Lecture Notes in Bioinformatics 3692

Edited by S. Istrail, P. Pevzner, and M. Waterman

Subseries of Lecture Notes in Computer Science

Rita Casadio Gene Myers (Eds.)

Algorithms in Bioinformatics

5th International Workshop, WABI 2005
Mallorca, Spain, October 3-6, 2005
Proceedings

Series Editors

Sorin Istrail, Celera Genomics, Applied Biosystems, Rockville, MD, USA
Pavel Pevzner, University of California, San Diego, CA, USA
Michael Waterman, University of Southern California, Los Angeles, CA, USA

Volume Editors

Rita Casadio
University of Bologna, Department of Biology/CIRB
Via Irnerio 42, 40126 Bologna, Italy
E-mail: casadio@alma.unibo.it

Gene Myers
Howard Hughes Medical Institute
4000 Jones Bridge Road, Chavy Chase, MD 20815-6789, USA
E-mail: gene@eecs.berkeley.edu

Library of Congress Control Number: 2005932938

CR Subject Classification (1998): F.1, F.2.2, E.1, G.1-3, J.3

ISSN 0302-9743
ISBN-10 3-540-29008-7 Springer Berlin Heidelberg New York
ISBN-13 978-3-540-29008-7 Springer Berlin Heidelberg New York

Springer is a part of Springer Science+Business Media

springeronline.com

Printed in Germany

Typesetting: Camera-ready by author, data conversion by Scientific Publishing Services, Chennai, India
Printed on acid-free paper SPIN: 11557067 06/3142 5 4 3 2 1 0

Preface

We are pleased to present the proceedings of the 5th Workshop on Algorithms in Bioinformatics (WABI 2005) which took place in Mallorca, Spain, October 3–6, 2005. The WABI 2005 workshop was part of the five ALGO 2005 conference meetings, which, in addition to WABI, included ESA, WAOA, IWPEC, and ATMOS. WABI 2005 was sponsored by EATCS (the European Association for Theoretical Computer Science), the ISCB (the International Society for Computational Biology), the Universitat Politècnica de Catalunya, the Universitat de les Illes Balears, and the Ministerio de Educación y Ciencia. See http://www.lsi.upc.edu/~wabi05/ for more details.

The Workshop on Algorithms in Bioinformatics highlights research work specifically developed to address algorithmic problems in biosequence analysis. The emphasis is therefore on statistical and probabilistic algorithms that address important problems in the field of molecular and structural biology. At present, given the enormous scientific and technical efforts in functional and structural genomics, the relevance of the problem is therefore constrained by the need for sound, efficient and specialized algorithms, capable of achieving solutions that can be tested by the biological community. Indeed the ultimate goal is to implement algorithms capable of extracting real features from real biological data sets. Therefore the workshop aims to present recent research results, including significant work in progress, and to identify and explore directions of future research.

Original research papers (including significant work in progress) or state-of-the-art surveys were solicited on all aspects of algorithms in bioinformatics, including, but not limited to: exact and approximate algorithms for genomics, genetics, sequence analysis, gene and signal recognition, alignment, molecular evolution, phylogenetics, structure determination or prediction, gene expression and gene networks, proteomics, functional genomics, and drug design. We received 94 submissions in response to our call for papers, and were able to accept 35 of these. In addition, WABI 2005 hosted a distinguished lecture by Dr. Marino Zerial of the Max Planck Institute for Molecular Cell Biology and Genetics in Dresden, given to the entire ALGO 2005 conference.

We would like to sincerely thank all the authors of submitted papers, and the participants of the workshop. We also thank the Program Committee and their sub-referees for their hard work in reviewing and selecting the papers for the workshop. The Program Committee consisted of the following 40 distinguished researchers:

Pankaj Kumar Agarwal (Duke University)
Tatsuya Akutsu (Kyoto University)
Amir Amihood (Bar-Ilan University)

Alberto Apostolico (Purdue University)
Craig Benham (University of California, Davis)
Gary Benson (MSSN, New York)
Mathieu Blanchette (McGill University)
Nadia El-Mabrouk (University of Montreal)
Olivier Gascuel (LIRMM, Montpelier)
Raffaele Giancarlo (University of Palermo)
Roderic Guigo (IMIM, Barcelona)
Michael Hallet (McGill University)
Daniel Huson (University of Tuebingen)
Gregory Kucherov (INRIA Nancy)
Michelle Lacey (Tulane University)
Jens Lagergren (KTH Stockholm)
Giuseppe Lancia (Univeristy of Udine)
Gad M. Landau (University of Haifa)
Thierry Lecroq (Université de Rouen)
Bernard Moret (University of New Mexico)
Shinichi Morishita (University of Tokyo)
Elchanan Mossel (Univeristy of California, Berkeley)
Vincent Moulton (University of Uppsala)
Lior Pachter (University of California, Berkeley)
Knut Reinert (Free University of Berlin)
Isidore Rigoutsos (IBM Watson)
Marie-France Sagot (INRIA Rhône-Alpes)
David Sankoff (University of Ottawa)
Sophie Schbath (INRIA Jouv-en-Josas)
Eran Segal (Rockefeller University)
Charles Semple (University of Canterbury)
Joao Carlos Setubal (Virginia Polytechnic Institute)
Roded Sharan (Tel Aviv Univeristy)
Steven Skiena (University of New York, Stony Brook)
Jens Stoye (University of Bielefeld)
Esko Ukkonen (University of Helsinki)
Lisa Vawter (Aventis Inc., USA)
Alfonso Valencia (CNB-CSIC, Spain)
Tandy Warnow (University of Texas)
Lusheng Wang (City Univeristy of Hong Kong)

Finally we would like to especially thank Bernard Moret, the de facto steering committee, for answering questions on history and precedence, for his advice on difficult protocol issues, and for setting up and hosting the EasyChair refereeing system used by the Program Committee.

July 2005

Rita Casadio and Gene Myers
WABI 2005 Program Co-chairs

Table of Contents

2. Tree Reconciliation

3. Clades and Haplotypes

Networks

Genome Rearrangements

1. Trasposition Model

2. Other Models

Sequences

1. Strings

2. Multi-alignment and Clustering

3. Clustering and Representation

Structure

1. Threading

2. Folding

Spectral Clustering Gene Ontology Terms to Group Genes by Function

Nora Speer, Christian Spieth, and Andreas Zell

University of Tübingen, Centre for Bioinformatics Tübingen (ZBIT),
Sand 1, D-72076 Tübingen, Germany
nspeer@informatik.uni-tuebingen.de

Abstract. With the invention of biotechnological high throughput methods like DNA microarrays, biologists are capable of producing huge amounts of data. During the analysis of such data the need for a grouping of the genes according to their biological function arises. In this paper, we propose a method that provides such a grouping. As functional information, we use Gene Ontology terms. Our method clusters all GO terms present in a data set using a Spectral Clustering method. Then, mapping the genes back to their annotation, genes can be associated to one or more clusters of defined biological processes. We show that our Spectral Clustering method is capable of finding clusters with high inner cluster similarity.

1 Introduction

In the past few years, high-throughput techniques like microarrays have become major tools in the field of genomics. In contrast to traditional methods, these technologies enable researchers to collect tremendous amounts of data, whose analysis itself constitutes a challenge. Since these techniques provide a global view on the cellular processes as well as on their underlying regulatory mechanisms, they are quite popular among biologists. After the analysis of such data, using filtering methods, clustering techniques or statistical approaches, researchers often end up with long lists of interesting candidate genes that need further examination. Then, in a second step, they categorize these genes by known biological functions.

In this paper, we address the problem of finding functional clusters of genes by clustering Gene Ontology terms. Based on methods originally developed for semantic similarity, we are able to compute a functional similarity between GO terms [13]. This information is fed into a spectral clustering algorithm [15]. This has the advantage, that after mapping the genes back to the GO terms, a gene with more than one associated term (function) can be present in more than one cluster which seems biologically plausible.

The organization of this paper is as follows: a brief introduction to the Gene Ontology is given in section 2. Related Work is discussed in section 3. Section 4 explains our method in detail. The experimental setup and the results on real world data sets are shown in section 5. Finally, in section 6, we conclude.

R. Casadio and G. Myers (Eds.): WABI 2005, LNBI 3692, pp. 1–12, 2005.

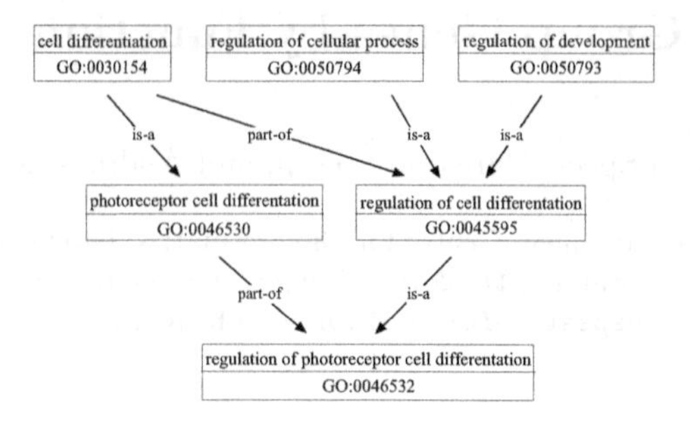

Fig. 1. Relations in the Gene Ontology. Each node is annotated with a unique accession number.

2 The Gene Ontology

The Gene Ontology (GO) is one of the most important ontologies within the bioinformatics community and is developed by the Gene Ontology Consortium [21]. It is specifically intended for annotating gene products with a consistent, controlled and structured vocabulary. Gene products are for instance sequences in databases as well as measured expression profiles. The GO is independent from any biological species. It represents terms in a Directed Acyclic Graph (DAG), covering three orthogonal taxonomies or "aspects": *molecular function*, *biological process* and *cellular component*. The GO-graph consists of over 18.000 terms, represented as nodes within the DAG, connected by relationships, represented as edges. Terms are allowed to have multiple parents as well as multiple children. Two different kinds of relationship exist: the "is-a" relationship (*photoreceptor cell differentiation* is, for example, a child of *cell differentiation*) and the "part-of" relationship that describes, for instance, that *regulation of cell differentiation* is part of *cell differentiation.*

Providing a standard vocabulary across any biological resources, the GO enables researchers to use this information for automatic data analysis done by computers and not by humans.

3 Related Work

While GO analysis is an increasingly important field, existing techniques suffer from some weaknesses: Many methods consider the GO simply as a list of terms, ignoring any structural relationships [2,7,17,23]. Others regard the GO primarily as a tree and convert the GO graph into a tree structure for determining distances between nodes [11]. Again others use a pseudo-distance that does not fulfill all metric conditions and relies on counting path lengths [3]. This is a delicate approach in unbalanced graphs like the GO those subgraphs have different degrees of detail.

Besides, the aim of some methods is primary either to use the GO as preprocessing [1] or as visualization tool [6]. Only few approaches utilize its structure for computation. Many methods are scoring techniques describing a list of genes annotated with GO terms [2,6,7,11,17,23]. But to our knowledge and apart from our earlier publications [20,19], there exists no automatic functional GO-based clustering method. One method is related to clustering and can be used to indicate which clusters are present in the data [3]. However, it suffers from the weaknesses that come along with using pseudo-distances as mentioned earlier.

4 Methodology

Our method consists of different steps that will be explained separately in this section: the mapping of the genes to the Gene Ontology, the calculation of functional similarities on GO terms, the spectral clustering algorithm and finally how the appropriate number of clusters is determined.

4.1 Mapping the Genes to the Gene Ontology

The functional similarity measure operates on pairs of GO nodes in a DAG, whereas in general, researchers are dealing with database ids of genes or probes. Therefore, a mapping M that relates the genes of a microarray experiment to nodes in the GO graph is required. Many databases (e.g. TrEMBL (GOA-project)) provide GO annotation for their entries and companies like Affymetrix provide GO mappings to their probe set ids as well. We used GeneLynx [8] to map the genes of dataset I. Hvidsten *et al.* [9] provide a mapping for dataset II.

4.2 Similarities Within the Gene Ontology

To calculate functional similarities between GO nodes, we rely on a technique that was originally developed for other taxonomies like WordNet to measure semantic similarities between words [12].

Following the notation in information theory, the information content (IC) of a term t can be quantified as follows [13]:

$$IC(t) = -\ln P(t) \tag{1}$$

where $P(t)$ is the probability of encountering an instance of term t in the data.

In the case of a hierarchical structure, such as the GO, where a term in the hierarchy subsumes those lower in the hierarchy, this implies that $P(t)$ is monotonic as one moves towards the root node. As the node's probability increases, its information content or its informativeness decreases. The root node has a probability of 1, hence its information content is 0. As the three aspects of the GO are disconnected subgraphs, this is still true if we ignore the root node "Gene Ontology" and take, for example, "biological process" as our root node instead.

To compute a similarity between two terms, one can use the IC of their common ancestor. As the GO allows multiple parents for each term, two terms can share ancestors by multiple paths. We take the minimum $P(t)$, if there is more than one ancestor. This is called P_{ms}, for *probability of the minimum subsumer* [13]. Thereby, it is guaranteed, that the most specific parent term is selected:

$$P_{\mathrm{ms}}(t_i, t_j) = \min_{t \in S(t_i, t_j)} P(t) \tag{2}$$

where $S(t_i, t_j)$ is the set of parental terms shared by both t_i and t_j. Based on Eq. 1 and 2, Lin extended the similarity measure, so that the IC of each single node was also taken into account [12,13]:

$$s(t_i, t_j) = \frac{2 \ln P_{ms}(t_i, t_j)}{\ln P(t_i) + \ln P(t_j)} \ . \tag{3}$$

Since $P_{ms}(t_i, t_j) \geq P(t_i)$ and $P_{ms}(t_i, t_j) \geq P(t_j)$, its value varies between 1 (for similar terms) and 0.

One should note, that the probability of a term as well as the resulting similarity between two terms differs from data set to data set, depending on the distribution of terms. Therefore, our clustering differs from a general clustering of the GO and a subsequent mapping of the genes to such a general clustering. Due to our approach, we are able to arrange the resulting cluster boundaries depending on the distribution of the GO terms either more specific (if the terms concentrate on a specific part of the GO) or more general (if the terms are widely spread).

4.3 Spectral Clustering

We decided to cluster GO terms, not genes, because of two reasons: first, we do not face the problem of combining different similarities per gene like in earlier publications [19,20] and second, after mapping the genes back to the GO, they can be present in more than one functional cluster which is biologically plausible, since they can also fulfill more than one biological function.

Recently, Spectral Clustering methods haven been growing in popularity. Several new algorithms have been published [22,18,14,15]. A set of objects (in our case GO terms) to be clustered will be denoted by T, with $|T| = n$. Given an affinity measure $A_{ij} = A_{ji} \geq 0$ for two objects i, j, the affinities A_{ij} can be seen as weights on the undirected edges ij of a graph G over T. Then, the matrix $A = [A_{ij}]$ is the real-valued adjacency matrix for G. Let $d_i = \sum_{j \in T} A_{ij}$ be called the degree of node i, and D be the diagonal matrix with d_i as its diagonal. A clustering $C = \{C_1, C_2, \ldots, C_K\}$ is a partitioning of T into the nonempty mutually disjoint subsets $C_1, C_2, \ldots, C_K$. In the graph theoretical paradigm a clustering represents a multiway cut in the graph G.

In general, all Spectral Clustering algorithms use Eigenvectors of a matrix (derived from the affinity matrix A) to map the original data to the K-dimensional vectors $\{\gamma_1, \gamma_2, \ldots, \gamma_n\}$ of the spectral domain $\Re^K$. Then, in a second step, these vectors are clustered with standard clustering algorithms. Here, we use K-means. We chose the newest Spectral Clustering algorithm by Ng *et al.* [15] and we will now review it briefly:

1. From the affinity matrix A and its derived diagonal matrix D, compute the Laplacian matrix $L = D^{-1/2}AD^{-1/2}$.
2. Find $v^1, v^2, \ldots, v^K$, the Eigenvectors of L, corresponding to the K largest Eigenvalues.
3. Form the matrix $V_{n \times k} = \left[v^1, v^2, \ldots, v^K\right]$ with these Eigenvectors as columns.
4. Form the matrix Y from V by renormalizing each of X's rows to have unit norm.
5. Cluster the rows of $Y = [\gamma_1, \gamma_2, \ldots, \gamma_n]$ as points in a K-dimensional space.
6. Finally assign the original object i to cluster j if and only if row γ_i of the matrix Y was assigned to j.

Since Spectral Clustering relies on the affinity matrix A, it is easy to apply it to any kind of data, where affinities can be computed. For numerical data, affinities are usually computed with a kernel function, e.g. $A_{ij} = \exp(\frac{-d(i,j)^2}{2\sigma^2})$, with $d(i,j)$ denoting the Euclidean distance between point i and j and σ denoting the kernel width. For non-numerical data, like GO terms, affinity can either be defined in the same way, given a distance measure d. This approach has the advantage of non-linearity, controlled by the kernel width σ, which allows for sharper separation between clusters. But it has also disadvantages: the question of how to deduce σ in a meaningful way arises and additionally, for many data types, especially the GO, similarity is much easier to define since it does not need to fulfill any metric conditions. As noted in [16], there is nothing magical about the definition of affinity. Therefore, we directly apply our similarity matrix as affinity matrix.

4.4 Cluster Validity

We selected the number of clusters K in our data according to the Davies-Bouldin index [5]. Given a clustering $C = \{C_1, C_2, \ldots, C_K\}$, it is defined as:

$$DB(C) = \frac{1}{K}\sum_{i=1}^{K} \max\left\{\frac{\Delta(C_i) + \Delta(C_j)}{\delta(C_i, C_j)}\right\} \tag{4}$$

where $\Delta(C_i)$ represents the inner cluster distance of cluster C_i and $\delta(C_i, C_j)$ denotes the inter cluster distance between cluster C_i and C_j. K is the number of clusters. Small values of $DB(C)$ indicate a good clustering.

$\Delta(C_i)$ and $\delta(C_i, C_j)$ are calculated as the sum of distances to the respective cluster mean and the distance between the centers of two clusters, respectively. Since we use similarities, not distances, and cannot compute means in the GO, we apply the DB-Index in the spectral domain $\Re^K$ (after the Eigenvector decomposition) where we are dealing with simple numerical data.

5 Computational Experiments

5.1 Data Sets

One possible scenario where researchers would like to group a list of genes according to their function is when they received lists of up- or down-regulated genes from the analysis of an DNA microarray experiment. Thus, we chose two publicly available microarray data sets, annotated the genes with the GO and used them for functional clustering. We only use the taxonomy *biological process*, because we are mainly interested in gene function in a more general sense. However, our method can be applied in the same way for the other two taxonomies.

The authors of the first data set examined the response of human fibroblasts to serum on cDNA microarrays in order to study growth control and cell cycle progression. They found 517 genes whose expression levels varied significantly, for details see [10]. We used these 517 genes for which the authors provide NCBI accession numbers. The GO mapping was done using GeneLynx [8]. After mapping to the GO, 238 genes showed one or more mappings to *biological process* or a child term of *biological process*. These 238 genes were used for the clustering.

In order to study gene regulation during eukaryotic mitosis, the authors of the second data set examined the transcriptional profiling of human fibroblasts during cell cycle using microarrays [4]. Duplicate experiments were carried out at 13 different time points ranging from 0 to 24 hours. Cho *et al.* [4] found 388 genes whose expression levels varied significantly. Hvidsten *et al.* [9] provide a mapping of the data set to GO. 233 of the 388 genes showed at least one mapping to the GO *biological process* taxonomy and were thus used for clustering.

5.2 Experimental Design

In the experiments, we had the problem of how to compare our method to other known clustering algorithms, because to our best knowledge, there is no clustering method that does a clustering only due to a similarity matrix. Instead, most algorithms need distances. Beside that, most clustering techniques were originally developed for numerical data and therefore utilize means during the clustering process which we cannot compute in the GO. Only linkage methods work on a proximity matrix, although this is also usually a distance matrix. Average Linkage clustering is known to be its most robust, non-means based representative. Therefore, we compare our approach to a modified version of an Average Linkage algorithm that joins the most similar clusters, instead of joining those with the smallest distance. Inner cluster similarity of cluster C_i is computed as follows:

$$s(C_i) = \frac{1}{|C_i|(|C_i - 1|)} \sum_{t_i, t_j \in C_i, t_i \neq t_j} s(t_i, t_j) \quad (5)$$

with $s(t_i, t_j)$ denoting the similarity between term t_i and t_j and $|C_i|$ denoting the number of terms in cluster C_i.

For Spectral Clustering, K-means was carried out 25 times and the solution with the minimum distortion was taken as proposed in [15]. For both algorithms, we performed runs for different values of K, ranging from $K = 5, 6, \ldots, 25$.

5.3 Results

Fig. 2 shows the average inner cluster similarity for Average Linkage and Spectral Clustering for both data sets and different numbers of K. It is clearly visible that except for one exception ($K = 5$, data set I), Spectral clustering always shows a much higher inner cluster similarity than Average Linkage clustering.

Additionally, we wanted to evaluate the best solutions generated by Spectral Clustering in more detail. Since inner cluster similarity is not independent from the number of clusters K, we chose the best solution according to the Davies-Bouldin index (Eq. 4) that was calculated after the Eigenvalue decomposition in the spectral domain $\Re^K$. Fig. 3 shows the Davies-Bouldin index for the cluster numbers $K = 5, ..., 25$ for data set I and II, respectively. For data set I, the best clustering was achieved with 10 clusters and for data set II with 9 clusters. These two solutions (indicated by an arrow in Fig. 3) were then used for further examination.

Figure 4 shows the Euclidean distance matrix calculated after the Eigenvector decomposition for data set I (left) and II (right). Higher values are indicated

Table 1. Cluster 5 of dataset I. This cluster contains mainly GO terms associated with mitosis

Term Acc.	GO Term Name
GO:0007050	cell cycle arrest
GO:0000074	regulation of cell cycle
GO:0008151	cell growth and/or maintenance
GO:0007049	cell cycle
GO:0007095	mitotic G2 checkpoint
GO:0000079	regulation of CDK activity
GO:0008284	positive regulation of cell proliferation
GO:0008283	cell proliferation
GO:0006878	copper ion homeostasis
GO:0008285	negative regulation of cell proliferation
GO:0006260	DNA replication
GO:0006874	calcium ion homeostasis
GO:0008156	negative regulation of DNA replication
GO:0006269	DNA replication, priming
GO:0007093	mitotic checkpoint
GO:0007096	regulation of exit from mitosis
GO:0006298	mismatch repair
GO:0000080	G1 phase of mitotic cell cycle
GO:0007088	regulation of mitosis
GO:0000067	DNA replication and chromosome cycle
GO:0007089	start control point of mitotic cell cycle
GO:0000085	G2 phase of mitotic cell cycle
GO:0007079	mitotic chromosome movement
GO:0000089	mitotic metaphase
GO:0007080	mitotic metaphase plate congression
GO:0006261	DNA dependent DNA replication

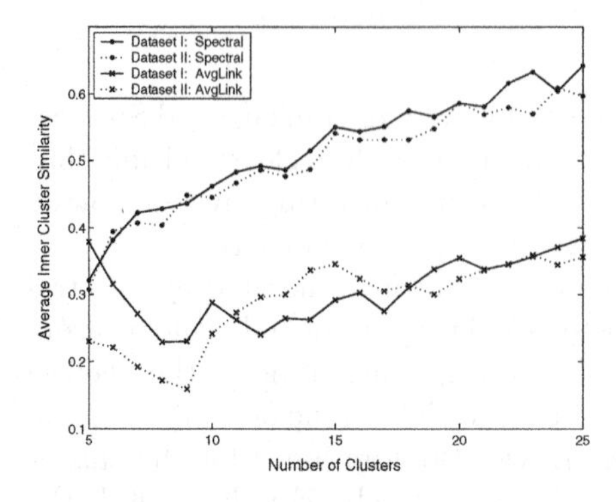

Fig. 2. Average Inner Cluster Similarity for Average Linkage and Spectral clustering for data set I and II

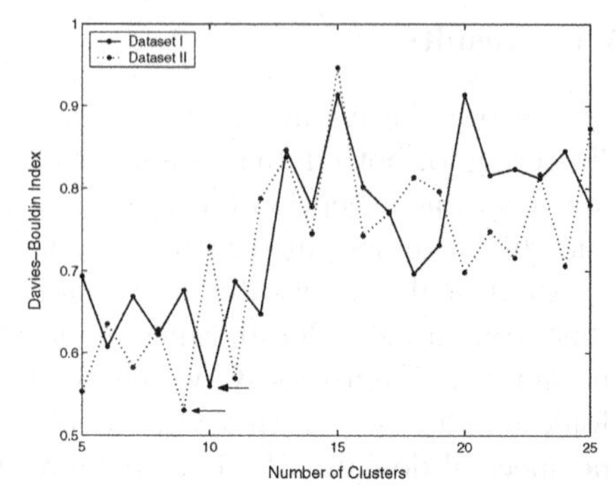

Fig. 3. Davies-Bouldin index in the spectral domain $\Re^K$ for Spectral Clustering of data set I and II

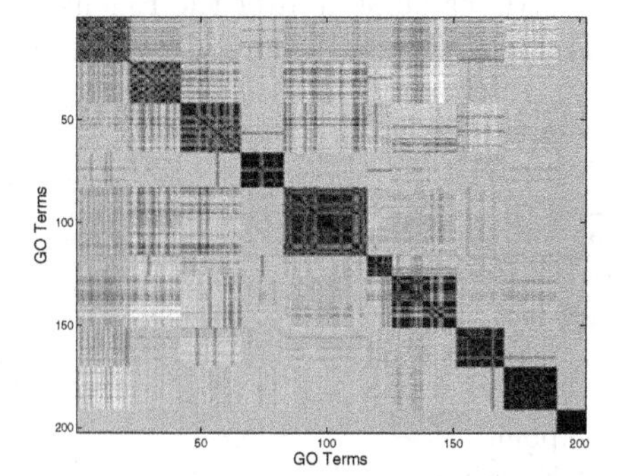

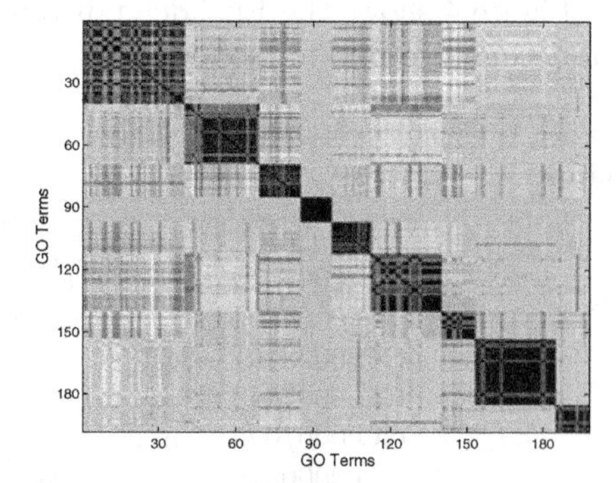

Fig. 4. Distance matrices in the spectral domain $\Re^K$ (after the Eigenvector decomposition) of data set I (left) and II (right): The 10 (left) and 9 (right) clusters are clearly visible.

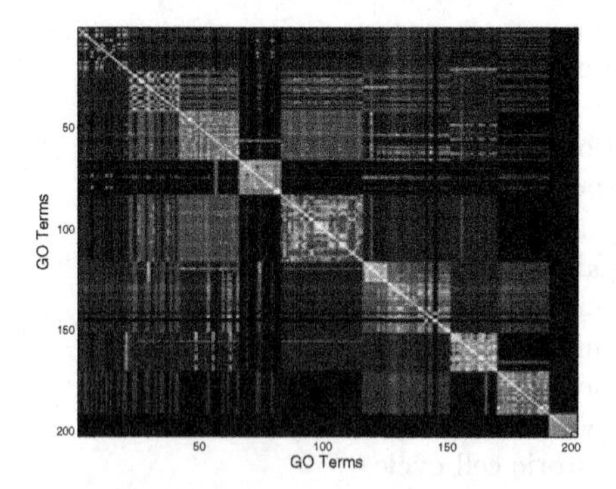

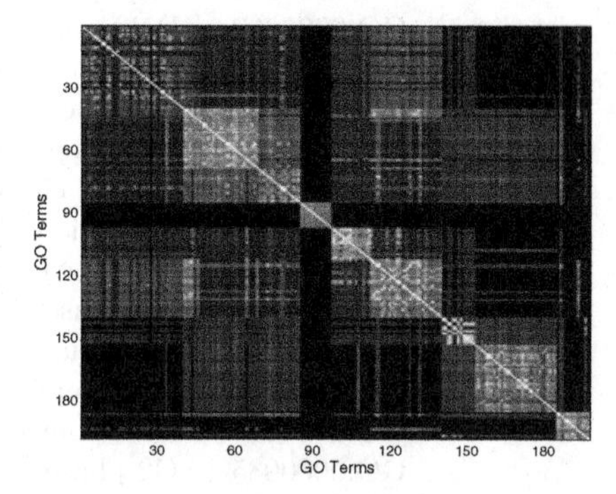

Fig. 5. The original similarity matrix of data set I (left) and II (right): Again the 10 (left) and the 9 (right) clusters are clearly visible.

Table 2. Cluster 8 of dataset I: this cluster contains mainly GO terms associated with signal transduction

Term Acc.	GO Term Name
GO:0000188	inactivation of MAPK
GO:0008277	regulation of G-protein coupled receptor protein signaling pathway
GO:0007165	signal transduction
GO:0007267	cell-cell signaling
GO:0007166	cell surface receptor linked signal transduction
GO:0007200	G-protein signaling, coupled to IP3 second messenger (phospholipase C activating)
GO:0007186	G-protein coupled receptor protein signaling pathway
GO:0007181	TGFbeta receptor complex assembly
GO:0007155	cell adhesion
GO:0008038	neuronal cell recognition
GO:0007179	TGFbeta receptor signaling pathway
GO:0007156	homophilic cell adhesion
GO:0007229	integrin-mediated signaling pathway
GO:0007178	transmembrane receptor protein serine/threonine kinase signaling pathway
GO:0007160	cell-matrix adhesion
GO:0007268	synaptic transmission
GO:0007173	EGF receptor signaling pathway
GO:0000165	MAPKKK cascade
GO:0000187	activation of MAPK
GO:0007169	transmembrane receptor protein tyrosine kinase signaling pathway
GO:0007243	protein kinase cascade

by a light color and lower values by a dark color. Thus, the 10 squares (left) and the 9 squares (right) indicate regions of small distances corresponding to the 10 and clusters, respectively. Figure 4 demonstrates that the clusters in the spectral domain $\Re^K$ have small inner cluster distances and high distances between them. The original affinity (or similarity) matrices for both data sets are visualized in Fig. 5. Again, light colors indicate higher values, thus, in this case a higher similarity. The 10 (left) and 9 (right) clusters are still clearly visible as regions of high inner cluster similarity compared to the similarity between the clusters.

Additionally, we examined clusters of a solution in more detail, but due to space limitations, we cannot show all clusters of both data sets. Therefore, we confine ourselves to show three selected clusters of data set I: cluster 5, 8 and 9. Tab. 1 - Tab. 3 show the GO terms of each of these clusters, respectively. A closer study of the GO term names reveals that our method produces from each other distinct functional clusters each containing GO terms that belong to a defined biological process. The GO terms of cluster 5 (Tab. 1) are mainly related to mitosis like cell cycle regulation or CDK activity regulation and DNA replication. In Tab. 2, the GO terms of cluster 8 are listed. They are mostly related to processes associated with signal transduction pathways like the TGF-

Table 3. Cluster 9 of dataset I: this cluster contains mainly GO terms associated with metabolism

Term Acc.	GO Term Name
GO:0006101	citrate metabolism
GO:0015936	coenzyme A metabolism
GO:0006629	lipid metabolism
GO:0006768	biotin metabolism
GO:0006633	fatty acid biosynthesis
GO:0006564	L-serine biosynthesis
GO:0006729	tetrahydrobiopterin biosynthesis
GO:0006048	UDP-N-acetylglucosamine biosynthesis
GO:0006631	fatty acid metabolism
GO:0016042	lipid catabolism
GO:0005989	lactose biosynthesis
GO:0006096	glycolysis
GO:0006700	C21-steroid hormone biosynthesis
GO:0008203	cholesterol metabolism
GO:0008202	steroid metabolism
GO:0006695	cholesterol biosynthesis
GO:0008299	isoprenoid biosynthesis
GO:0006694	steroid biosynthesis
GO:0006529	asparagine biosynthesis
GO:0006541	glutamine metabolism
GO:0006635	fatty acid beta-oxidation
GO:0006809	nitric oxide biosynthesis
GO:0006559	phenylalanine catabolism
GO:0006520	amino acid metabolism
GO:0006563	L-serine metabolism
GO:0006636	fatty acid desaturation
GO:0006004	fucose metabolism
GO:0006099	tricarboxylic acid cycle
GO:0006693	prostaglandin metabolism
GO:0006207	'de novo' pyrimidine base biosynthesis
GO:0006780	uroporphyrinogen III biosynthesis

β pathway or G-protein coupled signaling and these GO terms form cluster 8. Finally, cluster 9 (Tab. 3) contains GO terms associated with metabolic processes like amino acid synthesis, lipid metabolism or fatty acid biosynthesis, just to name a few.

6 Discussion

In this paper, we presented a clustering method for GO terms that can be used to cluster genes or any other gene products that can be annotated with the Gene Ontology. We showed that the clusters produced by our method have a higher average inner cluster similarity than those produced by a similarity-based variant of Average Linkage Clustering. Beside that, we showed for the best two solutions in detail that their GO terms have a much higher similarity to each other than to those in the other clusters. This is not only true for the data in the spectral domain $\Re^K$, but also for the original affinity matrix. Furthermore, we evaluated three clusters in more detail and could show that the GO terms in each cluster belong to a defined and separated biological process.

The Spectral Clustering technique enables us to cluster those objects, like GO terms, where it is easy to calculate similarities but more difficult to calculate distances or even means, that are needed by many popular clustering methods. In contrast to these methods, Spectral Clustering is able to produce a clustering only due to an affinity matrix. To be suitable for clustering, the affinity matrix only needs to reflect the natural relationships of the data.

Additionally, the fact that we are using GO terms for clustering and not genes like in our previous publications has the advantage that now, one gene can belong to more than one cluster. This makes also biologically sense, since one gene can also have more than one function. Thus, our method facilitates the functional analysis of high throughput data.

Acknowledgment

This work was supported by the National Genome Research Network (NGFN) of the Federal Ministry of Education and Research in Germany under contract number 0313323. We thank Holger Fröhlich for fruitful discussions.

References

1. B. Adryan and R. Schuh. Gene Ontology-based clustering of gene expression data. *Bioinformatics*, 20(16):2851–2852, 2004.
2. T. Beißbarth and T. Speed. GOstat: find statistically overexpressed Gene Ontologies within groups of genes. *Bioinformatics*, 20(9):1464–1465, 2004.
3. A. Flmer C.A. Joslyn, S.M. Mniszewski and G. Heaton. The gene ontology categorizer. *Bioinformatics*, 20(Suppl. 1):i169–i177, 2004.
4. R.J. Cho, M. Huang, M.J. Campbell, H. Dong, L. Steinmetz, L. Sapinoso, G. Hampton, S.J. Elledge, R.W. Davis, and D.J. Lockhart. Transcriptional regulation and function during the human cell cycle. *Nature Genetics*, 27(1):48–54, 2001.
5. J.L. Davies and D.W. Bouldin. A cluster separation measure. *IEEE Transactions on Pattern Analysis and Machine Intelligence*, 1:224–227, 1979.
6. S.W. Doniger, N.Salomonis, K.D. Dahlqusi, K. Vranizan, S.C. Lawlor, and B.R. Conklin. MAPPFinder: using Gene Ontology and GenMAPP to create a global gene-expression profile from microarray data. *Genome Biology*, 4(1):R7, 2003.
7. I. Gat-Viks, R. Sharan, and R. Shamir. Scoring clustering solutions by their biological relevance. *Bioinformatics*, 19(18):2381–2389, 2003.
8. Gene Lynx. http://www.genelynx.org, 2004.
9. T.R. Hvidsten, A. Laegreid, and J. Komorowski. Learning rule-based models of biological process from gene expression time profiles using Gene Ontology. *Bioinformatics*, 19(9):1116–1123, 2003.

10. V.R. Iyer, M.B. Eisen, D.T. Ross, G. Schuler, T. Moore, J.C.F. Lee, J.M. Trent, L.M. Staudt, J Hudson Jr, M.S. Boguski, D. Lashkari, D. Shalon, D. Botstein, and P.O. Brown. The transcriptional program in response of human fibroblasts to serum. *Science*, 283:83–87, 1999.
11. S.G. Lee, J.U. Hur., and Y.S. Kim. A graph-theoretic modeling on go space for biological interpretation on gene clusters. *Bioinformatics*, 20(3):381–388, 2004.
12. D. Lin. An information-theoretic definition of similarity. In Morgan Kaufmann, editor, *Proceedings of the 15th International Conference on Machine Learning*, volume 1, pages 296–304, San Francisco, CA, 1998.
13. P.W. Lord, R.D. Stevens, A. Brass, and C.A. Goble. Semantic similarity measures as tools for exploring the gene ontology. In *Proceedings of the Pacific Symposium on Biocomputing*, pages 601–612, 2003.
14. M. Meila and J. Shi. Learning segmantation by random walks. In T.K. Leen, T.G. Dietterich, and V. Tresp, editors, *Advances in Neural Information Processing Systems*, volume 13, pages 873–879, 2001.
15. A.Y. Ng, M.I. Jordan, and Y. Weiss. On spectral clustering: Analysis and an algorithm. In T. G. Dietterich, S. Becker, and Z. Ghahramani, editors, *Advances in Neural Information Processing Systems*, volume 14, pages 849–856, Cambridge, MA, 2002. MIT Press.
16. P.Perona and W. Freeman. A factorization approach to grouping. In *Lecture Notes in Computer Sience*, 1406, pages 655–670. Springer, 1998.
17. P.N Robinson, A. Wollstein, U. Böhme, and B. Beattie. Ontologizing gene-expression microarray data: characterizing clusters with gene ontology. *Bioinformatics*, 20(6):979–981, 2003.
18. J. Shi and J. Malik. Normalized cuts and image segmentation. *IEEE Transactions on Pattern Analysis and Machine Intelligence*, 22(8):888–905, 2000.
19. N. Speer, H. Fröhlich, C. Spieth, and A. Zell. Functional grouping of genes using spectral clustering and gene ontology. In *To appear in Proceedings of the IEEE International Joint Conference on Neural Networks*, 2005.
20. N. Speer, C. Spieth, and A. Zell. A memetic clustering algorithm for the functional partition of genes based on the Gene Ontology. In *Proceedings of the IEEE Symposium on Computational Intelligence in Bioinformatics and Computational Biology*, pages 252–259, 2004.
21. The Gene Ontology Consortium. The gene ontology (GO) database and informatics resource. *Nucleic Acids Research*, 32:D258–D261, 2004.
22. Y. Weiss. Segmentation using eigenvectors: a unifying view. In *Proceedings of IEEE International Conference on Computer Vision*, volume 2, pages 975–982, 1999.
23. B.R. Zeeberg, W. Feng, G. Wang, and A.T. Fojo *et al.* GOminer: a resource for biological interpretation of genomic and proteomic data. *Genome Biology*, 4(R28), 2003.

Dynamic De-Novo Prediction of microRNAs Associated with Cell Conditions: A Search Pruned by Expression

Chaya Ben-Zaken Zilberstein and Michal Ziv-Ukelson

Dept. of Computer Science, Technion - Israel Institute of Technology, Haifa 32000, Israel
{chaya, michalz}@cs.technion.ac.il

Abstract
Biological background: Plant microRNAs (miRNAs) are short RNA sequences that bind to target genes (mRNAs) and change their expression levels by redirecting their stabilities and marking them for cleavage. In *Arabidopsis thaliana,* microRNAs have been shown to regulate development and are believed to impact expression both under many conditions, such as stress and stimuli, as well as in various tissue types.
Methods: *mirXdeNovo* is a novel prototype tool for the de-novo prediction of microRNAs associated with a given cell condition. The work of *mirXdeNovo* is composed of two off-line preprocessing stages, which are executed only once per genome in the database, and a dynamic online main stage, which is executed again and again for each newly obtained expression profile. During the preprocessing stages, a set of candidate microRNAs is computed for the genome of interest and then each microRNA is associated with a set of mRNAs which are its predicted targets.

Then, during the main stage, given a newly obtained cell condition represented by a vector describing the expression level of each of the genes under this condition, the tool will efficiently compute the subset of microRNA candidates which are predicted to be active under this condition. The efficiency of the main stage is based in a novel branch-and-bound search of a tree constructed over the microRNA candidates and annotated with the corresponding predicted targets. This search exploits the monotonicity of the target prediction decision with respect to microRNA prefix size in order to apply an efficient yet admissible pruning. Our testing indicates that this pruning results in a substantial speed up over the naive search.
Biological Results: We employed *mirXdeNovo* to conduct a study, using the plant *Arabidopsis thaliana* as our model organism and the subject of our "hunt for microRNAs". During the preprocessing stage, 2000 microRNA precursor candidates were extracted from the genome. Our study included the 3'UTRs of 5800 mRNAs. 380 different conditions were analyzed including various tissues and hormonal treatments. This led to the discovery of some interesting and statistically significant newly predicted microRNAs, annotated with their potential condition of activity.

1 Introduction

The DNA of an organism determines all the RNA and protein molecules constituing its cells. The DNA sequence by itself, however, can not explain the whole picture of

R. Casadio and G. Myers (Eds.): WABI 2005, LNBI 3692, pp. 13–26, 2005.

how the organism functions and survives. In order to understand that, we need to know how the various genes are used. The cellular expression levels of genes typically vary according to cell conditions. A condition can be, for example, a specific tissue in which the gene is expressed or an exposure to certain environmental stress such as bacterial pathogens, heat, etc. The cellular expression levels of genes are largely influenced by their transcription rates, as well as by the degradation rates of their mRNAs. The degradation rates of mRNA molecules can vary by 100-fold or more between different cell conditions [3, 24]. These rates are affected by a wide variety of stimuli and cellular signals, including: specific hormones [24, 25], iron [4, 27], cell cycle progression [12], cell differentiation [13, 15], and viral infection [26]. In this paper we focus on specific regulatory signals denoted microRNAs which, in plants, are known to induce quick mRNA degradations. These microRNAs can change their activity with varying conditions and, thus, contribute to the differential expression of the genes at the post transcriptional level. Note that there are other factors that influence gene expression, including transcription factors which regulate the transcription rates of the RNA molecules, and proteins which mediate mRNA degradations as well, however, in this paper we will focus on microRNAs.

MicroRNAs are endogenous $\sim$ 20 nucleotide RNAs, which are initially transcribed as much longer RNA precursors that contain imperfect hairpins from which the mature microRNAs are excised by Dicer-like enzymes. Each mature microRNA derives from the double-stranded portion of the hairpin and is initially excised as a duplex comprising two RNAs of size $\sim 20nt$ each, one of which is the mature microRNA. MicroRNAs recognize their targets through base pairing and bind to them to form a microRNA:mRNA duplex. In animals, microRNAs often display limited complementarity to multiple sites in the 3'UTR of the target mRNA and act to repress its productive translation. In plants, microRNAs generally display near-perfect complementarity to a single site within the target mRNA and can direct the cleavage at this site [8, 19, 23, 9, 7].

To date, cloning has identified over 200 microRNAs from diverse eukaryotic organisms. Despite their success such biochemical approaches are skewed towards identifying abundant microRNAs. In order to identify microRNA candidates, bioinformatics methods take advantage of the properties of known microRNA precursors, including their hairpin structure (typically $\sim$ 70 nt), the length of their hairpin stem (typically $\sim 20nt$), and their tendency to be found in intergenic regions [1, 16]. Some improved predictions can be achieved by focusing on microRNA precursors that appear conserved across species or within a species [2, 11, 18]. However, none of these computational approaches provide rigorous statistics to evaluate the real functionality of their predictions in vivo. In this paper we provide a framework which allows the statistical evaluation of some of these predictions.

Here we propose a practical bioinformatic approach for discovering new and unknown microRNAs active in down-regulation of mRNA levels, by marking the mRNAs for cleavage and degradation, under a specific condition of interest. We focus on plant microRNAs. However, since recent evidence indicates that microRNAs may cause mRNA cleavage in animals too [10], our approach holds the potential to be expanded to these cases as well.

Let us first explain how we predict whether a *previously known* microRNA with *known* mRNA targets could induce quick mRNA degradations under a specific condition, given the expression profile of that condition, *i.e.* a vector describing the expression level of each gene under this condition. Later, we will show how this approach is extended to address *new* microRNA candidates with *unknown targets*. The following notation will be used.

- S denotes a set of microRNA candidates.
- $P \in S$ denotes a microRNA.
- Q denotes an expression condition.
- Z is a statistical score for associating P with Q, to be formulated below.
- θ denotes a user-specified threshold on Z.
- δ denotes a user pre-defined threshold on the free energy of a potential duplex to be formed between a microRNA and its predicted mRNA target.

Associating a known microRNA with a given condition: suppose that mRNAs t_1, $t_2 \ldots, t_n$ are targets of microRNA P, yet other mRNAs $x_1, x_2, \ldots, x_m$ clearly do not bind to P. Then if there is evidence that $x_1, x_2, \ldots, x_m$ are highly expressed under some given condition Q, yet $t_1, t_2 \ldots, t_n$ are expressed in low levels under the very same condition, then it may be possible to statistically assert that microRNA P is active under condition Q, contributing to the degradation of mRNAs $t_1, t_2 \ldots, t_n$. In practice, we use the full set of the mRNAs of the studied species as a background set, instead of using the specific subset $x_1, x_2, \ldots, x_m$. We compute μ the mean, and σ^2 the variance, of the expression levels of that population. Then we calculate $\bar{t}$ the mean of the expression levels of $t_1, t_2 \ldots, t_n$ and conduct a statistical test to assess whether $\bar{t}$ is significantly lower than μ. To do that we compute the following statistic to which we can associate a significance *p-value*, as will be explained in section 4.1.

Theorem 1. Let x be a vector of observations of size n and let $\overline{x}$ denote the mean of x. If x was sampled from a normal population with a mean μ and a variance σ^2 then Z has an approximate standard normal distribution.

$$Z = \frac{\overline{x} - \mu}{\sigma / \sqrt{n}} \tag{1}$$

Clearly, the lower the statistical score Z, the higher the probability that microRNA P is active, inducing mRNA cleavage and quick degradations under the given condition. Note that this approach was introduced in [29] where it was applied to a study associating known microRNAs with various expression conditions.

Associating new microRNA candidates with unknown targets in the context of activity conditions: here, the above approach is expanded as we spring forward in a hunt for new, previously undiscovered (de-novo) microRNAs. Our microRNA candidates are $\sim 20nt$-sized sequences comprising the duplexes of the hairpin structures of the potential precursors discovered by any of the above informatics approaches. One can predict the potential targets of each of these candidates, by identifying mRNAs which have the potential to form a stable duplex with this candidate. This potential is assessed based on the analysis of the sequences only, as explicitly described in section 3.2. Note that, in order to discover new microRNAs which are active under one or more conditions of interest, one could use the following naive three-stage search. An important input of this search are the expression profiles covering various cell conditions, where each expression profile is usually genome-wide measured using microarray techniques. The objective is to find pairs (P, Q) such that the corresponding score $Z < \theta$. Such a pairing would indicate that P is a predicted microRNA which is potentially active under cell condition Q.

1. Use the whole genome sequence of the organism of interest to collect a set S of microRNA candidates. This is done by scanning the genome for stretches which can be folded into stable hairpin structures with features similar to know microRNA precursors, and extracting the two $\sim 20nt$ sequences from their double-stranded portion, as described in section 3.1.
2. For each string $P \in S$: Predict the set of P's potential mRNA targets, by looking for mRNAs with potential to form a stable duplex with P. This potential is evaluated based on sequence analysis (local hybridization with score bounded by δ), as described in section 3.2.
3. For each condition Q under consideration and for each string $P \in S$: Examine the set of expression levels associated with P's set of predicted targets under Q, using the input expression profile for condition Q, and compute the corresponding score Z.
   If, for a given condition/microRNA pair (P, Q), the computed $Z < \theta$ for some user-specified θ: Report P as a potentially active microRNA under condition Q.

In the context of an effort to make our prototype tool available online to the bioinformatics community, we note that the relevant genomic sequences are pretty much static (we are, after all, in the midst of the "post genomic" era). Expression data, on the other hand, is quite dynamic, and new expression matrices are constantly discovered and made available to the public. We also observed that in order for our tool to be useful to the bioinformatic user, the statistical threshold θ often needs to be iteratively "fine tuned" to the subject organism as part of the set-up process. Clearly, the lower the predefined threshold θ, the more statistically significant the results reported to the user. On the other hand, using too low a θ should be avoided as it could result in the "loss" of active microRNAs.

Practically this means that for each genome which participates in our database the hunt for precursors which are potential microRNAs (step 1) only needs to be applied once per genome. Similarly, the computation of the set of mRNAs which are putative targets of each microRNA candidate (step 2) can also be computed once during a preprocessing stage. The need arises, however, for an efficient statistical computation to associate a dynamically changing expression data (and perhaps a dynamically changing θ), collected for specific conditions, with a set of correlated microRNAs (step 3). Therefore, we re-formalize the work described in step 3 of our framework as the following search problem.

Definition 1. *The "one (condition) against all (microRNA candidates)" association search problem is, given a cell condition Q and its expression profile, as well as an expression threshold θ: to find all microRNA candidate strings P such that the Z score computed for P and Q is upper bounded by θ.*

The challenge of speeding up the computation of the "one against all" association search is met in this paper by an online search engine that applies a simple yet powerful admissible pruning in order to speed up the statistical computations without loss of output optimality.

1.1 Our Results

In this paper we describe a novel framework for the de-novo discovery of new microRNAs which are predicted as regulators of gene expression levels (via mRNA degradations) under dynamically specified conditions of interest.

mirXdeNovo is a novel prototype tool for the de-novo prediction of microRNAs associated with a given cell condition. The work of *mirXdeNovo* is composed of two offline preprocessing stages and a dynamic online main stage. During the preprocessing stages, first a set of candidate microRNAs is computed for the genome of interest and then each microRNA is associated with a set of genes which are its predicted targets. Note that the preprocessing stages are executed only once per each genome which is to be included in the database, as the genomic sequences are quite static by now.

The main stage, on the other hand, is dynamically executed again and again for each newly obtained expression profile for a given condition. During this stage, given a newly obtained cell condition plus a vector describing the expression level of each of the genes under this condition, our tool will efficiently compute the subset of microRNA candidates which are predicted to be active under this condition. The efficiency of the main stage is based on a branch-and-bound search on a tree constructed from the microRNA candidates and annotated with the corresponding microRNA targets. The search is strongly yet admissibly pruned by exploiting the monotonicity of the target prediction decision with respect to microRNA prefix size.

When comparing the time invested in the "one against all condition association" stage (step 3) by the two methods, *mirXdeNovo* seems to be substantially faster than the naive search (see section 5.1 and Figure 2) on a wide range of practical testing setups. Thus, the contribution of this paper, in addition to its interesting biological results, is in suggesting a simple yet efficient "pruning by expression" method to accelerate this heavy computation.

Note that our tool can be used to further assert the relevance of microRNA candidates which were previously discovered by other methods, this by "plugging in" the output of these engines as a replacement to our first preprocessing stage. Discovering conditions in which a previously predicted candidate microRNA gets a significant Z score is a further validation to the function of this candidate as a microRNA in vivo. Furthermore, this enables us to associate $p\text{-}values$ to such previously predicted microRNAs.

Also note that, even though we focus on plants here, our approach holds the potential to be used for a wide variety of organisms including animals.

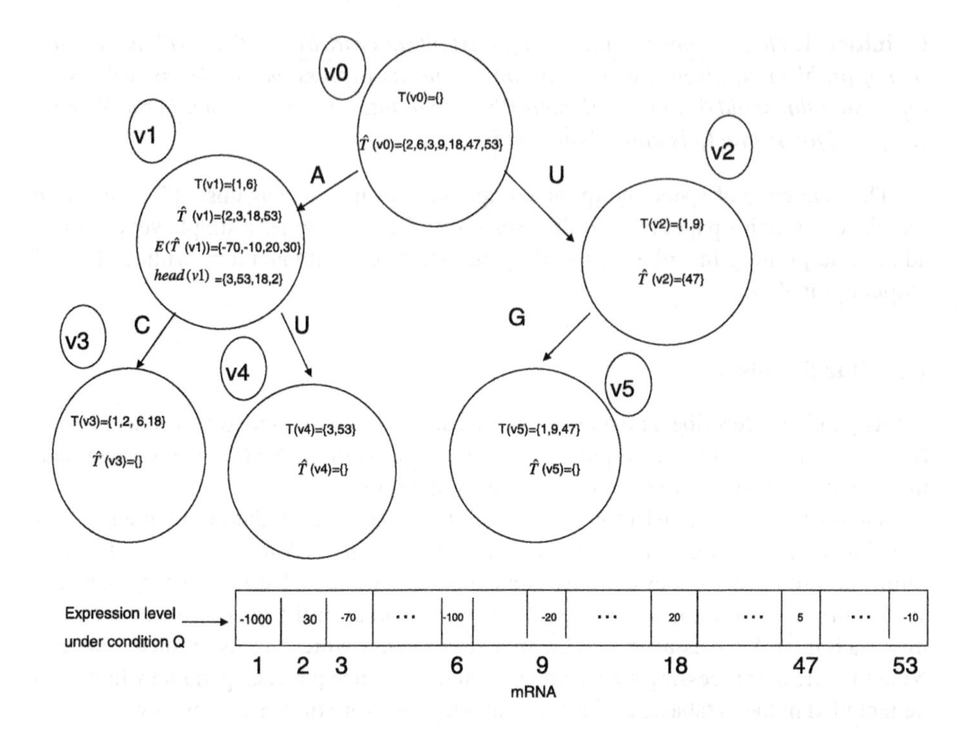

Fig. 1. An example of a trie $\mathcal{T}_S$ and an expression profile corresponding to a given condition

We employed *mirXdeNovo* to conduct a study, using the *A. thaliana* plant as our study subject. Our study included 2000 microRNA candidates and the 3'UTRs of 5800 mRNAs. 380 different conditions were analyzed including various tissues and hormonal treatments. Section 5.2 depicts some of the newly discovered microRNAs predicted by our engine, annotated with their potential condition of activity.

2 Search Tree Preliminaries and Formalism

We refer the reader to Figure 1 for an exemplification of the following definitions. We point out in advance that this example is only for demonstration purposes and that in practice the length of the strings corresponding to the leaves should be 20.

- Let $\mathcal{T}_S$ denote a prefix trie over the set of microRNA candidates with each leaf corresponding to one candidate $P \in S$.
- Let v denote any vertex in $\mathcal{T}_S$.
- Let $\mathcal{T}_S^v$ denote the subtree of $\mathcal{T}_S$ rooted at v.
- Let $L(v)$ denote the set of leaves in $\mathcal{T}_S^v$.
- Let $T(v)$ denote a set of mRNAs predicted to be targets of the string corresponding to the path from the root of $\mathcal{T}_S$ to v. These targets are predicted according to Definition 2 and have the potential to form a stable duplex, *i.e.* a hybridization duplex

with energy score $< \delta$, with with this string. (In the example of Figure 1, $T[v_4]$, the targets predicted for the potential microRNA prefix "AU" are $\{3, 53\}$.)

- Let $E(T(v))$ denote the set of expression levels corresponding to $T(v)$ under the input condition Q as reflected by the corresponding expression profile. (According to the expression profile at the bottom of Figure 1, $E(T(v_4)) = \{-70, -10\}$, since these are the expression levels which correspond to $T[v_4] = \{3, 53\}$.)
- $\widehat{T}(v) = (\bigcup_{x \in L(v)} T(x)) \setminus T(v)$ denotes the set of mRNAs which are targets of at least one of the leaves of the subtree rooted in v, but not of v itself. (In the example of Figure 1, $\widehat{T}(v_1) = \{2, 3, 53, 18\}$ since these target numbers occur either in $T(v_3)$ or $T(v_4)$ but not in $T(v_1)$ and $L(v_1) = \{v_3, v_4\}$.)
- Let $diff(x, v) = (|T(x) \setminus T(v)|)$ where v is a node in τ_S and $x \in L(v)$.
- Let $head_i(\widehat{T}(v))$ represent the i first members of the sorted-by-increasing-expression-value $\widehat{T}(v)$.
- Let $s = |S|$ denote the number of microRNA candidates under consideration.
- Let g denote the number of genes (mRNA sequences) in the database for the organism under consideration.

Since τ_S is constructed over the set S of microRNA strings of size $20nt$ each, the number of nodes in τ_S is $O(s)$. Since each node of τ_S contains $O(g)$ information, $|\tau_S| = O(s\, g)$.

Note that, in contrast to the rest of the information which is associated with the nodes of τ_S, both $E(T(v))$ and $head(\widehat{T}(v))$ are only available dynamically online and can not be computed via off-line pre-processing.

3 The Preprocessing Stage

3.1 Extracting Precursors

To extract candidate microRNAs, we scanned the genome of *A. Thaliana* for potential precursors: stretches of length $70 - 100nt$, which can be folded into a stable hairpin structure with a stem of length $20nt$. In this paper, we only considered cases with almost perfect stems, having at most 3 mismatches with no loops or bulges allowed. However, note that our tool is modular and that the set of candidate microRNAs is supplied to the main stage engine as the input computed during a pre-processing stage. Therefore, one could alternatively "plug in" any of the existing microRNA precursor hunters (see for example [1, 14, 16, 17, 2, 28]) by supplying our engine with an input file of candidate microRNAs generated by these alternative tools.

3.2 Predicting the microRNA Targets

An important basic module of *mirXdeNovo* is the microRNA target prediction engine. Work to date on microRNA target prediction consists of methods that either rely solely on edit distance [9], or combine sequence similarity information (such as measured by Smith-Waterman alignment) with secondary structure prediction by energy minimization [5, 6, 21]. In this paper we predict microRNA targets similarly to [22], using the nearest-neighbor thermodynamics approach developed for predicting RNA secondary structure, by Zuker *et al.* [20, 30]. Thus, we use dynamic programming to find the energetically most favorable duplex between a small microRNA and a large mRNA. Since microRNAs in plants are known to have almost perfect complementarity to their targets, we restrict the size of the allowed gap both in the microRNA and in the target to 3.

Note that the target prediction score reflects, in essence, the existence of a local alignment between a microRNA P and a target mRNA T with a score bounded by δ.

Definition 2. *An mRNA sequence $T = t_1, t_2, \ldots, t_n$ is a $\overline{\text{target}}$ of a microRNA candidate $P = p_1, p_2, \ldots, p_m$ iff $min_{0 \leq i \leq n; 0 \leq j \leq m} DP[i, j] \leq \delta$, where DP denotes the dynamic programming table for computing the existence, bounded by δ, of a local alignment between P and T.*

Observation 1. *Note that Definition 2 is monotone with respect to an increase in the prefix size of P. In other words, assume that Y is a prefix of P, and $Y\sigma$ is an extension of Y by some symbol. Then, the set of targets computed for $Y\sigma$ will always contain the subset computed for Y. Thus, if v_i is a parent of v_j in $\mathcal{T}_S$, $T(v_i) \subseteq T(v_j)$.*

3.3 Constructing the microRNA Candidate Tree $\mathcal{T}_S$

The search tree $\mathcal{T}_S$ is constructed during the preprocessing stage in time and space $O(s\ g)$.

4 The Online Main Stage

In this section we will describe a practical solution to the "one (condition) against all (microRNA candidates)" association search problem (see Definition 1). We will address this problem equipped with the tree $\mathcal{T}_S$ which was constructed off-line. Given as input the new expression profile which corresponds to the specific condition Q and a predefined threshold θ, we will search $\mathcal{T}_S$ to find all leaves $v \in L(root(\mathcal{T}_S))$ such that the Z score for microRNA P represented by leaf v under condition Q is smaller than θ.

Suppose we were to naively compute the set of expression levels associated with each of the leaves of $\mathcal{T}_S$ in order to find the leaves whose associated Z score is below θ. The time complexity of such a task would be $O(s\ g)$ (this is the time complexity of naively solving the "all against one" problem without constructing $\mathcal{T}_S$). Note that the number of nodes in $\mathcal{T}_S$ is $20 \times s$. Therefore, constructing the tree $\mathcal{T}_S$ scales the size of the problem input up only by a constant factor as each node contains $O(g)$ information. We will show in section 5.1 that this initial investment will pay off later during the online computations, since using $\mathcal{T}_S$ to represent the search space will allow us to apply a very strong pruning which will pretty much confine the search to the top of $\mathcal{T}_S$ so that only a subset of the leaves which are likely candidates will actually be reached. In section 4.2 we will assert that this pruning is admissible, i.e. it can be proven than none of the leaves with a $Z \leq \theta$ will be missed.

The search itself is quite simple and is implemented as a pruned pre-order traversal of $\mathcal{T}_S$. At each traversed internal node v we compute a score $Zmin(v)$ which is a lower bound to the Z scores of all the leaves in the sub-tree rooted by v. Recall, that the lower the Z-score, the more significant a candidate is. Thus, two cases can occur when reaching an inner node v in $\mathcal{T}_S$:

- if $Zmin(v) > \theta$ we abandon i.e. prune, the sub-tree rooted by v, knowing that no appropriate candidate resides in its leaves.
- otherwise, we recursively continue the pre-order traversal on the sub-tree rooted at v. The method for computing $Zmin(v)$ is explicitly described in section 4.2.

Due to this simple yet powerful pruning, for each condition, only a small subset r of the nodes of $\mathcal{T}_S$ is explored. In section 4.2 we will show that the work invested by the search engine in computing the pruning decision for each traversed node is $O(g)$. Therefore, the total work invested by the search engine is $O(r\ g)$. In comparison to the $O(s\ g)$ complexity of the naive algorithm we get an $O(r/s)$ speedup potential. Theoretically, s could reach 4^{20}. The expected number of traversed nodes r, on the other hand, remains small since the search is statistically confined to the top (near the root) of $\mathcal{T}_S$. We refer the reader to Section 5.1 and Figure 2 for a demonstration of the pruning power of our search engine in practice.

4.1 Computing the Z-Score of a Leaf

Let v be a microRNA candidate i.e. a leaf of $\mathcal{T}_S$, let $T(v)$ be its set of predicted mRNA targets and $E(T(v))$ the set of expression levels corresponding to $T(v)$ under the input condition (see Figure 1). In order to check whether v induces mRNA cleavage and quick degradation, we want to reject the hypothesis that $E(T(v))$ was sampled from the population of all mRNAs and, therefore, accept the hypothesis that the binding of v to its target induces cleavage and reduces the corresponding mRNA levels, making $E(T(v))$ a population of lower expression levels. Thus, we use the mean and variance of the expression levels of the entire mRNA population as μ and σ^2 and calculate $\Phi(Z)$, the corresponding level of significance i.e. *p-value*, associated with $Z = \frac{\overline{E(T(v))}-\mu}{\sigma/\sqrt{|T(v)|}}$. **If $\Phi(Z)$ is significantly small and $\overline{E(T(v))} \leq \mu$, we predict that v is active under the given input condition**. Note that if one wishes to use a threshold over $\Phi(Z)$, one can use a threshold over Z instead. Thus, by applying a low negative θ to *mirXdeNovo* we achieve two goals: first, we upper bound $\Phi(Z)$, and second, we ensure that $\overline{E(T(v))} \leq \mu$.

4.2 An Admissible Pruning Based on $Zmin(v)$

In this section we describe the work associated with each node v traversed by the search, and show how to admissibly compute $Zmin(v)$, the lower bound to the Zs of all the leaves in the sub-tree rooted by v. We show that the work per node is $O(g)$ and assert that the search is indeed admissible.

Note that, by Observation 1, any target which appears in the set $T(v)$ will also participate in the set of targets $T(x)$ for any node $x \in L(v)$. This allows us to incrementally compute the bound on the Z-potential of $L(v)$, as follows.

When computing the potential of $L(v)$ we aim to exploit all the information which is available at this point in order to efficiently (in $O(g)$) compute a lower bound on

$\overline{E(T(x))}$ for any $x \in L(v)$ (see Section 4.1). We know, by Observation 1, that all members of $T(v)$ participate in $\overline{E(T(x))}$ for any $x \in L(v)$. Furthermore, at this point we can consider the targets of v in $O(g)$ but do not wish to consider the targets of all the leaves $x \in L(v)$ as this may sum up to $O(s\,g)$.

Therefore, the potential score for $\overline{E(T(x))}$ could be lower bounded by considering a best-case scenario in which the additional targets of x which are not included in $T(v)$ are the lowest-scoring members of $E(\widehat{T}(v))$. In the best-case scenario, the $diff(x, v)$ targets in $T(x)$ which are not included in $T(v)$ are the lowest-scoring members of $E(\widehat{T}(v))$. Therefore, for each leaf $x \in L(v)$ we can compute the lowest possible score obtained by adding to $T(v)$ the $diff(x, v)$ lowest-valued members of $E(\widehat{T}(v))$. Note, however, that when using this approach there is no need to consider all $O(s)$ leaves in $L(v)$, as $diff(x, v)$ can only assume up to g distinct values. Thus, the binding-scenario described above need only be computed for each of the $O(g)$ possible distinct $diff(x, v)$ values among all $x \in L(v)$. Based on Equation 1, $Zmin(v)$ will thus be computed as follows.

$$Zmin(v) = min_{(i|\exists x \in L(v)\,:\, diff(x,v)=i)}\langle \frac{\overline{E(T(v)) \cup E(head_i(\widehat{T}(v)))} - \mu}{\sigma/\sqrt{i + |T(v)|}} \rangle \tag{2}$$

Example: consider node v_1 in the example presented in Figure 1. When considering whether it is safe to prune $\mathcal{T}_S^{v_1}$ we wish to assess whether any of the leaves in $L(v_1)$ may yield a $Z < \theta$. Note that $v_3 \in L(v_1)$ and that $diff(v_3, v_1) = 2$, since v_3 has four targets while v_1 has two targets. Therefore, when computing the pruning decision for v_1, the term contributed by v_3 to the pruning decision minimization, which should be a lower bound on $\overline{E(v_3)}$, will be based on $\overline{E(T(v_1)) \cup E(head_2(\widehat{T}(v_1))} = \overline{E(\{1,6\} \cup \{3,53\})} = \overline{\{-1000, -100, -70, -10\}} = -295$. The real $\overline{E(T(v_3))}$, on the other hand, is $\overline{E(\{1,2,6,18\})} = \overline{\{-1000, -100, 20, 30\}} = -262.5$, which is indeed above the -295 lower bound computed by the pruning decision.

Lemma 1. *$Zmin(v)$, as computed in equation 2, is a lower bound to the Z-scores of any of the leaves in $L(v)$.*

Computing the Pruning Decision Per Node in $O(g)$ time. In order to efficiently compute equation 2, the dynamic set $E(\widehat{T}(v))$ must be ordered by increasing expression value for each traversed node v (this will allow efficient $head_i$ queries). Therefore, the work per node v consists of updating the sorted list $E(\widehat{T}(v))$ and then computing $Zmin(v)$. Each of the two tasks can be done in $O(g)$ time, as follows.

1. By Observation 1 it is clear that the sorted $E(\widehat{T}(v))$ can be incrementally obtained from the sorted list of its parent in $O(g)$ time by removing one or more of its items. Thus, the set of expression values needs only be sorted once at the root and then incrementally updated by inheritance in $O(g)$ time.
2. Computing $Zmin(v)$ in $O(g)$ time: this is achieved by incrementally computing weighted averages.

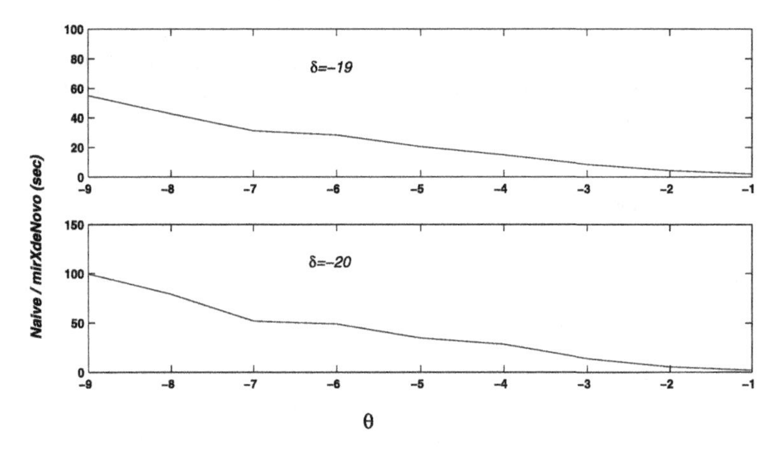

Fig. 2. The running time of *mirXdeNovo* compared to that of the naive search for $\delta = -19, -20$ and $\theta = -9 \ldots -1$

5 Experimental Results

5.1 Benchmarking mirXdeNovo

We used the sequences of the 3'UTRs of 5800 *A. thaliana* mRNAs with 2000 microRNA candidates to assemble $\mathcal{T}_S$. Then, we ran *mirXdeNovo* and the naive search each over 380 different conditions and recorded the run-times [1].

The results, which are summarized in Figure 2, indicate the advantage of *mirXdeNovo* over the naive search in a wide range of practical setups. We hypothesize that the speed of *mirXdeNovo* results from the strong pruning which occurs at high (*i.e.* close to the root) levels of $\mathcal{T}_S$. This pruning clearly become stronger as θ (see the x-axis) decreases and δ increases.

5.2 Associating New Potential *A. Thaliana* microRNAs with Various Activity Conditions

Different expression patterns are part of how a plant grows, develops, and adapts to environmental changes. In this section we use our approach to shed light on the contributions of various new microRNAs to these expression patterns in *A. thaliana*. We assembled $\mathcal{T}_S$ over the 3'UTRs of 5800 mRNAs and 3000 microRNA candidates and used it to analyze the expression profiles corresponding to 380 different conditions. We used the same data as described before, setting $\delta = -19$ and $\theta = -3$.

Table 1 depicts some significant relations discovered in our study. Our approach enables us to discover interesting relationships, *e.g.*:

[1] The microarray data describing the genome-wide expression profile for each of the conditions, as well as the 3'UTRs of the mRNA sequences were both retrieved from the *TAIR* database `http://www.Arabidopsis.org/`. The microRNA candidate were discovered as described in section 3.1.

Table 1. Newly discovered microRNAs and their hypothesized condition of activity. $E(T(microRNA))$ represents the mean of the expression levels corresponding to the set of the mRNA predicted targets of the candidate.

microRNA	***Sequence***	***Condition***	$\Phi(Z)$	$E(T(microRNA))$	μ	***Target Number***
microRNA-1	GGAGAGAGAGAGAGAUGAAAAA	Iron deficiency	10^{-7}	-0.02	0.11	389
microRNA-2	GGGAGUUUAUUUAUAUAUAU	Flowers	10^{-6}	-2.7	-0. 008	3
microRNA-3	GGGAGAGAGAGAGAUGAAAA	Zinc deficiency	10^{-6}	0.03	0.13	460
microRNA-4	AUUUUUAAUUUUGGUUAAACC	Auxin Response	0.005	-0.23	0	9
microRNA-5	CUUAUCAUUCUUCUUCCACU	Chlorophyll Starvation	0.0005	-0.2	0.13	61
microRNA-6	UAACAAUACCAGUUGUUUAA	Inhibition of the mitochondrial electron transport chain	10^{-5}	-0.04	0.02	7
microRNA-7	AGAAGAAUGAUAAGAAGACG	Infection response	0.00052	-0.18	0.05	18
microRNA-8	UUUAAAAAUUUAACACUUAUC	Mechano Stimulation	10^{-12}	-2	0	2
microRNA-9	AAAUUAUAUUUAGUUAUAAAU	Cell Death	0.0004	-0.7	-0.003	2
microRNA-10	GAUUUAUUUUAUUGUAUUUUU	Shade Avoidance	0.005	-0.5	0.03	3
microRNA-11	UAGGAACUAAAAAGAUAAUUA	Potassium Nitrate Excess	0.001	-0.34	-0.03	12

1. MicroRNA-4 was found to be active in response to Auxin. Auxin is a plant hormone involved in growth and development. It is well known that the response of cells to hormones usually involves global transcriptional changes. Correspondingly, we found that four out of the nine (44%) of the predicted targets of MicroRNA-4 are transcription factor genes. The significance of this observation is highlighted by the fact that, according to the data in the TAIR database, only about 2% of the Arabidopsis genes are indeed transcription factors.
2. MicroRNA-11 was found to be active under Potassium-Nitrate excess. Concurrently, two out of 12 (16%) of the predicted targets of microRNA-11 are involved in transport (compared to $\sim$ 4.8% of the general Arabidopsis gene population). We speculate that microRNA-11 destroys the mRNAs of these targets in order to limit the entrance of Potassium-Nitrate into the cell under excess condition.

Acknowledgments. Many thanks to Zohar Yakhini (Agilent Technologies and Technion) for inspiration as well as for very helpful discussions and comments. The authors are also grateful to Ron Y. Pinter (Technion) for fruitful discussions as well as for encouragement and support. Additional thanks to the anonymous WABI referees for their helpful detailed comments.

References

1. A. Adai, C. Johnson, S. Mlotshwa, S. Archer-Evans, V. Manocha, V. Vance, and V. Sundaresan. Computational prediction of miRNAs in arabidopsis thaliana. *Proc. Natl. Acad. Sci. U S A*, 15:78–91, 2005.
2. E. Bonnet, J. Wuyts, P. Rouze, and Y. Van de Peer. Detection of 91 potential conserved plant microRNAs in Arabidopsis thaliana and Oryza sativa identifies important target genes. *Proc. Natl. Acad. Sci. U S A*, 101:11511–11516, 2004.
3. C.V. Cabrera, J.J. Lee, J.W. Ellison, R.J. Britten, and E.H. Davidson. Regulation of cytoplasmic mRNA prevalence in sea urchin embryos. Rates of appearance and turnover for specific sequences. *J Mol Biol*, 174(1):85–111, 1984.
4. J.L. Casey, M.W. Hentze, D.M. Koeller, S.W. Caughman, T.A. Rouault, R.D. Klausner, and J.B. Harford. Iron-responsive elements: regulatory RNA sequences that control mRNA levels and translation. *Science*, 240:924–928, 1988.
5. A.J. Enright and et al. MicroRNA targets in drosophila. *Genome Biol., Pubmed*, 5(1), 12 2003.
6. A. Stark et al. Identification of Drosophila microRNA targets. *Plos. Biol., Pubmed*, 1(3), 2003.
7. G. Tang et al. Framework for RNA silencing in plants. *Genes Dev.*, 17:49–63, 2003.
8. K.D. Kasschau et al. P1/HC-Pro, a viral suppressor of RNA silencing, interferes with Arabidopsis development and miRNA function. *Dev. Cell*, 4:205–217, 2003.
9. M.W. Rhoades et al. Prediction of plant microRNA targets. *Cell*, 23:513–520, 2002.
10. S. Yekta et al. MicroRNA-directed cleavage of HOXB8 mRNA. *Science*, 304:594–596, 2004.
11. Y. Grad, J. Aach, G.D. Hayes, B.J Reinhart, G.M. Church, G. Ruvkun, and J. Kim. Computational and experimental identification of C. elegans microRNAs. *Mol Cell*, 11:1253–1263, 2003.
12. N. Heintz, H.L. Sive, and R.G. Roeder. Regulation of human histone gene expression: kinetics of accumulation and changes in the rate of synthesis and in the half-lives of individual histone mRNAs during the hela cell cycle. *Mol Cell Biol*, 3(4):539–550, 1983.
13. H.M. Jack and M. Wabl. Immunoglobulin mRNA stability varies during B lymphocyte differentiation. *EMBO J.*, 7(4):1041–1046, 1988.
14. B.T. Zhang J.W. Nam, W.J. Lee. Computational methods for identification of human microrna precursors. *Lecture Notes in Artificial Intelligence*, 3157:732–741, 2004.
15. A. Krowczynska, R. Yenofsky, and G. Brawerman. Regulation of messenger RNA stability in mouse erythroleukemia cells. *J Mol Biol*, 181(2):231–239, 1985.
16. E.C. Lai. Predicting and validating microRNA targets. *Genome Biol*, 5:115, 2004.
17. M. Legendre, A. Lambert, and D. Gautheret. Profile-based detection of microrna precursors in animal genomes. *Bioinformatics*, 21(7):841–845, 2005.
18. L.P. Lim, N.C. Lau, E.G. Weinstein, A. Abdelhakim, S. Yekta, M.W. Rhoades, C.B. Burge, and D.P. Bartel. The microRNAs of C. elegans. *Genes and Development*, 17:991–1008, 2003.
19. C. Llave, Z. Xie, K.D. Kasschau, and J.C. Carrington. Cleavage of scarecrow-like mRNA targets directed by a class of Arabidopsis miRNA. *Science*, 297:2053–2056, 2002.
20. D.H. Mathews, J. Sabina, M. Zuker, and D.H. Turner. Expanded sequence dependence of thermodynamic parameters improves prediction of RNA secondary structure. *J. Mol. Biol.*, 288:911–940, 1999.
21. N. Rajewsky and N.C. Socci. Computational identification of microRNA targets. *Genome Biology*, 5, 2004.

22. M. Rehmsmeier, P. Steffen, M. Hochsmann, and R. Giegerich. Fast and effective prediction of microRNA/target duplexes. *RNA*, 10:1507–1517, 2004.
23. B.J. Reinhart, F.J. Slack, M. Basson, A.E. Pasquinelli, J.C. Bettinger, A.E. Rougvie H.R., Horvitz, and G. Ruvkun. The 21-nucleotide let-7 RNA regulates developmental timing in caenorhabditis elegans. *Nature*, 403:901–906, 2000.
24. J Ross. mRNA stability in mammalian cells. *Microbiol Rev*, 59(3):423–450, 1995.
25. J Ross. Control of messenger RNA stability in higher eukaryotes. *Trends. Genet.*, 12(5):171–175, 1996.
26. C.M. Sorenson, P.A. Hart, and J. Ross. Analysis of herpes simplex virus-induced mRNA destabilizing activity using an in vitro mRNA decay system. *Nucleic Acids Res.*, 19:4459–4465, 1991.
27. A.M. Thomson, J.T. Rogers, and P.J. Leedman. Iron-regulatory proteins, iron-responsive elements and ferritin mRNA translation. *Int J Biochem Cell Biol*, 31(10):1139–1152, 1999.
28. P.E. Warburton, J. Giordano, F. Cheung, Y. Gelfand, and G. Benson. Inverted repeat structure of the human genome: the x-chromosome contains a preponderance of large, highly homologous inverted repeats that contain testes genes. *Genome Res*, pages 1861–1869, 2004.
29. C. Zilberstein, M. Ukelson, R. Pinter, and Z. Yakhini. A high-throughput approach for associating micrornas with their activity conditions. In *The ninth annual international conference on research in computational molecular biology*, 2005.
30. M. Zuker. Mfold web server for nucleic acid folding and hybridization prediction. *Nucleic Acids Res.*, 31(13):3406–3415, 2003.

Clustering Gene Expression Series with Prior Knowledge

Laurent Bréhélin

Laboratoire d'Informatique, Robotique et Microélectronique de Montpellier,
161, rue Ada, 34392 Montpellier Cédex 5, France
brehelin@lirmm.fr

Abstract. Microarrays allow monitoring of thousands of genes over time periods. Recently, gene clustering approaches specially adapted to deal with the time dependences of these data have been proposed. According to these methods, we investigate here how to use prior knowledge about the approximate profile of some classes to improve the classification result. We propose a Bayesian approach to this problem. A mixture model is used to describe and classify the data. The parameters of this model are constrained by a prior distribution defined with a new type of model that can express both our prior knowledge about the profile of classes of interest and the temporal nature of the data. Then, an EM algorithm estimates the parameters of the mixture model by maximizing its posterior probability.

Supplementary Material:
http://www.lirmm.fr/~brehelin/WABI05.pdf

1 Introduction

Technological advances such as microarrays allow us to simultaneously measure the level of expression of thousands of genes in a given tissue over time —for example along the cell cycle [1]. In the following, such a series of gene expression measurements is called an *expression series*. One common problem of gene expression data analysis is the identification of co-regulated genes. This problem naturally turns into a gene clustering problem. Until recently, expression series have been analyzed with methods that do not take the time dependences into account. Such methods include hierarchical clustering with Euclidean distance [2], k-means approaches [3,4] and the Self Organizing Maps [5,6]. Since these methods are unable to explicitly deal with the data order, permuting two or more time points in all series does not change the clustering result. A few methods specially adapted to expression series have recently been proposed. These methods involve probabilistic modeling of the data. For example, [7] use autoregressive models of order p. [8] use cubic splines with a probabilistic component to model the classes, while [9] model each class of gene with Hidden Markov Models (HMMs) [10].

R. Casadio and G. Myers (Eds.): WABI 2005, LNBI 3692, pp. 27–38, 2005.

Our aim here is to investigate how to explicitly use *rough* prior knowledge about the general shape of interesting classes. By *general shape*, we mean elementary and potentially incomplete information about the evolution of the mean expression level of the classes over time. This can, for example, be knowledge like: *"Classes with increasing expression level"*, *"Classes with bell curve shapes"*, *"Classes with high expression level in the beginning of the series"*, etc. Of course we do not know the profile of *all* the gene classes, but sometimes we are more concerned with one or more classes. For example, in the study of [1] on the Yeast cell cycle, the authors are interested in finding the cycle-regulated genes, and thus look for sinusoidal shape classes. In a similar way, we sometimes search for genes which tend to be quickly over- (or under-) expressed at the beginning of the series —in response to a given treatment, for example. A problem of importance that arises when the awaited classes are sparse —i.e., there are few interesting genes with regards to all the other ones— is that standard methods can completely omit these classes. This results in a final clustering where the interesting genes are lost among many other genes, in one or more classes that do not show the desired profile.

The approach we propose here tackles this problem. When information about one or several class shapes are available, these are directly integrated into the model, thus favoring classes with the desired profiles, and putting the other genes in separate classes. On the other hand, when no a priori information is available, the method allows a classical clustering of the series that deals with the temporal nature of the data in a very intuitive way.We use a Bayesian approach for this purpose. The approach involves two types of models. The first one is a probabilistic mixture model used to describe and classify the expression series. Parameters of this model are unknown and have to be estimated for the clustering. A second model, close to the HMMs and called *HPM* —for *Hidden Phase Model*—, is used to express our a priori knowledge (or simply the temporal feature of the data). We define two types of HPMs which can be used according to the situation: probabilistic and non-probabilistic HPMs. These models are completely specified by the user, and their parameters do not have to be estimated. They are used to define a prior probability distribution over the parameters of the mixture model. These parameters are estimated by maximizing the posterior probability of the model through an EM algorithm [11].

The next section presents our method, the mixture model, the two types of HPMs and the learning algorithm. In Section 3 we evaluate our method. We conclude in Section 4.

2 Method

2.1 Principle

Let $\mathcal{X}$ be a set of N expression series of length T. We assume that the data arise from a mixture model [12] with C components. We denote π_c as the prior probability of component c, and we have $\sum_{c=1}^{C} \pi_c = 1$. We assume that conditionally to component c, expression values at each time $t \in [1, T]$ are independent and follow a Gaussian distribution of mean μ_{ct} and variance σ_{ct}^2. The shape of component c is defined by the sequence of means $\mu_{c1} \dots \mu_{cT}$. We then have a probabilistic model of parameters $\Theta = (\pi_1, \dots, \pi_C, \theta_1, \dots, \theta_C)$ with $\theta_c = (\mu_{c1}, \dots, \mu_{cT}, \sigma_{c1}^2, \dots, \sigma_{cT}^2)$. The probability of an expression series $X = x_1 \dots x_T$ in this model is

$$P(X|\Theta) = \sum_{c=1}^{C} \pi_c \prod_{t=1}^{T} P(x_t|\mu_{ct}, \sigma_{ct}^2),$$

with $P(x_t|\mu_{ct}, \sigma_{ct}^2) = \mathcal{N}(x_t; \mu_{ct}, \sigma_{ct}^2)$. Under the assumption that series of $\mathcal{X}$ are independent, the likelihood of Θ is given by

$$L(\Theta|\mathcal{X}) = P(\mathcal{X}|\Theta) = \prod_{X \in \mathcal{X}} P(X|\Theta). \tag{1}$$

In a clustering task, the standard approach to classify a set of expression series $\mathcal{X}$ involves estimating parameters Θ that maximize Formula (1) (Maximum Likelihood Principle), and then assigning the most probable component c_{MAP} (*MAP* stands for *maximum a posteriori*) to each series $X \in \mathcal{X}$:

$$c_{\mathrm{MAP}} = \operatorname*{argmax}_{c=1 \dots C} P(c|X, \Theta) = \operatorname*{argmax}_{c=1 \dots C} \pi_c P(X|c, \Theta) \tag{2}$$

Note that finding parameters Θ that maximize (1) is a difficult task. However, approximate solutions can be inferred with EM algorithms [11].

The above mixture model does not explicitly take into account the potential dependences between times, nor any prior knowledge about the profile of the most interesting classes. Our aim is to constraint one or some components to follow a given profile, while leaving the other components free of constraints so that they can "collect" the expression series that do not have the desired profile. For example, if we are looking for classes with bell curves, we would build a 10 component model, with 5 bell-constrained and 5 unconstrained components. We thus propose to use a Bayesian approach, which introduces knowledge by way of a prior distribution of Θ —see for example [13] for a general introduction to Bayesian theory. Simply speaking, our idea is to define a prior distribution $P(\Theta)$ which is merely the product of the prior probability of the sequences of means $\mu_{c1} \dots \mu_{cT}$ associated with each component. Moreover, we want the prior probability of a given mean sequence for component c as follows: (i) the more the sequence agrees with the constraints associated with c, the higher its prior probability; (ii) sequences that disagree with the constraints have probability zero.

With a prior, we can write the posterior probability of Θ as

$$P(\Theta|\mathcal{X}) = \frac{P(\mathcal{X}|\Theta)P(\Theta)}{P(X)} \propto P(\mathcal{X}|\Theta)P(\Theta). \tag{3}$$

In this Bayesian framework, parameters Θ are estimated by maximizing the posterior probability —Equation (3)— instead of the likelihood —Expression (1).

However, maximizing the posterior probability is generally more difficult than maximizing the likelihood. For example, the classical re-estimation formulae of the EM algorithm do not directly apply and, depending on the form of the chosen prior distribution, it may be hard to perform the task in reasonable time.

In our case, we first discretize the space of the means μ_{ct} in order to be able to introduce various bits of knowledge and constraints about the profiles, as well as to efficiently estimate the parameters of the model. Since we know the maximal and minimal expression values taken by the series in $\mathcal{X}$ (say $x_{\max}$ and $x_{\min}$), we already know an upper and lower bound of the space of the means. Now we discretize this space in M equidistant steps, so that the lower and higher steps are equal to $x_{\min}$ and $x_{\max}$, respectively. Of course M is chosen to be sufficiently large (e.g. $M = 30$) to allow accurate representation of the data. Steps are named by their number, so M is the highest step. In this discretized mean space, our probabilistic model is re-defined as $\Theta = (\pi_1, \ldots, \pi_C, \theta_1, \ldots, \theta_C)$ with $\theta_c = (l_{c1}, \ldots, l_{cT}, \sigma^2_{c1}, \ldots, \sigma^2_{cT})$, with $l_{ct} \in \{1, \ldots, M\}$. We denote $m : \{1, \ldots, M\} \to [x_{\min}, x_{\max}]$ as the map function that associates step l with its expression level. The probability of an expression series $X \in \mathcal{X}$ is rewritten as

$$P(X|\Theta) = \sum_{c=1}^{C} \pi_c \prod_{t=1}^{T} P(x_t | l_{ct}, \sigma^2_{ct}),$$

with $P(x_t|l_{ct}, \sigma^2_{ct}) = \mathcal{N}(x_t; m(l_{ct}), \sigma^2_{ct})$ that follows a Gaussian distribution of mean equal to the level of expression associated with step l_{ct}, and variance σ^2_{ct}. In the following, the step sequence $l_{c1} \ldots l_{cT}$ associated with class c —and which defines its shape— is denoted as L_c. Note finally that the discretization only involves the means of the model, and not the space of the expression levels of the data. These, as well as the model variances σ^2_{ct}, remain in a continuous space.

2.2 Defining the Prior Distribution

Fist we define a new type of model called *Hidden Phase Models* (or *HPMs*), close to models like HMMs and finite automata [14]. These HPMs are used to express the desired profiles of the components, and each component c is then associated with a given HPM H_c. We define two types of HPMs: probabilistic and non-probabilistic HPMs. We next show how to derive the prior distribution of Θ from the HPMs.

Hidden Phase Models. The general assumption behind HPMs is that the genes of a given component pass through *phases* or *biological states* over the time. This means that, for a given component, we assume that some ranges of consecutive times actually correspond to the same biological state. These phases are hidden, but they affect the mean expression level evolution of the component. For example, some phases induce an increase in the mean level expression level while others tend to decrease or stabilize the level. In the same manner, the increase (or decrease) can be high for some phases and low for others, etc.

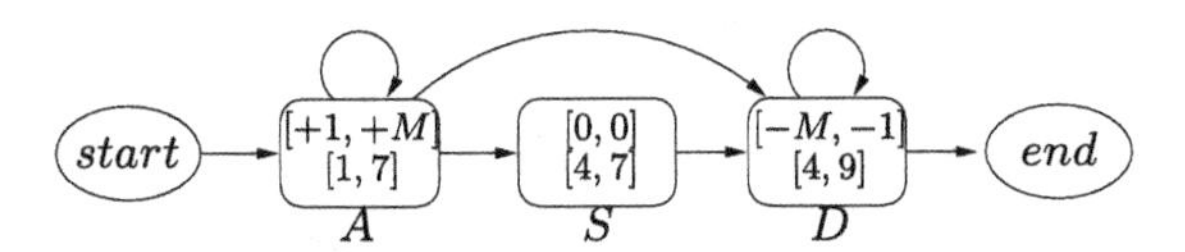

Fig. 1. An HPM for clustering 9-time expression series. In each state, upper and lower intervals represent the step-difference and time intervals associated with the state, respectively. This HPM induces bell curve shapes.

A (non-probabilistic) HPM is defined by a quadruplet $(\mathcal{S}, \delta, \epsilon, \tau)$, where

- $\mathcal{S}$ is a set of states representing the different phases; $\mathcal{S}$ contains two special states, *start* and *end*, which are used to initiate and conclude a sequence, respectively.
- $\delta : \mathcal{S} \times \mathcal{S} \mapsto \{0, 1\}$ is a function describing the authorized transitions between states. We denote $Out(s)$ as the set of states that can be reached from s.
- ϵ is a function that associates each state $s \in \mathcal{S}$ with an interval of integers defining the minimal and maximal differences of steps that can be observed between times t and $t-1$ when genes are in state s at time t. For example, if $\epsilon(s) = [1, 3]$, this means that if the genes of the component are in phase s at time t then the step difference $(l_t - l_{t-1})$ is between 1 and 3 (so phase s increases the expression level).
- τ is a function that associates each state $s \in \mathcal{S}$ with the interval of time the state can be reached. For example, if $\tau(s) = [3, 5]$ then the genes can be in state s between times 3 and 5 included.

An HPM example is depicted in Figure 1.

Now we can see how to express our prior knowledge with an HPM. Actually an HPM defines a set of *compatible* step sequences. We say that a step sequence $L = l_1 \dots l_T$ is compatible with an HPM H if there is a state sequence $s_0 \dots s_{T+1}$ —with $s_0 = start$ and $s_{T+1} = end$— in H, which is compatible with L. And we say that a state sequence $s_0 \dots s_{T+1}$ is compatible with L *iff* for each time $1 \leq t \leq T$ we have: i) t included in the time interval $\tau(s_t)$; ii) $\forall t \geq 2$, $(l_t - l_{t-1})$ included in $\epsilon(s_t)$ —for $t = 1$, as we do not know l_0, the genes can be in any phase so s_1 can be any state. Considering the step sequence on the top of Figure 2, a compatible phase sequence in the HPM of Figure 1 is, for example, *start* $- A - A - A - A - S - D - D - D - D -$ *end*. For the step sequence on the right, there is no compatible phase sequence in this HPM. In brief, building an HPM involves designing an HPM such that the compatible sequences have the desired profile. For example, the HPM of Figure 1 is well suited for the discovery of bell curve classes.

Probabilistic HPMs. Non probabilistic HPMs can be used to express strong constraints. They are generally sufficient to express knowledge about simple or well defined profiles. For more complex knowledge, or when we do not have any information about profiles and just want to express the fact that we are dealing with series data, these models can be unsuitable. Then probabilistic HPMs can be more suitable.

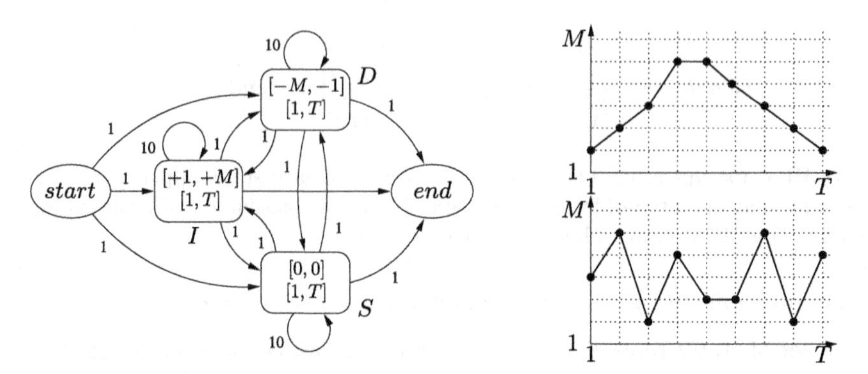

Fig. 2. Left, a probabilistic HPM for clustering expression series without prior knowledge about the form of the profiles. Right, two examples of step sequences.

A probabilistic HPM is defined by a quintuplet $(\mathcal{S}, \delta, \epsilon, \tau, w)$, where $\mathcal{S}$, δ, ϵ, and τ are the same as for non-probabilistic HPMs, and $w : \mathcal{S} \times \mathcal{S} \mapsto \mathbb{R}^+$ is a function associating a weight with each authorized transition. These weights are used to compute the transition probabilities from state to state. Due to the time constraints associated with the states by way of the τ function, transition probabilities are time dependent, so we cannot simply label transitions with a probability as is done for classical HMMs. In contrast, the probability, denoted as $P(s|s', t)$, to reach state s from state s' at time t is computed as follows:

$$P(s|s',t) = \begin{cases} 0 \text{ if } t \notin \tau(s); \\ w(s)/\left(\sum_{s'' \in Out(s') \mid t \in \tau(s'')} w(s'')\right) \text{ else.} \end{cases} \quad (4)$$

One example of probabilistic HPM is depicted in Figure 2.

Probabilistic HPMs also define compatible step sequences. Moreover, all compatible sequences do not have the same probability. Let H be a probabilistic HPM and $S = s_0, s_1 \ldots s_T, s_{T+1}$ a state sequence in this HPM. The probability of this sequence given H is defined by

$$P(S|H) = \prod_{t=1}^{T+1} P(s_t|s_{t-1}, t). \quad (5)$$

This model enables us to introduce more knowledge about the desired classes. For example, when we do not have any information about interesting profiles, the only thing we know is that we have to classify expression series. This means that we are seeking relatively "regular" profiles, in contrast to chaotic spiky profiles as that depicted on the bottom of Figure 2. This knowledge can be easily expressed with the probabilistic three-states HPM of Figure 2: one state (I) induces increasing steps, one (D) induces a decrease, and the last (S) induces stability. Moreover, it is assumed that, at each time, the probability of staying in the same state is higher than the probability of departure from it (weights on loops are higher than on other transitions). This HPM is compatible with any step sequence of length 9. However all sequences do not have the same probability, and spiky sequences involving many state changes are not favored.

Note that given a step sequence L, there are potentially many state sequences compatible with L. In reference to the HMM literature, the sequence of phases compatible with L which has the highest probability is called the *Viterbi sequence* of L [10], and is denoted as $V^L = v_0^L \dots v_{T+1}^L$. For example, the Viterbi sequences of the two step sequences of Figure 2 in the HPM of Figure 2, are $start-I-I-I-I-S-D-D-D-D-end$ and $start-I-I-D-I-D-S-I-D-I-end$, respectively.

Defining prior with HPMs. First we assume that prior probabilities of parameters π_c, L_c and σ_{ct}^2 are independent, as well as the C sets of parameters L_c and $(\sigma_{c1}^2, \dots, \sigma_{cT}^2)$, i.e., the probability distribution can be written as:

$$P(\Theta) = P(\pi_1, \dots, \pi_C) \prod_{c=1}^{C} P(L_c) \prod_{c=1}^{C} P(\sigma_{c1}^2, \dots, \sigma_{cT}^2).$$

Next we assume that distributions $P(\pi_1, \dots, \pi_C)$ and $P(\sigma_{c1}^2, \dots, \sigma_{cT}^2)$ are uninformative and that probabilities $P(L_c)$ are the only ones that express our knowledge.

Let c be a component and H_c a non probabilistic HPM associated with this class. A prior distribution of parameters L_c can be defined with H_c by assuming that the step sequences incompatible with H_c have probability zero while compatible sequences have all the same probability, i.e.,

$$P(L|H_c) = \begin{cases} 0 \text{ if } L \text{ is incompatible with } H_c; \\ K_c \text{ else,} \end{cases} \tag{6}$$

with K_c such that $\sum_{L \in \mathcal{L}_T} P(L) = 1$, with $\mathcal{L}_T$ being the set of length T sequences.

For probabilistic HPM, we want the prior probability of a step sequence L to be proportional to the Viterbi sequence of L in H_c. Then, we set

$$P(L|H_c) = \begin{cases} 0 \text{ if } L \text{ is incompatible with } H_c; \\ K'_c \cdot P(V^L|H_c) \text{ else,} \end{cases} \tag{7}$$

with K'_c such that $\sum_{L \in \mathcal{L}_T} P(L) = 1$. For example, for the HPM of Figure 2, the prior probabilities of the two step sequences are proportional to $1/3 \cdot .7 \cdot .7 \cdot .7 \cdot .1 \cdot .1 \cdot .7 \cdot .7 \cdot .7 \cdot .1 \sim 3.9 \cdot 10^{-5}$ and $1/3 \cdot .7 \cdot .1 \cdot .1 \cdot .1 \cdot .1 \cdot .1 \cdot .1 \cdot .1 \cdot .1 \sim 2.3 \cdot 10^{-10}$, respectively. The spiky sequence is then less likely than the other one, which agrees with our prior intuition.

A prior distribution of the step sequences of length T can then be defined with a probabilistic or a non-probabilistic HPM. In practice, one or more components can be associated with a given HPM (e.g. that of Figure 1), and the other ones with a less informative HPM like that of Figure 2. We then have

$$P(\Theta) \propto \prod_{c=1}^{C} P(L_c|H_c). \tag{8}$$

2.3 Learning

Here we briefly describe the learning algorithm used to estimate parameters Θ of the mixture model. A more detailed version can be found in the supplementary information material[1]. It is an EM algorithm that searches for parameters that maximize Expression (3). We only give the algorithm used for probabilistic HPMs, since that for non-probabilistic ones can be easily adapted.

Let us first define the *complete-data* likelihood. Likelihood of Expression (1) is actually the incomplete-data likelihood, since the real components of series $X \in \mathcal{X}$ are unknown. Under the assumption that this set of components $\mathcal{C} = \{c_X \in \{1, \ldots, C\}, \forall X \in \mathcal{X}\}$ is known, the complete-data likelihood can be written as

$$L(\Theta|\mathcal{X}, \mathcal{C}) = P(\mathcal{X}, \mathcal{C}|\Theta) = \prod_{X \in \mathcal{X}} \pi_{c_X} \prod_{t=1}^{T} P(x_t; l_{c_X t}, \sigma^2_{c_X t}).$$

The EM algorithm is an iterative algorithm that starts from an initial set of parameters $\Theta^{(0)}$, and iteratively reestimates the parameters at each step of the process. Let $Q(\Theta, \Theta^{(i)})$ denote the expectation, on the space of the hidden variables $\mathcal{C}$, of the logarithm of the complete-data likelihood, given the observed data $\mathcal{X}$ and parameters $\Theta^{(i)}$ at step i:

$$Q(\Theta, \Theta^{(i)}) = \sum_{\mathcal{C} \in \mathbf{C}} \log P(\mathcal{X}, \mathcal{C}|\Theta) P(\mathcal{C}|\mathcal{X}, \Theta^{(i)}),$$

with $\mathbf{C}$ being the space of values $\mathcal{C}$ can take. From [11], one can maximize Expression (3) by searching at each step of the algorithm for parameters π_c^*, L_c^* and $\sigma^{2\,*}_{ct}$ that maximize the quantity

$$Q(\Theta, \Theta^{(i)}) + \log P(\Theta). \tag{9}$$

Since $P(\Theta)$ is not related to the parameters π_c, after some calculus, an expression can be derived for π_c^* that maximizes Expression (9):

$$\pi_c^* = \frac{1}{|\mathcal{X}|} \sum_{X \in \mathcal{X}} P(c|X, \Theta^{(i)}). \tag{10}$$

Now, due to our independence assumptions, one can estimate the L_c and σ^2_{ct} that maximize Expression (9) for each component c independently. As for parameters π_c, σ^2_{ct} are not involved in the expression of $P(\Theta)$. Moreover, since the σ^2_{ct} associated with time t is independent of all the other times, the expression of $\sigma^{2\,*}_{ct}$ that maximizes (9) depends solely on the step l^*_{ct} in L^*_c:

$$\sigma^{2\,*}_{ct} = \frac{\sum_{X \in \mathcal{X}} (x_t - m(l^*_{ct}))^2 P(c|X, \Theta^{(i)})}{\sum_{X \in \mathcal{X}} P(c|X, \Theta^{(i)})}. \tag{11}$$

[1] `http://www.lirmm.fr/~brehelin/WABI05.pdf`

For L_c the situation is quite different since it is involved in the expression of $P(\Theta)$. The L_c that maximizes Expression (9) depends both on the data and on its Viterbi path in H_c and hence the different steps l^*_{ct} of L^*_c cannot be estimated independently. However, the step space is of finite size, so the space of the step sequences of length T is also finite. One way to compute the new L_c would be to enumerate all possible step sequences and then select the one that maximizes Expression (9). However, as the total number of length T sequences is equal to M^T, enumerating them all is clearly not suitable. Instead, we use a dynamic programming approach that iteratively computes the best sequence without enumerating all the solutions. Briefly, for each step l and each time t, we compute iteratively, from $t = 1$ to T, the best sequence —with regard to Expression(9)— that ends on step l at time t. At each iteration and for each step l, this best sequence is computed using the results of the previous iteration, and at the end of the process the best sequence L^*_c has thus been computed in polynomial time.

The learning algorithm is depicted in Figure 3. When no better solution is available, the initial parameter values can be set randomly. Thanks to the EM properties, the posterior probability $P(\Theta|\mathcal{X})$ —and hence $P(\mathcal{X}|\Theta)P(\Theta)$— increases at each loop of the algorithm, until a local optimum is reach. Then it continues to increase but to a much lesser extent. A practical way to detect the convergence is to check the increase at each loop and to stop the algorithm when this value goes under a given boundary.

1 Set parameters to initial values
2 **repeat**
3 | **for** $c = 1$ to C **do**
4 | | compute π^*_c with Formula (10)
5 | | Find the optimal step sequence $L^*_c = l^*_{c1} \dots l^*_{cT}$ with the dynamic programming algorithm
6 | | **foreach** time t **do** compute $\sigma^{2\,*}_{ct}$ from l^*_{ct} with Formula (11)
7 | Compute $P(\mathcal{X}|\Theta)P(\Theta)$
8 **until** convergence

Fig. 3. Learning algorithm

The total time complexity of the learning algorithm is $O(BCTM^2R^2N)$ —see supplementary information for details—, with B, C, T, M, R and N the maximal number of loops of the EM algorithm, the number of components of the mixture model, the number of time points of the data, the size of the step space, the maximal number of states of the HPMs, and the number of expression series to classify, respectively. In practice, N is potentially high (some thousands), T and R are relatively low (ten or less), M is around thirty, and less than one hundred loops are generally sufficient to ensure convergence. For the experiments in the next section for example, computing times on a 2 GHz Pentium 4, range from 20 seconds to 3 minutes according to the dataset, the type of HPMs and the number of components.

3 Experiments

In order to quantify the advantages of using prior knowledge to recover a particular class of genes, we first conducted some experiments on a dataset made up of the original Fibroblast dataset of [15] (see Supplementary Information for more details), along with some additional synthetic series that form a new artificial class. Briefly, we use a probabilistic model involving two Gaussian distributions to generate the expression levels of the artificial expression series: one Gaussian distribution is used to independently generate the gene expression levels of the first three times, while the other is used for the last nine times of the series. The mean of the first one is higher than the second, so the shape of the artificial class looks like a descending step. Figure 4 shows an example of synthetic series generated with this model. We conducted several experiments to recover the synthetic class among all other series, with the proportion of synthetic data ranging from 2% to 16% of the total data.

We use two quantities to measure the ability to recover the artificial class in the final clustering: *Recall* is the highest proportion of this class that can be found in a single cluster —so a recall of 100% is achieved when all the artificial series are in the same cluster—, and *precision* represents the proportion of artificial series in this cluster —so a precision of 100% indicates that all the series in the cluster containing most artificial series are actually artificial. For each proportion of

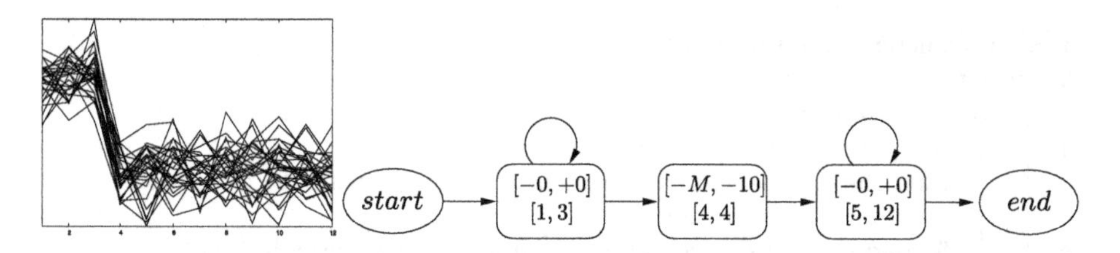

Fig. 4. Left, examples of synthetic expression series added to the fibroblast dataset. Right, the HPM designed to find the synthetic class among the "real" biological classes in the fibroblast dataset.

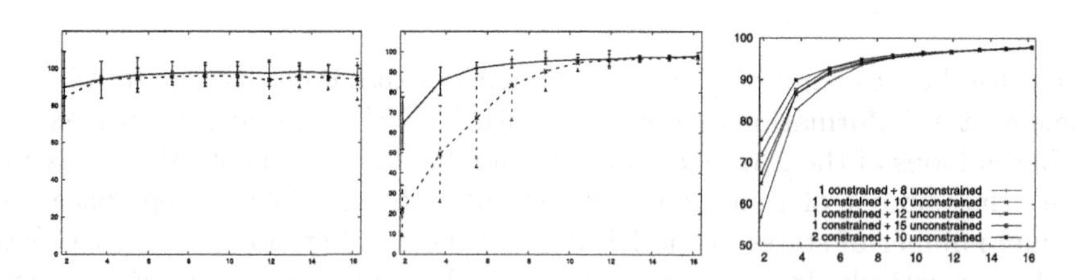

Fig. 5. Recall (left) and precision (middle) achieved with (solid lines) and without (dashed lines) prior knowledge about the class of interest. The x-axes denote the proportion (in percent) of this class among all the expression series. Right, precision achieved using different number of components.

synthetic data, we run a clustering of 11 components with two different methods. The first one does not use any prior knowledge about the class of interest, i.e., its components are completely unconstrained —this method can be viewed as a kind of k-means clustering. The second method makes use of the HPM of Figure 4 to constrain the first class, leaving the 10 others unconstrained. The experiments were repeated 100 times for each proportion of synthetic data and the results are reported in Figure 5.

Both methods achieve quite good recall, even when the proportion of the class of interest is low. Using prior knowledge gives only slightly better results. Concerning the precision, however, there is a clear difference between the two methods, and we can see that the lower the proportion of interesting class, the higher the benefit of our method. When the proportion is 2%, for example, the precision achieved with no prior knowledge is only about 21% —vs. 65% when using prior knowledge—, so the interesting series are lost among many other series, leading to a class that does not show the desired profile.

Next we investigated the sensitivity of the method to the number of components. Determining the number of clusters is a difficult task for all clustering methods. However, when the aim is to recover a particular class of genes rather than to infer a global clustering of the data, the problem is less acute. To illustrate this, we computed, in 100 runs, the precision and recall achieved with various numbers of constrained and unconstrained components, with the proportion of synthetic data ranging from 2% to 16% of the total data. We tried 1 constrained with 8, 10, 12 and 15 unconstrained components, and 2 constrained with 10 unconstrained components. All trials gave recall of up to 80% for all proportions of synthetic data (data not shown), and quite good precision —see the right hand curves in Figure 4. Actually the best results are achieved with the highest numbers of components, so giving a sufficiently high number of components seems to be a good strategy to efficiently recover the clusters of interest.

Experiments to find "real" classes have also been carried out. We used the datasets of [15] and [1] with the aim to uncover classes that show a quick overexpression at the beginning of the series and classes with sinusoidal shape, respectively. Due to space limitations, these experiments have been included in Supplementary Material.

4 Conclusions

We proposed a Bayesian approach for the clustering of gene expression series. This approach allows the user to easily integrate prior knowledge about the general profile of the classes of interest.

We experimentally observed on a mixture of natural and synthetic data that the benefit of the method increases when the number of expression series composing the classes of interest decreases with regard to the total number of series, and that it can be really interesting when this number is very low.

Many improvements seem possible on this basis. Indeed, other knowledge can be integrated in the HPMs. For example, knowledge about the desired mean expression level —and not about the *evolution* of the expression has it is done— could be easily added. Another improvement would be to introduce long-range dependences, i.e., to constrain differences of expression not only between consecutive times but also between separate times. For example, this would allow us to stipulate that the profiles should achieve their maximum at a specific time t.

Acknowledgements

I thank Olivier Martin, Gilles Caraux and Olivier Gascuel for their help and comments on this work.

References

1. Spellman, P.T., Sherlock, G., Zhang, M.Q., Iyer, V.R., Anders, K., Eisen, M.B., Brown, P.O., Botstein, D., Futcher, B.: Comprehensive identification of cell cycle-regulated genes of the yeast saccharomyces cerevisiae by microarray hybridization. Mol Biol Cell **9** (1998) 3273–3297
2. Eisen, M.B., Spellman, P.T., Brown, P.O., Botstein, D.: Cluster analysis and display of genome-wide expression patterns. Proc Natl Acad Sci U S A **95** (1998) 14863–14868
3. Lloyd, S.: Least squares quantization in PCM. IEEE Trans. Info. Theory **IT-2** (1982) 129–137
4. Herwig, R., Poustka, A.J., Muller, C., Bull, C., Lehrach, H., O'Brien, J.: Large-scale clustering of cDNA-fingerprinting data. Genome Res **9** (1999) 1093–105
5. Kohonen, T.: Self-Organizing Maps. Springer (1997)
6. Tamayo, P., Slonim, D., Mesirov, J., Zhu, Q., Kitareewan, S., Dmitrovsky, E., Lander, E.S., Golub, T.R.: Interpreting patterns of gene expression with self-organizing maps: methods and application to hematopoietic differentiation. Proc Natl Acad Sci U S A **96** (1999) 2907–2912
7. Ramoni, M.F., Sebastiani, P., Kohane, I.S.: Cluster analysis of gene expression dynamics. Proc Natl Acad Sci USA **99** (2002) 9121–9126
8. Bar-Joseph, Z., Gerber, G.K., Gifford, D.K., Jaakkola, T.S., Simon, I.: Continuous representations of time-series gene expression data. J Comput Biol **10** (2003) 341–356
9. Schliep, A., Schonhuth, A., Steinhoff, C.: Using hidden markov models to analyze gene expression time course data. Bioinformatics **19 Suppl 1** (2003) 255–263
10. Rabiner, L.R.: A tutorial on hidden Markov models and selected applications in speech recognition. Proceedings of the IEEE **77** (1989) 257–285
11. Dempster, A.P., Laird, N.M., Rubin, D.B.: Maximum likelihood from incomplete data via the EM algorithm. J. Royal Stat. Soc. B **39** (1977) 1–38
12. McLachlan, G., Krishnan, T.: Finite mixture models. John Wiley (2000)
13. Duda, R., Hart, P., Stork, D.: Pattern Classification. John Wiley (2001)
14. Casacuberta, F.: Some relations among stochastic finite state networks used in automatic speech recognition. IEEE Transactions on Pattern Analysis and Machine Intelligence **12** (1990) 691–695
15. Iyer, V.R., Eisen, M.B., Ross, D.T., Schuler, G., Moore, T., Lee, J.C., Trent, J.M., Staudt, L.M., Hudson, J.J., Boguski, M.S., Lashkari, D., Shalon, D., Botstein, D., Brown, P.O.: The transcriptional program in the response of human fibroblasts to serum. Science **283** (1999) 83–87

A Linear Time Biclustering Algorithm for Time Series Gene Expression Data

Sara C. Madeira[1,2,3] and Arlindo L. Oliveira[1,2]

[1] INESC-ID, Lisbon, Portugal
[2] Technical University of Lisbon, IST, Lisbon, Portugal
[3] University of Beira Interior, Covilhã, Portugal
smadeira@di.ubi.pt, aml@inesc-id.pt

Abstract. Several non-supervised machine learning methods have been used in the analysis of gene expression data obtained from microarray experiments. Recently, biclustering, a non-supervised approach that performs simultaneous clustering on the row and column dimensions of the data matrix, has been shown to be remarkably effective in a variety of applications. The goal of biclustering is to find subgroups of genes and subgroups of conditions, where the genes exhibit highly correlated behaviors. In the most common settings, biclustering is an NP-complete problem, and heuristic approaches are used to obtain sub-optimal solutions using reasonable computational resources.

In this work, we examine a particular setting of the problem, where we are concerned with finding biclusters in time series expression data. In this context, we are interested in finding biclusters with consecutive columns. For this particular version of the problem, we propose an algorithm that finds and reports all relevant biclusters in time linear on the size of the data matrix. This complexity is obtained by manipulating a discretized version of the matrix and by using string processing techniques based on suffix trees. We report results in both synthetic and real data that show the effectiveness of the approach.

1 Introduction

Recent developments in DNA chips enabled the simultaneous measure of the expression level of a large number of genes (sometimes all the genes of an organism) for a given experimental condition (sample) [11]. The samples may correspond to different time points, different environmental conditions, different organs or even different individuals. Extracting biologically relevant information from this kind of data, widely called (gene) expression data, is a challenging and very important task.

Most commonly, gene expression data is arranged in a data matrix, where each gene corresponds to one row and each condition to one column, as in Fig. 1(a). Each element of this matrix represents the expression level of a gene under a specific condition, and is represented by a real number, which is usually the logarithm of the relative abundance of the mRNA of the gene under

R. Casadio and G. Myers (Eds.): WABI 2005, LNBI 3692, pp. 39–52, 2005.

the specific condition. Gene expression matrices have been extensively analyzed in both the gene dimension and the condition dimension. These analyses correspond, respectively, to the analysis of the expression patterns of genes and to the analysis of the expression patterns of samples. A number of different objectives are pursued when this type of analysis is undertaken. Among these, relevant examples are the classification of genes, the classification of conditions and the identification of regulatory processes. Clustering techniques have been extensively applied towards these objectives. However, applying clustering algorithms to gene expression data runs into a significant difficulty: many activation patterns are common to a group of genes only under specific experimental conditions. In fact, our general understanding of cellular processes leads us to expect subsets of genes to be co-regulated and co-expressed only under certain experimental conditions, but to behave almost independently under other conditions. Discovering such local expression patterns may be the key to uncovering many genetic mechanisms that are not apparent otherwise [1]. Researchers have therefore moved past this simple idea of row or column clustering and have turned to biclustering [2], a technique that when applied to gene expression matrices identifies subgroups of genes that show similar activity patterns under a specific subset of the experimental conditions.

Many approaches to biclustering in expression data have been proposed to date [9]. In its general form, this problem is known to be NP-complete [14], and almost all the approaches presented to date are heuristic and obtain only approximate results. In a few cases, exhaustive search methods have been used, but limits are imposed on the size of the biclusters that can be found, in order to obtain reasonable runtimes. There exists, however, a particular restriction to the problem that is very important but has not been extensively explored before, and that leads to a tractable problem. This restriction is applicable when the gene expression data corresponds to snapshots in time of the expression level of the genes. Under this experimental setup, the researcher is particularly interested in biclusters with contiguous columns, that correspond to samples taken in consecutive instants of time. In this context, we show that there exists a linear time algorithm that finds all maximal contiguous column biclusters.

2 Definitions and Related Work

2.1 Biclusters in Gene Expression Data

Let A' be an n row by m column matrix, where A'_{ij} represents the expression level of gene i under condition j. In this work, we are interested in the case where the gene expression levels can be discretized to a set of symbols of interest, Σ, that represent distinctive activation levels. In the simpler case, Σ may contain only three symbols, $\{D, U, N\}$ meaning *DownRegulated*, *UpRegulated* or *NoChange*. However,in other applications, the values in matrix A' may be discretized to a larger set of values.

After the discretization process, matrix A' is transformed in matrix A and $A_{ij} \in \Sigma$ represents the discretized value of the expression level of gene i under condition j. Figure 1(b) represents a possible discretization of the gene expression values in Fig. 1(a). In this example, an expression level was considered as *NoChange* if it falls in the range $[-0.3 : 0.3]$. The matrix A is defined by its set of rows, R, and its set of columns, C. Let $I \subseteq R$ and $J \subseteq C$ be subsets of the rows and columns, respectively. Then, $A_{IJ} = (I, J)$ denotes the sub-matrix of A that contains only the elements A_{ij} belonging to the sub-matrix with set of rows I and set of columns J. We will use A_{iC} to denote row i of matrix A and A_{Rj} to denote column j of matrix A.

Definition 1. *A bicluster is a subset of rows that exhibit similar behavior across a subset of columns, and vice-versa. The bicluster $A_{IJ} = (I, J)$ is thus a subset of rows and a subset of columns where $I = \{i_1, ..., i_k\}$ is a subset of rows ($I \subseteq R$ and $k \leq n$), and $J = \{j_1, ..., j_s\}$ is a subset of columns ($J \subseteq C$ and $s \leq m$), and can be defined as a k by s submatrix of the matrix A.*

The specific problem addressed by biclustering algorithms can now be defined. Given a data matrix, A', or its discretized version, A, the goal is to identify a set of biclusters $B_k = (I_k, J_k)$ such that each bicluster B_k satisfies some specific characteristics of homogeneity. The exact characteristics of homogeneity vary from approach to approach, and will be studied in Section 2.2.

2.2 Bicluster Types and Merit Functions

Biclustering approaches may identify many types of biclusters by analyzing directly the values in matrix A or using its discretized version [9]. However, in this paper we will deal with biclusters that exhibit coherent evolutions, characterized by a specific property of the symbols present in the discretized matrix. In particular, we are interested in finding column coherent biclusters satisfying the following definition:

Definition 2. *A column coherent bicluster (cc-bicluster), $A_{IJ} = (I, J)$, is a subset of rows $I = \{i_1, \ldots, i_k\}$ and a subset of columns $J = \{j_1, \ldots, j_s\}$ from the matrix A such that $A_{ij} = A_{lj}$ for all $i, l \in I$ and $j \in J$.*

Although interesting biclusters can be identified in the discretized matrix A, they are usually ranked using merit functions computed over the original,

	C1	C2	C3	C4	C5
G1	0.07	0.73	-0.54	0.45	0.25
G2	-0.34	0.46	-0.38	0.76	-0.44
G3	0.22	0.17	-0.11	0.44	-0.11
G4	0.70	0.71	-0.41	0.33	0.35

(a)

	C1	C2	C3	C4	C5
G1	N	U	D	U	N
G2	D	U	D	U	D
G3	N	N	N	U	N
G4	U	U	D	U	U

(b)

	C1	C2	C3	C4	C5
G1	N1	U2	D3	U4	N5
G2	D1	U2	D3	U4	D5
G3	N1	N2	N3	U4	N5
G4	U1	U2	D3	U4	U5

(c)

Fig. 1. Toy example. (a) represents the original expression matrix, (b) the discretized matrix and (c) the discretized matrix after alphabet transformation (section 3.1).

non-discretized version of the matrix, A'. To understand these metrics, consider a bicluster $A_{IJ} = (I, J)$ and let a'_{iJ} represent the mean of the ith row in the bicluster, a'_{Ij} the mean of the jth column in the bicluster and a'_{IJ} the mean of all elements in the bicluster.

Depending upon the application, it may be helpful to characterize biclusters by the degree of fluctuation in gene expression level as well as the similarity in behavior. For example, a bicluster with a low *mean squared residue* (1), $MSR(I, J)$, where $r(A'_{ij})$ are the residues (2), indicates that the expression levels fluctuate in unison [2]. This includes, however, flat biclusters with no fluctuation. In order to remove flat biclusters or identify biclusters with high degree of fluctuation in expression levels is beneficial to use the bicluster *variance* (3), $VAR(I, J)$, the *average row variance* (4), $ARV(I, J)$, and the *average column variance* (5), $ACV(I, J)$. A low $VAR(I, J)$ identifies a constant bicluster. A bicluster with high $ARV(I, J)$ and low $ACV(I, J)$ has high fluctuation on the rows and coherent columns while a bicluster with low $ARV(I, J)$ and high $ACV(I, J)$ is a bicluster with high fluctuation on the columns and coherent rows. Since a low value of $MSR(I, J)$ identifies a bicluster with coherent values [2], if the value of $ARV(I, J)$ is high and the value of $MSR(I, J)$ is low we can also identify a bicluster with high fluctuation on the rows and coherent columns.

$$MSR(I, J) = \frac{1}{|I||J|} \sum_{i \in I, j \in J} r(A'_{ij})^2 \tag{1}$$

$$r(A'_{ij}) = A'_{ij} - a'_{iJ} - a'_{Ij} + a'_{IJ} \tag{2}$$

$$VAR(I, J) = \frac{1}{|I||J|} \sum_{i \in I, j \in J} (A'_{ij} - a'_{IJ})^2 \tag{3}$$

$$ARV(I, J) = \frac{1}{|I||J|} \sum_{i \in I, j \in J} (A'_{ij} - a'_{iJ})^2 \tag{4}$$

$$ACV(I, J) = \frac{1}{|I||J|} \sum_{i \in I, j \in J} (A'_{ij} - a'_{Ij})^2 \tag{5}$$

Many heuristic approaches have been proposed for the selection of biclusters that minimize directly this type of merit functions [9]. However, the inherent difficulty of this problem when dealing with the non-discretized matrix A' and the great interest in finding coherent behaviors regardless of the exact numeric values in the data matrix, has led many authors to a formulation based on a discretized version of the gene expression matrix. Since most versions of the problem addressed by these authors are NP-complete [1,5,6,7,13,15,16] the solutions proposed are heuristic and are not guaranteed to find optimal solutions.

A different approach, from Ji and Tan [4], aims at finding time-lagged biclusters in time series expression data. As in the present work, they are also interested in identifying biclusters formed by consecutive columns. They propose to use a naive algorithm that has a complexity $O(|R||C|^3)$, if all consecutive column biclusters are to be found. With an appropriate implementation (not described in the paper) their sliding window approach can have its complexity reduced to $O((|R||C|^2)$, a complexity that is still of the order of $|C|$ higher than our proposed approach.

2.3 Biclusters in Time Series Expression Data

Finding a set of maximal biclusters that satisfy the coherence property defined in Def. 2 remains an NP-complete problem. As such, most biclustering algorithms use heuristic approaches that are not guaranteed to find optimal solutions. However, we are interested in the analysis of time series expression data, and this leads to an important restriction, on which relies the linear time algorithm we propose.

When analyzing time series expression data, with the objective of isolating coherent activity between genes in a subset of conditions, we want to restrict the attention to biclusters with contiguous columns. Other authors have already pointed out the importance of biclusters that span consecutive columns [4], and their importance in the identification of gene regulatory processes. In fact, the activation of a set of genes under specific conditions corresponds, in many cases, to the activation of a particular biological process. As time goes on, biological processes start and finish, leading to increased (or decreased) activity of sets of genes that can be identified because they form biclusters with contiguous columns, as illustrated in Fig. 2(a). In this figure, the existence of three processes (P1, P2 and P3) leads to increased activity of different sets of genes, represented by three biclusters. Note that, although the columns of each of the biclusters are contiguous, the rows are in arbitrary positions, and are represented as contiguous for P1 and P2 only for convenience. Overlapping is also allowed.

Time series expression data are often used to study dynamic biological systems and gene regulatory networks since their analysis can potentially provide more insights about biological systems [8]. In this context, the identification of biological processes that lead to the creation of biclusters, together with their relationship, is crucial for the identification of gene regulatory networks and for the classification of genes. This leads us to the definition of the type of biclusters that are of interest in this work.

Definition 3. *A contiguous column coherent bicluster (ccc-bicluster),* $A_{IJ} = (I, J)$*, is a subset of rows* $I = \{i_1, \ldots, i_k\}$ *and a* ***contiguous*** *subset of columns* $J = \{r, r+1, \ldots, s-1, s\}$ *from matrix* A *such that* $A_{ij} = A_{lj}, \forall i, l \in I$ *and* $j \in J$*.*

For the remainder of this work, we will refer to a contiguous column coherent bicluster simply as a ccc-bicluster. By definition, each row in matrix A is a ccc-bicluster. These are **trivial biclusters** and will not be of interest, in general. The biclusters with only one row or only one column will also be considered as trivial.

In this settings, each ccc-bicluster defines a string S that is common to every row in the ccc-bicluster, between columns r and s of matrix A. Figure 2(b) illustrates two ccc-biclusters that appear in the expression matrix in Fig. 1(a). These two ccc-biclusters are maximal, in the sense that they are not properly contained in any other ccc-biclusters. This notion will be defined more clearly later on.

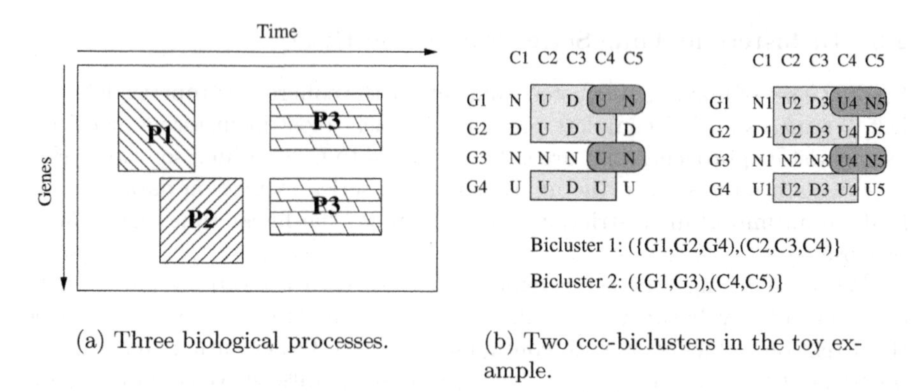

(a) Three biological processes.

(b) Two ccc-biclusters in the toy example.

Fig. 2. Biclusters in time series gene expression data

2.4 Suffix Trees

A *string* S is an ordered list of characters written contiguously from left to right [3]. For any string S, $S[i..j]$ is the (contiguous) *subtring* of S that starts at position i and ends at position j. In particular, $S[1..i]$ is the *prefix* of S that ends at position i and $S[i..|S|]$ is the *suffix* of S that starts at position i, where $|S|$ is the number of characters in S. A *suffix tree* is a data structure built over all the suffixes of a string S that exposes its internal structure. This data structure has been extensively used to solve a large number of string processing problems.

Definition 4. *A* suffix tree *of a $|S|$-character string S is a rooted directed tree with exactly $|S|$ leaves, numbered 1 to $|S|$. Each internal node, other than the root, has at least two children and each edge is labeled with a nonempty substring of S. No two edges out of a node have edge-labels beginning with the same character. The key feature of the suffix tree is that for any leaf i, the label of the path from the root to the leaf i exactly spells out the suffix of S that starts at position i.*

In order to enable the construction of a suffix tree obeying this definition when one suffix of S matches a prefix of another suffix of S, a character terminator, that does not appear nowhere else in the string, is added to its end. For example, the suffix tree for the string S=TACTAG is presented in Fig. 3(a). The suffix tree construction for a set of strings, called a *generalized suffix tree*, can be easily obtained by consecutively building the suffix tree for each string of the set. The leaf number of the single string suffix tree is now converted to two numbers: one identifying the string and other the starting position (suffix) in that string.

Suffix trees can be built in time that is linear on the size of the string, using several algorithms. Generalized suffix trees can be built in time linear on the sum of the sizes of the strings. Ukkonen's algorithm [18], used in this work, uses *suffix links* to achieve a linear time construction.

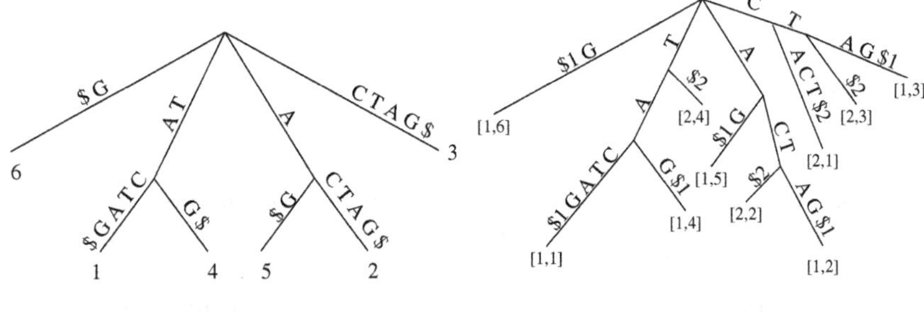

(a) Suffix tree for S=TACTAG.

(b) Generalized suffix tree for S_1=TACTAG and S_2=CACT.

Fig. 3. Example of a suffix tree for the string S=TACTAG and a generalized suffix tree for the strings S_1=TACTAG and S_2=CACT

Definition 5. *There is a suffix link from node v to node u, (v, u), if the path-label of node u represents a suffix of the path-label of node v and the length of the path-label of u is exactly equal to the length of the path-label of v minus 1.*

3 Biclustering Time Series Expression Data

3.1 Biclusters and Suffix Trees

We can now introduce the major results of this work, that lead to the linear time biclustering algorithm. We first introduce the concept of maximal ccc-bicluster.

Definition 6. *A ccc-bicluster $A_{IJ} = (I, J)$ is maximal if no other ccc-bicluster exists that properly contains A_{IJ}, that is, if for all other ccc-biclusters $A_{LM} = (L, M)$, $I \subseteq L$ and $J \subseteq M \Rightarrow I = L \wedge J = M$.*

We will also call a ccc-bicluster *right-maximal* if it cannot be extended to the right by adding one more column at the end, and *left-maximal* if it cannot be extended to the left by adding one more column at the beginning. Stated more plainly, a ccc-bicluster is maximal if no more rows nor contiguous columns (either at the right or at the left) can be added to it while maintaining the coherence property in Def. 3.

We will now consider a new alphabet $\Sigma' = \Sigma \times \{1 \ldots m\}$, where each element Σ' is obtained by concatenating one symbol in Σ and one number in the range $\{1 \ldots m\}$. In order to do this alphabet transformation we use a function f : $\Sigma \times \{1 \ldots m\}$ defined by $f(a, k) = a|k$ where $a|k$ represents the character in Σ' obtained by concatenating the symbol a with the number k. For example, if $\Sigma = \{U, D, N\}$ and $m = 3$, then $\Sigma' = \{U1, U2, U3, D1, D2, D3, N1, N2, N3\}$. For this case, $f(U, 2) = U2$ and $f(D, 1) = D1$.

Consider now the set of strings $S = \{S_1, \ldots, S_n\}$ obtained by mapping each row A_{iC} in matrix A to string S_i such that $S_i(j) = f(A_{ij}, j)$. Each of these strings has m characters and corresponds to the symbols in a row of matrix A after the above alphabet transformation. After this transformation, the matrix in Fig. 1(a) corresponding to the discretized matrix in Fig. 1(b), becomes the matrix in Fig. 1(c).

Let T be the generalized suffix tree obtained from the set of strings S. Let v be a node of T and let $P(v)$ be the path-length of v, that is, the number of characters in the string that labels the path from the root to node v. Additionally, let $B(v)$ be the branch-length of v and let $L(v)$ denote the number of leaves in the sub-tree rooted at v, in case v is an internal node. It is easy to verify that every internal node of the generalized suffix tree T corresponds to one ccc-bicluster of the matrix A. This is so because an internal node v in T corresponds to a given substring that is common to every row that has a leaf rooted in v. Therefore, node v defines a ccc-bicluster that has $P(v)$ columns and a number of rows equal to $L(v)$. It is also true that all the leaves except the ones whose path label is simply a terminator also identify ccc-biclusters.

Since these ccc-biclusters may not be maximal, we will now present with only sketches of proofs the two lemmas that lead to our the main theorem.

Lemma 1. *Every right-maximal ccc-bicluster corresponds to one node in T.*

Proof. Let B be a ccc-bicluster that cannot be extended to the right by adding a column at the right, that is, a right-maximal ccc-bicluster. Since B is a ccc-bicluster, every row in B shares the substring that defines B. Since B is right maximal, at least one of the rows in B must have a character that differs from the character in the other rows, in the first column to the right that is not in B. Therefore, there is a node in T that matches B and the path-label of that node is the string that defines B. □

Lemma 2. *Let node v_1 correspond to a ccc-bicluster B_1 and node v_2 correspond to a ccc-bicluster B_2. Then, if there is a suffix link from node v_1 to node v_2, bicluster B_2 contains one less column than bicluster B_1.*

Proof. Follows directly from the definition of suffix links. □

From these lemmas, we can now derive the theorem that is our main result.

Theorem 1. *Let v be a node in the generalized suffix tree T. If v is an internal node, then v corresponds to a maximal ccc-bicluster iff $L(v) > L(u)$ for every node u such that there is a suffix link from u to v. If v is a leaf node, then v corresponds to a maximal bicluster iff the path-length of v, $P(v)$, is equal to $|S_i|$ and the label of the branch that leads to v has characters other than the terminator, that is, $B(v)$ is greater than one. Furthermore, every maximal ccc-bicluster in the matrix corresponds to a node v satisfying one of these conditions.*

Proof. Let B be a maximal ccc-bicluster and S the string that defines B. Now, S must lead to a node v (by Lemma 1). If node v is an internal node and does not have an incoming suffix link, the conditions of the theorem are met. Since B is also left-maximal, every node u that defines a bicluster B' with one more column than B (by Lemma 2) must have $L(v) > L(u)$, since B' cannot contain all the rows in B (otherwise, B would not be left-maximal). Therefore, it is sufficient to check that every internal node u that has a suffix link directed at v has $L(u) < L(v)$ to ensure that node v corresponds to a maximal ccc-bicluster.

If node v is a leaf node the conditions of the theorem are met. In fact, if B is maximal the path-length of v must be equal to $|S|$, otherwise, B would not be left-maximal since it could be extended to the left until all the characters at its left in S had been added. Furthermore, the path-label of v cannot be only a string terminator, that is, $B(v)$ must be greater than one, otherwise B would also not be maximal. In one hand, if the parent of v was the root, then B would not be left maximal since it could be extended to the left by adding at least the last character of S. On the other hand, if the parent of v was an internal node, then B would not be maximal either since it could be extended by adding to it the rows that correspond to the remaining leaves of the sub-tree rooted at the internal node that is the parent of v. □

Figure 4 illustrates the generalized suffix tree obtained from the strings that correspond to the rows of the matrix in Fig. 1(c). For clarity, this figure does not contain the leaves that represent string terminators that are direct daughters of the root. Each non-terminal node, other than the root, is labeled with the value of $L(v)$, the number of leaves in its subtree. Also shown in this tree are the suffix links between nodes. For clarity, the suffix links that end at the root are not shown.

This figure also shows that there are six internal nodes, other than the root. Each one of these nodes corresponds to one ccc-bicluster. Furthermore, each leaf node with branch-length greater than one is also a ccc-bicluster. However, some of these ccc-biclusters are trivial, since they represent biclusters with only one column (nodes with branch-labels $N1$, $U4$ and $N5$, since they have branch-labels with only one character), or are represented by leaves (trivial ccc-biclusters with only one row even when they are maximal). Others are non-maximal (nodes with branch-labels $D3U4$ and $N5$), since they have an incoming suffix link from a node with the same number of leaves. As such, only the internal nodes with branch-labels $U2D3U4$ and $U4N5$ identify maximal, non-trivial ccc-biclusters. These

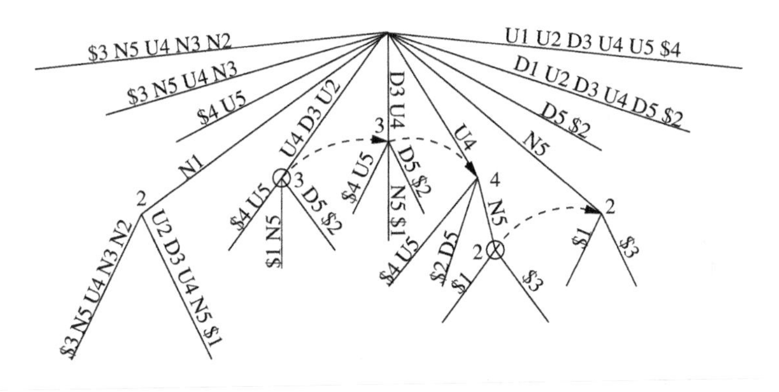

Fig. 4. Generalized suffix tree for the matrix in Fig. 1(c)

nodes correspond to the maximal ccc-biclusters $(\{G1, G2, G4\}, \{C2, C3, C4\})$ and $(\{G1, G3\}, \{C4, C5\})$ in Fig. 2(b).

Note that the rows in each ccc-bicluster are obtained from the terminators in the leaves in the subtree of each node v, while the columns in each ccc-bicluster are obtained from the value of $P(v)$ and the information on the branch-label that connects v to its parent. In fact, the value of $P(v)$ and the first character of the path label of v is needed to identify the set of columns that belong to the bicluster.

3.2 A Linear Time Algorithm for Finding and Reporting ccc-Biclusters

Theorem 1 directly implies that there is an algorithm that finds and reports all maximal ccc-biclusters in a discretized and transformed gene expression matrix A in time linear on the size of the matrix (see Alg. 1). With appropriate data structures at the nodes, the suffix tree construction is linear on the size of the input matrix, using Ukkonen' algorithm [18]. The remaining steps of our algorithm are also linear since they are performed using depth first searches (*dfs*) on the suffix tree. A more detailed analysis shows that the increase in the alphabet size does not have an impact on this linear time complexity. In fact, only two types of nodes have more than $|\Sigma|$ children: the root node and nodes that have as children only leaf nodes. In both cases, it is easy to devise a data structure that enables constant time manipulation of these nodes.

Algorithm 1. Algorithm for finding and reporting all maximal ccc-biclusters

1: **procedure** FIND AND REPORT ALL MAXIMAL CCC-BICLUSTERS(A)
2: Map each row i in matrix A, A_{iC}, to a string S_i using function f: $S_i \leftarrow f(A_{iC})$.
3: Build a generalized suffix tree, T, for the set of strings S.
4: **for all** nodes $v \in T$ **do**
5: Compute the path-length and the branch-length of v: $P(v)$ and $B(v)$.
6: Mark v as "Valid".
7: **end for**
8: **for all** internal nodes $v \in T$ **do**
9: Compute the number of leaves in the sub-tree rooted at v: $L(v)$.
10: **end for**
11: **for all** nodes $v \in T$ **do**
12: **if** (v is an internal node and there is a suffix link (v, u) and $L(v) >= L(u)$) **or** (v is a leaf node and $(P(v)! = |S_i|$ or $B(v) = 1)$) **then**
13: Mark u as "Invalid".
14: **end if**
15: **end for**
16: **for all** nodes $v \in T$ **do**
17: **if** v is "Valid" **then**
18: Report the ccc-bicluster that corresponds to v.
19: **end if**
20: **end for**
21: **end procedure**

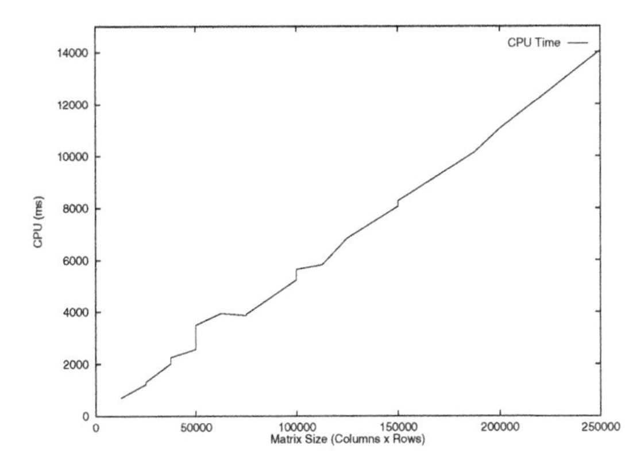

Fig. 5. CPU time versus size of the synthetic input data

4 Experimental Results

In order to validate the approach, we performed experiments with both synthetic and real data, using a prototype implementation of the algorithm, coded in Java, and a 3GHz Pentium-4 machine, running Linux with 1GB of memory.

To evaluate the efficiency of the algorithm, and validate experimentally the predicted linear time complexity, we generated matrices with random values, on which 10 biclusters were hidden, with dimensions ranging from $15-25$ rows and $8-12$ columns. The size of the matrices varied from 250×50 (rows $\times$ columns) up to 1000×250. We used a three character alphabet, $\Sigma = \{U, D, N\}$. In all cases, we recovered the *planted* ccc-biclusters, together with a large number of artifacts that result from random coincidences in the data matrix. Figure 5 shows a plot of the variation of the CPU time with the size of the input data matrix. A clear linear relationship over several orders of magnitude is apparent from this plot. It is also clear that the algorithm runs in less than 15 seconds even in the larger synthetic matrices used.

To validate the approach with real data, we used time-series from the yeast cell-cycle dataset described by Tavazoie et al. [17] which contains the expression profiles of more than 6000 yeast genes measured at 17 time points over two complete cell cycles. We used 2884 genes selected by Cheng and Church [2] as in [17] and removed the genes with missing values. The matrix with the remaining 2268 genes was discretized by gene to an alphabet $\Sigma = \{D, U, N\}$ using an equal bin frequency discretization.

The resulting matrix was then processed by our algorithm and 14728 maximal non-trivial ccc-biclusters were reported in 13.5 seconds. From these 825 had more than 4 conditions and at least 25 genes. Since we were interested in ccc-biclusters with high values of $ARV(I,J)$ and low values of $MSR(I,J)$, this last set was then ordered in descending order according to the value of $(ARV(I,J) - MSR(I,J))$ (see Sec. 2.2). The computation of the values of the metrics and the ordering of the biclusters cannot be done, in general, in time linear on the size of

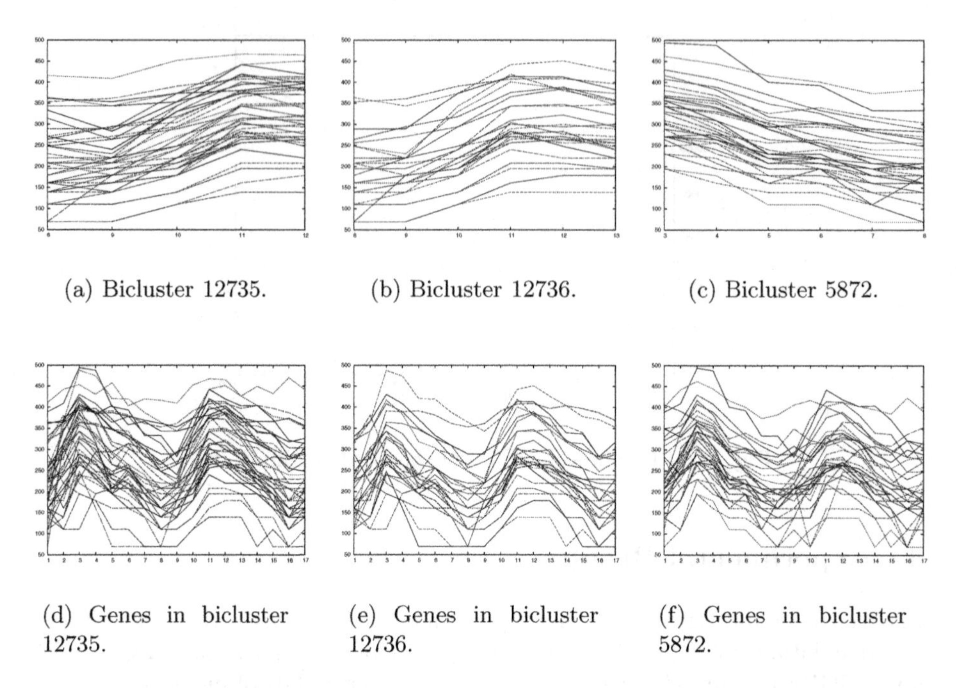

(a) Bicluster 12735. (b) Bicluster 12736. (c) Bicluster 5872.

(d) Genes in bicluster 12735. (e) Genes in bicluster 12736. (f) Genes in bicluster 5872.

Fig. 6. Expression level of genes in the top three ccc-biclusters

Table 1. Biological relevance of the top three ccc-biclusters

Bicluster ID	*Genes*	*Conditions*	*GO-Category*	*GO-Level*	*P-Value*
12735	41	5 (8-12)	Cell Cycle	4	2.9E-05
12736	26	6 (8-13)	DNA replication	6,7	9.2E-03
5872	33	6 (3-8)	DNA metabolism	6,5	2.0E-05
			DNA replication	7,6	2.5E-05
			DNA repair	6,7	9.1E-05
			response to DNA damage stimulus	9	2.4E-04
			response to endogenous stimulus	9	2.7E-04
			biopolymer metabolism	4	4.3E-03
			negative regulation of DNA transposition	9,10	8.0E-03
			regulation of DNA transposition	8,9	8.0E-03
			lagging strand elongation	9,10	8.8E-03

the matrix. In practice, we observed that in real data these steps take less time than the bicluster generation steps.

We present some preliminary evidence of the biological significance of these results by analyzing (for lack of space) only the top three ccc-biclusters according to this criterion. To access the biological relevance of the biclusters we used the Gene Ontology (GO) and the p-values obtained from the hypergeometric distribution to model the probability of observing at least k genes, from a bicluster with $|I|$ genes, by chance in a category containing c genes from the 2268 genes in the dataset. For this, we used the functions from the three GO categories, biological process, molecular function and cell component at level higher than 3. Table 1 shows p-values computed using the GOToolBox [10] and the utilities from the YEASTRACT database [12].

In order to show that the generated ccc-biclusters have biological significance, shown by statistically significant enrichment in one or more GO categories, we report the categories in which the (Bonferroni corrected [11]) p-values are below 0.01. Figure 6 presents the expression levels of these ccc-biclusters and shows how biclustering is able to identify highly correlated expression patterns of genes, under a given subset of conditions. Note that the highly correlated activity under this subset of columns does not necessarily translate into highly correlated activity under all conditions.

5 Conclusions

In this work, we presented a linear time algorithm for the identification of all maximal contiguous column biclusters in time series expression data. By discretizing the gene expression values, and manipulating the strings that correspond to each row using string processing techniques, we have been able to demonstrate that there is a correspondence between the maximal ccc-biclusters and the nodes of the generalized suffix tree that represents the rows (genes) in the matrix. This simple correspondence lead to a very efficient algorithm for the extraction of ccc-biclusters, that runs in a few seconds even for matrices with thousands of genes and hundreds of conditions. We have demonstrated the correctness of the algorithm and sketched the complexity analysis. We have also presented experimental results with synthetic data and preliminary results with real data from yeast. This work opened several promising directions for future research. Among these are the discovery of imperfect ccc-biclusters (ccc-biclusters allowing a given number of errors) and the development of methods for the identification of regulatory networks.

Acknowledgements. This work was partially supported by projects POSI/SRI/47778/2002, BioGrid and POSI/EIA/57398/2004, DBYeast, financed by FCT, Fundação para a Ciência e Tecnologia, and the POSI program.

References

1. A. Ben-Dor, B. Chor, R. Karp, and Z. Yakhini. Discovering local structure in gene expression data: The order–preserving submatrix problem. In *Proc. of the 6th International Conference on Computacional Biology*, pages 49–57, 2002.
2. Y. Cheng and G. M. Church. Biclustering of expression data. In *Proc. of the 8th International Conference on Intelligent Systems for Molecular Biology*, pages 93–103, 2000.

3. D. Gusfield. *Algorithms on strings, trees, and sequences: computer science and computational biology.* Cambridge University Press, 1997.
4. L. Ji and K. Tan. Identifying time-lagged gene clusters using gene expression data. *Bioinformatics*, 21(4):509–516, 2005.
5. M. Koyuturk, W. Szpankowski, and Ananth Grama. Biclustering gene-feature matrices for statistically significant dense patterns. In *Proc. of the 8th Annual International Conference on Research in Computational Molecular Biology*, pages 480–484, 2004.
6. J. Liu, W. Wang, and J. Yang. Biclustering in gene expression data by tendency. In *Proc. of the 3rd International IEEE Computer Society Computational Systems Bioinformatics Conference*, pages 182–193, 2004.
7. S. Lonardi, W. Szpankowski, and Q. Yang. Finding biclusters by random projections. In *Proc. of the 15th Annual Symposium on Combinatorial Pattern Matching,Springer*, pages 102–116, 2004.
8. Y. Luan and H. Li. Clustering of time-course gene expression data using a mixed-effects model with b-splines. *Bioinformatics*, 19(4):474–482, 2003.
9. S. C. Madeira and A. L. Oliveira. Biclustering algorithms for biological data analysis: a survey. *IEEE/ACM Transactions on Computational Biology and Bioinformatics*, 1(1):24–45, Jan-Mar 2004.
10. D. Martin, C. Brun, E. Remy, P. Mouren, D. Thieffry, and B. Jacq. Gotoolbox: functional investigation of gene datasets based on gene ontology. *Genome Biology*, 5(12):R101, 2004.
11. G. McLachlan, K. Do, and C. Ambroise. *Analysing microarray gene expression data.* Wiley, 2004.
12. P. Monteiro, M. C. Teixeira, P. Jain, S. Tenreiro, A. R. Fernandes, N. Mira, M. Alenquer, A. T. Freitas, A. L. Oliveira, and Isabel Sá-Correia. Yeast search for transcriptional regulators and consensus tracking (yeastract). *http://www.yeastract.com*, 2005.
13. T. M. Murali and S. Kasif. Extracting conserved gene expression motifs from gene expression data. In *Proc. of the Pacific Symposium on Biocomputing*, volume 8, pages 77–88, 2003.
14. R. Peeters. The maximum edge biclique problem is NP-complete. *Discrete Applied Mathematics*, 131(3):651–654, 2003.
15. Q. Sheng, Y. Moreau, and B. De Moor. Biclustering microarray data by Gibbs sampling. In *Bioinformatics*, volume 19 (Suppl. 2), pages 196–205, 2003.
16. A. Tanay, R. Sharan, and R. Shamir. Discovering statistically significant biclusters in gene expression data. In *Bioinformatics*, volume 18 (Suppl. 1), pages S136–S144, 2002.
17. S. Tavazoie, J. D. Hughes, M. J. Campbell, R. J. Cho, and G. M. Church. Systematic determination of genetic network architecture. *Nature Genetics*, 22:281–285, 1999.
18. E. Ukkonen. On-line construction of suffix trees. *Algorithmica*, 14:249–260, 1995.

Time-Window Analysis of Developmental Gene Expression Data with Multiple Genetic Backgrounds
[Extended Abstract]

Tamir Tuller[1,*], Efrat Oron[2], Erez Makavy[1], Daniel A. Chamovitz[2], and Benny Chor[1]

[1] School of Computer Science, Tel-Aviv University, Tel-Aviv 69978, Israel
{tamirtul, erezmak, benny}@post.tau.ac.il
[2] Department of Plant Sciences, Tel-Aviv University, Tel-Aviv 69978, Israel
{oronefra, dannyc}@post.tau.ac.il

Abstract. We study gene expression data, derived from developing tissues, under multiple genetic backgrounds (mutations). Motivated by the perceived behavior under these background, our main goals are to explore *time windows questions*:

1. Find a large set of genes that have a similar behavior in two different genetic backgrounds, under an appropriate time shift.
2. Find a model that approximates the dynamics of a gene network in developing tissues at different continuous time windows.

We first explain the biological significance of these problems, and then explore their computational complexity, which ranges from polynomial to NP-hard. We developed algorithms and heuristics for the different problems, and ran those on synthetic and biological data, with very encouraging results.

1 Introduction

A major goal of systems biology is to infer the relationships among genes and proteins in the cell and organism. A large number of works have tried to identify genes that appear to be coexpressed, in an approach known as "guilt by association". These works come in roughly three major flavors - clustering (*e.g.* [3]), biclustering (*e.g.* [6]), and methods for model inferring (*e.g.* [7,9]). The problems we deal with in this paper include ingredients from all three. The inputs to our problem include gene expression datasets from two genetic backgrounds. In the first set of problems, the goal is to identify sets of genes with a similar behavior in two equisize subsets of the conditions in the two datasets. Biologically, we are interested in the case where one dataset was generated when a component of the system underwent mutation, while the other dataset represents the wildtype. We

* Corresponding author.

R. Casadio and G. Myers (Eds.): WABI 2005, LNBI 3692, pp. 53–64, 2005.

focus on developmental gene expression datasets [2], where the conditions are ordered in time, and the subset of conditions of interest are continuous time windows. This natural restriction makes some variants of our problems polynomial. In our experiments, a few key mutations were induced (separately) in a component of a central protein complex. These mutations caused various changes in the behavior of many genes. Some of these changes are best described across contiguous time intervals. This motivates us to define and explore such time interval questions to better understand the functional relations among participating genes. We call these problems "time windows problems", since we want to find a set of genes and a time window such that the genes' behavior during this time window in the wildtype is similar to their behavior in the mutant in a different second time window, namely each mutation causes a "time shift" in the expression levels when compared to wildtype. For example, suppose a mutation inhibits the expression of a set of genes, such that it remains 0 in all time points. This phenomenon can cause time shift in the expression level of another gene set. Figure 1 illustrates such a hypothetical example, where a mutation in one gene causes a shift in the expression level of another gene. Gene g is regulated by genes pg_1 and pg_2 according to the table in figure 1A. A mutation causes gene pg_1 to stay at level 0. According to the regulation table, the expression level of gene g at later developmental stages in the mutation (figure 1C) is similar to its expression level in earlier developmental stages in the wildtype (figure 1B). By grouping together genes which exhibit similar shifts in the mutant gene expression compared to the wildtype gene expression, and by combining information about the functionality of some of these genes, one can conclude about the functionality of a mutated gene network (or a mutated protein complex), and the way such a network may regulate directly or indirectly these shifted genes. Thus part of our goals is to find subsets of genes with the same GO annotations [1] and a similar shift.

The second problem of interest is finding a model approximating the dynamics of a gene network in certain continuous time windows. We want to find the regulatory rules of genes by other genes in these time windows. For example, in this work, for the mutant dataset we have the expression levels of several thousand genes at just three time points, while for the wildtype dataset, we have the expression levels of several thousand genes at a few dozen time points. We want to understand the dynamics of the genes in continuous time windows in the wildtype around the time points sampled in the mutant. We employ linear models, and develop heuristic for solving this problem, while being careful to avoid over-fitting.

A number of works have examined analyzing developmental gene expression datasets and comparing gene expression in multiple genetic backgrounds. Arbeitman *et al.* [2] report the gene expression patterns for nearly one-third of all Drosophila genes during a complete time course of development. We use this dataset here. Chang *et al.* [5] performed a quantitative inference of dynamic regulatory pathways via microarray data. They used a second order model of differential equations with many parameters, combined with maximum

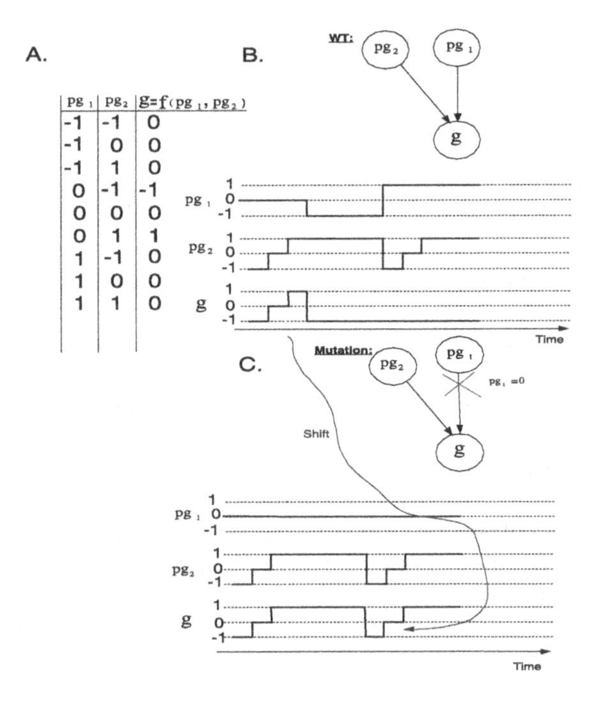

pg 1	pg 2	g=f(pg 1, pg 2)
-1	-1	0
-1	0	0
-1	1	0
0	-1	-1
0	0	0
0	1	1
1	-1	0
1	0	0
1	1	0

Fig. 1. Illustration of a gene which was shifted by a mutation. A. The expression level of gene g as a function of its regulators, the genes pg_1 and pg_2, where -1 denote under-expression, 0 denote normal expression, and 1 denote over-expression. B. Hypothetical developmental expression level of gene g in the wildtype. C. Hypothetical developmental expression level of gene g when a mutation causes gene pg_1 to be stuck in level $'0'$.

likelihood methods for inferring a regulatory pathways. McDonal and Rosbash [10], developed methods for identification of genes with cyclic behavior while studying circadian rhythms. D'haeseleer and Fuhrman [8] suggested modelling a gene network by a linear model. We note that the main drawback in their approach is the large number of parameters (compared to the size of the dataset) they used, which may lead to over-fitting.

2 Time-Windows Problems: Definition and Mathematical Properties

In this section we present the time shift problems, and deal with their mathematical properties. For lack of space, the proofs are deferred to the full version of this paper. Let S be a set of genes. Let M_1 and M_2 be two gene expression datasets for S, and m_1 and m_2 denote the number of conditions in M_1 and M_2, respectively. Let $d_S : \mathcal{M} \times \mathcal{M} \rightarrow R^{\geq 0}$, where $\mathcal{M}$ is the space of datasets over S, denote a measure for the *dissimilarity* of the expression pattern of the gene set S in M_1 vs. M_2. The problems in this work have the following general structure:

Problem 1. **Input:** Two gene expression datasets, M_1 and M_2, over a gene set S; a positive number, δ, and a dissimilarity measure $d_S(M_2, M_1)$.

Task: Find a maximum subset of genes, $S' \subseteq S$, such that $d_{S'}(M_2, M_1) < \delta$.

For a specific example, suppose M_1 and M_2 have the same number of conditions. For every gene g, $M_1(g)$ and $M_2(g)$ are real vectors of the same length. Let $\| \|_p$ denote the ℓ_p norm. Then $d_{S,p}(M_1, M_2) = \sum_{g \in S} \|M_1(g) - M_2(g)\|_p$ is such a dissimilarity measure. A second example is parameterized by an integer k. For every choice of k conditions C_1, C_2 from M_1 and M_2, respectively, we look at $M_{1,|C_1}$, $M_{2,|C_2}$ (the restriction of each dataset to the respective k conditions). Then $d_{S,k,p}(M_1, M_2) = \min_{|C_1|=|C_2|=k} \sum_{g \in S} \|M_{1,|C_1}(g) - M_{2,|C_2}(g)\|_p$ is a dissimilarity measure. In this paper, we deal with problems where the conditions are ordered, usually by time, so that this order has a biological meaning. One example of such order is developmental gene expression dataset, where the i-th condition (column) refers to time t_i, and $i > j \iff t_i > t_j$. We emphasize that t_i in the two dataset need not be the same. In the first problem we use $d_{S,k,2}(M_1, M_2)$ as the dissimilarity measure. We further restrict here C_1 and C_2 to be continuous time windows in M_1 and M_2, respectively, and $k = m_2$ is the total number of the conditions in the smaller dataset (say M_2). We denote this dissimilarity measure $D^1_S(M_1, M_2)$. For $S = \emptyset$ we define $D^1_S(M_1, M_2) = 0$.

Problem 2. Time shift.

Input: Two ordered gene expression datasets, M_1 and M_2, where the number of conditions in M_1 is larger than in M_2 ($m_1 \geq m_2 = k$), and a positive number, δ.

Task: Find a maximum set of genes S' and a continues time window in M_1, $W_1 = i_1, .., i_k$, such that the expression of the gene in this window is similar (error less than δ) to their expression level in all the conditions of M_2. Quantitatively $D^1_{S'}(M_2, M_1) < \delta$.

A generalization of problem 2 with allows k (the size of the windows) to be *smaller* than both m_1 and m_2. The windows should still be continuous. In this case, we denote the dissimilarity measure $D^2_{S,k}(M_2, M_1)$. In the second problem we want to infer models, which describe the behavior of the genes in M_1 (the larger dataset). We allow different models for different time windows in M_1, while we focus on "interesting" time windows, that contain the time points in M_2 (each time window in M_1 is around different time point in M_2). The simplest such model is a linear model.

D'haeseleer and Fuhrman [8] were the first to suggest the use of linear models for analyzing gene networks. Regulation of genes can be described as a function of the expression levels of other genes by a differential equations. These equations can be approximated by difference equations which can be described by an equivalent set of linear equations. More generally, the behavior of many dynamic models at time $t + \Delta t$ can be approximated by a linear function of the model's parameters at time t. The relative error is the ratio between the error and the average gene expression level. In our model we got better results in terms of relative error when we worked with the logarithm of the expression value. Taking logarithms when working with gene expression level is justified for example in [12]. So we aimed to express the log of the expression level of a gene at time point i by a linear combination of the log expression levels of a subset of the genes at time $i-1$ (the "parents" of the gene), such that the expression level of gene g_n at time i equals approximately: $g_n(t) \approx \sum_{g_j \in pa(g_n)} w_{n,j} \cdot g_j(t-1)$, where $pa(g_n)$ denotes the set of "parents" of gene g_n, and $w_{n,j}$ are constants. For each gene we need to find a *different* set of parents and weights, $w_{n,j}$ but these are fixed for all the conditions. The error of a gene according to such model is the Euclidean distance between the predicted vector and the actual vector across the window, W: Let $\hat{g}_n(t) = \sum_{g_j \in pa(g_n)} w_{n,j} \cdot g_j(t-1)$ be the "predicted" value, then the error for gene g_n in window W is: $e_{g_n,W} = \sqrt{\sum_{i \in W}(g_n(t) - \hat{g}_n(t))^2}$. This roughly describes the model of [8]. We deal here with a more general model: First, we want to find different linear models for different continuous time windows of development, in that we are looking for a phase, or window, dependent network. We define the error of a set of genes in a time window to be the error (e_{g_n}) of the gene with the largest error in this set in the time window. As defined, this problem has too many degrees of freedom in choosing the parents' sets of each genes, resulting in models that are often meaningless [13]. Thus we want to bound the maximum in-degree in the linear network. The bound should be smaller than the number of conditions in that window.

Thus by [13] if we want to describe an ℓ dimensional vector by a linear model where its parent set are of ℓ or more vectors in the ℓ-dimensional space we can get zero error by using random vectors with probability approaching 1 as $k \to \infty$. If the number of parents is smaller than ℓ, this phenomenon does not occur. Thus if our vector describes a gene expressions, such model may reveal a true relation between a gene and its parents. This motivates us to restrict the number of parents of each gene in the model to be smaller than the number of time points in the window. By using this upper bound, we prevent over-fitting a model to a window. Thus, we are interested in the following problem:

Problem 3. Linear approximation with bounded in-degree.

Input: Two ordered gene expression datasets M_1 and M_2, a positive number, ε, two positive integer h and W.

Task: For every point in M_2 find a linear model with in-degree less than h for the set of genes in M_1, and for the window of size W around the point. When the expression level of the set in M_1 is described by this model, its error is required to be smaller than ε for the window.

The restriction of bounded in degree makes the problem computationally hard. In the decision version of problem 3 we have the same inputs, and have to decide if there is a model with in degree less than h that approximate the genes in M_1 in windows around the points of M_2 with error less than ε.

Lemma 1. *The decision version of problem 3 is NP-hard.*

For a given h (if h is not an input of the problem) problem 3 is practical only for $h \leq 3$ (by exhaustive search in complexity $O(n^h)$).

The following lemma indicates that if our dataset was sampled from a linear model, when using this dataset for inferring a linear model in a time window where many genes change slowly, we should expect that the parents' set of a gene will contain genes which are not directly connected to the gene in the real model (but are close to it in the real model). It is known that other model inferring methods suffer from similar problems. However, when our method missed edges of a gene's "real parents", it often replaces them by edges from the "grandparents generation". This phenomenon is tolerable since we get sets of genes that are relatively close to a gene's parental set. Let $gp(g_n)$ denote the set of grandparents of the gene g_n in a linear model. We say that a gene g_k changes slowly if $(1-\varepsilon) \cdot g_k(t-1) \leq g_k(t) \leq (1+\varepsilon) \cdot g_k(t-1)$, where $0 < \varepsilon << 1$. The following lemma explains this phenomena. By recursively using the arguments of lemma 2 we can get similar results for the connection between the expression level of a gene and its ancestral of depth d. In this case we will get the approximating factors $(1-\varepsilon)^d$ and $(1+\varepsilon)^d$ instead of $(1-\varepsilon)$ and $(1+\varepsilon)$, which increase the error exponentially (with d). It easy to see that when the average in-degree in the net is larger than 1, replacing the real parents of a gene by its ancestors implies an increase of its in-degree.

Lemma 2. *For the linear model with bounded degree, if the gene's grandparents change slowly, the optimal set of parents for a gene can be well approximated by the set of its grandparents.*

3 Algorithms and Heuristics

The time shift problems have two stages: Finding a set of shifted genes, and identifying a subset with functional enrichment (GO annotation) in each such a set. Let $M_{r,|C_{j,i}}(g)$ denote the i-th sample of gene g in time window C_j of dataset M_r. A direct calculation of the cost function for the two variants of the time shift problems for a gene g, when comparing the time windows C_2 and C_1 (of size k) in M_1 and M_2 respectively, is:

$\|M_{2,|C_2}(g) - M_{1,|C_1}(g)\|_2 = \sqrt{\sum_{i=1}^{k}(M_{2,|C_{2,i}}(g) - M_{1,|C_{1,i}}(g))^2}$. Subtraction and addition are much cheaper processor operations, compared to squaring and square root operations. This calculation involve performing k squaring and one square root operations, namely $k+1$ "expensive" operations. In the naive way of finding solutions to the time shift problem, we separately calculate for each gene the cost function using the above equation, and attribute it to the pair of windows C_1, C_2 if this function is less than δ. In other words, for each gene we need to calculate the cost function $(m_1-k-1)\cdot(m_2-k-1)$ times, for each pair of time windows $C_1 \subseteq M_1$ and $C_2 \subseteq M_2$. In total we have $(m_1-k-1)\cdot(m_2-k-1)\cdot(k+1)$ expensive operations for just one gene.

Let $M_{r,i}(g)$ denote the expression level of gene g in dataset r in time point i. In order to speed up the process, we do the following: For each gene, we first calculate a table of size $m_1 \cdot m_2$, where the (i,j) entry in the table contains the value $(M_{2,i}(g) - M_{1,j}(g))^2$. We use the fact that each entry in the table is used for many pairs of windows, and use the values in the table for calculating the cost function for different pair of windows for g. Since calculating the table costs us $m_1 \cdot m_2$ expensive operations and by using the table we only need to perform one expensive operation (square root operation) for the calculation of the cost for a pair of windows, we now perform a total of $m_1 \cdot m_2 + (m_1 - k - 1) \cdot (m_2 - k - 1)$ expensive operations. Asymptotically (for large m_1 and m_2) this is $\Theta(k)$ faster.

In the next stage we searched for functional enrichment in the solution set, to better understand possible biological meanings of our results. We used GO annotations [1] which attribute genes to cellular functions. We are interested in sets of genes that have both a similar time shift and a common function, as determined by GO annotation. Let G denote a bound for the maximum number of GO annotations for a gene. We calculated the number of genes with each GO annotation in our dataset by one pass over all the genes, and for each gene we checked at most G annotations (a gene may have more than one annotation, and we assume no more than G). We generated a table with the number of genes with each GO annotations. The overall complexity of this stage $n \cdot G$. Given a set of size $|S|$ of shifted genes, we generated a similar "small" table only for the set, in time complexity $G \cdot |S|$. These tables enable us to calculate the enrichment's p-values, using the standard formula of hypergeometric distribution.

We now turn to the linear model problem. Since the bounded linear model problem is NP-hard, we used variations of the following heuristic. Let m_1 denote the number of time samples in M_1. Check all the $m_1 - k - 1$ continuous windows of size k, the specified window size. For each such window, perform the following greedy heuristic for each gene g_n:

1. Start with an empty set of "parents" for g_n.
2. At step r, by exhaustive search, find the gene that causes the largest decrease in the prediction error of gene g_n when adding it to the $r-1$ current parents of the gene g_n, add it to the parents set of the gene.
3. Stop if the decrease is less than $\alpha \cdot \hat{\varepsilon}_{r,n}$ or the number of parent is larger than m_1.

The number α is a parameter to our algorithm, and $\varepsilon_{r,n}$ denotes the average decrease in the error of gene g_n with $r-1$ parent that were found by the greedy algorithm, when adding a random r-th parents to the gene. Let $\hat{\varepsilon}_{r,n}$ denote an estimation of $\varepsilon_{r,n}$. For each r $(1 \leq r < k)$ and gene, g_n, in the dataset M_1, $\varepsilon_{r,n}$ is estimated empirically during the above exhaustive search, by averaging the decrease in the error in r-th step. *I. e.* when adding a gene to the parent set of size $r-1$ of g_n. In multiple regression we fit the coefficient $w_{n,j}$, of the gene's "parents" in the linear model. This is done by finding $w_{n,j}$, which minimize the error e_{g_n} (by differentiating and comparing it to zero). In stage 2 of the algorithm, we perform multiple regression for each candidate set of parents. Let

k (the size of the window) denote an upper bound on the number of parents for each gene in the model. Let $C_{mk}(k)$ denote the time complexity for calculating multiple regression with k variables. In our case this equals the complexity of inverting a $k \times k$ matrix, which is $O(k^3)$ practically. Let n denote the number of genes in the dataset. The overall time complexity of the algorithm for a given time window is $O(k \cdot C_{mk}(k) \cdot n^2) = O(k^4 \cdot n^2)$. As in the previous problems, here we also used the fact that models of close windows are similar. For each gene we kept the parents set which our algorithm found for the closest previous window and tried to find a better one. To avoid local maxima, for each gene we tried to optimize its parents set by checking different subsets from the set of genes which give large decrease in the error in the initial stages of the algorithm.

4 Results

Our data consist of developmental gene expression datasets of the fruit fly *Drosophila melanogaster.* One was wildtype dataset and others were from different mutants in the Cop9 signalosome. The Cop9 signalosome (CSN) is a highly conserved protein complex, conserved across different organisms and known to be essential for development of plants and animals. CSN has eight subunits that regulate multiple signal transduction pathways [4]. These subunits are inter-related, and some are found in multiple configurations [11]. Consequently, the biological roles of the complex as a whole, and of individual subunits, are not completely understood. To clarify this situation, we are employing transcriptional profiling on Drosophila *csn* mutants. Four mutants in different CSN subunits were analyzed at three developmental time points: 60, 72, and 96 hours after egg deposit (AED). This is the first global comparison between multiple CSN mutants in animals, and as such we expect it to shed light on CSN involvement in unknown processes, and lead to new and improved models for the role of CSN and its subunits. We analyzed our data together with publicly available wildtype samples of Arbeitman *et. al* [2], containing 80 time points.

4.1 Detecting Putative Time-Shifted and Partially Time-Shifted Genes

Some of the results for putative time-shifted genes compared to all the three time points in the mutants (problem 2) are summarized in table 4.1. In table 4.1 a gene is attributed to a shift bin $[A, B]$ if it there is at least one $A \leq \Delta \leq B$ such the gene exhibits at time t in the mutant expression pattern as in time $t - \Delta$ in the wildtype. For the four mutants analyzed, 120 sets of genes with suspected time shifts were identified. In a global look at the data, we fist notice that while in mutants 3 and 4 the number of negative and positive shifts was equal ($14 : 16$ and $16 : 14$ respectively), in mutant 2, and especially mutant 1, most of the shifts were negative ($21 : 9$ and $25 : 4$, respectively). This may suggest that mutations 1 and 2 caused late-acting genes to be induced earlier. Table 4.1 shows a sample of these sets with their predicted shifts and accompanying P-value. The functional analysis of these genes indicates that most of these sets are involved in various aspects of development and cellular regulation, such as regulation of DNA structure and integrity (rows 3, 5, 6, 8) and signal transduction (rows 1, 2, 9, 12, 14). Only a few sets are obviously involved in "house keeping" functions. For example, three genes, whose gene products are all involved in glycolysis, were found to have a bin shift of $[-4, 4]$ (line 4). These genes comprise 1/3 of the genes with a similar bin shift, but only 0.49% of total genes used in the analysis. The $[-4, 4]$ bin basically represents genes for whom no significant shift is found. Glycolysis, the breakdown of glucose to usable energy forms, is the one metabolic pathway that occurs in all living cells and is the starting point for aerobic respiration and fermentation. As such, no effect of the CSN on glycolysis was expected *a priori*, and indeed, this is illustrated in this example. Interestingly, mutants 3 and 4 both show bin shifts with genes encoding subunits of the proteasome (lines 11, 15, 17), and the proteasome regulatory lid in particular (lines 11 and 17). The proteasome is a large multiprotein complex that degrades proteins in regulated fashion. That these genes are regulated by the CSN is interesting as the proteasome lid is evolutionarily related to CSN, and the lid and CSN interact physically to regulate similar processes. As yet, there is no in depth understanding on the cross-talk between these two complexes, nor is the regulation on the lid clearly understood. However, finding that the regulated expression of these genes is shifted in a COP9 signalosome mutant provides further evidence for the mutual dependence of these complexes.

In the next stage we deal with windows of size 1. We observed that in all mutants at time 60 AED, more genes were up- than down regulated in relation to the wild type. This expression pattern may be explained by the corresponding mutations causing loss of function of transcriptional repressors. We hypothesized that these genes are either up regulated in the mutant before they would normally be so in the wildtype (that is in the wildtype they should be up regulated at $t = 60+$ shift), or alternatively that these genes are normally upregulated early in development, but then not repressed in the mutant (that is in the will type they should be upregulated at $t = 60-$ shift). Our results for putative time-shifted genes compared to only one time point 60 in the mutants (problem 2, where the conditions of the mutants include only time point 60) are summarized in table 2. Table 2 shows that both types of behaviors were identified. The first two rows of Table 2 show genes that are up regulated in two CSN mutants at 60 hours, while in wildtype these genes peak during early or late embryogenesis $(0 - 24$ hrs AED). At the other extreme, the last two rows in Table 2 show sets of genes that are induced in a CSN mutant at 60 hours, while in the wildtype, these genes normally peak either during metamorphosis $(149 - 161$ hrs AED) or in old adults $(527 - 827$ hours from hatching).

4.2 Intermittent Linear Model: Synthetical and Biological Inputs

To evaluate our method, we first checked our algorithm on small nets (containing a few dozen genes and conditions). We sampled known nets and tried to reconstruct them by our and by D'haeseleer's algorithms. We counted the number of real edges each algorithm missed, and the number of of edges that do

Table 1. Representative results from time-shift analysis where the threshold = 0.1. Bin Shift Range shows range of the shift in hours - the difference between the original time point and the shifted time point, for example if the shift is −100 the shifted time point are 100 hours after the original ones, i.e. the mutation caused the gene expression of time $T + 100$ hours to be expressed in time T. Shift Bin Size: Number of genes in this shift. Func Bin Size: Number of genes with this GO-ID in this shift. Genes with this GO-ID: Total no of genes with this GO-ID on the chip, out of 2869 genes with GO-ID. P-value.

No	*mutant*	*GO-ID*	*Bin Shift Range*	*Shift Bin Size*	*Func Bin Size*	*Genes with this Func*	*P-value*
1	1	7274	[-16 -8]	29	2	3	$2.933 \cdot 10^{-4}$
2	1	19221	[-68 -64]	14	2	11	0.0012
3	1	6333	[20 28]	23	2	13	0.0044
4	1	6096	[-4 -4]	33	3	14	$4.499 \cdot 10^{-4}$
5	2	6398	[-12 -8]	26	2	6	0.0011
6	2	5730	[-104 -100]	40	3	10	$2.73 \cdot 10^{-4}$
7	2	3779	[52 56]	7	2	49	0.0056
8	2	6281	[-64 -48]	16	2	25	0.0078
9	3	8523	[8 16]	30	2	2	$1.057 \cdot 10^{-4}$
10	3	15144	[-96 -92]	28	2	5	$8.94 \cdot 10^{-4}$
11	3	5838	[52 60]	17	2	9	0.0011
12	3	5099	[-104 -100]	82	2	2	$8.07 \cdot 10^{-4}$
13	4	4559	[52 56]	27	2	2	$8.53 \cdot 10^{-5}$
14	4	8195	[24 32]	56	2	2	$3.74 \cdot 10^{-4}$
15	4	8540	[12 20]	48	3	5	$4.26 \cdot 10^{-5}$
16	4	9993	[-16 -12]	14	2	15	0.0022
17	4	5838	[12 16]	53	4	9	$1.2 \cdot 10^{-5}$

Table 2. Representative results from partial time-shift analysis. Bin Shift Range shows the wildtype expression I nduction range (in hours) for the genes upregulated at 60 hrs AED in the mutants. Shift Bin Size: Number of genes in this shift. Func Bin Size: Number of genes with this GO-ID in this shift. Genes with this GO-ID: Total no of genes with this GO-ID among the 242 genes up-regulated at 60 hrs AED that have a GO-ID and are also present in the wildtype data set.

No	*Bin Shift Range*	*GO-ID*	*Shift Bin Size*	*Func Bin Size*	*Genes with Func*	*P-value*
1	[8 9]	6139	60	5	7/242	0.0102
2	[19 20]	4702	32	5	9/242	0.0024
3	[149 161]	4674	37	4	6/242	0.0053
2	[527 827]	8248	50	3	3/242	0.0084

not exist in the real model and each algorithm adds The simple least square fit method was substantially worse than our method, it missed 42% more real edges than our method, and it add 425% more false edges compared to our method. We then ran our procedure on real dataset of Arbeitman *et al.* [2] with 4000 genes, and generated linear models for three time windows as output. The first window was for the times window 24 – 105, the second for times 19 – 57, and the third was for times 67 – 113 (around the time points in the datasets of our

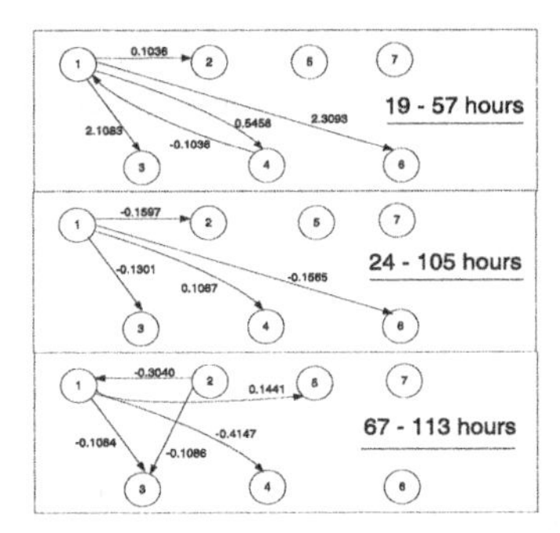

Fig. 2. Description of the dynamics of a gene set at three time windows around time 60 - 96. The figure describes only the sub-model for this set, edges from/to other genes were omitted.

mutants). Each window contained ten samples. For lack of space, we describe here the results for only one sub graph constructed model. Further analysis is deferred to the full version of this paper. Figure 2 describes the sub graph of the constructed models for a small set of chosen genes. Figure 2 shows an analysis of a small gene network, where gene 1 encodes a transcription factor known to be involved in a developmental process, gene 4 encodes a hormone receptor involved in this process, and genes 2, 3, 5, and 7 are known to be regulated in this process, though their connection to 1 and 4 is unknown. Our linear modeling correctly identifies gene 1 as a key node in this network, where it regulates the other members of this network, with the exception of gene 7, which appears as an "orphan". Interestingly, early in development we identify a putative feedback inhibition loop between the transcription factor (1) and the activating receptor (4). In late development, gene 2, whose biochemical and developmental function are unknown, has an inhibitory effect on the network, negatively affecting both the transcription factor (1) and another gene (3).

5 Conclusions and Further Research

In this work we investigated problems originating from developmental gene expression datasets of multiple genetic backgrounds. We defined two major questions, explained their biological significance, and their mathematical properties. One of the problems is polynomial, while the other is NP-hard. We developed algorithms for solving two variants of first one, and a heuristic for the other. We implemented and ran them on synthetic and biological inputs. Our methods generated many interesting biological results, some exhibiting agreement with the acceptable biological knowledge. This supports the underlying reasoning of our approach. There are many open questions and directions we are considering. Here we describe two of them. First, more biological experiments are underway, in order to achieve a richer dataset for our mutants (a dataset with more time points). Such datasets will enable us to infer linear models for the mutants and compare these models to the one inferred for the wildtype. This way, we could explore a new time window problem, where we compare models from different genetic backgrounds in different time windows. Another direction involves inferring linear models from datasets of multiple species, an approach that may help filtering noise and avoid over-fitting.

Acknowledgements

We wish to thank Drs. Bruce Edgar and Ling Li of the Fred Hutchinson Cancer Research Center for providing the facilities for and assistance in carrying out the microarray hybridizations and Dr. Daniel Yekutieli from Tel Aviv University for helpful discussions. This work was partially supported by grants from the Manna Institute (EO) and Israel Science Foundation (DAC, BC).

References

1. GO annotation guide. http://www.geneontology.org/go.annotation.
2. Arbeitman, N.M., Furlong, M.E.E., Imam, F., Johnson, E., Null, H.B., Baker, S.B., Krasnow, A.M., Scott, P.M., Davis, W.R., White P.K. Gene expression during the life cycle of drosophila melanogaster. *Science*, 297:2270–2275, 2002.
3. Ben-Dor, A., Shamir, R., Yakhini, Z. Clustering gene expression patterns. *Journal of Computational Biology*, 6:281–297, 1999.
4. Chamovitz DA., Yahalom, A. A systems approach to the cop9 signalosome. *Plant Physiol*, 132:426–427, 2003.
5. Chang, W.C., Li, C.W., Chen, B.S. Quantitative inference of dynamic regulatory pathways via microarray data. *BMC Bioinformatics*, 6(44), 2005.
6. Cheng, Y., Church, G.M. Biclustering of expression data. *In Proc. ISMB'00*, pages 93–103, 2000.
7. Chor, B., Tuller, T. Adding hidden node to gene network. *WABI2004*, 2004.
8. D'haeseleer, P., Fuhrman, S. Gene network inference using a linear, additive regulation model. *Bioinformatics*, 2000.
9. Friedman, N., Linial, M., Nachman, I., Pe'er, D. Using bayesian network to analyze expression data. *Journal of Computational Biology*, 7:601–620, 2000.
10. McDonald, M.J., Rosbash, M. Microarray analysis and organization of circadian gene expression in drosophila. *Cell*, 107(5):567–578, 2001.
11. Orian, A., Van Steensel, B., Delrow, J., Bussemaker, H.J., Li, L., Sawado, T., Williams, E., Loo, L.W., Cowley, S.M., Yost, C., Pierce, S., Edgar, B.A., Parkhurst, S.M., Eisenman, R.N., Genomic binding by the drosophila myc, max, mad/mnt transcription factor network. *Genes Dev*, 17:1101–14, 2003.
12. Rocke, D.M., Durbin, B. Approximate variance-stabilizing transformations for gene-expression microarray data. *Bioinformatics*, 19(8):966–972, 2003.
13. Tao, T., Vu, V. On the singularity probability of random bernoulli matrices. *submited*, 2005.

A Lookahead Branch-and-Bound Algorithm for the Maximum Quartet Consistency Problem

Gang Wu, Jia-Huai You, and Guohui Lin*

Department of Computing Science, University of Alberta,
Edmonton, Alberta T6G 2E8, Canada
Tel: (780) 492-3737, Fax: (780) 492-1071
{wgang, you, ghlin}@cs.ualberta.ca

Abstract. A lookahead branch-and-bound algorithm (LBnB) is proposed for solving the Maximum Quartet Consistency (MQC) Problem where the input is a complete set of quartets on the taxa and the goal is to construct a phylogeny that satisfies the maximum number of given quartets. It integrates a number of previous efforts on exact algorithms, heuristics, and approximation algorithms for the NP-hard MQC problem, and a few improved search techniques, especially a lookahead scheme, to solve the problem optimally. The theoretical running time analysis of the LBnB algorithm is provided, and an extensive simulation study has been well designed to compare the algorithm to previous existing exact algorithms and a best heuristic Hypercleaning. The experimental results on both synthetic and real datasets show that LBnB outperformed other exact algorithms, and it was competitive to Hypercleaning on many datasets.

1 Introduction

With the availability of more and more genomic data, there are more needs in fast phylogenetic analysis to facilitate biological applications. As a concrete example, the Influenza Sequence Database (http://www.flu.lanl.gov/) has a deposit of about 5318 whole genomes for Avian Influenza viruses as of January 16, 2005. The phylogenetic analysis on these viruses, at both the chronological level and the territorial level, to carry out the evolutionary relationships is crucial to fast understanding of the emergence of new variants and fast vaccination strategy design. There are a number of models as well as algorithms that have been proposed for the phylogenetic analysis, each assumes the same amount of biological data associated with every taxon under investigation. In practice, however, the data disparity problem exists which means the availability of biological data for analysis is different for every taxon. This data disparity problem either forces the phylogenetic analysis to use a much less amount of (but common) data for all taxa or limits the phylogenetic analysis to consider a smaller number of taxa. Subsequently, it raises the question of how confident we

* Corresponding author.

R. Casadio and G. Myers (Eds.): WABI 2005, LNBI 3692, pp. 65–76, 2005.

should trust the analytical results that might be obtained using different data for different subsets of taxa. One way of resolving this issue is to use different phylogenetic analysis methods to handle different subsets of taxa where the methods find most applicable, and then to assemble a global phylogenetic pattern out of the achieved sub-patterns for the subsets. The promise of this approach is that, since every piece of phylogenetic analysis on subsets is of high confidence, the global analytical results must also be of high confidence, though there might be needs to resolve potential conflicts among the sub-patterns. The quartet-based phylogeny construction methods can be classified into such efforts to construct a (global) phylogeny for a set of taxa associated with different biological data.

1.1 Quartet-Based Phylogeny Construction

In our discussion of phylogeny construction for a set of taxa, the phylogeny is an unrooted binary tree whose leaves bijectively map to the set of taxa and every internal node in the tree has degree 3. In quartet-based phylogeny construction methods, researchers try to build a phylogeny for every (or most) quartet, which is a subset of 4 taxa, called a *quartet topology*, and then assemble a global phylogeny for the whole set of taxa to satisfy all the quartet topologies that have been built, or if not at all possible, to satisfy as many of them as possible. For a quartet, there are 3 possible phylogenies associated with it. For example, Figure 1 shows the 3 topologies for a quartet $\{s_1, s_2, s_3, s_4\}$. For simplicity, we use $[s_1, s_2|s_3, s_4]$ to denote the topology in which the path connecting s_1 and s_2 doesn't intersect the path connecting s_3 and s_4, as shown in Figure 1(a).

Given a phylogeny T for a set of taxa S, for every quartet X, we can derive a topology for X by computing the induced subtree of T on X. Such a set of $\binom{n}{4}$ induced quartet topologies is denoted as Q_T. Conversely, given a set Q of quartet topologies (which can be built by various quartet inference approaches), for every quartet, Q contains at most one topology for it (i.e. no ambiguity). If there exists one global phylogeny T for S such that a quartet topology $q \in Q$ for a quartet is the same as the one derived from T, then T *satisfies* q or q is *consistent* with T. If there exists one global phylogeny T satisfying all quartet topologies in Q, i.e. $Q = Q_T$, then Q is *compatible* and T is the phylogeny *associated* with Q.

The recognition problem, called the *Quartet Compatibility Problem* (QCP), is to determine whether a given set Q of quartet topologies on a set of taxa S is compatible or not, or equivalently if there is a phylogeny T on S satisfying

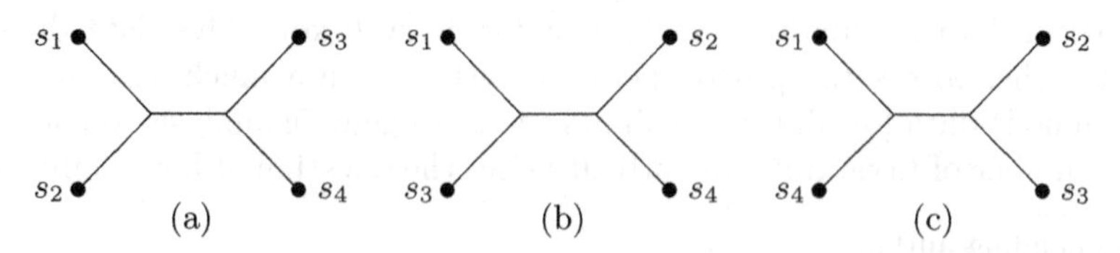

Fig. 1. Three possible topologies for quartet $\{s_1, s_2, s_3, s_4\}$: (a) $[s_1, s_2|s_3, s_4]$, (b) $[s_1, s_3|s_2, s_4]$, and (c) $[s_1, s_4|s_2, s_3]$.

all the quartet topologies in Q. If Q contains exactly one topology for every quartet, then Q is *complete*; Otherwise, Q is *incomplete*. It has been known that when Q is complete, the QCP problem can be solved in $O(n^4)$ time, where n is the size of taxa set S; furthermore, if Q is compatible, then the associated phylogeny T is unique and can be constructed within the same time [7]. The situation changes when Q is incomplete, where the recognition problem becomes NP-complete [14].

The more interesting computational problem is the optimization problem where Q (either complete or incomplete, but in this paper we only consider the complete case) is not compatible and the goal is to construct a phylogeny to satisfy as many quartet topologies as possible. This is the so-called *Maximum Quartet Consistency Problem* (MQC). A dual minimization problem to the MQC problem, the *Minimum Quartet Inconsistency Problem* (MQI) with the same input, is to construct a phylogeny to minimize the number of inconsistent quartet topologies. Despite the fact that the MQC and the MQI problems have the same optimal solution(s), their approximabilities differ a lot. The MQC problem is NP-hard [4] and it admits a *Polynomial Time Approximation Scheme* (PTAS) [12]; The MQI problem is NP-hard [4] too but the best approximation ratio so far is $O(n^2)$ where n is the size of taxa set [11].

A few attempts have been made to solve the MQC (or MQI) problem optimally. Ben-Dor et al. [2] presents a dynamic programming algorithm that evaluates the number of quartet topologies that are consistent with a bipartition of the taxa set and thereby determines a phylogeny to satisfy the maximum number of quartet topologies. The running time of the algorithm is $O(3^n n^4)$. The inconsistent quartet topologies with a bipartition are referred to as *quartet errors* (across the bipartition), which are subject to be determined and changed in order to be compatible to other quartets. If the number of quartet errors is known ahead of time, then the fixed-parameter algorithm proposed by Gramm and Niedermeier [9], which for simplicity is referred to as the GN algorithm subsequently, would be able to detect and correct them and return an associated phylogeny. The GN algorithm has a running time $O(4^k n + n^4)$, where k is the number of quartet errors. We note that the algorithm could be modified to solve the general MQC problem with unknown number of quartet errors. In fact, the branch-and-bound algorithm we developed in this paper takes advantage of many technical steps in the GN algorithm, and many technical steps in another exact algorithm for solving the MQC problem through Constraint Programming, which we developed in a previous work [15]. It also extends some ideas embedded in a class of heuristics and approximation algorithms called *quartet cleaning* for the MQC and/or MQI problems [3,4,6,10,12] on identifying those edges that must be in the optimal phylogeny, among which *hypercleaning* performs the best with its running time complexity $O(n^5 f(2m) + n^7 f(m))$, where $f(m) = 4m^2(1+2m)^{4m}$ and m denotes the greatest value that the set $Best(Q, m)$ has to be computed during the execution.

1.2 The Organization

In the next section, we present a number of facts associated with the quartet topologies to provide the theoretical foundations for the design of our branch-and-bound algorithm. Some of these facts might be attached to existing algorithms such as the GN algorithm and the hypercleaning algorithm, others might be independent on any algorithm. We present the lookahead branch-and-bound algorithm in Section 3 where the meaning of "lookahead" would become clear. We have implemented our algorithm, as well as some other algorithms mentioned above, and tested them on various synthetic and real datasets. Section 4 summarizes the experimental results and the comparisons we have made. We conclude the paper in Section 5 with some remarks.

2 Theoretical Foundations for the BnB Algorithm

Our branch-and-bound algorithm takes advantage of many technical steps of the GN algorithm [9], which is designed for the MQC problem where the number of quartet errors is known to be exactly k. In the following we include some propositions from [9] that are found useful for the design of our algorithm. The interested readers should refer to [9] for more details. The most important idea in the GN algorithm is to resolve global quartet conflicts through resolving local quartet conflicts [5,1], referred to as *local conflicts.* A local conflict is a set of 3 incompatible quartet topologies on a subset of exactly 5 taxa. For example, $\{[a,b|c,d],[a,c|b,e],[a,c|d,e]\}$ is a local conflict.

Theorem 1. [9] *Given a complete set of quartets Q over a set of taxa S and some taxon $e \in S$, Q is compatible iff there exists no local conflict whose taxa set includes e.*

Notice that with a fixed taxon e, there are $\binom{n-1}{4}$ subsets of 5 taxa containing e and for each such subset there are $\binom{5}{3} = 10$ potential local conflicts. Therefore, Theorem 1 tells us that testing if Q is compatible can be reduced to testing if none of the $O(n^4)$ subsets of 3 quartet topologies is a local conflict. Since testing the compatibility of each subset of 3 quartet topologies takes constant time, this implies an $O(n^4)$ algorithm for the QCP problem. With a fixed quartet topology q, a similar argument gives that there are at most $6(n-4)$ local conflicts containing q. Subsequently, if we decide to change the topology for a quartet, then at most $6(n-4)$ new local conflicts would be generated and at most $6(n-4)$ previous local conflicts would be resolved.

Theorem 2. *Given a complete set of quartets Q over a set of taxa S and some taxon $e \in S$, let m denote the number of local conflicts whose taxa set includes e. Then there are at least $\frac{m}{6(n-4)}$ quartet errors in Q.*

The GN algorithm begins with building a conflict list that contains all the local conflicts (involving a fixed taxon). At each step, it randomly chooses one local conflict and tries to resolve it through changing some quartet topology in the local conflict. It is proven in [9] that it is sufficient to consider at most four ways of resolving a local conflict and the GN algorithm picks one in some order. The new topology becomes fixed for that specific quartet in the rest of the algorithm execution unless there is no solution along the way and the algorithm goes back to undo the changing. The algorithm subsequently updates the conflict list. When there is no further local conflict left, the algorithm terminates and assembles a phylogeny in another $O(n^4)$ time. It is proven that the main effort at every node in the search tree associated with the GN algorithm is the conflict list updating, which takes $O(n)$ time. Therefore, the overall running time of the algorithm is $O(4^k n + n^4)$. In [9], some efforts have been made to improve the running time in practice through determining the quartet topologies that must be changed, as stated in the following Theorem 3; some other efforts have been made to improve the running time in practice but potentially dropping the optimality, through fixing some quartet topologies (and bipartitions) along the computation.

Theorem 3. [9] *For a quartet topology $q \in Q$, if there are more than $3k$ distinct local conflicts that contain q, then q must be changed in the optimal solution.*

We remark that the GN algorithm is designed for a special case of the MQC problem where the number of quartet errors is known to be exactly k. It therefore terminates when it finds the first solution. The algorithm can be modified to return all solutions in $O(4^k n + n^4)$ time. Such a modification would also be able to solve the MQC problem when the number of quartet errors is only known to be not exceeding some constant. Our branch-and-bound algorithm is designed to solve the general MQC problem where the number of quartet errors is unknown ahead of time. The rest of this section is devoted to some propositions which can fast determine (more) quartet topologies that need to be fixed and (more) quartet topologies that need to be changed. Some of these speedups are novel and make the branch-and-bound algorithm significantly outperform the existing exact algorithms. The readers might find the technical proofs of these propositions in [16].

Let (X, Y) be a bipartition of the taxa set S and $|X| = \ell$ and $|Y| = n - \ell$. It follows that $|Q_{(X,Y)}| = \binom{\ell}{2}\binom{n-\ell}{2}$. Let $p_1 = |Q_{(X,Y)} - Q|$. Fixing three taxa from Y, a subset of ℓ quartet topologies from Q, where each quartet contains these three taxa and one taxon from X, is called an ℓ-subset with respect to (X, Y). There are in total $\binom{n-\ell}{3}$ such ℓ-subsets. For an ℓ-subset, if ignoring the difference of the taxa from X gives rise to one unique quartet topology, then the ℓ-subset is *exchangeable* on X; Otherwise, it is *nonexchangeable* on X. Let p_2 and p_3 denote the number of nonexchangeable ℓ-subsets on X and the number of nonexchangeable $(n - \ell)$-subsets on Y, respectively.

Theorem 4. [16] *Let Q be a complete set of quartets on S of n taxa. For a bipartition (X, Y) of S where $|X| = \ell$, let p_1 be the number of quartet errors in Q across (X, Y), p_2 be the number of nonexchangeable ℓ-subsets on X, and p_3 be the number of nonexchangeable $(n-\ell)$-subsets on Y. If $2p_1 + (\ell-1)p_2 + (n-\ell-1)p_3 \leq (\ell - 1)(n - \ell - 1)$, then bipartition (X, Y) must be in the optimal phylogeny.*

Our extensive experimental results show that around 5 edges out of 100 can be discovered by Theorem 4, which consequently reduces the overall running time several orders of magnitude.

In [15], we have developed a scheme that transforms the computing of a phylogeny to satisfy the maximum number of quartet topologies to the computing of an ultrametric matrix, which essentially is about the computing of the least common ancestor for every pair of taxa such that a maximum number of quartet topologies are satisfied. The computing of a desired ultrametric matrix is then formulated into a constraint programming that can be readily solved by calling to Smodels (`http://www.tcs.hut.fi/Software/smodels/`). Though the work is mainly regarded as a reformulation of an old problem, we have imposed in the Smodels a number of deduction rules which speed up the computation. These deduction rules are also found useful in determining the quartet topologies that must be fixed according to a set of already fixed quartet topologies. Subsequently, at every node in the branch-and-bound search tree we might be able to shorten the conflict list and identify the illegal ways of resolving a local conflict. The following Theorem 5 lists some of the rules that have been implemented into our algorithm, which help reduce the overall running time significantly.

Theorem 5. [8,15] *For a set of* 5 *taxa* $\{a, b, c, d, e\}$, *if* $[a, b|c, d]$ *and* $[a, b|c, e]$ *are fixed, then* $[a, b|d, e]$ *must be fixed too; if* $[a, b|c, d]$ *and* $[a, c|d, e]$ *are fixed, then* $[a, b|c, e]$, $[a, b|d, e]$, *and* $[b, c|d, e]$ *must be fixed too.*

3 A Lookahead Branch-and-Bound Algorithm

Our branch-and-bound algorithm (LBnB) is designed to solve the MQC problem optimally regardless whether or not the number of quartet errors is known ahead of time. Typically, in the unknown case, we can use some other phylogeny construction method(s) to infer a phylogeny and determine subsequently its associated quartet topology set. This provides us a first upper bound on the number of quartet errors, denoted as k.

LBnB might also be regarded as an improvement over the GN algorithm, with several newly designed speedup techniques. In fact, most of the basic operations in the GN algorithm have been adopted into LBnB, though they are put together in different places and rearranged in different orders. Some significant differences between the two algorithms are detailed in the following: In LBnB, at the root node of the search tree, a global search is performed to identify quartet topologies need to be fixed using Theorem 4 (in $O(n^4)$ time). It then applies Theorem 5 to deduce more quartet topologies need to be fixed (in $O(n^4)$ time). Another global search is performed to identify need-to-be-changed quartet topologies according to Theorem 3 (in $O(nk)$ time). Subsequently, and similarly as in the GN algorithm, it then builds a conflict list (in $O(n^4)$ time). In LBnB, besides this list, two other lists, one contains the fixed quartet topologies and the other contains the need-to-be-changed quartet topologies, are also maintained. In fact, every node in the search tree is associated with these three lists. At every node, the conflict list is further partitioned into two parts, one consists of local conflicts each contains a need-to-be-changed quartet topology and the other consists of local conflicts each contains no need-to-be-changed quartet topology. The LBnB algorithm puts priorities on need-to-be-changed quartets. For every need-to-be-changed quartet topology, it 1) changes the topology to resolve some (at least one) local conflict; 2) updates the the conflict list; 3) updates the fixed quartet topology list; and 4) updates the need-to-be-changed quartet topology list. The difference between the size of the new conflict list and the size of the conflict list before quartet topology changing is defined to be the *contribution* of this need-to-be-changed quartet topology. The LBnB algorithm picks the need-to-be-changed quartet topology achieving the largest contribution to proceed the search. In this sense, the LBnB algorithm uses the contribution to guide its search. In the case that there is no need-to-be-changed quartet topology, a similar treatment is done for every way of resolving a local conflict, and then the algorithm picks the way achieving the largest contribution to proceed. The algorithm updates the upper bound k if a solution is found and in the solution a less number of quartet errors were found.

At every search node, the LBnB algorithm also checks the lower bound on the number of quartet errors using Theorem 2. If adding this lower bound to the number of already determined quartet errors is greater than the upper bound k, then the search node is cut off from further consideration. Notice that our algorithm LBnB has the mechanism to look one step forward to identify the best way to resolve a local conflict, through examining the contributions of all possible branches. For this reason, we call it the *lookahead* branch-and-bound algorithm. It should be noted that although it appears that at every node LBnB spends more time than the GN algorithm ($O(n^2k)$ vs. $O(n)$), LBnB does not do any extra computation compared to the GN algorithm. Therefore, the overall running time

1.	At every node in the search tree,
1.1.	Use Theorem 2 to decide to cut the node or not;
1.2.	Use Theorem 4 to determine need-to-be-fixed quartets;
1.3.	Use Theorem 5 to deduce need-to-be-fixed quartets;
1.4.	Use Theorem 3 to determine need-to-be-changed quartets;
1.5.	Build a conflict list and partition it into two parts;
1.6.	If there are need-to-be-changed quartets,
1.6.1.	For a need-to-be-changed quartet, calculate its contribution;
1.6.2.	Pick the need-to-be-changed quartet achieving the largest contribution;
1.7.	Else,
1.7.1.	For every way of resolving a local conflict, calculate its contribution;
1.7.2.	Pick the resolvement way achieving the largest contribution;
1.8.	If k quartets have been changed but the conflict list is nonempty,
1.8.1.	Kill the node and return to the parent node;
1.9.	If the conflict list is empty,
1.9.1.	Update k, update the best solution, and return to the parent node;
1.10.	Continue on search at the picked node;

Fig. 2. A high-level description of the LBnB algorithm for the MQC problem

of LBnB is also $O(4^k n + n^4)$. Nonetheless, with the embedded greedy lookahead, LBnB runs much faster in practice, mostly because the depth of its search tree is much smaller than that for the GN algorithm. This is demonstrated true by the experimental results in Section 4. One high-level description of the LBnB algorithm at every search node is depicted in Figure 2.

Theorem 6. *The lookahead branch-and-bound algorithm solves the MQC problem and runs in $O(4^k n + n^4)$ time, where n is the size of the taxa set and k is an upper bound on the number of quartet errors.*

4 Experimental Results

We have designed four experiments to compare the performance of LBnB to existing algorithms, among which three are on synthetic datasets and the last one is on a real dataset of 30 taxa with about 8.3% quartet errors (i.e. $k = 2272$). Note that the algorithms tested are all quartet-based phylogeny construction algorithms. Therefore, the comparisons were for the purpose of showing the speed of LBnB (except the comparison made to the hypercleaning algorithm). In all experiments, we said an algorithm could not solve an instance if the algorithm does not terminate in 48 hours. We chose to implement all the algorithms in C/C++ (either by ourselves or courtesy to the original authors). All experiments were done on an IBM P690 computer with a 1.7 GHz processor and 32 GB main memory (shared by 16 CPUs).

Similar to [9], to provide a common test bench for all the algorithms, we generated artificial datasets. Again we want to emphasis that such datasets were for the purpose of testing the computing power of the algorithms. Therefore, some of our datasets were very hard for all algorithms. For every pair of (n, k), where n refers to the number of taxa and k refers to the upper bound of the quartet errors, we generated a random phylogeny by recursively inserting one taxon onto an arbitrary edge in the existing sub-phylogeny. After all taxa have been added to the phylogeny, which is an unrooted binary tree, we derived the set of quartets induced by the phylogeny. We then arbitrarily picked k out of the $\binom{n}{4}$ quartets and altered the topologies. We remark that this process only guarantees the number of quartet errors is upper bounded by k, but not necessarily equal to k since some combinations of quartet altering might give rise to a new compatible set of quartets. For every pair of (n, k) we generated 10 datasets, and the following reported results are the average of test run on them. Different algorithms have their own computing limits and we have different datasets for them, to be detailed in the following sections.

4.1 Experiment 1

This experiment was intended to make comparisons among the exact algorithms proposed for the general MQC problem, including the dynamic programming algorithm by Ben-Dor et al. [2] (denoted as DP), the constraint programming

Table 1. The running times for four exact algorithms: DP, CP, GN-opt, and LBnB-opt

Problem Size		DP	CP	GN-Opt	LBnB-Opt
$n = 10$	$p = 1\%$	2 secs	1 sec	5 secs	1 sec
	$p = 5\%$	2 secs	1 sec	6 secs	1 sec
	$p = 10\%$	2 secs	1 sec	15 secs	1 sec
	$p = 15\%$	2 secs	1 sec	30 secs	1 sec
	$p = 20\%$	2 secs	1 sec	35 secs	1 sec
	$p = 30\%$	2 secs	1 sec	2 mins	1 sec
$n = 15$	$p = 1\%$	2 mins	1 sec	20 secs	1 sec
	$p = 5\%$	2 mins	1 sec	10 mins	1 sec
	$p = 10\%$	2 mins	1 sec	5 hrs	10 secs
	$p = 15\%$	2 mins	1 sec	–	30 secs
	$p = 20\%$	2 mins	1 sec	–	1 min
	$p = 30\%$	2 mins	1 sec	–	2 mins
$n = 20$	$p = 1\%$	40 hrs	10 mins	20 mins	1 sec
	$p = 5\%$	40 hrs	40 mins	–	2 mins
	$p = 10\%$	40 hrs	6 hrs	–	10 mins
	$p = 15\%$	40 hrs	6 hrs	–	40 mins
	$p = 20\%$	40 hrs	8 hrs	–	3 hrs
	$p = 30\%$	40 hrs	10 hrs	–	10 hrs
$n = 25$	$p = 1\%$	–	20 mins	9 hrs	5 secs
	$p = 5\%$	–	10 hrs	–	10 mins
	$p = 10\%$	–	–	–	1 hr
	$p = 15\%$	–	–	–	3 hrs
	$p = 20\%$	–	–	–	18 hrs
	$p = 30\%$	–	–	–	52 hrs

approach in our previous work [15] (denoted as CP), the modified GN algorithm (denoted as GN-Opt), and our LBnB algorithm (denoted as LBnB-Opt). Since DP and CP do not take k as an input and in fact their running times are independent on k, we generated datasets defined by a pair (n, p) where for each dataset p records the percentage of quartet errors in the given complete quartet set. For GN-Opt and LBnB-Opt, we used $k = \binom{n}{4} \times p$ as the first upper bound on the number of quartet errors. We used quartet error percentage $p = 1\%, 5\%, 10\%, 15\%, 20\%, 30\%$. The largest value set for n was 25, since LBnB-Opt, which appeared to run the fastest, failed to terminate in 100 hours for pair $(30, 30\%)$. Some of the running time results for these four algorithms are summarized in Table 1, where a '–' indicates that an algorithm didn't terminate in 48 hours. It is surprising to see that GN-Opt performed inferior to DP and CP on most of the datasets. It is very encouraging to see that LBnB-Opt outperformed the other three algorithms on all datasets.

4.2 Experiment 2

This experiment was designed to make sole comparison between the GN algorithm and our LBnB algorithm. As we have seen in Experiment 1 that in terms of finding the optimal solution, LBnB-Opt outperformed GN-Opt significantly.

Table 2. The running time comparison between GN-1st and LBnB-1st

Problem Size		GN-1st	LBnB-1st
$n = 10$	$k = 40$	1 sec	1 sec
	$k = 50$	15 secs	1 sec
$n = 20$	$k = 40$	3 secs	1 sec
	$k = 50$	10 secs	1 sec
	$k = 100$	35 mins	20 secs
	$k = 200$	–	2 mins
$n = 30$	$k = 40$	1 sec	1 sec
	$k = 50$	6 secs	2 secs
	$k = 100$	21 mins	35 secs
	$k = 200$	–	5 mins
$n = 40$	$k = 40$	1 sec	1 sec
	$k = 50$	2 secs	2 secs
	$k = 100$	15 mins	1 min
	$k = 200$	–	10 mins
$n = 50$	$k = 40$	1 sec	1 sec
	$k = 50$	3 secs	3 secs
	$k = 100$	6 mins	2 mins
	$k = 200$	–	20 mins

Therefore, in this experiment we only compared them in finding the first solution, in which the number of quartet errors is at most k. We denote these two algorithms as GN-1st and LBnB-1st, respectively. We generated 10 datasets for each pair (n, k), where $n = 10, 20, 30, 40, 50$ and $k = 5, 10, 20, 30, 40, 50, 100, 200$. For every value of n, we found both GN-1st and LBnB-1st terminated within a second on the datasets corresponding to some small values of k; and only when $k \geq 40$ their running times start to depart. Another observation is that for each dataset in the experiment, surprisingly, the number of quartet errors in the solution by GN-1st is equal to the number of quartet errors in the solution by LBnB-1st, and is equal to k. For these reasons, we chose to report their running times only, and only for datasets with $k \geq 40$, in Table 2. We set the time limit to be 24 hours in this experiment because of time constraint. From the table, we see that LBnB-1st outperformed GN-1st on all datasets, slightly for small values of k but significantly for large values of k.

4.3 Experiment 3

The hypercleaning algorithm is known as a best heuristic for the MQC problem (at the writing of this paper). The largest instance reported in [3] contains 18 taxa. This third experiment was designed to compare the performances of our algorithm and the hypercleaning algorithm in terms of both running time and the quality of the returned solution. For this purpose, we generated datasets for $n = 30, 40, 50$ with various bounds on the number of quartet errors $k = 100, 200, 300$. Note that smaller values of n and larger values of k do not make sense since they correspond to either too easy datasets or too difficult datasets. Our algorithm was run to return both the first solution and the optimal solution (that is, by LBnB-1st and LBnB-Opt, respectively). It is again interesting to note that all three solutions for a dataset contain a same number of quartet errors. (The value of m in Hypercleaning on these datasets was set 3 in order to output solution phylogenies.) Subsequently, we chose to report the running time only in Table 3. From the statistics, it is not surprising to see that LBnB-Opt required much more time to search a whole tree than LBnB-1st which terminated

Table 3. The running time comparison among three algorithms: Hypercleaning, LBnB-1st, and LBnB-opt

Problem Size		Hypercleaning	LBnB-1st	LBnB-Opt
$n = 30$	$k = 100$	40 secs	35 secs	6 mins
	$k = 200$	55 secs	5 mins	36 mins
	$k = 300$	80 secs	45 mins	2 hrs
$n = 40$	$k = 100$	10 mins	1 min	20 mins
	$k = 200$	12 mins	10 mins	3 hrs
	$k = 300$	18 mins	135 mins	18 hrs
$n = 50$	$k = 100$	50 mins	2 mins	2 hrs
	$k = 200$	55 mins	20 mins	12 hrs
	$k = 300$	62 mins	3 hrs	–

at the time one solution was found. Nonetheless, it is encouraging to see that LBnB-1st beats hypercleaning in some "easy" datasets. We set the time limit to be 24 hours in this experiment.

4.4 A Real Dataset

We have obtained a real dataset which contains 30 Avian Influenza viruses with their pairwise evolutionary distances measured by the Complete Composition Vector defined on the whole set of proteins for the viruses [17]. We applied the Four-Point Method [7] on this 30×30 distance matrix to infer a topology for every quartet. We ran the Neighbor-Joining algorithm [13] on the distance matrix to produce a phylogeny on these 30 taxa and use the set of induced quartet topologies to estimate an upper bound on the quartet errors, which was 2272. Unfortunately GN-1st did not terminate in 100 hours. Hypercleaning finished in 62 minutes and the returned phylogeny resolves 1922 quartet errors. Our algorithms LBnB-1st and LBnB-Opt finished in 21 hours and 29 hours, respectively, and the returned phylogenies both resolve 1871 quartet errors.

5 Concluding Remarks

Among the exact algorithms for solving the MQC problem, we found the dynamic programming (DP) the least efficient and our lookahead branch-and-bound algorithm (LBnB-Opt) seemingly the most powerful. Nonetheless, the performance of DP (and CP) is not affected by the number of quartet errors and thus DP might be superior for datasets containing a small number of taxa.

The fixed-parameter algorithm solves the MQC problem in the special case where the number of quartet errors is known exactly. This special case is unlikely to happen, rather than that, the more often we find it easy to obtain an upper bound on the number of quartet errors. The fixed-parameter algorithm can be modified to solve the general problem, however, our experiments showed that the running time grows too fast to be feasible. Our LBnB algorithm might be regarded as an improvement over the fixed-parameter algorithm. It outperforms in general significantly in both finding the first solution and in finding the optimal solution. One of our future work is to design more techniques to speedup the search, including faster determination of need-to-be-changed quartets and need-to-be-fixed quartets. Our expectation is that the algorithm has a competitive running time to hypercleaning on datasets of modest size, say 50 taxa, and thus to prepare us to construct a global phylogenetic pattern for the 5318 Avian Influenza viruses.

Lastly, we remark that all experiments were done for the purpose of testing the power of the LBnB algorithm. Though on all the synthetic datasets every algorithm found the optimal solution, we haven't compare the quality of the phylogenies for the real dataset (but both LBnB-1st and LBnB-opt solve the instance optimally).

Acknowledgment

GW was supported in part by CFI and NSERC; JY was supported in part by NSERC; GL was supported in part by CFI, NSERC, and NNSF Grant 60373012.

References

1. H. Bandelt and A. Dress. *Advance in Applied Mathematics*, 7:309–343, 1986.
2. A. Ben-Dor, B. Chor, D. Graur, R. Ophir, and D. Pelleg. In *Proceedings of RECOMB'98*, pages 9–19, 1998.
3. V. Berry, D. Bryant, T. Jiang, P. Kearney, M. Li, T. Wareham, and H. Zhang. In *Proceedings of SODA'00*, pages 287–296, San Francisco, California, January 9–11, 2000.
4. V. Berry, T. Jiang, P. E. Kearney, M. Li, and H. T. Wareham. In *Proceedings of ESA'99*, LNCS 1643, pages 313–324, 1999.
5. H. Colonius and H. H. Schultze. *British Journal of Mathematical and Statistical Psychology*, 34:167–180, 1981.
6. G. Della Vedova and H. T. Wareham. *Bioinformatics*, 18:1297–1304, 2002.
7. P. Erdos, M. Steel, L. Szekely, and T. Warnow. *Random Structures and Algorithms*, 14:153–184, 1999.
8. P. L. Erdos, M. A. Steel, L. A. Szekely, and T. J. Warnow. *Computers and Artificial Intelligence*, 16:217–227, 1997.
9. J. Gramm and R. Niedermeier. *Journal of Computer and System Science*, 67:723–741, 2003.
10. T. Jiang, P. E. Kearney, and M. Li. In *Proceedings of FOCS'98*, pages 416–425, 1998.
11. T. Jiang, P. E. Kearney, and M. Li. *Journal of Algorithms*, 34:194–201, 2000.
12. T. Jiang, P. E. Kearney, and M. Li. *SIAM Journal on Computing*, 30:1942–1961, 2001.
13. N. Saitou and M. Nei. *Molecular Biology and Evolution*, 4:406–425, 1987.
14. M. Steel. *Journal on Classification*, 9:91–116, 1992.
15. G. Wu, G.-H. Lin, and J. You. In *Proceedings of ICTAI'04*, pages 612–619, 2004.
16. G. Wu, J. You, and G.-H. Lin. Technical Report TR05-05, Department of Computing Science, University of Alberta, January 2005.
17. X. Wu, X. Wan, G. Wu, D. Xu, and G.-H. Lin. Technical Report TR05-06, Department of Computing Science, University of Alberta, January 2005.

Computing the Quartet Distance Between Trees of Arbitrary Degree

Chris Christiansen[1], Thomas Mailund[2],
Christian N.S. Pedersen[1,2], and Martin Randers[1]

[1] Department of Computer Science, University of Aarhus,
Aabogade 34, DK-8200 Århus N, Denmark
{chrisc, cstorm, u002155}@daimi.au.dk
[2] Bioinformatics Research Center, University of Aarhus,
Høegh-Guldbergsgade 10, Bldg. 090, DK-8000 Århus C, Denmark
{mailund, cstorm}@birc.au.dk

Abstract. We present two algorithms for computing the quartet distance between trees of arbitrary degree. The quartet distance between two unrooted evolutionary trees is the number of quartets—sub-trees induced by four leaves—that differs between the trees. Previous algorithms focus on computing the quartet distance between binary trees. In this paper, we present two algorithms for computing the quartet distance between trees of arbitrary degrees. One in time $O(n^3)$ and space $O(n^2)$ and one in time $O(n^2d^2)$ and space $O(n^2)$, where n is the number of species and d is the maximal degree of the internal nodes of the trees. We experimentally compare the two algorithms and discuss possible directions for improving the running time further.

1 Introduction

The evolutionary relationship for a set of species is conveniently described by a tree in which the leaves correspond to the species, and the internal nodes correspond to speciation events. The true evolutionary tree for a set of species is rarely known, so inferring it from obtainable information is of great interest. Many different methods have been developed for this, see e.g. [8] for an overview.

Some methods infer rooted trees, where the most recent common ancestor of the set of species is represented by a root, and the direction of evolution is from the root to the leaves, while other methods infer unrooted trees, relying on an out group for inferring the direction of evolution. Most methods aim at fully resolving the evolutionary tree into a binary tree, while other methods, e.g. the Buneman [6, 2] and refined Buneman [9, 4], construct fully resolved binary trees only if this is well supported by the input data. Different methods often yield different inferred trees for the same set of species, and even the same method can give rise to different evolutionary trees for the same set of species when applied to different information about the species, e.g. different genes. To study such differences in a systematic manner, one must be able to quantify differences between evolutionary trees using well-defined and efficient methods.

R. Casadio and G. Myers (Eds.): WABI 2005, LNBI 3692, pp. 77–88, 2005.

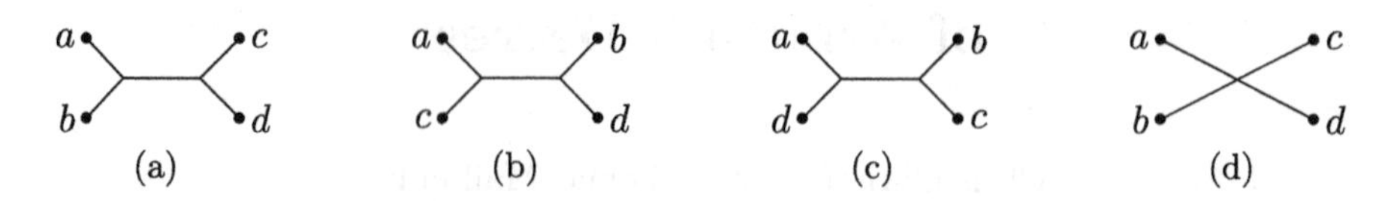

Fig. 1. The four possible quartet topologies of species a, b, c, and d. Topologies (a): $ab|cd$, (b): $ac|bd$, and (c): $ad|bc$ are *butterfly* quartets, while topology (d): $\substack{a \\ b} \times \substack{c \\ d}$, is a *star* quartet.

One approach for comparing evolutionary trees is to define a distance measure between trees and compare two trees by computing this distance. Several distance measures have been proposed, e.g. the symmetric difference metric [10], the nearest-neighbour interchange metric [13], the subtree transfer distance [1], the Robinson and Foulds distance [11], and the quartet distance [7]. Each distance measure has different properties and reflects different aspects of biology.

This paper is concerned with calculating the quartet distance. For an evolutionary tree, the *quartet topology* of four species is determined by the minimal topological subtree containing the four species. The four possible quartet topologies of four species are shown in Fig. 1. Given two evolutionary trees on the same set of n species, the *quartet distance* between them is the number of sets of four species for which the quartet topologies differ in the two trees.

Previous algorithms for computing the quartet distance all focus on comparing *binary* trees and therefore avoid star quartets. Steel and Penny in [12] present an algorithm for computing the quartet distance in time $O(n^3)$. Bryant *et al.* in [5] present an algorithm that computes the quartet distance in time $O(n^2)$. Brodal *et al.* in [3] present an algorithm that computes the quartet distance in time $O(n \log n)$. In this paper, we present two algorithms that compute the quartet distance between two trees of *arbitrary* degrees, i.e. trees that can contain star quartets. The first algorithm runs in time $O(n^3)$ and space $O(n^2)$, the second in time $O(n^2d^2)$ and space $O(n^2)$, where d is the maximal degree of inner nodes in the trees.

The rest of the paper is organised as follows. In Sect. 2, we present our algorithms for computing the quartet distance between two unrooted evolutionary trees of arbitrary degree. In Sect. 3 we experimentally compare the running time of the algorithms and confirm that the actual running times concur with the theoretical results. In Sect. 4 we draw our conclusions.

2 Algorithms

In this section we present three algorithms for counting the quartet distance between two non-rooted trees of arbitrary degrees over the same set of n species.

Comparing two quartets means comparing the topology of these two quartets in the two input trees. We say that a quartet is *shared* if it has the same topology in both trees. Two star quartets always have the same topology, but butterfly quartets can have three different topologies as shown in Fig. 1.

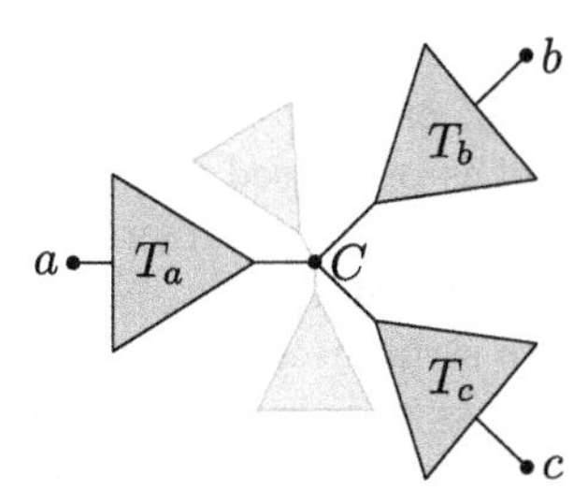

Fig. 3. The three subtrees containing the leaves a, b and c are called T_a, T_b and T_c, respectively. Any remaining trees (shown in lighter grey) are collectedly called T_{rest}.

The most obvious way to compute the quartet distance between two non-rooted trees over the same set of n species would be to explicitly compare each of the $\binom{n}{4}$ pairs of quartets. With only the input trees available, the straightforward approach for finding the topology of one quartet takes linear time, which leads to a total time usage of $O(n^5)$.

We first present an algorithm that runs in time $O(n^4)$ and space $O(n)$, which we modify to run in time $O(n^3)$ but space $O(n^2)$. Then we establish some terminology and present a third algoritm, using $O(n^2d^2)$ time and $O(n^2)$ space.

2.1 General Quartet Distance in Time $O(n^4)$ and Space $O(n)$

Given two input trees T and T' over the same set of n species, we can compute the quartet distance in time $O(n^4)$, if the quartet topology for each set can be found in constant time in both input trees. We will show how to achieve this by focusing on the *centers* between triplets of leaves: Given leaves a, b, and c, there is a unique inner node C, the center of a, b and c, in which the paths from a to b, a to c and b to c are joined, see Fig. 2.

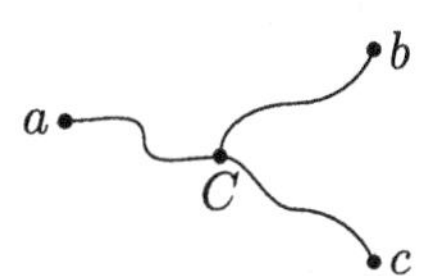

Fig. 2. The *center*, C, of a, b and c

Given leaves $a, b, c \in T$ with center C, let the subtree containing a be denoted T_a, similarly for leaves b and c. Any remaining subtrees of C are collectedly denoted T_{rest}, see Fig. 3. For each leaf x different from a, b and c, a quartet is defined, and its topology in T can be easily determined from the center: if $x \in T_a$ then the topology of the quartet is $ax|bc$, and if $x \in T_b$, the quartet is $bx|ac$ and if $x \in T_c$ the quartet is $ab|cx$. If $x \in T_{\text{rest}}$, then the topology is $\frac{a}{b} \times \frac{c}{x}$. Similarly, the quartet topologies for a, b, c and x can be determined in T' given the center of a, b, and c in T'.

To determine the quartet distance between two trees, each of the $\binom{n}{3} \in O(n^3)$ triplets of leaves is processed sequentially in both trees: The centers can be found, and the leaves in the different subtrees accumulated, in linear time. Assuming leaves are numbered $0, \ldots, n-1$, the topology of each of the $n-3$ quartets containing a, b and c can be stored in an array for each tree, where entry i contains the topology of the quartet a, b, c and the ith leaf. In this way, the topology of the quartets can be compared directly. Each of the $O(n^3)$ triplets can be processed in linear time, making the algorithm run in time $O(n^4)$ and space $O(n)$. Since each quartet is processed four times—once for each triplet it contains, the number of quartets that have different topologies must be divided by four to get the correct quartet distance.

The algorithm uses no data structures other than the trees and arrays, little memory and no complex methods, so it is very easy to implement. It is also very easy to extend it to count the number of *shared* quartet topologies between k trees over the same set of n species in time $O(kn^4)$ and space $O(kn)$.

2.2 General Quartet Distance in Time $O(n^3)$ and Space $O(n^2)$

Instead of counting quartets with different topologies in the input trees, we now count the number of shared quartets and then subtract this number from the total number of quartets, $\binom{n}{4}$, to get the quartet distance.

Given rooted subtrees T_x and T'_x of T and T' respectively, we can pre-compute the size of the intersection of leaf sets, denoted $|T_x \cap T'_x|$. There are $O(n^2)$ pairs of such subtrees, and all intersection sizes can be computed and stored in time and space $O(n^2)$, see e.g. [5]. Using these precomputed sizes, we can improve the running time of the above algorithm to $O(n^3)$.

Consider leaves a, b and c. Let centers and subtrees in T and T' be defined as above. Use prime (′) to denote the center and subtrees in T', and no prime for the center and subtrees in T. Any leaf, x, in T_a gives a butterfly quartet of the form $ax|bc$ in T, and any leaf x' in T'_a gives a butterfly quartet of the form $ax'|bc$ in T'. Therefore, any leaf x, $x \neq a$, in both T_a and T'_a, represents a shared butterfly quartet in the two trees. The same applies for b and c. The number of shared butterfly quartets containing a, b and c can therefore be computed by the expression:

$$|T_a \cap T'_a| + |T_b \cap T'_b| + |T_c \cap T'_c| - 3\,.$$

This number can be computed in constant time, since the sizes of the intersections are precomputed. Any leaf x in T_{rest} gives a star quartet in T, and any leaf x' in T'_{rest} gives a star quartet in T'. It follows that the number of shared star quartets containing a, b and c can be computed by the expression:

$$|T_{\text{rest}} \cap T'_{\text{rest}}|\,.$$

Since T_{rest} and T'_{rest} potentially consist of a large number of subtrees, it is not clear how to compute $|T_{\text{rest}} \cap T'_{\text{rest}}|$ in constant time. However, $|T_{\text{rest}} \cap T'_{\text{rest}}|$ can be expressed in terms of T_a, T_b, T_c, T'_a, T'_b and T'_c. Any leaf in T'_{rest} is either in T_a, T_b, T_c or T_{rest}, and never in two of those at the same time, since they are disjoint. Thus:

$$|T_{\text{rest}} \cap T'_{\text{rest}}| = |T'_{\text{rest}}| - (|T'_{\text{rest}} \cap T_a| + |T'_{\text{rest}} \cap T_b| + |T'_{\text{rest}} \cap T_c|)\,.$$

Any element in T_a that is not in T'_a, T'_b or T'_c is in T'_{rest}. The same applies for elements in T_b and T_c. This gives the equations:

$$|T_a \cap T'_{\text{rest}}| = |T_a| - (|T_a \cap T'_a| + |T_a \cap T'_b| + |T_a \cap T'_c|)\,,$$
$$|T_b \cap T'_{\text{rest}}| = |T_b| - (|T_b \cap T'_a| + |T_b \cap T'_b| + |T_b \cap T'_c|)\,,$$
$$|T_c \cap T'_{\text{rest}}| = |T_c| - (|T_c \cap T'_a| + |T_c \cap T'_b| + |T_c \cap T'_c|)\,.$$

Since there are n leaves in the trees, it follows directly that:

$$|T'_{\text{rest}}| = n - (|T'_a| + |T'_b| + |T'_c|)\,.$$

Combining all these equations give an expression for $|T_{\text{rest}} \cap T'_{\text{rest}}|$ that can be computed in constant time. So given a center of three leaves in each tree, the number of shared topologies of quartets containing these leaves can be found in constant time. Assuming the centers can be found in constant time, the algorithm will run in time $O(n^3)$, since there are $O(n^3)$ different triplets of leaves.

Finding the centers in constant time is done by finding a linear number of centers in linear time: For each pair of leaves, a and b, the path from a to b can be found in linear time by a single traversal of the tree. Each inner node in the path is the center of all triplets a, b, c where c can be reached from the node via an edge not in the the path, see Fig. 4 for an illustration.

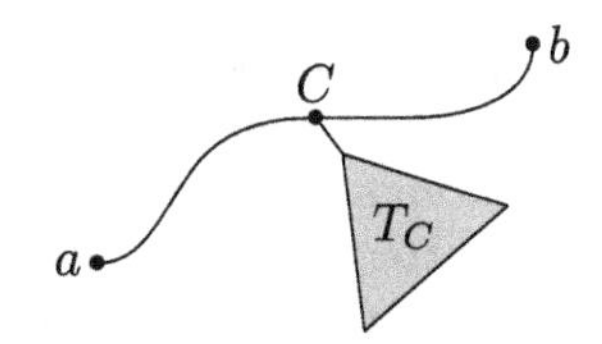

Fig. 4. For node C on the path from a to b, any subtree T_C, rooted in C and not on the path, defines a set of leaves, c, for which C is the center of a, b, and c

Assuming leaves are numbered $0, \ldots, n-1$, the center of each of the $n-2$ triples containing a and b can be put in an array, where entry i contains the center of a, b and the ith leaf. All internal nodes on the path from a to b is a covering and disjoint set of centers for all triplets containing a and b. Therefore, filling all array entries can be done by a single traversal of the tree. Using these arrays, and the equations above, the number of shared quartets containing a fixed pair of leaves can be found in linear time. Since there are $O(n^2)$ pairs of leaves, the algorithm runs in time $O(n^3)$. The space consumption is $O(n^2)$, since all the intersection sizes of induced subtrees must be stored. As before, all quartets are counted four times, so the final count must be divided by four.

2.3 General Quartet Distance in Time $O(n^2 d^2)$ and Space $O(n^2)$

Given four leaves and their quartet topology in T and T', there are four possible cases: Case 1: The quartet has a star topology in both trees (shared quartet). Case 2: The quartet has an equal butterfly topology in both trees (shared quartet). Case 3: The quartet has a different butterfly topology in the trees. Case 4: The quartet has a butterfly topology in one tree, and a star topology in the other tree. For $i = 1, 2, 3, 4$, let Q_i be the number of quartets in case i above. The quartet distance $\text{qdist}(T, T')$ between T and T' is $Q_3 + Q_4$. Let BQ and BQ'

be the number quartets that have butterfly topologies in T and T', respectively. The main observation of our approach is that $Q_4 = BQ + BQ' - 2(Q_2 + Q_3)$, which gives the following expression:

$$\mathrm{qdist}(T, T') = Q_3 + Q_4 = BQ + BQ' - 2Q_2 - Q_3$$

Surprisingly, this means that the general quartet distance can be calculated without directly considering the star quartets. What we need are three algorithms: one that can count the number of quartets with butterfly topology in a single general tree (for computing BQ and BQ'), one that can count the number of quartets that share the same butterfly topology in two general trees (for computing Q_2), and one that can count the number of quartets that have different butterfly topologies in two general trees (for computing Q_3).

The first algorithm can be implemented using the second: The number of quartets with the same butterfly topology in two instances of the same tree is the total number of butterfly quartets in that tree. Therefore we only need to consider the second and third algorithm. Both can be made by extending the algorithm described in [5].

The algorithm for counting butterfly quartets with the same topology uses the concepts of *directed edges*, *directed quartets*, and *claims*, that we define below:

A butterfly quartet $ab|cd$ is *induced* by edges which separate a, b from c, d, note that the inducing edge implies the topology of the quartet. Since several edges may induce the same butterfly quartet, it is convenient to look at directed edges: every edge e defines two rooted subtrees T_1 and T_2; instead of viewing e as a single undirected edge, we can view it as two directed edges e_1 and e_2, and we say that T_1 is *in front* of e_1 and T_2 is *behind* e_1 and similarly for e_2: T_2 is in front of e_2 and T_1 is behind e_2, see Fig. 5.

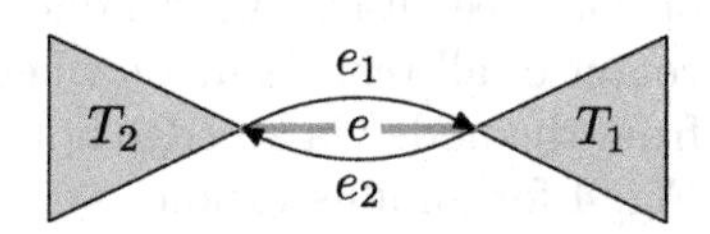

Fig. 5. Edge e defines rooted trees T_1 and T_2 and directed edges e_1 and e_2

A butterfly quartet $ab|cd$ defines two *directed* quartets: $ab \rightarrow cd$ and $ab \leftarrow cd$; for undirected edge e, inducing quartet $ab|cd$, the corresponding directed edges e_1 and e_2 induces the directed quartets $ab \rightarrow cd$ and $ab \leftarrow cd$.

To each directed quartet, $ab \rightarrow cd$, we can uniquely associate a directed edge, e_1 such that a and b are leaves in the tree *behind* e_1, and such that c and d are leaves in *different* subtrees of the root of the tree *in front* of e_1, see Fig. 6. We call such a tree substructure a *claim*, written $T_1 \xrightarrow{e_1} (T_2, T_3)$, and say that the edge e_1 *claims* the directed quartet $ab \rightarrow cd$ and we also say that an edge e_1 claims an undirected quartet $ab|cd$ if it claims one of its directed quartets.

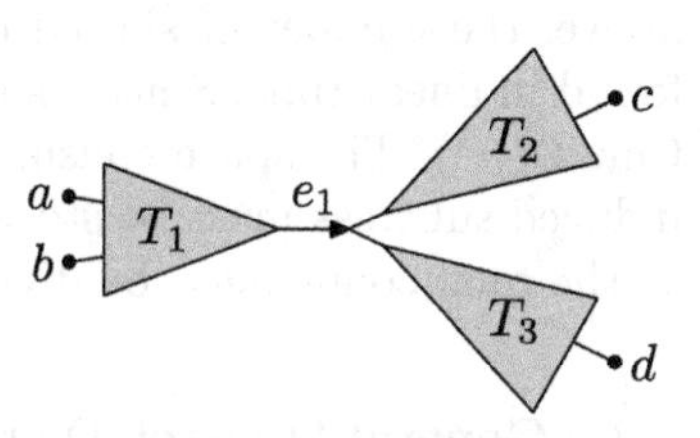

Fig. 6. The directed edge e_1 claims all directed quartets $ab \rightarrow cd$ where $a, b \in T_1$, $c \in T_2$ and $d \in T_3$

Since each (undirected) butterfly quartet defines exactly two directed quartets, and each directed quartet is claimed by exactly one directed edge, each butterfly quartet is claimed by exactly two directed edges.

Now, consider a pair of claims, $cl = T_1 \xrightarrow{e} (T_2, T_3)$ and $cl' = T'_1 \xrightarrow{e'} (T'_2, T'_3)$ in T and T', respectively. The number of oriented butterfly quartets shared by the claims can be computed by the expression:

$$\text{count}(cl, cl') = \binom{|T_1 \cap T'_1|}{2} \cdot (|T_2 \cap T'_2| \cdot |T_3 \cap T'_3| + |T_2 \cap T'_3| \cdot |T_3 \cap T'_2|) \quad (1)$$

and the total number of shared oriented butterfly quartets between T and T' is obtained by summing over all pairs of claims from the two trees:

$$2Q_2 = \sum_{cl \in T} \sum_{cl' \in T'} \text{count}(cl, cl') \quad (2)$$

Since each butterfly quartet corresponds to exactly two oriented butterfly quartets, the number of shared butterfly quartets is obtained by dividing the sum with two. Since the size of the intersection of subtrees is precomputed, $\text{count}(cl, cl')$ can be computed in constant time, and the sum can thus be computed in time proportional to the number of claims in T times the number of claims in T'.

For binary trees, there is a $1-1$ correspondence between claims and directed edges pointing to internal nodes, and therefore $O(n)$ claims per tree, resulting in a $O(n^2)$ time algorithm. For general trees, a directed edge e pointing to a node of degree d will be part of $\binom{d-1}{2}$ claims, see Fig. 7. Thus, the straightforward application of the sum in (2) result in an $O(n^2d^4)$ time algorithm for $d-$airy trees, since each edge in each tree can lead to $O(d^2)$ claims.

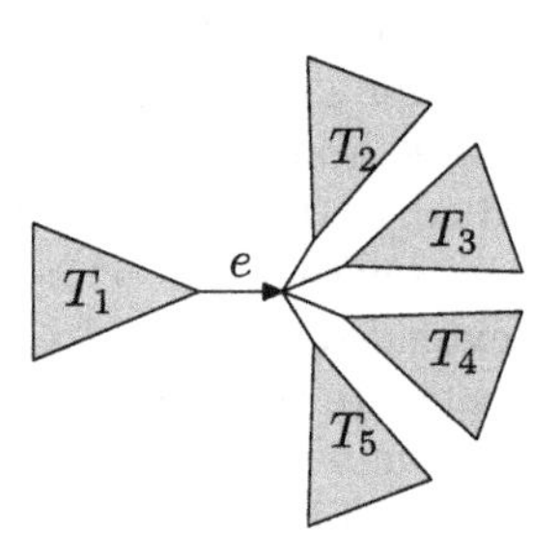

Fig. 7. The edge e is not part of a unique claim, but part of all claims $T_1 \xrightarrow{e} (T_i, T_j)$ for $i, j = 2, 3, 4, 5, i \neq j$

The reduction in time to $O(n^2d^2)$ is achieved by a transformation of the input trees T and T' into binary trees, annotated with information about the original trees, from which the butterfly quartets shared in the original trees can be calculated.

Expanding two trees of arbitrary degree is done by expanding every node with degree higher than three to a number of binary nodes. This adds a linear number of edges and nodes to the trees, so it does not change their sizes asymptotically. The two expanded trees induce all the butterfly quartets in the two original non-expanded input trees, but they also induce additional butterfly quartets due to the newly added nodes and edges: each star quartet in the original trees will be translated into butterfly quartets by the introduction of new edges, when resolving high-degree nodes, see Fig. 8. Thus, summing the contribution of all pairs of claims using (2) on the expanded trees will count too many butterfly quartets.

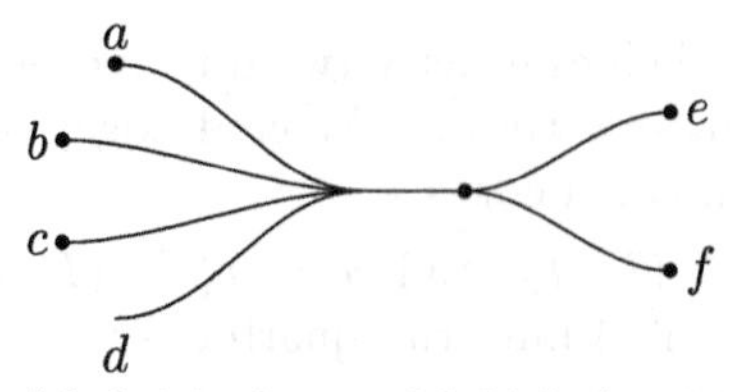

(a) Original tree with high-degree node.

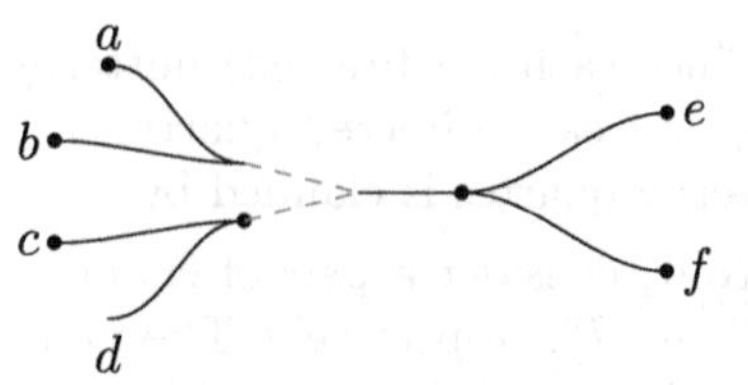

(b) Expanded tree, new edges shown dashed in grey.

Fig. 8. A tree with a high-degree node (a), and an expanded tree for this tree (b). The star quartets in the tree on the left becomes butterfly quartets on the right, due to the newly introduced edges, e.g. $\genfrac{}{}{0pt}{}{a}{b} \times \genfrac{}{}{0pt}{}{c}{e}$ becomes $ab|ce$.

Consider an edge, e_1, present in the original tree, and any edge e_2 resulting from the expansion of the tree, where e_2 is reachable from e_1 through newly introduced edges exclusively, see Fig. 9. These two edges, *together*, claim a set of butterfly quartets found in both the extended and the original tree: for leaves $a, b \in T_1$ behind e_1 and $c \in T_2$ and $d \in T_3$ in front of e_2 have, by construction, quartet topology $ab|cd$ in the extended tree, but also in the original tree since e_1 separates a and b from c and d in the original tree.

Formally, we let $\dashrightarrow^*$ denote the reflexive and transitive closure of newly introduced edges, that is, for $e_1 \dashrightarrow^* e_2$, e_2 is either e_1 or e_2 is reachable from e_1 following only newly introduced edges. If e_2 form a claim in the extended tree, we say that $e_1 \dashrightarrow^* e_2$ is an *extended claim*. For extended claims $ecl \in T$ and $ecl' \in T'$ we can calculate the number of shared quartets using (1), but with the sub-trees from the extended tree, defined by the extended claims.

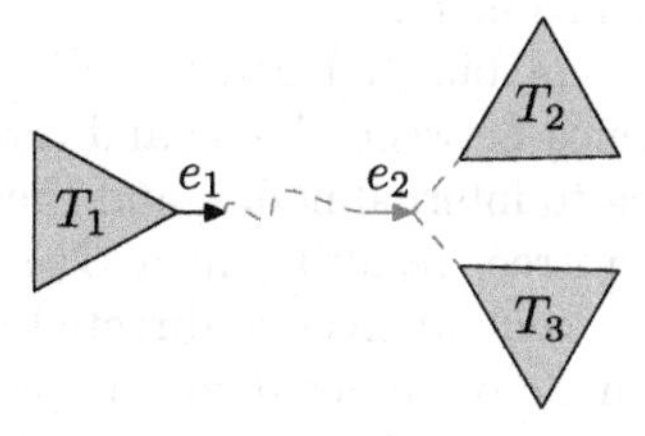

Fig. 9. An extended claim from original edge e_1 through newly introduced edge e_2

The following propositions establish that we can count the ordered butterfly quartets shared in the original trees using the extended claims in the extended trees, ET and ET', which gives us.

$$2Q_2 = \sum_{ecl \in ET} \sum_{ecl' \in ET'} \text{count}(ecl, ecl') \tag{3}$$

Proposition 1. *Each directed butterfly quartet, $ab \to cd$ in the original tree corresponds to exactly one extended claim in the extended tree.*

Proof. Consider the quartet $ab \to cd$ in the original tree, and corresponding claim $T_{ab} \xrightarrow{e} (T_c, T_d)$ in the original tree. The node separating trees T_{ab}, T_c, and T_d is either a binary node (degree 3) or a node of higher degree.

In the first case, where the node has degree 3, it will not have been expanded in ET and the claim will also be an extended claim in ET, and the only extended claim in ET that claims $ab \to cd$.

In the second case, where the node has degree higher than 3, it will be expanded into a sub-tree of newly introduced nodes and edges. Among the newly introduced nodes, there is a node that separates the subtrees containing c and d, since they were separated by the high degree node that was expanded. Let e' denote the edge pointing to this node, not from the trees containing c and d; the situation is then as this:

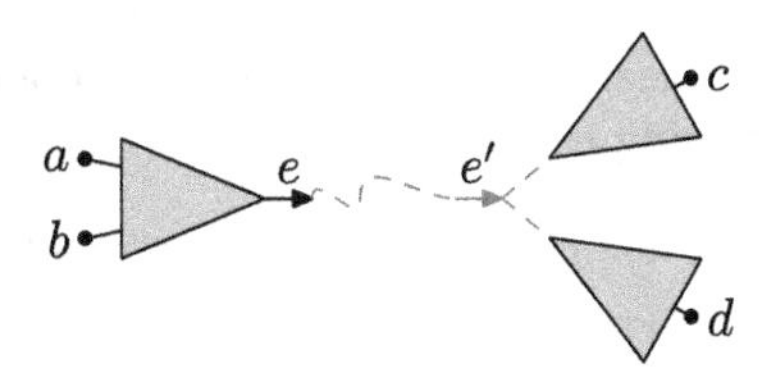

This is an extended claim, claiming the oriented quartet $ab \rightarrow cd$ in ET. Any other extended claim, claiming $ab \rightarrow cd$, must contain e', by definition, and an original edge e'' such that a and b are found in the tree behind e'' and such that $e'' \dashrightarrow^* e'$. Assume that such an e'' exist, and that $e'' \neq e$. Since e separates a and b from the expanded node that originally separated ab from c and d, any edge separating ab from cd must either be on the path from e to e', or be behind e. Assume e'' is on the path. Since $e \dashrightarrow^* e'$ the path only contain new edges, which contradicts that e'' is an original edge. Assume that e'' is behind e, which means that e is on the path from e'' to e'. Since $e'' \dashrightarrow^* e'$ by assumption, this path consist of new edges, which contradicts that e is an original edge. The conclusion is the edge e'' cannot exist, thus the extended claim is unique. □

Proposition 2. *All directed quartets $ab \rightarrow cd$ claimed by extended claims in the extended tree correspond to directed quartets in the original tree.*

Proof. If $ab \rightarrow cd$ is claimed by an extended claim, through original edge e_1 and new edge e_2, in ET, then, by the construction of ET, the original edge e_1 separates ab from cd in T. Thus e_1 induces, but does not necessarily claim, the quartet $ab \rightarrow cd$ in T. □

Using precomputed sizes of intersecting subtrees, as before, we can calculate count(ecl, ecl') in constant time, so to achieve the desired complexity for computing (3) we need to shown that we can calculate all pairs of extended claims in time $O(n^2d^2)$; we do this by showing how we can calculate all extended claims for a single tree in time $O(nd)$.

When building the extended tree, we can tag each edge with a flag specifying if it is an edge in the original tree or if it is a newly introduced edge. With these tags, we can iterate through all original edges in the extended tree by a simple tree traversal in linear time. For each original edge, e_1, we can find all e_2 such that $e_1 \dashrightarrow^* e_2$ by a simple traversal starting at e_1 following only newly introduced edges. Since the newly introduced edges reachable from e_1 found in this way are all edges in a tree expanded from a node of degree at most d, this search can be done in time $O(d)$ giving a total time for finding all extended claims in a single tree of $O(nd)$.

For counting butterfly quartets with different topologies we use an algorithm that works almost in the same way as the one just described, but it uses an alternative version of (1), which counts the number of butterfly quartets that have different topologies instead of the ones with the same topology. Similar to (1), it can be calculated in constant time.

Observe that when comparing two extended claims *ecl* and *ecl'*, butterfly quartets associated to these claims, that have different topologies, are those quartets where exactly one leaf is in $T_1 \cap T_1'$. The number of such quartets can be counted using the following expression:

$$\begin{aligned}\text{dcount}(ecl, ecl') = \; & |T_1 \cap T_1'| \cdot |T_1 \cap T_3'| \cdot |T_2 \cap T_2'| \cdot |T_3 \cap T_1'| + \\ & |T_1 \cap T_1'| \cdot |T_1 \cap T_2'| \cdot |T_2 \cap T_3'| \cdot |T_3 \cap T_1'| + \\ & |T_1 \cap T_1'| \cdot |T_1 \cap T_2'| \cdot |T_2 \cap T_1'| \cdot |T_3 \cap T_3'| + \\ & |T_1 \cap T_1'| \cdot |T_1 \cap T_3'| \cdot |T_2 \cap T_1'| \cdot |T_3 \cap T_2'|\end{aligned}$$

Counting the total number of butterfly quartets with different topologies in the two trees is done by comparing all pairs of extended claims.

$$4Q_3 = \sum_{ecl \in ET} \sum_{ecl' \in ET'} \text{dcount}(ecl, ecl') \tag{4}$$

Since each undirected butterfly quartet is claimed twice in each tree, the ones with different topologies are counted four times—once for each combination of claims. Therefore, the number of butterfly quartets with different topology is obtained by dividing the sum with four.

3 Experiments

We have implemented the above algorithms in Java (the implementation is available upon request) and performed a set of experiments to validate the theoretical time complexities of our algorithms. The performance of the $O(n^2d^2)$ time algorithm has been verified by running the algorithm on trees where all nodes have the same degree d. To verify the running time with respect to d, we have run the algorithm for various degrees d but fixed numbers of leaves n, . Similarly, to verify the running time with respect to n, we have kept d constant while changing n.

Fig. 10 shows the running time of the $O(n^2d^2)$ time algorithm as a function of d. For fixed n, the running time evolves as $O(d^2)$, which verifies the theoretical time complexity with respect to d. Fig. 11 shows the running time of the algorithms as a function of the number of leaves n. For fixed d the running time evolves as $O(n^2)$ and $O(n^3)$ respectively, which agrees with the theoretical time complexities. As expected the $O(n^3)$ does not depend on d, however for $d = 3$ the algorithm is slightly slower than for larger values of d. This can be explained by the longer paths between every pair of leaves in such trees, since the algorithm is based on processing paths between pairs of leaves.

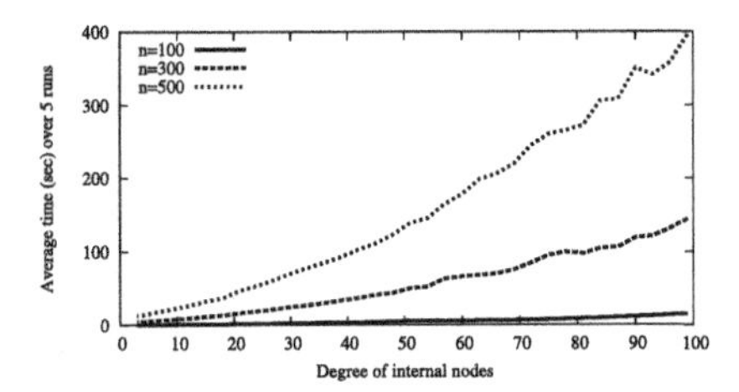

Fig. 10. Running time of the $O(n^2d^2)$ algorithm as a function of the degree d of the internal nodes for three fixed numbers of leaves n

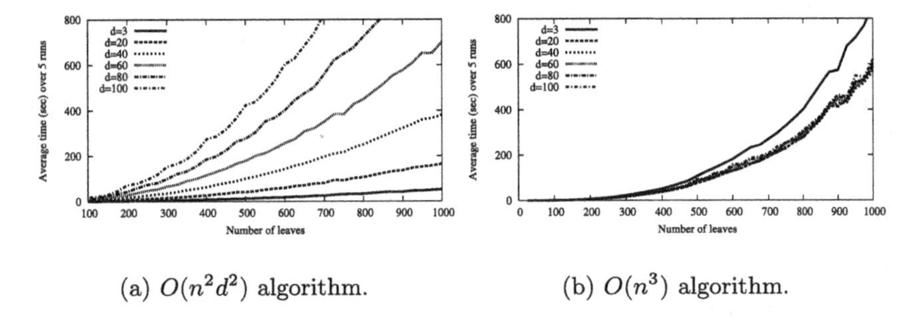

(a) $O(n^2d^2)$ algorithm. (b) $O(n^3)$ algorithm.

Fig. 11. Running time of the $O(n^2d^2)$ and $O(n^3)$ algorithms as a function of the number of leaves n for six fixed internal node degrees d. For every d it is verified that the running time as a function of n is $O(n^2)$ and $O(n^3)$ respectively.

Fig. 11 also show that the choice of algorithm for computing the quartet distance should depend on both n and d as expected. Note that all trees in our testing scenario have many internal nodes with a high degree, whereas trees generated using e.g. the Buneman [6, 2] and refined Buneman [9, 4] methods (which might reconstruct non-binary trees) usually contain only a few nodes of high degree. On such trees, the $O(n^2d^2)$ time algorithm can be expected to run faster than in our tests. More experiments are needed to investigate the performance of the algorithms on such trees. One can view the graphs in Fig. 11 as a worst case guideline of when to use the $O(n^3)$ time algorithm. Fig. 12 shows a direct comparison of the running times of the two algorithms. Since Fig. 11(b)

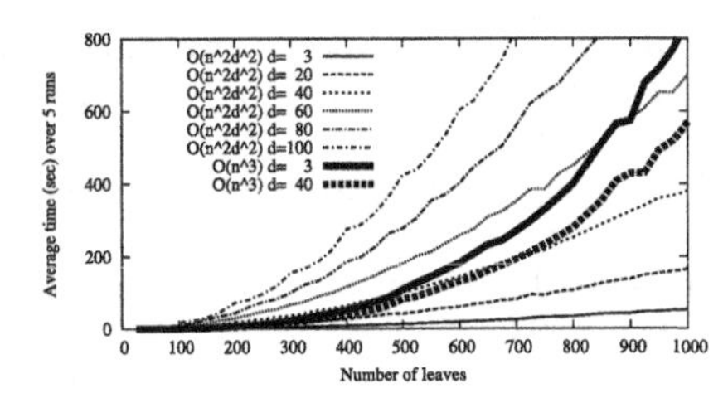

Fig. 12. Comparison of the running times

shows no significant variance in the running time for $d \geq 20$, Fig. 12 includes only two different plots for the $O(n^3)$ algorithm.

4 Conclusions

The contributions of this paper are two algorithms for computing the quartet distance between trees of arbitrary degrees. Earlier algorithms focus on binary trees and does not work on general trees. In this paper we have also shown that the quartet distance between two general trees can be expressed in terms not involving star quartets. The use of leaf intersection sizes in our algorithms prevents them from having a running time faster than $O(n^2)$. Consequently work on improving the $O(n^2d^2)$ time algorithm should be directed at removing one or both factors of d. Another approach for reducing the time complexity is to try to adapt the ideas from the $O(n \log n)$ time algorithm in [3] to trees of arbitrary degree.

References

1. B. L. Allen and M. Steel. Subtree transfer operations and their induced metrics on evolutionary trees. *Annals of Combinatorics*, 5:1–13, 2001.
2. V. Berry and D. Bryant. Faster reliable phylogenetic analysis. In *Proc. 3rd International Conference on Computational Molecular Biology (RECOMB)*, 1999.
3. G. S. Brodal, R. Fagerberg, and C. N. S. Pedersen. Computing the quartet distance between evolutionary trees in time $O(n \log n)$. *Algorithmica*, 38:377–395, 2003.
4. D. Bryant and V. Moulton. A polynomial time algorithm for constructing the refined buneman tree. *Applied Mathematics Letters*, 12:51–56, 1999.
5. D. Bryant, J. Tsang, P. E. Kearney, and M. Li. Computing the quartet distance between evolutionary trees. In *Proceedings of the 11th Annual Symposium on Discrete Algorithms (SODA)*, pages 285–286, 2000.
6. P. Buneman. The recovery of trees from measures of dissimilarity. In F. Hodson, D. Kendall, and P. Tautu, editors, *Mathematics in Archaeological and Historical Sciences*, pages 387–395. Edinburgh University Press, 1971.
7. G. Estabrook, F. McMorris, and C. Meacham. Comparison of undirected phylogenetic trees based on subtrees of four evolutionary units. *Syst. Zool.*, 34:193–200, 1985.
8. J. Felsenstein. *Inferring Phylogenies.* Sinauer Associates Inc., 2004.
9. V. Moulton and M. Steel. Retractions of finite distance functions onto tree metrics. *Discrete Applied Mathematics*, 91:215–233, 1999.
10. D. F. Robinson and L. R. Foulds. Comparison of weighted labelled trees. In *Combinatorial mathematics, VI (Proc. 6th Austral. Conf)*, Lecture Notes in Mathematics, pages 119–126. Springer, 1979.
11. D. F. Robinson and L. R. Foulds. Comparison of phylogenetic trees. *Mathematical Biosciences*, 53:131–147, 1981.
12. M. Steel and D. Penny. Distribution of tree comparison metrics–some new results. *Syst. Biol.*, 42(2):126–141, 1993.
13. M. S. Waterman and T. F. Smith. On the similarity of dendrograms. *Journal of Theoretical Biology*, 73:789–800, 1978.

Using Semi-definite Programming to Enhance Supertree Resolvability

Shlomo Moran[1], Satish Rao[2], and Sagi Snir[3]

[1] Computer Science dept., Technion, Haifa 32000, Israel[†]
moran@cs.technion.ac.il
[2] Computer Science dept. University of California, Berkeley, CA 94720, USA[‡]
satishr@cs.berkeley.edu
[3] Mathematics dept. University of California, Berkeley, CA 94720, USA[§]
ssagi@math.berkeley.edu

Abstract. Supertree methods are used to construct a large tree over a large set of taxa, from a set of small trees over overlapping subsets of the complete taxa set. Since accurate reconstruction methods are currently limited to a maximum of few dozens of taxa, the use of a supertree method in order to construct the tree of life is inevitable.

Supertree methods are broadly divided according to the input trees: When the input trees are unrooted, the basic reconstruction unit is a quartet tree. In this case, the basic decision problem of whether there exists a tree that agrees with all quartets is NP-complete. On the other hand, when the input trees are rooted, the basic reconstruction unit is a rooted triplet, and the above decision problem has a polynomial time algorithm. However, when there is no tree which agrees with all triplets, it would be desirable to find the tree that agrees with the maximum number of triplets. However, this optimization problem was shown to be NP-hard. Current heuristic approaches perform mincut on a graph representing the triplets inconsistency and return a tree that is guaranteed to satisfy some required properties.

In this work we present a different heuristic approach that guarantees the properties provided by the current methods and give experimental evidence that it significantly outperforms currently used methods. This method is based on divide and conquer where we use a semi-definite programming approach in the divide step.

1 Introduction

The study of evolution and the construction of phylogenetic (evolutionary) trees are classical subjects in biology. DNA sequences from a variety of organisms are rapidly accumulating, providing large amounts of data to a number of sequence

[†] This research was supported by the Technion VPR-fund and by the Bernard Elkin Chair in Computer Science.
[‡] Supported by NSF Award–0331494.
[§] This reasearch was supported by National Institutes of Health Grant R01-HG02362-02.

R. Casadio and G. Myers (Eds.): WABI 2005, LNBI 3692, pp. 89–103, 2005.

based approaches for phylogenetic trees reconstruction. The goal behind the "tree of life" project is to construct the tree representing the evolutionary history of over a million and a half different species. This task cannot be achieved by today's substitution based phylogenetic reconstruction methods. Therefore, the need to design methods capable to amalgamate data taken from different sources is emerging.

Phylogeny reconstruction methods are broadly divided into *character-based* and *distance-based* methods. Distance based methods start by computing "evolutionary distances" between pairs of taxa. Then a tree with weighted edges whose pairwise tree distances approximate the evolutionary distances is sought, typically by some version of the *neighbor joining* clustering paradigm [SN87]. In contrast, character based methods work directly on character data. The best known and most widely used character-based methods are *maximum parsimony* [Fit81] and *maximum likelihood* [Fel81]. Maximum parsimony (MP) is a non-parametric combinatorial method, while maximum likelihood (ML) is a parametric statistical method. In MP, the tree sought is such that minimizes the total number of state changes on all edges for all the characters. MP was proved to be NP-hard already at 1982 by [FG82] while ML only recently by [CT05]. Nevertheless, for data comprise of up to few dozens of taxa, there exist good heuristics that are heavily used in practice.

The *supertree reconstruction problem* (or for short, the supertree problem) (to be defined rigorously later) is as follows: given a set of phylogenetic trees over overlapping non identical sets of taxa, find a tree over the union of the given taxa sets that *represents the best* the input subtrees. This output tree is denoted as the *supertree.* There are few criteria of how to measure the quality of the supertree. These criteria differ by the type of the input trees.

Phylogenetic trees are divided into *rooted* and *unrooted* trees. In unrooted trees, the decision problem of whether there exists a supertree that resolves all the input subtrees was shown to NP-hard by [Ste92]. However, in the same paper, it is shown that if the trees are rooted then this problem is solvable in polynomial time by a simple divide and conquer algorithm of Aho *et. al.* , devised originally in the setting of relational data bases. Rooted trees have other attractive properties. In [SDB00] three required basic properties from a supertree method are listed:

- The method can be applied to any unordered set of input trees.
- if we rename all the species and then apply the method to the new input trees, we get the same old output tree under the renaming performed.
- if the input trees are consistent, the supertree returned should resolve all the input trees.

In the same work, it is shown that if the input tree are unrooted, there is no supertree method that can satisfy these three requirements simultaneously. However, when the setting is changed to a rooted setting, these requirements are achievable, in addition to polynomial running time of the method. They suggested that the rooted setting was perhaps superior to the unrooted one in the context of supertree construction. Based on these requirements, [SS00] devised the min cut (MC) supertree method and showed it satisfies all of them. A later modification by [Pag02] to the (MC) algorithm, denoted the *modified min cut* (MMC) algorithm, guaranteed that the resulting supertree satisfied additional desirable properties. Page has made code available that implements this algorithm and maintains a server where one can run the MMC algorithm (at http://darwin.zoology.gla.ac.uk/ rpage/supertree/).

Briefly, both MC and MMC proceed recursively by finding a minimum cut in a graph built from the current set of triplets. The triplets in this minimum cut will then be discarded so that the remaining triplets can be partitioned into subtrees that are combined. The discarded triplets correspond to the unsatisfied ones. The algorithms are a bit narrow in their view in that they proceed cautiously in order to violate a minimum number of triplets at each step. At each step there are also triplets that are satisfied. They do not include this phenomena in their divide step.

In this paper, we extend the above approach and apply a more inclusive and less greedy criterion in the divide step. In short, instead of minimizing the number of triplets that are violated at every step, we aim at maximizing the ratio between satisfied to violated triplets. Unfortunately, maximizing this ratio is, itself, an NP-complete problem (by reduction from the sparsest cut problem for general demands). Thus, we develop a very fast heuristic that is motivated by semi-definite programming (inspired in part by work of Goemans and Williamson [GW95]) that works quite well for this application. This approach leads to practically much better results in terms of violated triplets without interfering with any of the properties guaranteed by these algorithms. Indeed, our methods are significantly better and significantly faster than implementations of MMC. Moreover, the implementation effort was quite modest.

We also compare our performance to a supertree method that is far slower but has significantly better performance for the triplet problem. This method proceeds by reducing an instance of a triplet problem to an instance of the matrix representation parsimony problem (MRP), and then using state of the art parsimony engines to solve the parsimony problem. This method previously produced far better results than the triplet based methods at the cost of much larger running times [ECB+04]. Our methods also outperform these methods in terms of satisfying triplets. On one of our input distributions, MRP appears to do better on a fit measure involving the model tree, called maximum agreement subtree (MAST). On the other, we appear to do better. Also significantly, while for small sets of taxa we improve only modestly, as the number of taxa grows our advantage increases a bit. The computational requirements of the MRP approach, do not allow us to obtain performance numbers for MRP for very large number of taxa. We do report on results with our algorithms, however without comparing to other methods.

The use of semi-definite programming in the context of phylogenetic reconstruction is not new. Ben-Dor *et. al.* [BDCG+98] have used semi-definite programming to construct a tree over a given set of quartets. However, we stress that our method and the one of [BDCG+98] resemble only in the use of the semi-definite programming technique. While [BDCG$^+$98] use the embedding on the sphere once to resolve the whole tree and then use a neighbor joining type algorithm to reconstruct a complete tree, we use it at every step in the recursion only as an intermediate step to obtain an optimal partition of the taxa while the main algorithm is the divide and conquer Aho *et. al.* algorithm.

The remainder of this paper is organized as follows. In the next section we present the notations used and define the maximum triplet consistency method. In Section 3 we survey some existing methods for rooted supertrees and discuss some of their properties. Section 4 describes the pitfalls of current triplet based supertree methods, outlines our contribution, the algorithmic challenges involved and their resolutions. Section 5 describes experimental results comparing our method with two other representative methods on two types of synthetic data and also results on one instance of real data. We conclude in Section 6 with discussion and future research directions.

2 Preliminaries

For a tree $T = (V, E)$, we denote by $\mathcal{L}(T)$ the set of leaves of T. A phylogenetic tree T over a set of taxa $\mathcal{X}$ is a tree for which there is a bijection between $\mathcal{X}$ and $\mathcal{L}(T)$. Henceforth, we will identify the taxa set with the leaves they are mapped to. A tree T is said to be *rooted* if the set of edges E is directed and there is a single distinguished internal vertex r with in-degree 0. Let u and v be two vertices in a rooted tree T. We say that u is a *descendant* of v and v an *ancestor* of u, if there is a (directed) path from v to u. For $u, v \in V$, the *least common ancestor* of u and v, or $lca(u, v)$, is a vertex w that is an ancestor of both u and v and there is no descendant of w, w' that is also an ancestor of both u and v. From now on, we will restrict our attention to phylogenetic trees solely. Let T be a tree and $A \subseteq \mathcal{L}(T)$. We denote by $T|_A$ the tree induced by the sub set of leaves A where all internal vertices with degree two contracted. When T is rooted, the contraction is done at vertices with out-degree one. For two trees T and T', we say that T *satisfies* T', and T' is *satisfied* by T, if $\mathcal{L}(T') \subseteq \mathcal{L}(T)$ and $T|_{\mathcal{L}(T')} = T'$. Otherwise, T' is *violated* by T. In [BS00] it is shown that this requirement is equivalent to the condition that for every $u, v, w \in \mathcal{L}(T')$, $LCA(u, v)$ is a descendant of $LCA(u, w)$ in T if and only if $LCA(u, v)$ is a descendant of $LCA(u, w)$ in T'. This observation forms the basis to the triplet approach in supertree construction. For a set of trees $\mathcal{T} = \{T_1, \ldots, T_k\}$ with possibly overlapping leaves, we say that $\mathcal{T}$ is *consistent* if there exists a tree T^* over the set of leaves $\bigcup_i \mathcal{L}(T_i)$ that satisfies every tree $T_i \in \mathcal{T}$. Otherwise, $\mathcal{T}$ is *inconsistent.* When $\mathcal{T}$ is inconsistent, it is desirable to find a tree T^* over $\bigcup_i \mathcal{L}(T_i)$ that minimizes some objective function. T^* is denoted a *supertree* and the problem of finding T^* is the *supertree problem.*

A *rooted triplet*, or for short a triplet, is a rooted tree over three leaves u, v, w. We write a rooted triplet over u, v, w as $u, v|w$ if $LCA(u, v)$ is a descendant of $LCA(u, w)$. The *triplet score* between a tree T_i and T^* is the sum of $u, v, w \in \mathcal{L}(T_i)$ such that $T_i|_{\{u,v,w\}} = T^*|_{\{u,v,w\}}$. The triplet score between $\mathcal{T}$ and T^* is the sum of the triplet scores between T^* and every $T_i \in \mathcal{T}$. When $\mathcal{T}$ is a set of rooted triplets, this score is just the number of triplets satisfied by T^*. We refer to this problem as the *maximum triplet consistency* (MTC) problem.

Another score between a set of trees is the *maximum agreement sub tree* (MAST). This is defined as the largest set of leaves A common to all $T_i \in T$ such that $T_i|_A = T_j|_A$ for every $T_i, T_j \in \mathcal{T}$.

3 Supertree Methods

In this section we discuss competitive procedures for the supertree problem; the minimum cut methods and maximum parsimony and related methods.

3.1 Triplets Methods

We now describe minimum cut algorithms for triplet based reconstruction. The first algorithm solves the problem when all the triples are consistent and was developed in the context of relational databases by Aho *et. al.* [ASSU81]. Steel presented the algorithm for use in phylogenetics in [Ste92]. The algorithm uses two generic partitioning rules, of which only one is used in our case. It proceeds recursively on the set of taxa by applying the partitioning rule to produce a tree. The algorithm is described below:

Aho *et. al.* $(V, \mathcal{T})$

1. Let V be the set of taxa.
2. If $\mathcal{T} = \emptyset$ return a tree of depth 1 with all V as sister taxa.
3. Build the connectivity graph $G_C = (V', E')$ as follows:
   - $V' \leftarrow V$,
   - for every triplet $i, j|k \in \mathcal{T}: \ (i,j) \in E'$
4. Let c be the number of connected components in G_C.
5. If $c = 1$, return NULL, no tree is consistent with all triplets.
6. else
   - create an internal vertex u in T.
   - For every connected component C_i in G,
     - $T_i \leftarrow$ Aho *et. al.* $(V(C_i), \{(i,j|k) \in \mathcal{T}: \ i,j,k \in V(C_i)\})$.
     - make T_i a child of u.
7. return T_u.

It is clear that the algorithm finds a tree T over $\mathcal{X}$ that satisfies all the input triplets if such a tree exists. The algorithm runs in time $O(|\mathcal{T}| \cdot n)$ and a later improvement by Henziger *et. al.* [HKW96] reduced it to $min\left\{O(|\mathcal{T}| \cdot n^{0.5}),\right.$ $\left.O(|\mathcal{T}| + n^2 \log n)\right\}$.

As biological data suffers from noise it is very likely that $\mathcal{T}$ will be inconsistent, i.e. there will be no tree T that satisfies all the triplets in $\mathcal{T}$. Therefore it is desirable to find a tree that maximizes (minimizes) the number of satisfied (violated) triplets. The Aho et. al. algorithm above will, thus, typically fail upon encountering such an inconsistency. Still, one can easily prove that it does not hurt to run it up to this point. That is, the following lemma can be proved.

Lemma 1. *Given an inconsistent set of triplets $\mathcal{T}$, but the connectivity graph G_C (constructed at step 3 of Aho* et. al. *algorithm) induced by $\mathcal{T}$ is not connected. Then, any optimal tree T^* for $\mathcal{T}$, will have a different subtree for every component in the connectivity graph.*

The min-cut algorithm of Semple an Steel [SS00] was the first to cope with this problem of inconsistent input triplets. Their algorithm handles rooted sub trees of arbitrary size and is centered around the following idea: Apply Aho *et. al.* algorithm as long as possible. if at a certain stage in the recursion of the algorithm, step 3 yields a connected graph (i.e a single connected component), perform the following modification to the algorithm:

- convert the connectivity graph G of step 3 into an edge weighted graph $G' = (V, E, w)$ where for $e = (i, j)$, $w(e) = |\{(i, j|k) \in \mathcal{T} : k \in X\}|$.
- Compute min-cut on the graph G'.
- Apply the Aho et.al. algorithm on the two subproblems induced by the two resulted components.

Although [SS00] did not prove any bound for the quality of their solution, they did claim that subtrees that are shared by all input subtrees, are resolved by the output supertree. An elegant extension of this idea was presented by Page [Pag02] to maintain all sub trees that are not in disagreement by any input subtree. The algorithm, denoted *modified min cut* (MMC) applies the min cut criterion but on a somewhat different graph that of [SS00].

3.2 Character Based Supertree Methods

We now describe a family of supertree methods that is based on solving a more general problem. The family of character based supertree methods contains these two methods: Matrix Representation using Parsimony (MRP) and Matrix Representation using Flipping (MRF). MRP [Bau92, Rag92] is the most widely used supertree method by practitioners. It has been found to have good performance [ECB$^+$04]. In this method, the input subtrees are encoded in a $\{0, 1\}$ matrix in the following way: As every edge e in an input subtree T_i induces a partition on $\mathcal{L}(T_i)$, (A, B), e is encoded as binary character C where for each $s \in A$, $C(s) = 0$ and for $s \in B$, $C(s) = 1$. For $s \in \mathcal{X} \setminus (A \cup B)$, $C(s) =?$ indicating a *missing state.* The method tries to find the maximum parsimonious tree w.r.t that matrix. It is not confined to a rooted setting and hence loses some amount of its power when the input trees are rooted. However, to code for rooted trees, the following idea is employed: augment the taxa set with an artificial species s_r. Now let T_i be some input sub tree with root r_i and let e be an edge in T_i inducing the partition (A, B) over $\mathcal{L}(T_i)$. w.l.o.g assume that $T|_A$ contains r_i. Then $C(r_i) = 0$ and for all $s \in A$, $C(s) = 0$ and for $s \in B$, $C(s) = 1$. This is a reduction from the MTC problem to the MRP problem in the following sense.

Observation 1. *Let T be a tree over $\mathcal{X}$ and M an $n \times m$ matrix as defined above. Then for every character C in M, the parsimony score of C w.r.t. T is 1 if the triplet coded for C is satisfied by T. Otherwise it is 2.*

Corollary 1. *The parsimony score of M (or alternatively, the MRP score) is m plus the number of triplets violated by T.*

Since the MTC is NP-complete, Corollary 1 implies that solving the MRP problem even for rooted triplets is NP-hard and therefore, some heuristics need to be employed. In practice, this leads to the somewhat confounding results that even when given a consistent set of triples MRP methods do not obtain a consistent solution as do the triplet methods.

The other character based supertree method is the Matrix Representation using Flipping (MRF)[CEFBS02, ECB+04]. MRF starts with the same matrix as MRP however, its objective function is somewhat different: It seeks for the minimum number of states flipping at the characters in order to make all characters compatible. In that case, the supertree is convex on all characters. This problem is somewhat a restricted version of the "big convex recoloring" problem introduced at [MS04], as the characters are restricted to two states only. By similar lines to Observation 1 the following observation is derived:

Observation 2. *Given a set of triplets $\mathcal{T}$ and let M be the partial matrix representing $\mathcal{T}$, the MRF score for M is exactly the MTC score for $\mathcal{T}$.*

It was found that MRP and MRF perform similarly in experiments [ECB+04]. Since MRF runs much slower, we compare our method just to MRP.

4 MAX Cut Tree Construction

In this section we describe our algorithm, MAX CUT triplets. Next we show by an example how it improves over the local algorithms MC and MMC. Our algorithm proceeds along the divide and conquer strategy of Aho *et. al.* algorithm identically to MC and MMC. However, it differs from MC and MMC by the action performed when a set of inconsistent triplets is encountered.

We first observe that the algorithm terminates since at any divide step, a nontrivial cut is produced, identically to the other algorithms. It is easy to verify the following observation:

Observation 3. *For every triplet $u, v|w$ such that a good edge (u, w) or (v, w) is in the cut but (u, v) is not, the triplet $u, v|w$ is satisfied by the algorithm.*

Therefore, it is plausible that the algorithm should maximize the ratio between good to bad edges in every divide step of the algorithm. We remark here that all triplet based algorithms do that maximization implicitly in cases when the triplets are consistent. At that time, the ratio is infinity.

MAX CUT triplets $(V, \mathcal{T})$

1. Let V be the set of taxa.
2. If $\mathcal{T} = \emptyset$ return a tree of depth 1 with all V sister taxa.
3. For every triplet $i, j|k \in \mathcal{T}$: $(i, j) \in E$
4. Let c be the number of connected components in $G = (V, E)$.
5. If $c = 1$
   - Denote the edges created in step 3 as *bad edges*
   - For every triplet $i, j|k \in \mathcal{T}$, augment two *good edges* (i, k) and (j, k) to E.
   - Find a cut $(C, \bar{C})$ in G such that the ratio of good edges versus bad edges in the cut is maximized.
6. create an internal vertex u.
7. For every connected component C_i in G,
   - $T_i \leftarrow$ MAX CUT triplets $(V(C_i), \{(i, j|k) \in \mathcal{T} : i, j, k \in V(C_i)\})$.
   - make T_i a child of u.
8. return T_u.

Heuristic for Optimizing the Ratio. Unfortunately, in general, finding a cut that maximizes the ratio of good edges to bad edges (that we use in step 5 of the algorithm) is NP-complete as well (for example, one can reduce from the max cut or sparsest cut problems.) Several semi-definite programming based approximation algorithms, however, have been suggested for related problems [GW95, ARV04]. Based on these approaches we developed the following heurstic.

The heuristic proceeds embedded vertices of the graph onto the surface of a 3 dimensional sphere by locally moving vertices to minimize the function

$$\sum_{\text{good edges} e=(i,j)} w(e)d(i,j) - \alpha \sum_{\text{bad edges} e=(i,j)} w(e)d(i,j), \tag{1}$$

for various values of the parameter α. Essentially, we search for an α where the value is less than 0. The intuition is that for a good cut, $(C, \overline{C})$ where the ratio of good edges to bad is at most α, mapping all the points in C to a vector v and all the points in $\overline{C}$ to $-v$, yields a negative value for the function 1. Finding the minimal such value, makes it where other higher ratio cuts don't have good embeddings. Thus, the minimal value of α where the embedding problem has a nontrivial solution is a lower bound on the cut ratio for all cuts. The embedding can be computed with semidefinite programming packages for embedding into n dimensinos. We, however, found that a local heuristic that embedded onto a 3 dimensional sphere sufficed for the results we present here. It is very fast, and it appears to be effective.

Once, we obtain the embedding, we produce a cut as follows, partitioned the vertices into two sets by selecting a random hyperplane to partition the points and producing the corresponding cut.

Finally, we used a version of the Fiduccia-Matheyes bisection improvement algorithm [KL70, FM82] to improve the resulting cut.

We remark that with a provably accurate optimization procedure that we could binary search for an optimal value of α will yield an optimal ratio cut. For our inputs, our heuristic worked quite well with only a few values of α.

We also remark that with a semi-definite solution, one could could actually get a bound on how far from optimal the cut procedure is from optimal. Since this is a side issue, we leave that for the future.

4.1 Example

We now illustrate our algorithm on an example and compare it to the local algorithms MC and MMC. Consider the set of triplets $\{(1,2|4),(2,3|5),(2,5|3),(4,5|2),(5,6|3)\}$. It can easily be shown that their connectivity graph contains a single component (depicted in Figure 1). Moreover, as every subset of leaves appears in at most one input tree, the weight of every edge in the connectivity graph is one. In this case, the local algorithms will apply the min cut criterion in order to partition the taxa and continue in the divide and conquer approach. However, a naive application of the min cut criterion is indifferent of where to partition the graph and pathological example is the removal of the edge $(1,2)$ violating the triplet $(1,2|4)$, then the edge $(4,5)$ violating the triplet $(4,5|2)$, then $(5,6)$ violating $(5,6|3)$ and eventually $(3,2)$ violating $(2,3|5)$. This gives a total of three violated triplets.

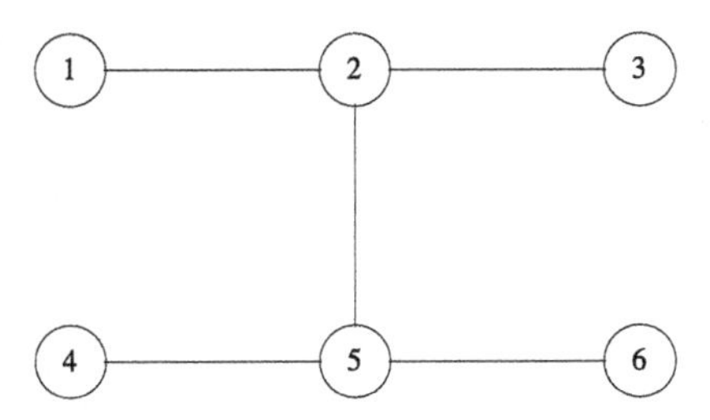

Fig. 1. A connectivity graph induced by the set of triplets $\{(1,2|4),(2,3|5),(2,5|3),(4,5|2),(5,6|3)\}$

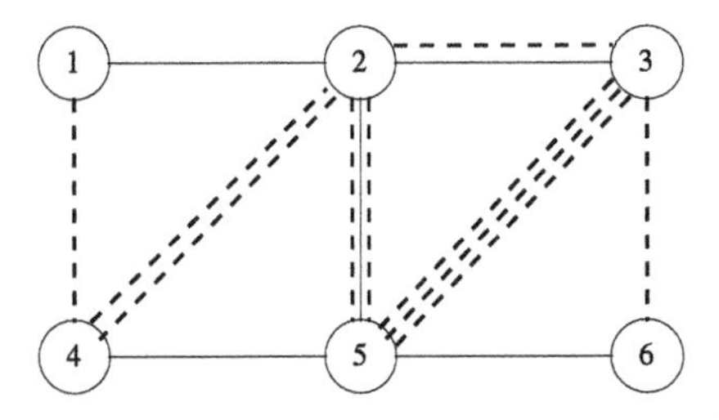

Fig. 2. The graph created by MAXCUT triplets algorithm for the set of triplets $\{(1,2|4),(2,3|5),(2,5|3),(4,5|2),(5,6|3)\}$. Good edges are drawn by dashed line.

In contrast, our algorithm constructs the graph depicted in Figure 2. It can easily be seen that the cut maximizing the ratio between good and bad edges is the cut $(\{1,2,3\},\{4,5,6\})$. This cut violates the triplet $2,5|3$ but the rest of the triplets would be satisfied recursively by the algorithm.

5 Experiments and Discussion

In order to test our method, we conducted experiments that compared our method versus two previously mentioned methods. As we mentioned before, further results on these are reported on in [ECB+04]. Again, it appears that MRP is substantially more accurate than both MC and MMC but in the cost of a much longer running time. They also study a heuristic based on MRF as we previously mentioned. Here, we only test MRP and MMC versus our method since MRF was reported to have similar behavior to MRP with much longer running times.

We generated our triplets from some model tree. In particular, we generate triplets from a given model tree and output some percentage of incorrect triplets. The properties we wanted to measure are the accuracy of the supertree returned in terms of the number of triplets satisfied and the resemblance of that tree to the model tree.

We conducted two types of experiments differing by the way the species of any triplet were chosen.

1. **uniform:** The species are chosen according to a uniform distribution from the the species set.
2. **geometric:** Triplets over species with distance d were chosen with probability $\frac{1}{d}$. This introduces locality into the process of triplet generation.

5.1 Uniform Distribution Results

Our results in terms of triplet score, MAST score, and running times are reported in Table 1. We note that the gap in performances and running time between MMC and MRP in these experiments, is as reported in [ECB+04], while MRP and our procedure perform in a similar fashion. Indeed, the latter methods typically output a tree that is *better* than the model tree. We also observe our method is usually a bit better than MRP, and as the problem size grows our advantage increases. For example, for the largest problem where we could get scores for both, we obtain a score of 75.7 while MRP achieves 67.3. Moreover, the running of MRP (and even MMC) became prohibitive far before the limits of our methods. It appears that MRP outperforms our method in terms of MAST fit, however, as an evidence of the orthogonality of the two measures, it can be noted that even when the triplets were consistent (100% triplet fit for our method and MMC), still a higher MAST fit was obtained by MRP, although the tree it returned did not satisfy all the triplets.

Moreover, MXC is much faster than MRP and a fair bit faster than MMC. For example, MRP takes more than two hours and a half to solve a problem with four hundred triplets where MXC took 9 seconds. This is a factor of a thousand better. MMC also took more than twenty five minutes to solve a problem with

Table 1. Data from experiments of uniform distribution of triplet selection. "-" denotes that the problem took too long.

#taxa	#triplets	%correct	% Triplet fit			MAST fit			Running Time		
			MRP	MXC	MMC	MRP	MXC	MMC	MRP	MXC	MMC
50	400	80	81	81.5	40	17	14	8	13	1	4
50	1000	80	79	80.9	38.4	18	22	7	18	2	14
50	1000	100	99	100	100	27	26	26	12	1	6
100	100	50	96	100	100	11	8	8	74	1	1
100	400	70	70	77.7	52.7	14	12	8	152	3	17
100	1000	70	72.3	73.1	33.6	23	18	8	642	4	16
100	2000	50	49.6	54.5	34.1	19	18	8	402	7	63
150	100	70	96	100	100	14	6	6	290	1	1
150	1000	70	70	73.7	43.7	16	17	8	1712	4	35
150	1000	90	89.2	90.7	42.3	25	16	8	2628	5	29
150	2000	90	87.9	90.3	34.8	31	24	8	4790	17	108
200	400	70	81.2	81.5	65.7	17	10	9	9047	9	25
200	1000	70	67.3	75.7	44.1	15	17	9	8581	23	200
200	4000	50	-	54.2	34.5	-	25	200	-	53	546
400	10000	50	-	53.5	36.1	-	31	10	-	80	1603
400	50000	50	-	50.6	-	-	55	-	-	320	-
600	4000	50	-	62.4	-	-	17	-	-	38	-
600	50000	50	-	51	-	-	56	-	-	961	-
800	20000	50	-	53.5	-	-	34	-	-	132	-
1000	50000	50	-	51.4	-	-	49	-	-	339	-
2000	50000	50	-	53.5	-	-	-	-	-	383	-

ten thousand triplets and four hundred taxa where MXC took eighty seconds. It appears that MMC has some implementation problems as it crashed on twenty thousand triplets. The largest experiment we performed in this distribution was with fifty thousand triplets on two thousand taxa. The running time on this data was a bit more than six minutes (383 seconds). [1]

5.2 Geometric Distribution Results

In this type of experiment, we gave preferences to "close" over "far" triplets. That means that a triplet whose species are of distance d was chosen with probability $\frac{1}{d}$. We denote that as the geometric distribution. The reasoning for using that distribution is that, in general, we are less certain about the order of speciation of distantly related species, than of more closely related species, and therefore we "weigh" these distant triplets less. We believe this type of distribution is more realistic than the uniform distribution.

Table 2 depicts a sample of our experiments under the geometric distribution. It can be seen that both MRP and our method maintain the same superiority

[1] We note further that our code is hardly optimized in that while the divide step (the max ratio cut) is implemented in C, the recursive algorithm is written in perl with system calls to the max cut code. Thus, we are hopeful that this approach can work with even much larger datasets.

Table 2. Statistics on experiments of geometric distribution of triplet selection

#taxa	#triplets	%correct	% Triplet fit			MAST fit			Running Time		
			MRP	MXC	MMC	MRP	MXC	MMC	MRP	MXC	MMC
50	400	50	65.7	62	50.2	11	14	9	25	58	12
50	1000	70	65.3	70.3	43	21	29	13	44	7	26
50	2000	90	88.2	89.3	65.6	34	36	26	46	22	48
100	400	50	74.2	74	53	15	21	9	499	45	30
100	1000	50	62	63.4	38.1	17	24	12	806	30	78
100	1000	70	66.6	69.8	46.5	29	28	12	775	18	111
100	1000	90	82.7	85.5	0.49	39	45	15	411	10	53
150	1000	50	64.8	66.9	41.5	16	33	15	2564	65	43
150	2000	50	61.6	60	35	21	22	13	7553	1596	243
150	400	90	80.5	83.5	71.75	24	20	18	1114	4	13
200	1000	70	68.3	71.3	44.1	22	25	13	6317	11	66
200	400	90	83	86	73	19	19	15	3053	3	13
200	2000	90	83.9	86.35	38.65	84	93	15	10980	17	76
250	100	70	98	100	100	17	6	6	2036	2	3
250	2000	70	66.1	71.7	42	33	52	18	25114	29	92
250	1000	90	75.8	77.4	57.6	30	24	19	13626	38	217
300	2000	70	65	67.3	44.8	32	29	16	40176	38	302
300	400	90	88.5	97.2	91.7	25	22	20	27183	2	14
300	2000	90	71	74.9	49.8	37	46	19	35555	18	153

in terms of triplet fit score over MMC with average of 2% advantage to MXC over MRP. MXC is still much faster than MRP although in this distribution we needed to check for more values of α what increased the running time in some cases (e.g. the line with 150 species, 2000 triplets and 50% correctness). A somewhat interesting phenomenon, is that with this distribution, our method strictly outperforms MRP in terms of MAST score, in contrast to the uniform distribution where it appears MRP has some advantage. It is notable that even under this distribution MRP has better MAST score when the triplets are sparse, so MXC and MMC satisfy all the triplets but leave an unresolved tree, as implied by Aho *et. al.* algorithm (e.g. the line with 250 species, 100 triplets and 70% correctness). Perhaps this distribution of triplets along with the MXC algorithm is better for learning the true tree as compared to the uniform distribution of triplets.

5.3 Experiment on Real Data

Although our method was designed to handle input in a form of triplet trees, we wanted to test its behavior on real data. Real data comes normally in a form of trees over more than three species. When coming to apply a triplet based method, a separate task is to generate a "representative set" of triplets that will lead to the best construction of a tree. Since this task is beyond the scope of this paper, we used a rather naive way and generated *all* the triplets induced by *any* subtree. Of course this approach is biased and gives more representation to higher ancestral vertices over more recent ancestral vertices.

The real data we used for our experiment is composed of 158 source trees with 267 marsupial species from 107 published studies [CBEBP04]. The source trees were based on a wide range of data types, including molecular sequences, DNA hybridization, karyotypes, and immunological, morphological and behavioral data. The average number of species per tree is about 16.4. The number of triplets generated is 2,380,724. Although the true tree is assumed to be known and is found in **TREEBASE** [PSDW], we compared the results we obtained to either the set of triplets (triplet fit) or to the input subtrees (Robinson-Foulds).

In this experiment we used the Robinson-Foulds (RF) topological distance between two trees [RF81]. This distance equals the number of splits existing at exactly one tree. In addition, since MRP returns a fully resolved tree, whereas a triplet based method (seeks to) return the minimally resolved tree that satisfies most of the triplets, we measured a more MRP-favorable measure which is the number of common edges between the two trees. This is simply derived from the latter by $\frac{E_1+E_2-RF}{2}$, where E_i is the number of edges in tree i and RF is the Robinson-Foulds distance.

The results on this data using this method of triplets generation show strict advantage to MRP. The triplet fit achieved by MRP was 98.2% versus 96% by MXC. Both the RF distance and the number of common edges between the supertree and each of the subtrees were normalized by the number of species in the subtree. The final scores are the sum of the latter over all subtrees. For these measures, MRP achieved an average of 0.49 different edges per species and average of 0.95 common edges per species. MXC in turn, achieved inferior results of 0.65 different edges per species and 0.78 common edges per species.

6 Conclusion and Further Work

In this work we described a novel idea of using the semi-definite technique for the purpose of constructing triplet based supertrees. We introduced a less greedy partition criterion for the cases where the triplets are inconsistent. Moreover, we showed experimentally that although optimizing this criterion implies solving a NP-hard problem, a simple heuristic suffices to provide good performance. This in turn yields a very fast algorithm that outperforms in terms of running time even the theoretically efficient algorithms of mincut. Moreover, we showed that for the type of inputs we studied here, the performance of triplet based methods can exceed those of the heavier character based methods. These results pose semi-definite programming as a promising direction in the supertree field.

We want to emphasize that there are few major questions to be answered in this direction:

- While triplet based method try to maximize the number of satisfied triplets, this does not necessarily goes along with optimizing the other measures such as MAST or RF distance. It will be valuable to study the difference between these measures and to try to come with some conclusions for which really measures trees similarity.

- We saw that on real data, MRP is still superior over naively taking all triplets and trying to solve this problem. Moreover, MRP outperforms MXC even on that criterion, (although it solves an apparently different problem). Therefore, a more insightful approach for selecting which triplets to include is requested.
- Our experiments showed that actually, an optimal solution to the ratio cut problem is unnecessary. It would be of interest to explore the influence of using different ratios on the quality of the trees inferred by the method.

Acknowledgments

We would like to thank very much David Fernandez Baca and Duhong Chen for a lot of help with the many technicalities involved. We also thanks David Bryant, Benny Chor, Oliver Eulenstein, Arie Freund, Dick Karp, Rod Page, Mauricio Resende, Dror Rawitz, and Mike Steel for helpful discussions.

References

[ARV04] S. Arora, S. Rao, and U. Vazirani. Expander flows, geometric embeddings and graph partitioning. In *Symposium on the Foundations of Computer Science*, 2004.

[ASSU81] A.V. Aho, Y. Sagiv, T.G. Szymanski, and J.D. Ullman. Inferring a tree from lowest common ancestors with an application to the optimization of relational expressions. *SIAM Journal of Computing*, 10(3):405–421, 1981.

[Bau92] B.R. Baum. Combining trees as a way of combining data sets for phylogenetic inference. *Taxon*, 41:3–10, 1992.

[BDCG+98] A. Ben-Dor, B. Chor, D. Graur, R. Ophir, and D. Pelleg. Constructing phylogenies from quarbcgoptets: Elucidation of eutherian superordinal relationships. *Jour. of Comput. Biology*, 5(3):377–390, 1998.

[BS00] D.J. Bryant and M.A. Steel. Extension operations on sets of leaf-labelled trees. *Advances in Applied Mathematics*, 16(4):425–453, 2000.

[CBEBP04] M. Cardillo, O. R. P. Bininda-Emonds, E. Boakes, and A. Purvis. A species-level phylogenetic supertree of marsupials. *Journal of Zoology*, 264(1):11–31, 2004.

[CEFBS02] D. Chen, O. Eulenstein, D. Fernandez-Baca, and M. Sanderson. Supertrees by flipping. In *COCOON*, 2002.

[CT05] B. Chor and T. Tuller. Maximum likelihood of evolutionary trees is hard. In *RECOMB*, 2005.

[ECB+04] O. Eulenstein, D. Chen, J.G. Burleigh, D. Fernandez-Baca, and M.J. Sanderson. Performance of flip supertrees with a heuristic algorithm. *Systematic Biology*, 53(2):299–308, 2004.

[Fel81] J. Felsenstein. Evolutionary trees from DNA sequences: A maximum likelihood approach. *J. Mol. Evol.*, 17:368–376, 1981.

[FG82] L.R. Foulds and R.L Graham. The steiner problem in phylogeny is NP-complete. *Advances in Applied Mathematics*, 3:43–49, 1982.

[Fit81] W. M. Fitch. A non-sequential method for constructing trees and hierarchical classifications. *Journal of Molecular Evolution*, 18(1):30–37, 1981.

[FM82] C.M. Fiduccia and R.M. Mattheyses. A linear time heuristic for improving network partitions. In *Design Automation Conference*, pages 175–181, 1982.

[GW95] M.X. Goemans and D.P. Williamson. Improved approximation algorithms for maximum cut and satisfiability problems using semidefinite programming. *Journal of the Association for Computing Machinery*, 42(6):1115–1145, November 1995.

[HKW96] M. R. Henzinger, V. King, and T. Warnow. Constructing a tree from homeomorphic subtrees, with applications to computational evolutionary biology. In *SODA*, pages 333–340, 1996.

[KL70] B. W. Kernighan and S. Lin. An ecient heuristic procedure for partitioning graphs. *The Bell System Technical Journal*, 29(2):291–307, 1970.

[MS04] S. Moran and S. Snir. Convex recoloring of strings and trees: Definitions, hardness results and algorithms. *submitted*, 2004.

[Pag02] R.D.M. Page. Modified mincut supertrees. In *R. Guigo and D Gusfield (Eds): WABI 2002, LNCS 2452*, 2002.

[PSDW] W. Piel, M. Sanderson, M. Donoghue, and M. Walsh. Treebase. http://www.treebase.org.

[Rag92] M.A. Ragan. Matrix representation in reconstructing phylogenetic-relationships among the eukaryotes. *Biosystems*, 28:47–55, 1992.

[RF81] D.R. Robinson and L.R Foulds. Comparison of phylogenetic trees. *Mathematical Biosciences*, 53:131–147, 1981.

[SDB00] M. Steel, A. Dress, and S. Boker. Simple but fundamental limitations on supertree and consensus tree methods. *Systematic Biology*, 49:363–368, 2000.

[SN87] N. Saitou and M. Nei. The neighbor-joining method: A new method for reconstructing phylogenetic trees. *Molecular Biology and Evolution*, 4, 1987.

[SS00] C. Semple and M. Steel. A supertree method for rooted trees. *Discrete Applied Mathematics*, 103:147–158, 2000.

[Ste92] M. Steel. The complexity of reconstructing trees from qualitative characters and subtress. *Journal of Classification*, 9(1):91–116, 1992.

An Efficient Reduction from Constrained to Unconstrained Maximum Agreement Subtree

Z.S. Peng and H.F. Ting

Department of Computer Science,
The University of Hong Kong, Hong Kong
{zspeng, hfting}@cs.hku.hk

Abstract. We propose and study the Maximum Constrained Agreement Subtree (MCAST) problem, which is a variant of the classical Maximum Agreement Subtree (MAST) problem. Our problem allows users to apply their domain knowledge to control the construction of the agreement subtrees in order to get better results. We show that the MCAST problem can be reduced to the MAST problem efficiently and thus we have algorithms for MCAST with running times matching the fastest known algorithms for MAST.

1 Introduction

Evolutionary trees, which are rooted trees with their leaves labeled by some unique species, are commonly used to capture the evolutionary relationship of the species in nature. Different theories capture different kinds of evolutionary relationships and induce different evolutionary trees. To find out how much these theories are in common, we compare the corresponding evolutionary trees and find some consensus of these trees.

One successful approach for finding consensus of different evolutionary trees is to construct their maximum agreement subtree (MAST), which is the largest evolutionary tree that is a topology subtree of the given trees. There are many algorithms proposed for constructing MAST; for example, [6,7,9,10,12,13,17], or more recently, [1,2,5,14].

A major problem of these algorithms is that it does not allow biologists to apply their knowledge to control the construction for getting better results. For example, the evolutionary relationship of many species is well understood. Any evolutionary tree including these species should be consistent with this commonly accepted relationship. With this additional constraint, MAST is not a good measure for comparing evolutionary trees. Let us consider the trees S and T in Fig. 1. Note that the maximum agreement subtree of S and T is large, and one would consider that the two trees are similar. However, the two trees agree on almost nothing if we insist that the agreement subtree must be consistent with the evolutionary relationship of e, f, h, which is given by the tree P. In fact, if P is a correct relationship, then S and T infer different evolutionary relationship for many other species. For example, for the species a, S suggests

R. Casadio and G. Myers (Eds.): WABI 2005, LNBI 3692, pp. 104–115, 2005.

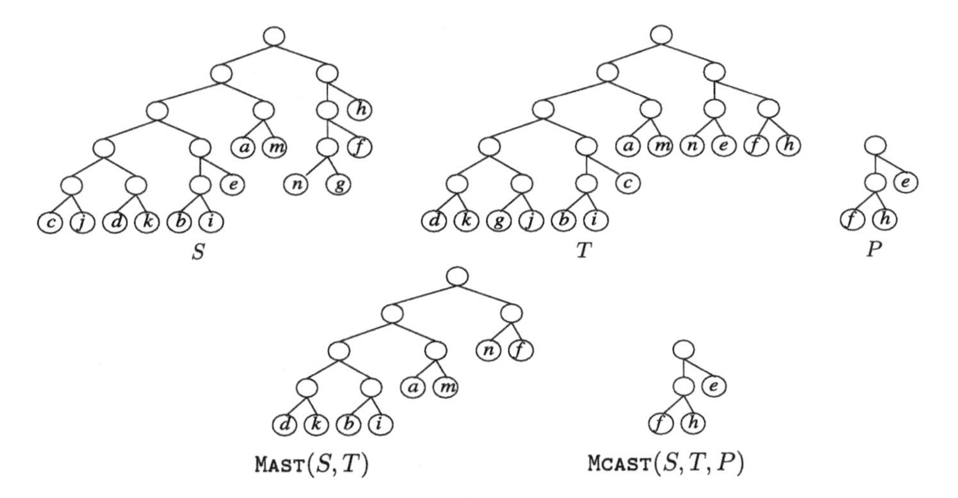

Fig. 1. Maximum agreement and maximum constrained agreement subtrees

that the least common ancestor of a and e is different from the least common ancestor of a and f, while T suggests they are the same.

To allow biologists to enforce such predefined relationship in the agreement subtree, we propose and study the *maximum constrained agreement subtree* (MCAST) problem, which is defined as follows:

> Let S and T be two evolutionary trees, and P be an agreement subtree of S and T. Find the largest agreement subtree of S of T that contains P as a subtree. We say that this agreement subtree is the *maximum constrained agreement subtree* of S and T with respect to P.

In [15], we gave an $O(n \log n)$ time recursive algorithm for this problem when the input trees are binary. However, it is difficult to generalize the algorithm for general trees. In this paper, we give a deeper analysis of the structure of the constrained agreement subtrees and show that the MCAST problem can be indeed reduced to the classical Maximum Agreement Subtree (MAST) problem. Note that this reduction is not surprising when P is empty or has only one leaf.

If P is the empty tree, our MCAST problem is just the MAST problem. If P has only one leaf κ, the problem is equivalent to finding a largest agreement subtree A of S and T that contains κ. By a simple trick, we can reduce the problem to the MAST problem as follows. Let $|S|$ and $|T|$ be the number of leaves in S and T, respectively. To find A, we simply replace the leaf κ in S and T by some large tree X of size at least $|S|+|T|$. Then, any maximum agreement subtree A' of the resulting trees must contain X. In other words, the role of X is the same as the role of κ in S and T. By replacing X in A' by κ, we get A.

The major contribution of this paper is to show that we have this reduction even for general P. To be more precise, let $n = |S| + |T|$ and $|P| = k$. In Theorem 3, we prove that given S, T and P, we can find in $O(n + k \log k)$ time subtrees $S_1, S_2, \ldots, S_m$ of S, $T_1, T_2, \ldots, T_m$ of T, and $P_1, \ldots, P_m$ such that

- $\sum_{1 \le i \le m} |S_i| \le 2|S|$ and $\sum_{1 \le i \le m} |T_i| \le 2|T|$;

Table 1. Time complexity of MAST and MCAST

	MAST	MCAST
Binary trees	$O(n \log n)$ [4]	$O(n \log n)$
Trees with constant degree d	$O(\sqrt{d}n \log n)$[16]	$O(\sqrt{d}n \log n)$
General trees	$O(n^{1.5})$ [11]	$O(n^{1.5})$

- each P_i has only one leaf; and
- to find a maximum constrained agreement subtree of S and T with respect to P, it suffices to find maximum constrained agreement subtrees of S_i and T_i with respect to P_i for $1 \leq i \leq m$.

Hence, our MCAST problem is reduced to a number of subproblems with input S_i, T_i and P_i where each P_i has one leaf. As mentioned above, these subproblems can be further reduced to the MAST problem with double input size (because of the subtree X). Therefore, if $\phi(h)$ is the worst case running time of an algorithm for finding a maximum agreement subtree of two trees with totally h leaves, then $T(n, k)$, the time complexity of the problem of finding a maximum constrained agreement subtree of S and T with respect to P, can be bounded as follows:

$$T(n,k) \leq \sum_{1 \leq i \leq m} \phi(2(|S_i| + |T_i|)) + O(n + k \log k). \tag{1}$$

We note that for all existing algorithms for MAST, their running times are upper bounded by some convex functions $\phi(m)$, and by Jensen's inequality [8], we have

$$\sum_{1 \leq i \leq m} \phi(2(|S_i| + |T_i|)) \leq \phi(\sum_{1 \leq i \leq m}(2(|S_i| + |T_i|)) = \phi(4(|S| + |T|)) = \phi(4n). \tag{2}$$

From (1) and (2), we can use an existing algorithm for MAST to solve MCAST without increasing the running time asymptotically. For a summary, Table 1 lists the running time of the MCAST problem by our reduction using the fastest known MAST algorithms for different kinds of input trees.

Remarks: Note that when P is large, i.e., when $k = \Omega(n)$, our reduction takes $O(n \log n)$ time. Recently, Berry [3] has shown that the time complexity can be reduced to $O(n)$ by devising an interesting algorithm to sort all nodes of a tree in linear time.

Our paper is organized as follows. In Section 2, we give the necessary definitions and notations for our discussion. We also prove some properties on agreement subtrees that help simplify our analysis. In Section 3 and 4, we analyze the structure of the agreement subtrees, and in Section 5, we give formally the reduction.

2 Preliminaries

A *labeled* tree S is a rooted tree with every leaf being labeled with a unique species. In this paper, we use the label of the leaf as its name. Let $\mathcal{L}(S)$ denote the set of leaves of S. For any two leaves a, b, let $\texttt{lca}_S(a, b)$ denote the *least common ancestor* of a, b in S. Given any subset $H \subseteq \mathcal{L}(S)$ of leaves, the *restricted subtree* of S on H, denoted as $S\|_H$, is the subtree of S whose nodes includes the set of leaves in H as well as the least common ancestors of any two leaves, and whose edges preserve the ancestor-descendent relationship of S. Intuitively, $S\|_H$ can be constructed as follows: Discard those leaves of S not in H, as well as those internal nodes whose degrees eventually become one; then contract every path whose intermediate nodes are each of degree two into an edge. The following fact comes directly from the definition.

Fact 1. *Suppose that $H \subseteq L \subseteq \mathcal{L}(S)$. Then, we have (i) for any two leaves $a, b \in H$, $\texttt{lca}_{S\|_H}(a,b) = \texttt{lca}_{S\|_L}(a,b)$, and (ii) $(S\|_L)\|_H = S\|_H$.*

Let T be another labeled tree. We say that S and T are *leaf-label preserving isomorphic* if (i) they have the same set of leaves (i.e., $\mathcal{L}(S) = \mathcal{L}(T)$) and (ii) there exits a bijection f from the nodes of S to the nodes of T such that for any pair of leaves a, b of S, $f(\texttt{lca}_S(a,b)) = \texttt{lca}_T(a,b)$. Note that for any leaf a, $f(a) = f(\texttt{lca}_S(a,a)) = \texttt{lca}_T(a,a) = a$; f maps every leaf in S to the leaf in T with the same label. We write $S = T$ if the two trees are leaf-label preserving isomorphic.

Observe that given any two trees S and T with the same set of leaves, we can always define a mapping f such that for any pair of leaves a, b, $f(\texttt{lca}_S(a,b)) = \texttt{lca}_T(a,b)$. However, the necessary and sufficient condition for f being bijective, and hence $S = T$, is that for any two pairs of leaves a, b and c, d (not necessarily distinct), we have

$$\texttt{lca}_S(a,b) = \texttt{lca}_S(c,d) \text{ if and only if } \texttt{lca}_T(a,b) = \texttt{lca}_T(c,d). \qquad (3)$$

The following lemma gives a somewhat simpler condition; it helps simplify our analysis given in the rest of this paper.

Lemma 1. *Following is a necessary and sufficient condition for $S = T$: for any three leaves a, b, c, we have*

$$\texttt{lca}_S(a,b) = \texttt{lca}_S(a,c) \iff \texttt{lca}_T(a,b) = \texttt{lca}_T(a,c). \qquad (4)$$

Proof. It suffices to prove that (3) is equivalent to (4). Obviously, (3) implies (4). To prove the other direction, suppose that (3) does not hold. In other words, there are four leaves a, b, c, d such that in one tree, say S, we have $\texttt{lca}_S(a,b) = \texttt{lca}_S(c,d)$, but $\texttt{lca}_T(a,b) \neq \texttt{lca}_T(c,d)$. Below, we identify three leaves from a, b, c, d that violate (4).

In T, since $\texttt{lca}_T(a,b) \neq \texttt{lca}_T(c,d)$, the two nodes cannot be descendent of each other at the same time. Thus, one of them, say $\texttt{lca}_T(a,b)$ is not a descendent of $\texttt{lca}_T(c,d)$, and this further implies either a or b, say a, is not a descendent of $\texttt{lca}_T(c,d)$. In other words, all of the ancestors of a are not $\texttt{lca}_T(c,d)$, and it follows that

$$\texttt{lca}_T(a,c) \neq \texttt{lca}_T(c,d) \text{ and } \texttt{lca}_T(a,d) \neq \texttt{lca}_T(c,d). \qquad (5)$$

In S, since $\texttt{lca}_S(a,b) = \texttt{lca}_S(c,d)$, a is a descendent of $\texttt{lca}_S(c,d)$. Let w be the least common ancestor of $\texttt{lca}_S(a,c)$ and $\texttt{lca}_S(a,d)$. Note that

- $\texttt{lca}_S(c,d)$ is the ancestor of a, c, d and hence it is an ancestor of w, and
- w is an ancestor of c and d, and hence is an ancestor of $\texttt{lca}_S(c,d)$.

It follows that $w = \texttt{lca}_S(c,d)$. Finally, since $\texttt{lca}_S(a,c)$ and $\texttt{lca}_S(a,d)$ are on the path from a to the root, their least common ancestor w must be one of them, i.e., $w = \mathtt{lca}_S(a,c)$ or $w = \mathtt{lca}_S(a,d)$, or equivalently,

$$\mathtt{lca}_S(a,c) = \mathtt{lca}_S(c,d) \text{ or } \mathtt{lca}_S(a,d) = \mathtt{lca}_S(c,d). \quad (6)$$

Taking (5) and(6) together, we conclude that (4) does not hold; the lemma follows. □

We say that a subset $K \subseteq \mathcal{L}(S) \cap \mathcal{L}(T)$ of leaves is an *agreement leaf subset* of S and T if $S\|_K = T\|_K$; the two restricted subtrees are called *agreement subtrees* of S and T. Suppose that K is an agreement leaf subset of S and T. A leaf subset $L \subseteq \mathcal{L}(S) \cap \mathcal{L}(T)$ is called a *constrained agreement leaf subset of S and T with respect to K* if (i) $K \subseteq L$ and (ii) L is an agreement leaf subset of S and T. The classical *maximum agreement subtree problem* asks to find the largest agreement leaf subset of S and T. In this paper, we study the *maximum constrained agreement subtree*, which asks for finding the maximum constrained agreement leaf subset of S and T with respect to K.[1] As shown in Fig. 1, the output of the two problems can be very different.

In the rest of the paper, we assume that $K \neq \emptyset$ and $S\|_K = T\|_K$. We define $\textsc{Cast}(S,T,K)$ to be the set of all agreement leaf subsets of S and T with respect to K, and define $\textsc{Mcast}(S,T,K) \subseteq \textsc{Cast}(S,T,K)$ to be the subset of those with maximum size. In the next two sections, we describe some structural properties on S, T and K, which help us design efficient algorithms for solving the maximum constrained agreement subtree problem, or equivalently, finding an element in $\textsc{Mcast}(S,T,K)$. Our analysis is divided into two cases. Let κ be a leaf in K. In the following section, we focus on the case when κ is a child of the root of both S and T; we call such leaf a *shallow* leaf. The existence of a shallow leaf in K greatly simplifies our analysis. We handle the other case when κ is not a shallow leaf in Section 4.

3 The Case When κ is a Shallow Leaf

In this section, we show that the existence of a shallow leaf imposes great restrictions on how a constrained agreement leaf subset can be formed. The following lemma describes one such restriction. We call the whole subtree rooted at some child of the root a *rooted subtree.*

Lemma 2. *Suppose that $L \in \textsc{Cast}(S,T,K)$. For any rooted subtrees S' of S and T' of T, if S' and T' have a common leaf in L (i.e., $L \cap \mathcal{L}(S') \cap \mathcal{L}(T') \neq \emptyset$), then $L \cap \mathcal{L}(S') = L \cap \mathcal{L}(T')$*

Proof. It suffices to prove that for any leaves $a, b \in L$, a, b are in different rooted subtrees of S if and only if a, b are in different rooted subtrees of T, or equivalently, $\mathtt{lca}_S(a,b)$ is the root of S if and only if $\mathtt{lca}_T(a,b)$ is the root of T.

[1] Note that our problem is somewhat different from the one we mentioned in Section 1; however, it should be clear that we can solve the problem of finding a MCAST of S and T with respect to P by solving our problem with $K = \mathcal{L}(P)$.

From Fact 1, $\texttt{lca}_S(a,b) = \texttt{lca}_{S\|_{\mathcal{L}(S)}}(a,b) = \texttt{lca}_{S\|_L}(a,b)$, and $\texttt{lca}_T(a,b) = \texttt{lca}_{T\|_{\mathcal{L}(S)}}(a,b) = \texttt{lca}_{T\|_L}(a,b)$. Since $\kappa \in K \subseteq L$, $a, b \in L$, and $S\|_L = T\|_L$, by Lemma 1, $\texttt{lca}_{S\|_L}(a,b) = \texttt{lca}_{S\|_L}(a,\kappa) \iff \texttt{lca}_{T\|_L}(a,b) = \texttt{lca}_{T\|_L}(a,\kappa)$. The lemma follows immediately because κ is a shallow leaf, and $\texttt{lca}_{S\|_L}(a,\kappa)$ and $\texttt{lca}_{T\|_L}(a,\kappa)$ are the root of S and T, respectively. □

Note that $K \in \textsc{Cast}(S,T,K)$. Lemma 2 asserts that for any rooted subtree S' of S, if S' have a leaf in K, then there is a rooted subtree T' of T such that $K \cap \mathcal{L}(S') = K \cap \mathcal{L}(T')$. Let $S_1, S_2, \ldots, S_m$ be all the rooted subtrees of S that contain some leaf in K, and $T_1, T_2, \ldots, T_m$ be the rooted subtrees of T where $K \cap \mathcal{L}(S_i) = K \cap \mathcal{L}(T_i)$. Suppose that S_m and T_m are the subtrees that contain the single shallow leaf κ. Define S_0 to be the tree obtained by removing $S_1, S_2, \ldots, S_{m-1}$ from S. Note that only S_m remains in S_0 and thus S_0 has a single leaf in K, namely κ. Define T_0 similarly. It should be clear that $K \cap \mathcal{L}(S_0) = K \cap \mathcal{L}(T_0) = \{\kappa\}$. We call $\langle (S_0, S_1, \ldots, S_{m-1}), (T_0, T_1, \ldots, T_{m-1}) \rangle$ the *κ-decomposition* of S and T with respect to K. The following lemma shows that κ-decomposition imposes some nice structure on any constrained agreement leaf subset.

Lemma 3. *Suppose that $L \in \textsc{Cast}(S,T,K)$. Then,*
$L_i = L \cap \mathcal{L}(S_i) \in \textsc{Cast}(S_i, T_i, K \cap \mathcal{L}(S_i))$ *for* $0 \le i \le m-1$.

Proof. Recall that for each $1 \le i \le m-1$, $K \cap \mathcal{L}(S_i) = K \cap \mathcal{L}(T_i) \neq \emptyset$, and since $K \subseteq L$, S_i and T_i have a common leaf in L and thus $L \cap \mathcal{L}(S_i) = L \cap \mathcal{L}(T_i)$ (Lemma 2). It follows that the remaining leaves of L in S and T are the same; in other words, $L \cap \mathcal{L}(S_0) = L \cap \mathcal{L}(T_0)$. Therefore, for each $0 \le i \le m-1$, $L_i = L \cap \mathcal{L}(S_i) = L \cap \mathcal{L}(T_i)$. Since $L_i \subseteq \mathcal{L}(S_i)$ and $L_i \subseteq \mathcal{L}(T_i)$, we have $S_i\|_{L_i} = S\|_{L_i}$ and $T_i\|_{L_i} = T\|_{L_i}$. We use this fact to prove that $S_i\|_{L_i} = T_i\|_{L_i}$, and hence $L_i = L \cap \mathcal{L}(S_i) \in \textsc{Cast}(S_i, T_i, K \cap \mathcal{L}(S_i))$ as follows.

Consider any leaves $a, b, c \in L_i$. We have

$$\begin{aligned}
\texttt{lca}_{S_i\|_{L_i}}(a,b) = \texttt{lca}_{S_i\|_{L_i}}(a,c) &\iff \texttt{lca}_{S\|_{L_i}}(a,b) = \texttt{lca}_{S\|_{L_i}}(a,c) \\
&\iff \texttt{lca}_{S\|_L}(a,b) = \texttt{lca}_{S\|_L}(a,c) \quad (\text{as } L_i \subseteq L) \\
&\iff \texttt{lca}_{T\|_L}(a,b) = \texttt{lca}_{T\|_L}(a,c) \\
&\qquad\qquad (\text{as } S\|_L = T\|_L) \\
&\iff \texttt{lca}_{T\|_{L_i}}(a,b) = \texttt{lca}_{T\|_{L_i}}(a,c) \\
&\iff \texttt{lca}_{T_i\|_{L_i}}(a,b) = \texttt{lca}_{T_i\|_{L_i}}(a,c).
\end{aligned}$$

By Lemma 1, we conclude that $S_i\|_{L_i} = T_i\|_{L_i}$. □

The following theorem shows that we can construct a maximum constrained agreement leaf set by combining some smaller ones constructed according to the κ-decomposition.

Theorem 1. *Let $H_0, H_1, \ldots, H_{m-1}$ be leaf sets such that $H_i \in \textsc{Mcast}(S_i, T_i, K \cap \mathcal{L}(S_i))$ for $0 \le i \le m-1$. Then, $H = \cup_{0 \le i \le m-1} H_i \in \textsc{Mcast}(S,T,K)$.*

Proof. Note that $K = \bigcup_{0 \le i \le m-1} (K \cap \mathcal{L}(S_i)) \subseteq \bigcup_{0 \le i \le m-1} H_i = H$. Below, we prove that $S\|_H = T\|_H$ and hence $H \in \textsc{Cast}(S,T,K)$. By Lemma 1, it suffices to prove that for any three leaves $a, b, c \in H$, we have

$$\texttt{lca}_{S\|_H}(a,b) = \texttt{lca}_{S\|_H}(a,c) \iff \texttt{lca}_{T\|_H}(a,b) = \texttt{lca}_{T\|_H}(a,c). \quad (7)$$

Note that if a, b, c are all in the same leaf set H_i, then

$$\texttt{lca}_{S\|_H}(a,b) = \texttt{lca}_{S\|_H}(a,c) \iff \texttt{lca}_{S\|_{H_i}}(a,b) = \texttt{lca}_{S\|_{H_i}}(a,c)$$

$$\iff \texttt{lca}_{T\|_{H_i}}(a,b) = \texttt{lca}_{T\|_{H_i}}(a,c) \iff \texttt{lca}_{T\|_H}(a,b) = \texttt{lca}_{T\|_H}(a,c),$$

and we have (7). Suppose that a, b, c are not in the same leaf set. Either a, b or a, c must be in different sets. Assume that a and c are in different sets H_i and H_j. Then, $\texttt{lca}_{S\|_H}(a,c)$ and $\texttt{lca}_{T\|_H}(a,c)$ are the root of S and T, respectively. Therefore, to prove (7), it suffices to prove that

$$\texttt{lca}_{S\|_H}(a,b) \text{ is the root of } S \iff \texttt{lca}_{T\|_H}(a,b) \text{ is the root of } T. \quad (8)$$

Note that if a, b are in different leaf sets, $\texttt{lca}_{S\|_H}(a,b)$ and $\texttt{lca}_{T\|_H}(a,b)$ are the roots of S and T, respectively. If a, b are in the same set H_i where $i \neq 0$, a, b are within the rooted subtrees S_i in S and subtree T_i in T; hence, $\texttt{lca}_{S\|_H}(a,b)$ and $\texttt{lca}_{T\|_H}(a,b)$ are not the root of S and T. For the case when $a, b \in H_0$, recall that $\kappa \in K \cap \mathcal{L}(S_0) \subseteq H_0$ and $S_0\|_{H_0} = T_0\|_{H_0}$. Thus, for the three leaves $a, b, \kappa \in H_0$, we have

$$\texttt{lca}_{S_0\|_{H_0}}(a,b) = \texttt{lca}_{S_0\|_{H_0}}(a,\kappa) \iff \texttt{lca}_{T_0\|_{H_0}}(a,b) = \texttt{lca}_{T_0\|_{H_0}}(a,\kappa). \quad (9)$$

Note that (9) is equivalent to (8) because (i) $\texttt{lca}_{S_0\|_{H_0}}(a,b) = \texttt{lca}_{S\|_{H_0}}(a,b) = \texttt{lca}_{S\|_H}(a,b)$, $\texttt{lca}_{T_0\|_{H_0}}(a,b) = \texttt{lca}_{T\|_{H_0}}(a,b) = \texttt{lca}_{T\|_H}(a,b)$, and (ii) $\texttt{lca}_{S_0\|_{H_0}}(a,\kappa)$ and $\texttt{lca}_{T_0\|_{H_0}}(a,\kappa)$ are the root of S and T, respectively. Hence, in all possible cases, we have (8), and hence (7). Therefore $S\|_H = T\|_H$ and $H \in \textsc{Cast}(S,T,K)$.

To see that $H \in \textsc{Mcast}(S,T,K)$, i.e., H is a largest element in $\textsc{Cast}(S,T,K)$, let us consider any $L \in \textsc{Cast}(S,T,K)$. Lemma 2 asserts that for $0 \leq i \leq m-1$, $L_i = L \cap \mathcal{L}(S_i) \in \textsc{Cast}(S_i, T_i, K \cap \mathcal{L}(S_i))$. Since $H_i \in \textsc{Mcast}(S_i, T_i, K \cap \mathcal{L}(S_i))$, we have $|L_i| \leq |H_i|$. Then, $|L| = \sum_{0 \leq i \leq m-1} |L_i| \leq \sum_{0 \leq i \leq m-1} |H_i| = |H|$. □

4 The Case When κ is Not a Shallow Leaf

In this section, we analyze the structure of the maximum agreement leaf subsets of S and T with respect to K under the assumption that κ is not a shallow leaf.

Consider the unique path from the root of S to κ. We call the nodes on this path κ-nodes of S. Given any two different κ-nodes u, u', we say that u *is higher than* u', denoted as $u \succ u'$, if u is nearer the root. We say $u \succeq u'$ if either $u = u'$ or $u \succ u'$. Note that κ itself is the lowest κ-node in S. For any leaf a of S, define κ-parent of a, denoted as $\kappa_S(a)$, to be the least ancestor of a that is κ-node. Let $\mathcal{L}_\kappa(u) = \{a \mid \kappa_S(a) = u\}$ be the set of leaves whose κ-parents are u. (Note that $\mathcal{L}_\kappa(\kappa) = \{\kappa\}$, and for other u, $\mathcal{L}_\kappa(u)$ includes all the leaf descendents of u except those that are in the subtree rooted at the unique κ-node child of u.) For any set I of κ-nodes, define $\mathcal{L}_\kappa(I) = \bigcup_{u \in I} \mathcal{L}_\kappa(u)$. For any κ-node u, we say that u is *precious* if $\mathcal{L}_\kappa(u)$ has at least one leaf in K, i.e., $K \cap \mathcal{L}_\kappa(u) \neq \emptyset$. Otherwise, we say that u is *ordinary*. We have similar definitions for T. The following lemma gives some structural property related to κ-nodes.

Lemma 4. *Suppose that* $L \in \textsc{Cast}(S,T,K)$. *For any two leaves* $a, b \in L$, *we have (i)* $\kappa_S(a) \succ \kappa_S(b) \iff \kappa_T(a) \succ \kappa_T(b)$*; and (ii)* $\kappa_S(a) \neq \kappa_S(b) \iff \kappa_T(a) \neq \kappa_T(b)$.

Proof. To prove (i), suppose that $\kappa_S(a) \succ \kappa_S(b)$. Since $\kappa \in L$ and $S\|_L = T\|_L$, the three leaves in L are related as follows:

$$\mathtt{lca}_{S\|_L}(a,b) = \mathtt{lca}_{S\|_L}(a,\kappa) \iff \mathtt{lca}_{T\|_L}(a,b) = \mathtt{lca}_{T\|_L}(a,\kappa). \qquad (10)$$

Note that among the ancestors of b that are on the path from b to $\kappa_S(a)$, there is only one node, namely $\kappa_S(a)$ that is an ancestor of a; hence $\mathtt{lca}_{S\|_L}(a,b) = \kappa_S(a) = \mathtt{lca}_{S\|_L}(a,\kappa)$ (because κ is the lowest κ-node and all κ-nodes are its ancestors). Together with (10), $\mathtt{lca}_{T\|_L}(a,b) = \mathtt{lca}_{T\|_L}(a,\kappa) = \kappa_T(a)$, or equivalently, we have $\kappa_T(a) \succ \kappa_T(b)$. The other direction of (i) can be proved symmetrically.

Note that (ii) follows from (i) directly. □

Let $u_1 \succ u_2 \succ \cdots \succ u_m$ be the sequence of precious κ-nodes on S. We define the *κ-decomposition* of S to be the sequence of sets $(I_1, I_2, \ldots, I_{2m})$ where

- $I_{2\ell}$ is a singleton containing the ℓth precious κ-node u_ℓ,
- I_1 contains all the κ-nodes higher than u_1, and
- for $2 \le \ell \le m$, $I_{2\ell-1}$ contains those κ-nodes between $u_{\ell-1}$ and u_ℓ.

Note that $I_{2m} = \{\kappa\}$. Since κ is a leaf, the κ-decomposition covers all the κ-nodes. (See Fig. 2 for an example.) We define the κ-decomposition $(J_1, J_2, \ldots, J_{2n})$ for T similarly.

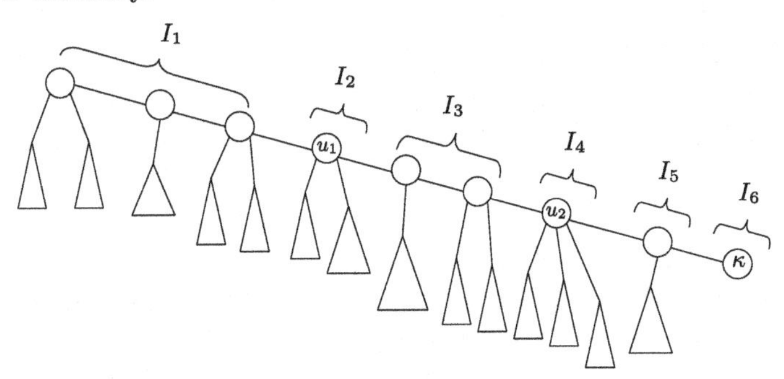

Fig. 2. The κ-decomposition of S

Recall that we assume $S\|_K = T\|_K$ and hence $\textsc{Cast}(S,T,K)$ is not empty. In the rest of the section, we study the structure of any $L \in \textsc{Cast}(S,T,K)$ according the κ-decompositions $(I_1, I_2, \ldots, I_{2m})$ and $(J_1, J_2, \ldots, J_{2n})$ of S and T, respectively. The following lemma shows that the two lists have the same length, i.e., $m = n$, and there is a one-one correspondence between the sets in the lists.

Lemma 5. *Given any $L \in \textsc{Cast}(S,T,K)$, $L \cap \mathcal{L}_\kappa(I_\ell) = L \cap \mathcal{L}_\kappa(J_\ell)$ for $1 \le \ell \le 2m$. Furthermore, we have $m = n$.*

Proof. We claim that for every $1 \le \ell \le \min\{m,n\}$, a leaf $a \in L$ is in $\mathcal{L}_\kappa(I_\ell)$ if and only if a is in $\mathcal{L}_\kappa(J_\ell)$. This implies $L \cap \mathcal{L}_\kappa(I_\ell) = L \cap \mathcal{L}_\kappa(J_\ell)$ for $1 \le \ell \le \min\{m,n\}$. Together with the fact that $\mathcal{L}_\kappa(I_{2m}) = \mathcal{L}_\kappa(J_{2n}) = \{\kappa\}$, we conclude $m = n$ and the lemma follows.

We prove our claim by induction. Note that by symmetry, we only need to prove that if a leaf $a \in L$ is in $\mathcal{L}_\kappa(I_\ell)$, then a is in $\mathcal{L}_\kappa(J_\ell)$. For the base case, suppose to the contrary that L has a leaf a in $\mathcal{L}_\kappa(I_1)$ but not in $\mathcal{L}_\kappa(J_1)$. Recall that J_2 has only one element, which is a precious κ-node v. It follows that $\mathcal{L}_\kappa(J_2)$ has a leaf $b \in K \subseteq L$, and $\kappa_T(b) = v \succeq \kappa_T(a)$ (because $a \notin \mathcal{L}_\kappa(J_1)$). On the other hand, by definition, $\mathcal{L}_\kappa(I_1)$ contains no leaf in K and thus $b \notin \mathcal{L}_\kappa(I_1)$ and $\kappa_S(a) \succ \kappa_S(b)$. Note that the κ-parent of the two leaves $a, b \in L$ have different relationships in S and T. By Lemma 4, $S\|_L \neq T\|_L$, a contradiction. Thus, the claim is true for $\ell = 1$.

Suppose that the claim is true for $1, 2, \ldots, \ell - 1$ and we consider ℓ. Assume that L has a leaf a in $\mathcal{L}_\kappa(I_\ell)$ but not in $\mathcal{L}_\kappa(J_\ell)$. Note that if ℓ is odd, the assumption will lead us to the contradictory conclusion that $S\|_L \neq T\|_L$ as in the base case $\ell = 1$. Suppose that ℓ is even. Then, J_ℓ has a single precious κ-node v, and there is a leaf $b \in K \subseteq L$ that is in $\mathcal{L}_\kappa(J_\ell)$. Together with induction hypothesis that $L \cap \mathcal{L}_\kappa(I_k) = L \cap \mathcal{L}(J_k)$ for $1 \leq k \leq \ell - 1$, we conclude that (i) $a \in \mathcal{L}_\kappa(I_\ell)$ and $a \in \mathcal{L}_\kappa(J_p)$ for some $p > \ell$, and (ii) $b \in \mathcal{L}_\kappa(J_\ell)$ and $b \in \mathcal{L}_\kappa(I_q)$ for some $q \geq \ell$. Therefore, $\kappa_S(a) \succeq \kappa_S(b)$ and $\kappa_T(b) \succ \kappa_T(a)$, and by Lemma 4, $S\|_L \neq T\|_L$, a contradiction. Thus the claim is also true for ℓ. □

Corollary 1. *Suppose that $L \in \textsc{Cast}(S, T, K)$. For any two leaves $a, b \in L$, if $a \in \mathcal{L}_\kappa(I_p)$ and $b \in \mathcal{L}_\kappa(I_q)$ where $p < q$ then*
(i) $\kappa_S(a) \succ \kappa_S(b)$ and $\mathtt{lca}_S(a, b) = \kappa_S(a)$, and
(ii) $\kappa_T(a) \succ \kappa_T(b)$ and $\mathtt{lca}_T(a, b) = \kappa_T(a)$.

Proof. (i) follows directly from definition. From Lemma 5, we have $L \cap \mathcal{L}_\kappa(I_p) = L \cap \mathcal{L}_\kappa(J_p)$ and $L \cap \mathcal{L}_\kappa(I_q) = L \cap \mathcal{L}_\kappa(J_q)$. Hence, $a \in \mathcal{L}_\kappa(J_p)$ and $b \in \mathcal{L}_\kappa(J_q)$ and we have (ii). □

We want to have lemmas like Lemma 3 and Theorem 1 that allow us to find the maximum agreement leaf subset by finding those for some partition of the leaves. However, in this case, every set in the partition needs to have some leaf in K. Thus, we extend the leaf set as follows: for $1 \leq \ell \leq 2m$, let $\mathcal{L}_\kappa{}'(I_\ell) = \mathcal{L}_\kappa(I_\ell) \cup \{\kappa\}$ and $\mathcal{L}_\kappa{}'(J_\ell) = \mathcal{L}_\kappa(J_\ell) \cup \{\kappa\}$. Note that $K \in \textsc{Cast}(S, T, K)$ and by Lemma 5, we have $K \cap \mathcal{L}_\kappa(I_\ell) = K \cap \mathcal{L}_\kappa(J_\ell)$, and hence $K \cap \mathcal{L}_\kappa{}'(I_\ell) = K \cap \mathcal{L}_\kappa{}'(J_\ell)$ for $1 \leq \ell \leq 2m$. It follows that $\textsc{Cast}(S\|_{\mathcal{L}_\kappa{}'(I_\ell)}, T\|_{\mathcal{L}_\kappa{}'(J_\ell)}, K \cap \mathcal{L}_\kappa{}'(I_\ell))$ is not empty.

Lemma 6. *Suppose that $L \in \textsc{Cast}(S, T, K)$. For $1 \leq \ell \leq 2m$, the leaf set $L_\ell = L \cap \mathcal{L}_\kappa{}'(I_\ell)$ is in $\textsc{Cast}(S\|_{\mathcal{L}_\kappa{}'(I_\ell)}, T\|_{\mathcal{L}_\kappa{}'(J_\ell)}, K \cap \mathcal{L}_\kappa{}'(I_\ell))$.*

Proof. Obviously $K \cap \mathcal{L}_\kappa{}'(I_\ell) \subseteq L_\ell$. Below, we show that $(S\|_{\mathcal{L}_\kappa{}'(I_\ell)})\|_{L_\ell} = (T\|_{\mathcal{L}_\kappa{}'(J_\ell)})\|_{L_\ell}$ and the lemma follows.

By Lemma 5, we have $L \cap \mathcal{L}_\kappa(I_\ell) = L \cap \mathcal{L}_\kappa(J_\ell)$ and hence $L \cap \mathcal{L}_\kappa{}'(I_\ell) = L \cap \mathcal{L}_\kappa{}'(J_\ell)$. Therefore, $L_\ell = L \cap \mathcal{L}_\kappa{}'(I_\ell) = L \cap \mathcal{L}_\kappa{}'(J_\ell)$ and $(S\|_{\mathcal{L}_\kappa{}'(I_\ell)})\|_{L_\ell} = S\|_{L_\ell}$ and $(T\|_{\mathcal{L}_\kappa{}'(J_\ell)})\|_{L_\ell} = T\|_{L_\ell}$. As in the proof of Lemma 3, we have, for any three leaves $a, b, c \in L_\ell$,

$$\mathtt{lca}_{(S\|_{\mathcal{L}_\kappa{}'(I_\ell)})\|_{L_\ell}}(a, b) = \mathtt{lca}_{(S\|_{\mathcal{L}_\kappa{}'(I_\ell)})\|_{L_\ell}}(a, c) \iff \mathtt{lca}_{S\|_{L_\ell}}(a, b) = \mathtt{lca}_{S\|_{L_\ell}}(a, c)$$

$$\iff \mathtt{lca}_{S\|_L}(a, b) = \mathtt{lca}_{S\|_L}(a, c) \iff \mathtt{lca}_{T\|_L}(a, b) = \mathtt{lca}_{T\|_L}(a, c) \iff$$

$$\mathtt{lca}_{T\|_{L_\ell}}(a, b) = \mathtt{lca}_{T\|_{L_\ell}}(a, c) \iff \mathtt{lca}_{(T\|_{\mathcal{L}_\kappa{}'(J_\ell)})\|_{L_\ell}}(a, b) = \mathtt{lca}_{(T\|_{\mathcal{L}_\kappa{}'(J_\ell)})\|_{L_\ell}}(a, c),$$

and by Lemma 1, $(S\|_{\mathcal{L}_\kappa{}'(I_\ell)})\|_{L_\ell} = (T\|_{\mathcal{L}_\kappa{}'(J_\ell)})\|_{L_\ell}$. □

The next theorem is similar to Theorem 1; it suggests a divide-and-conquer approach to find the maximum agreement leaf subset.

Theorem 2. *Let* $H_1, \ldots, H_{2m}$ *be leaf sets where* $H_\ell \in \texttt{MCAST}(S\|_{\mathcal{L}_\kappa{}'(I_\ell)}, T\|_{\mathcal{L}_\kappa{}'(J_\ell)}, K \cap \mathcal{L}_\kappa{}'(I_\ell))$ *for* $1 \le \ell \le 2m$. *Then,* $H = \bigcup_{1 \le \ell \le 2m} H_i \in \texttt{MCAST}(S, T, K)$.

Proof. Note that $K = \bigcup_{1 \le \ell \le 2m} K \cap \mathcal{L}_\kappa{}'(I_\ell) \subseteq \bigcup_{1 \le \ell \le 2m} H_\ell = H$. Below, we show that $S\|_H = T\|_H$, and hence $H \in \texttt{CAST}(S, T, K)$. By Lemma 1, it suffices to prove that for any three leaves $a, b, c \in H$, we have

$$\texttt{lca}_{S\|_H}(a,b) = \texttt{lca}_{S\|_H}(a,c) \iff \texttt{lca}_{T\|_H}(a,b) = \texttt{lca}_{T\|_H}(a,c) \tag{11}$$

Note that if a, b, c are all in the same leaf set H_ℓ, then,

$$\texttt{lca}_{S\|_H}(a,b) = \texttt{lca}_{S\|_H}(a,c) \iff \texttt{lca}_{S\|_{H_\ell}}(a,b) = \texttt{lca}_{S\|_{H_\ell}}(a,c)$$
$$\iff \texttt{lca}_{T\|_{H_\ell}}(a,b) = \texttt{lca}_{T\|_{H_\ell}}(a,c) \iff \texttt{lca}_{T\|_H}(a,b) = \texttt{lca}_{T\|_H}(a,c),$$

and we have (11). Suppose that a, b, c are not in the same leaf set. Then, either a, b or a, c, say a, b are in different leaf sets. Suppose $a \in H_p$ and $b \in H_q$. Note that κ is in all the leaf sets because $\kappa \in K \cap \mathcal{L}_\kappa{}'(I_\ell) \subseteq H_\ell$ for $1 \le \ell \le 2m$; hence a and b cannot be κ. We consider two cases.

Case 1: $p < q$. Since $a \in H_p \subseteq \mathcal{L}_\kappa{}'(I_p)$, $b \in H_q \subseteq \mathcal{L}_\kappa{}'(I_q)$ and a, b are not κ, we conclude that $a \in \mathcal{L}_\kappa(I_p)$ and $b \in \mathcal{L}_\kappa(I_q)$. Together with $p < q$, we have $\texttt{lca}_{S\|_H}(a,b) = \kappa_S(a)$ and $\texttt{lca}_{T\|_H}(a,b) = \kappa_T(a)$ (Corollary 1). To prove (11), it suffices to show that

$$\texttt{lca}_{S\|_H}(a,c) = \kappa_S(a) \iff \texttt{lca}_{T\|_H}(a,c) = \kappa_T(a). \tag{12}$$

Suppose that a, c are in the same leaf set, i.e., $a, c \in H_p$. Since $\kappa \in H_p$ and $S\|_{H_p} = T\|_{H_p}$, the three leaves a, c, κ are related by $\texttt{lca}_{S\|_{H_p}}(a,c) = \texttt{lca}_{S\|_{H_p}}(a,\kappa) = \kappa_S(a) \iff \texttt{lca}_{T\|_{H_p}}(a,c) = \texttt{lca}_{T\|_{H_p}}(a,\kappa) = \kappa_T(a)$. Then, we have (12) because $\texttt{lca}_{S\|_H}(a,c) = \texttt{lca}_{S\|_{H_p}}(a,c)$ and $\texttt{lca}_{T\|_H}(a,c) = \texttt{lca}_{T\|_{H_p}}(a,c)$.

Suppose that a, c are in the different leaf sets and let $c \in H_g \subseteq \mathcal{L}_\kappa{}'(I_g)$. Again, c cannot be κ and thus $c \in \mathcal{L}_\kappa(I_g)$. ¿From Corollary 1, if $g > p$, then $\texttt{lca}_{S\|_H}(a,c) = \kappa_S(a)$ and $\texttt{lca}_{T\|_H}(a,c) = \kappa_T(a)$; and if $g < p$, then $\texttt{lca}_{S\|_H}(a,c) = \kappa_S(c) \neq \kappa_S(a)$ and $\texttt{lca}_{T\|_H}(a,c) = \kappa_T(c) \neq \kappa_T(a)$. Therefore, regardless of where c is, we have (12), and hence (11).

Case 2: $p > q$. Similar to Case 1, we have $\texttt{lca}_{S\|_H}(a,b) = \kappa_S(b)$ and $\texttt{lca}_{T\|_H}(a,b) = \kappa_T(b)$. To prove (11), it suffices to prove that

$$\texttt{lca}_{S\|_H}(a,c) = \kappa_S(b) \iff \texttt{lca}_{T\|_H}(a,c) = \kappa_T(b). \tag{13}$$

Suppose $c \notin H_q$. Then neither a nor c are in $\mathcal{L}_\kappa(I_q)$ and thus their least common ancestor in S and T are not in I_q and J_q, respectively. Since $b \in \mathcal{L}_\kappa(I_q)$, $\kappa_S(b)$ and $\kappa_T(b)$ are in I_q and J_q respectively. Hence, $\texttt{lca}_{S\|_H}(a,c) \neq \kappa_S(b)$ and $\texttt{lca}_{T\|_H}(a,c) \neq \kappa_T(b)$ and we have (13).

Suppose $c \in H_q$. Since $b, c, \kappa \in H_q$ and $S\|_{H_q} = T\|_{H_q}$, the three leaves are related by $\texttt{lca}_{S\|_H}(c,\kappa) = \texttt{lca}_{S\|_H}(b,\kappa) \iff \texttt{lca}_{T\|_H}(c,\kappa) = \texttt{lca}_{T\|_H}(b,\kappa)$, or equivalently,

$$\kappa_S(c) = \kappa_S(b) \iff \kappa_T(c) = \kappa_T(b). \tag{14}$$

Since $p > q$, we have $\texttt{lca}_{S\|_H}(a,c) = \kappa_S(c)$ and $\texttt{lca}_{T\|_H}(a,c) = \kappa_T(c)$. Together with (14), we have (13) and hence (11).

In both cases, we have (11) and hence $S\|_H = T\|_H$ and $H \in \textsc{Cast}(S,T,K)$. Together with Lemma 6, we can prove easily that $H \in \textsc{Mcast}(S,T,K)$ as in the proof of Theorem 1. □

5 The Reduction

In this section, we describe a reduction for finding a maximum constrained agreement subtree of S and T with respect to K. To ease our discussion, we assume that the set of all possible leaves are totally ordered.

Let n be the total number of leaves in S and T. For every internal node u, we define the *classifying leaf* of u to be the smallest leaf descendent of u that is in K; if u has no such leaf, we define the classifying leaf of u to be $-\infty$. Let $C(u)$ be the set of classifying leaves of its children. Note that by performing a depth first search on S and T, we can decide the classifying leaves and hence $C(u)$ for every internal node in S or T. Then, taking another $O(\sum_{u\in S}|C(u)|\log|C(u)| + \sum_{u\in T}|C(u)|\log|C(u)|) = O(n\log n)$ time, we can sort the classifying leaves in every $C(u)$. Note that be a little bit more careful, all the sortings can actually be done in $O(n + |K|\log|K|)$ time.

Below, we explain how to use these sorted $C(u)$'s and apply the results of the previous sections to solve our problem. We pick a leaf κ in K.

- If κ is a shallow leaf, then by Theorem 1, we can reduce $\textsc{Mcast}(S,T,K)$ to the subproblems of $\textsc{Mcast}(S_0,T_0,K\cap\mathcal{L}(S_0)),\ldots,\textsc{Mcast}(S_{m-1},T_{m-1},K\cap\mathcal{L}(S_{m-1}))$. Note that by comparing the sorted $C(r_S)$ and $C(r_T)$ of the roots r_S and r_T, we can identify the S_i's and T_i's.
- If κ is not a shallow leaf, then by Theorem 2, we can reduce the problem to the subproblems $\textsc{Mcast}(S\|_{\mathcal{L}_\kappa'(I_\ell)}, T\|_{\mathcal{L}_\kappa'(J_\ell)}, K\cap\mathcal{L}_\kappa'(I_\ell))$ $(1\le\ell\le 2m)$. Note that by comparing the sorted $C(u)$ of those nodes along the paths from κ to the root of S and T, we can identify the $S\|_{\mathcal{L}_\kappa'(I_\ell)}$ and $T\|_{\mathcal{L}_\kappa'(J_\ell)}$ for $1\le\ell\le 2m$.

Observe that any two of the above subproblems share only one leaf, namely κ. For those subproblems with more than one leaf in K, we can recursively apply Theorems 1 and 2 to further divide them until we come up with only subproblems with only one leaf in K. It should be clear that the whole process takes $O(n+|K|\log|K|)$ time. The following theorem summarizes our discussion.

Theorem 3. *Consider any labeled trees S, T and a leaf subset K. Suppose that $S\|_K = T\|_K$. Let n be the total number of leaves of S and T. Then, using $O(n+|K|\log|K|)$ time, we can find subtrees $S_1,S_2,\ldots,S_m$ of S, $T_1,T_2,\ldots,T_m$ of T such that*

1. *given any $H_i \in \textsc{Mcast}(S_i,T_i,K\cap\mathcal{L}(S_i))$ for $1\le i\le m$, we have $\left(\bigcup_{1\le i\le m} H_i\right) \in \textsc{Mcast}(S,T,K)$;*
2. *$P_i = K\cap\mathcal{L}(S_i)$ has only one leaf; and*
3. *all the S_i's, as well as all the T_i's, have at most one leaf in common and hence $\sum_{1\le i\le m}|S_i| \le 2|S|$ and $\sum_{1\le i\le m}|T_i| \le 2|T|$.*

References

1. K. Amenta and F. Clarke. A linear-time majority tree algorithm. In *Proceedings of the 3rd International Workshop on Algorithms in Bioinformatics*, pages 216–227, 2003.
2. T.Y. Berger-Wolf. Online consensus and agreement of phylogenetic trees. In *Proceedings of the 4th International Workshop on Algorithms in Bioinformatics*, pages 350–361, 2004.
3. V. BERRY. Improving the reduction from the constrained to the unconstrained MAST. Technical Report 05041, LIRMM, 2005.
4. R. Cole, M. Farach, R. Hariharan, T. Przytycka, and M. Thorup. An $O(n \log n)$ algorithm for the maximum agreement subtree problem for binary trees. *SIAM Journal on Computing*, 30(5):1385–1404, 2000.
5. S. Dong and E. Kraemer. Calculation, visualization and manipulation of masts (maximum agreement subtrees. In *Proceedings of the IEEE Computational Systems Bioinformatics Conference*, pages 1–10, 2004.
6. M. Farach and M. Thorup. Optimal evolutionary tree comparison by sparse dynamic programming. In *Proceedings of the 35th Annual IEEE Symposium on Foundations of Computer Science*, pages 770–779, 1994.
7. M. Farach and M. Thorup. Fast comparison of evolutionary trees. In *Proceedings of the 5th Annual ACM-SIAM Symposium on Discrete Algorithms*, pages 481–488, 1995.
8. G.H. Hardy, J.E. Littlewood, and G. Pólya. *Inequalities*. Cambridge, 1952.
9. M.Y. Kao. Tree contractions and evolutionary trees. *SIAM Journal on Computing*, 27:1592–1616, 1998.
10. M.Y. Kao, T.W. Lam, W.K. Sung, and H.F. Ting. A decomposition theorem for maximum weight bipartite matchings with applications in evolution trees. In *Proceedings of the 7th Annual European Symposium on Algorithms*, pages 438–449, 1999.
11. M.Y. Kao, T.W. Lam, W.K. Sung, and H.F. Ting. An even faster and more unifying algorithm comparing trees via unbalanced bipartite matchings. *Journal of Algorithms*, 20(2):212–233, 2001.
12. D. Keselman and A. Amir. Maximum agreement subtree in a set of evolutionary trees– metrics and efficient algorithms. In *Proceedings of 35th Annual Symposium on the Foundations of Computer Sciences*, pages 758–769, 1994.
13. E. Kubicka, G. Kubicki, and F. McMorris. An algorithm to find agreement subtrees. *Journal of Classification*, 12:91–99, 1995.
14. A. Messmark, J. Jansson, A. Lingas, and E. Lundell. Polynomial-time algorithms for the ordered maximum agreement subtree problem. In *Proceedings of the 15th Annual Symposium on Combinatorial Pattern Matching*, pages 220–229, 2004.
15. Z.S. Peng and H.F. Ting. An $O(n \log n)$-time algorithm for the maximum constrained agreement subtree problem for binary trees. In *Proceedings of the 15th symposium on Algorithms and Computations*, pages 754–765, 2004.
16. T. Przytycka. Sparse dynamic programming for maximum agreement subtree problem. In *Mathematical Hierarchies and Biology*, pages 249–264. DIMACS series in Discrete Mathematics and Theoretical Computer Science, 1997.
17. M. Steel and T. Warnow. Kaikoura tree theorems: computing the maximum agreement subtree. *Information Processing Letters*, 48(2):77–82, 1994.

Pattern Identification in Biogeography*

Ganeshkumar Ganapathy[1], Barbara Goodson[2], Robert Jansen[2], Vijaya Ramachandran[1], and Tandy Warnow[1]

[1] Department of Computer Sciences, The University of Texas at Austin, TX 78712
{gsgk, vlr, tandy}@cs.utexas.edu

[2] Section of Integrative Biology, School of Biological Sciences, The University of Texas at Austin, TX 78712
{bgoodson, jansen}@mail.utexas.edu

Abstract. We develop and study two distance metrics for area cladograms (leaf-labeled trees where many leaves can share the same label): the *edge contract-and-refine* metric and the *MAAC* distance metric. We demonstrate that in contrast to phylogenies, the contract-and-refine distance between two area cladograms is not identical to the character encoding distance, and the latter is not a metric. We present a polynomial time algorithm to compute the MAAC distance, based on a polynomial-time algorithm for computing the largest common pruned subtree of two area cladograms. We also describe a linear time algorithm to decide if two area cladograms are identical.

1 Introduction

Biogeography is the study of the spatial and temporal distributions of organisms ([BL98, CKP03]). Biogeographers seek not only to understand ecological processes that influence the distribution of living organism over short periods of time (e.g., climatic stability, effect of area) but also to uncover events occurring in the distant past (e.g., continental drift, glaciation, evolution) which have resulted in the geographic distribution observed today.

Biogeography and Phylogeny. One of the ways of understanding the geographic distribution of species is by studying the *evolutionary history of the species* (see [CLW95, EO05, Jac04b] for instances of this approach). The evolutionary relationships are typically represented as branching tree structures called *phylogenetic trees*, or simply phylogenies. The branching structure of the phylogeny of a set of taxa can be used to differentiate between competing hypotheses concerning the observed geographic distribution of the set of taxa. Moreover, *a consistent pattern* observed in the phylogenies of species from different genera in the same geographic area will imply a stronger evidence for the particular hypotheses suggested by the pattern. As an example of this approach, consider a group of islands, each containing multiple ecological zones (for

* The research of Ganeshkumar Ganapathy was supported by NSF grants 0331453 and 0121680, Vijaya Ramachandran by NSF CCF-0514876, Tandy Warnow by NSF grants 0331453, 0312830, and 0121680, Barbara Goodson by NSF IGERT training grant 0114387, and Robert Jansen by NSF grant DEB 0120709.

R. Casadio and G. Myers (Eds.): WABI 2005, LNBI 3692, pp. 116–127, 2005.

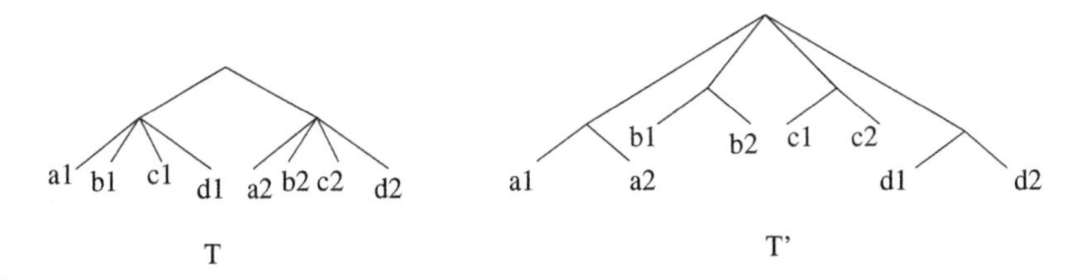

Fig. 1. Two hypothetical phylogenies on eight taxa on four islands (a, b, c, d) with two ecological zones each (1 and 2). T suggests dispersal, and T' suggests adaptive radiation.

example, each island can contain coastal and mountain ecological zones). Suppose our goal is to understand the observed geographic distribution of species on the islands. One hypothesis about the distribution could be that species dispersed from each ecological zone in each island to similar zones in other islands and then differentiated. This process is called *inter-island colonization*. Another hypothesis could be that dispersal *between* islands happened first followed by dispersal to the different ecological zones and differentiation into many species. This process is called *adaptive radiation* (see [JEOH00] for a discussion). The crucial idea is that we might be able to infer which of the above two hypotheses is responsible for the observed distribution: inter-island colonization is suggested by taxa on different islands but the same ecological zone forming a monophyletic group (rooted subtree), and adaptive radiation is suggested if species on the same island in different ecological zones form a monophyletic group (that is, form a rooted subtree in the phylogeny).

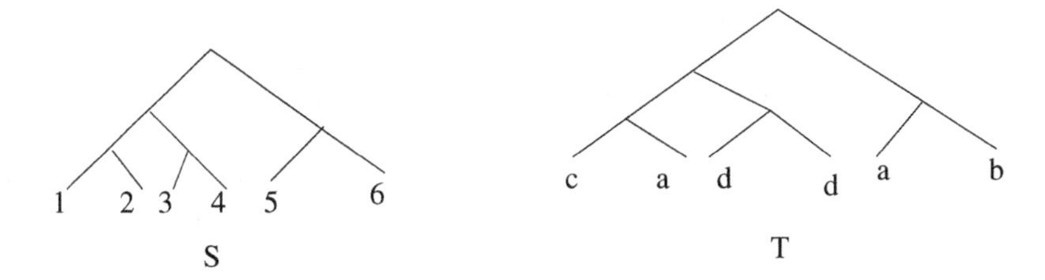

Fig. 2. A phylogeny S and its associated area cladogram T, assuming taxon 1 appears in area c; 2 appears in area a; 3 appears in area d; 4 appears in area d; 5 appears in area a; and 6 appears in area b.

Area Cladograms. Before looking for common patterns in the phylogenies of different sets of species in the same geographic area, the phylogeny for each set of species is converted to an *area cladogram*. Area cladograms are rooted or unrooted trees (as are phylogenies) whose leaves are labeled with *geographic areas* instead of taxa (see [Ros78, NP81]). To obtain the area cladogram for a set of species local to a set of areas, we start with the phylogeny for the set of species and, for each leaf, replace the taxon label with the label of the area in which the taxon is found. This process is illustrated in Figure 2. More formally, we define:

Definition 1. *Area Cladogram*
An area cladogram *is an unrooted or rooted leaf-labeled tree T. The leaves are labeled with areas, and many leaves may share the same label.*

In general, it might happen that a single taxon resides in more than one area (such taxa are called *widespread taxa*), and this would result in area cladograms with multiply-labeled leaves. We will develop our metrics and algorithms for area cladograms as in Definition 1, but we will show how to apply our results to more general cladograms where leaves can have multiple labels.

It should be noted that several methods have been proposed for obtaining area cladograms from phylogenetic trees (see [NP81, Pag88, Bro81, Pag94]). The methods "resolve" the issues of *widespread taxa* (single leaf being labeled by many areas), *redundant taxa* (many leaves being labeled by the same area), and *missing areas* to obtain a *resolved area cladogram* where the mapping between leaves and areas is one-one. Unresolved area cladograms are sometimes called *taxon area cladograms* in the literature.

Much of the prior work on area cladograms has focussed on suitable transformation that will result in resolved area cladograms, for which algorithms and metrics for phylogenetic trees apply.

In this paper, we address the problem of directly comparing two area cladograms. We develop distance metrics between area cladograms, and describe algorithms for computing a largest common pruned subtree of two area cladograms and for deciding if two given area cladograms are identical.

Prior Work. Inferring biogeographical history with species and areas is just one instance of the problem of inferring histories of two associated entities: the associated entities may be hosts and parasites, or genes and organisms [Pag94, PC98] (areas are analogous to hosts and organisms, and taxa in biogeography are analogous to parasites and genes). Hence, comparing area cladograms has a long history and a wide variety of applications (see [Jac04a, Jac04b, CLW95, GvVB02, Pag88] for example). Earlier work on comparing area cladograms has included pruning the cladograms until the two cladograms agree on the remaining leaves (see [Ros78, Pag88]), and using similarity metrics such as the *bipartition* metric (also called the *component* metric or the *character encoding* metric in the literature) and the *triplets* metric (see [Pag88]) between area cladograms (the triplets metric only applies when the area cladograms are rooted.)

All such methods apply only to resolved area cladograms. The methods of resolution differ in their interpretation of widespread taxa, redundant taxa and missing areas, and have been called *assumptions* 0, 1 *and* 2 in the literature (see [Pag88, vVZK99]). We will take a different approach to comparing area cladograms: we will compare them *without first resolving them so that the mapping between the leaves and labels is one-one*. This avoids the contentious issues ([Pag90]) surrounding the process of resolution.

Our Contributions. Our contributions are two-fold: we develop both metrics and algorithms for comparing area cladograms. More specifically,

- We show that the equivalence between the edge contract-and-refine metric ("RF-distance") and the bipartition metric ("character-encoding" metric) that holds for phylogenies *does not hold* for area cladograms. More specifically, we show that

the bipartition metric, when extended to area cladograms, is not a metric. For the edge contract-and-refine edit distance between two area cladograms we present a simple, but worst-case exponential-time algorithm. This edit distance can compare only area cladograms that are on the same number of leaves, and when each area labels the same number of leaves in both area cladograms (Section 3).
- We define another metric, the *MAAC* distance metric, for comparing two *rooted* area cladograms, which is based on the size of the largest common pruned subtree between the two area cladograms. The MAAC distance metric can compare two arbitrary trees that are not necessarily on the same number of leaves, which is particularly useful when comparing area cladograms (Section 3).
- We present a polynomial time algorithm for computing the MAAC distance between two rooted area cladograms. This algorithm is based on an algorithm we present for computing the largest common pruned subtree, the *maximum agreement area cladogram (MAAC)*, of two area cladograms. We also describe a faster, linear-time algorithm to decide if two area cladograms are identical (Section 4).

2 Phylogenies: Distance Metrics and Agreement Subsets

Character Encoding of Phylogenies. Tests for equality between phylogenies are based on the notion of the *character encoding* of phylogenies. Another notion crucial to the study of phylogenies is that of a *bipartition*: removing an edge e from a leaf-labeled tree T induces a bipartition π_e on its set of leaves.

Definition 2. *Character Encoding of a Phylogeny*
The character encoding *of a phylogeny T is the set $C(T) = \{\pi_e : e \in E(T)\}$, which represents the set of bipartitions induced by the edges of T.*

Theorem 1. *Character-Encoding Metric [Bun71]*
Let T and T' be two phylogenies on the same set of taxa. Then $|C(T)\triangle C(T')| = |(C(T)-C(T'))\cup(C(T')-C(T))|$ defines a distance metric.

By Theorem 1, two phylogenies T and T' are isomorphic (with the isomorphism preserving the leaf labels) if and only if $|C(T)\triangle C(T')| = 0$.

A *contraction* operation applied on an edge in a tree collapses that edge and identifies its two end points; a *refinement* operation applied at an unresolved node (i.e., an internal node with degree greater than three) expands that unresolved node into two nodes connected by an edge.

Definition 3. *Robinson-Foulds (RF) Distance*
The Robinson-Foulds distance *between two phylogenies T_1 and T_2 is defined as the number of contractions and refinements necessary to transform T_1 into T_2 (or vice-versa), and is denoted $RF(T_1,T_2)$.*

The RF distance naturally defines a metric since it is an edit distance.

Theorem 2. *[RF81] Let T_1 and T_2 be two phylogenies on the same set of taxa. Then $RF(T_1,T_2) = |C(T_1)\triangle C(T_2)|$.*

Finally, we define the maximum agreement subtree problem for phylogenies. The analogue of this problem for area cladograms is crucial to addressing the problems outlined in Section 1.

Definition 4. *Maximum Agreement Subset (MAST)*
Let $\{T_1, T_2, \ldots, T_k\}$ be a set of phylogenetic trees, on a set L of leaves. A maximum agreement subset *(MAST) of trees T_1 through T_k is a set of leaves $L' \subseteq L$ of maximum cardinality such that the restrictions of the trees $T_1, \ldots, T_k$ to the set L' are all isomorphic, with the isomorphism preserving leaf labels.*

The maximum agreement subset problem was introduced in [FG85], and has been studied thoroughly since then. The rooted and unrooted versions of MAST are polynomially related since the unrooted MAST problem can be solved by solving a polynomial number of rooted MAST problems. Computing a MAST is NP-hard for three or more trees [AK97]. A $O(n^{2+o(1)})$ time algorithm for the case of two trees on n leaves is given in [FCT94]. For two rooted binary trees, the best known algorithm takes $O(n \log^3 n)$ time ([FCPT95b, FCPT95a]); for two rooted trees which may not be binary, the best known algorithm takes $O(n^{1.5} c^{\sqrt{\log n}})$ time where c is a constant ([FCT94]). For computing a MAST of k rooted trees, an $O(kn^3 + n^d)$ algorithm (with d the maximum degree of a node in any tree) was presented in [FCPT95a].

3 Distance Measures Between Area Cladograms

In this section, we will develop distance metrics for the set of area cladograms. We will first show that the character encoding distance between two different area cladograms can be zero, and hence the character-encoding "distance" is not a metric on area cladograms, and in particular cannot be used as a test of isomorphism. We then propose a metric for comparing area cladograms that is based on computing the size of the largest common pruned subtree of the two area cladograms. We call this the *MAAC* metric, and show how to compute it in Section 4.

While the character-encoding metric for phylogenies does not extend to area cladograms, the contract-and-refine edit distance still defines a metric (because it is an edit distance). We present an algorithm to compute the edge contract-and-refine edit distance between area cladograms. This algorithm is efficient if there are few occurrences of widespread taxa, but it is exponential-time in general. For phylogenies this edit distance which is called the *Robinson-Foulds* distance, can be computed efficiently since it equals the character-encoding distance.

3.1 The Character Encoding Cannot Distinguish Between Area Cladograms

We first define the *extended character encoding* of an area cladogram.

Definition 5. *Let T be an area cladogram. The multi-set $\{\pi_e : e \in E(T)\}$ is called the* extended character encoding *of T, and will be denoted by $C(T)$. Here π_e denotes the bipartition of the multi-set of leaf labels induced by the edge e.*

Contrary to our experience with phylogenetic trees where the mapping between leaves and labels is 1-1, with two area cladograms T_1 and T_2, $C(T_1) = C(T_2)$ does not imply that T_1 and T_2 are isomorphic. We exhibit a pair of such of trees in Figure 3.

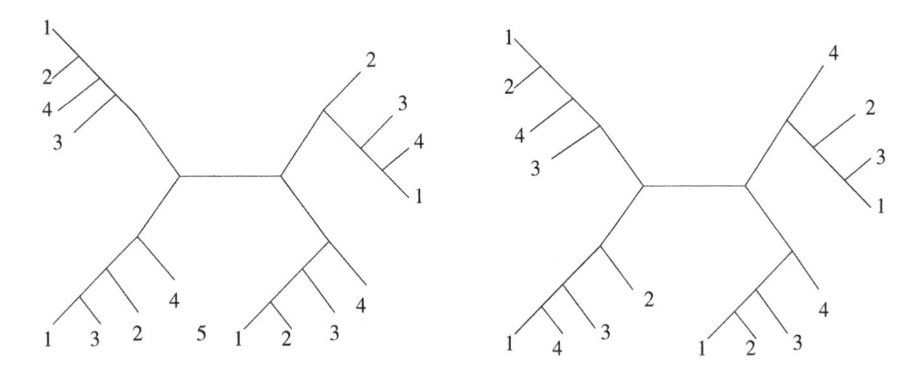

Fig. 3. Two different binary area cladograms that induce the same multi-set of partitions

3.2 The MAAC Distance Metric Between Area Cladograms

In this section we define the problem of computing the largest common pruned subtree of two rooted area cladograms and describe a distance metric based on the size of a largest common pruned subtree. We call a largest common pruned subtree a *Maximum Agreement Area Cladogram (MAAC)* (thus the MAAC is analogous to the maximum agreement subtree of two phylogenies).

Let T be an area cladogram on a set of leaves L. The *restriction* of T to a set of leaves L' is the cladogram obtained by deleting leaves in the set $L - L'$ from T and then suppressing internal nodes of degree two (except the root, if there is one).

We now define *a maximum agreement area cladogram (MAAC)* for a set of rooted area cladograms, and a distance measure between two rooted area cladograms that is based on the size of a MAAC of the two area cladograms.

Definition 6. *Maximum Agreement Area Cladogram (MAAC) and MAAC distance*

Let $\{T_1, T_2, \ldots, T_k\}$ be a set of rooted area cladograms, with L_i the leaf set of tree T_i, for $i = 1, 2, \ldots, k$. Let $\lambda_1 \subseteq L_1$ through $\lambda_k \subseteq L_k$ be sets of leaves of maximum cardinality such that the respective restrictions of the trees $T_1, \ldots, T_k$ to the sets $\lambda_1 \ldots \lambda_k$ are all isomorphic, with the isomorphisms preserving leaf labels. A restriction of any tree T_i to such a subset of leaves λ_i is a maximum agreement area cladogram *(MAAC) for the cladograms T_1 through T_k. The size of the MAAC is defined to be the number of leaves in the maximum agreement area cladogram, and is denoted by $size_{maac}(T_1, T_2, \ldots, T_k)$.*

The MAAC distance *between two trees T_1 and T_2 is $d_M(T_1, T_2) = max(n_1,\ n_2) - size_{maac}(T, T')$, where n_1 and n_2 are the number of leaves in T_1 and T_2 respectively.*

Note that in the above definition *we do not require that all the given set of trees contain the same number of leaves, or that they be labeled with the same set of areas, or even that they be consistent.* The MAAC distance can be viewed as a generalization of the maximum agreement subtree metric for phylogenies [GKK94], which for two phylogenies on the same set of n labeled leaves was defined as $n - size_{mast}$ where $size_{mast}$ is the size of a maximum agreement subset of the two phylogenies.

Handling Widespread Taxa. For comparing cladograms using maximum agreement area cladograms, leaves labeled by more than one area can be treated thus: each leaf

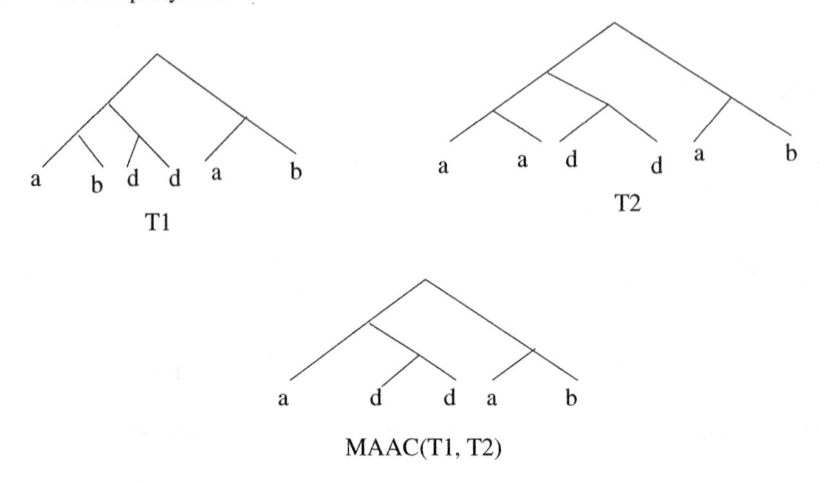

Fig. 4. Two area cladograms T1 and T2, and their MAAC

labeled by a group of areas can be split into many separate leaves (all having the same parent), each of which is labeled by a single unique area from the group of areas.

Due to space constraints, we state the following theorem without proof:

Theorem 3. *The MAAC distance d_M is a metric on the set of all area cladograms.*

Note that twice the MAAC distance between two cladograms is an upper bound on the number of insertions and deletions of leaves necessary to transform one of the cladograms to the other.

In Section 4, we present a polynomial-time algorithm for computing a maximum agreement area cladogram for two area cladograms.

3.3 Contract-and-Refine Distance Metric for Area Cladograms

Though the character-encoding distance fails to extend to area cladograms, the RF distance, being an edit distance, can be extended to unrooted area cladograms to provide a distance metric.

Definition 7. *Robinson-Foulds Distance Between Unrooted Area Cladograms*

The Robinson-Foulds distance *between two unrooted area cladograms T_1 and T_2 is defined to be the number of contractions and refinements necessary to transform T_1 to T_2 (or equivalently, T_2 to T_1).*

Note that if the number of leaves labeled l is different in T_1 and T_2 for some label l, then $RF(T_1, T_2)$ is undefined (i.e., there is no sequence of contractions and refinements that can transform T_1 into T_2). In such cases we define $RF(T_1, T_2)$ to be ∞.

Handling Widespread Taxa. Taxa endemic (resident) to more than one area would result in cladograms with leaves labeled by many areas. Our definition of the Robinson-Foulds distance applies to such cladograms as well: if a leaf is labeled with a set of areas, we can consider that set of areas to be the unique label for that leaf.

As shown in Section 3.1, for area cladograms, the RF distance will not be equal to the extended character-encoding distance. However, we can relate the RF distance between two area cladograms to the RF distance between two associated phylogenies, as we will show. We begin with some definitions.

Definition 8. *Full Differentiation of an Area Cladogram*

Let $T = (t, M)$ be an unrooted area cladogram, where t is an unlabeled tree and M is the mapping assigning labels to the leaves of t. Then, a full differentiation *of T is a leaf-labeled tree $T^* = (t, M^*)$ such that M^* is one-one. In other words, T^* has the same topology as T, but has its leaves labeled uniquely.*

Definition 9. *Consistent Full Differentiations*

Let $T_1 = (t_1, M_1)$ and $T_2 = (t_2, M_2)$ be two unrooted area cladograms with the same set L of leaf labels, and let $T_1^* = (t_1, M_1^*)$ and $T_2^* = (t_2, M_2^*)$ be full differentiations of T_1 and T_2 respectively. T_1^* and T_2^* are consistent full differentiations *if, for each label $l \in L$, the set of labels in assigned to leaves in T_1^* that were labelled l in T_1 is identical to the set of labels assigned to leaves in T_2^* that were labelled l in T_2. Mathematically, this is: $\forall l \in L, \{M_1^*(x) : M_1(x) = l\} = \{M_2^*(x) : M_2(x) = l\}$.*

Due to space constraints, we state the following theorem without proof:

Theorem 4. *Let T_1 and T_2 be two unrooted area cladograms. Then $RF(T_1, T_2)$ is equal to* $\max\{RF(T_1^*, T_2^*) : T_1^*$ *and T_2^* are mutually consistent full differentiations of T_1 and T_2, respectively*$\}$.

Note that the RF distance between two cladograms T_1 and T_2 is at most the RF distance between *any* consistent full differentiations of T_1 and T_2. Hence this provides a linear-time method for getting an upper bound on the RF distance between two area cladograms T_1 and T_2: we first compute two mutually consistent full differentiations, and then compute their RF distance.

Theorem 4 suggests the following trivial (but expensive) algorithm for computing the RF distance between two area cladograms T_1 and T_2: we simply compute the RF distance between all the possible consistent full differentiations of T_1 and T_2 (in $\Theta(n)$ time per pair, see [Day85]) and choose the minimum. Thus, we have the following theorem:

Theorem 5. *Let T_1 and T_2 be two unrooted area cladograms on n leaves on the same set of areas. For each area a_i appearing at the leaves of T_1 and T_2, let n_i be the number of leaves labeled with area a_i. Then, the RF distance between T_1 and T_2 can be calculated in $\Theta(n\Pi_{i=1}^{k}(n_i)!)$ time.*

4 An Algorithm for the Maximum Agreement Area Cladogram Problem

In this section we describe an algorithm for computing maximum agreement area cladogram (MAAC) of two given rooted area cladograms. The algorithm is based on a dynamic programming algorithm for the phylogenetic rooted maximum agreement subtree algorithm from [SW93]. We will first present the maximum agreement subtree algorithm. We will then observe that the basic recursion underlying the dynamic-programming algorithm will hold for the maximum agreement area cladogram algorithm as well though the mapping between leaves and their labels may not be one-one in area cladograms.

The MAST Algorithm from [SW93]. We now give a brief summary of the algorithm in [SW93] for computing the MAST of two rooted binary trees. In our description, the expression $MAST(T,T')$ denotes a maximum agreement subset of two given (rooted binary) phylogenies T and T'.

Let T and T' be two given binary phylogenies on n leaves. Let v be a node in T, and denote by T_v the subtree of T rooted at v. Similarly denote by T'_w the subtree of T' rooted at a node w in T'. The dynamic programming algorithm for MAST operates by computing $MAST(T_v,T'_w)$ for all pairs of nodes (v,w) in $V(T) \times V(T')$ "bottom-up". We now show how to reduce computing $MAST(T_v,T'_w)$ to computing a small number of smaller MAST computations $MAST(S,S')$ where S and S' are subtrees of T_v and T'_w respectively, with at least one of them being a proper subtree.

To begin with, the $MAST(T_v,T'_w)$ is easy to compute when either v or w are leaves. So in the following discussion assume neither v nor w is a leaf.

Let L^* be a MAST of T_v and T'_w, and let T^* be the corresponding MAST tree. Then there exist homeomorphisms mapping T^* to a rooted subtree of T_v and to a rooted subtree of T'_w. Let p be the (not necessarily proper) descendant of v such that the root of T^* is mapped to p. Similarly let q be the descendant of w in T' such that that the root of T^* is mapped to w. Then, $MAST(T_v,T'_w)$ is in fact equal to $MAST(T_p,T'_q)$.

The vertex p may be actually v or it might be a vertex below v. Similarly q may be w or some vertex below w. Based on the location of p and q, we have the following cases.

- *Vertex p is a proper descendent of v.* In this case, T_p is a proper subtree of T_v, and $MAST(T_v,T'_w)$ equals $MAST(T_p,T'_w)$.
- *Vertex q is a proper descendent of w.* In this case, $MAST(T_v,T'_w)$ equals $MAST(T_v,T'_q)$.
- *Vertex p equals v and vertex q equals w.*

In the first two cases, we have reduced the computation of $MAST(T_v,T'_w)$ to a MAST computation on a subproblem. In the last case, let v_1 and v_2 be the children of v, and let w_1 and w_2 be the children of w. Let T_1^* and T_2^* be the subtrees of the root of the MAST tree T^*. Then, T_1^* is homeomorphic to a subtree of T_{v_1} (or to a subtree of T_{v_2}; there is no loss of generality in assuming that it is homeomorphic to a subtree of T_{v_1}). Similarly, T_2^* is homeomorphic to a subtree of T_{v_2}. It cannot be homeomorphic to a subtree of T_{v_1}, since then T^* would be homeomorphic to a subtree of T_{v_1}, contradicting the assumption that there is no proper descendent p of v such that root of T^* is mapped p. Arguing similarly, we can conclude that T_1^* and T_2^* are homeomorphic to subtrees of T'_{w_1} and T'_{w_2} respectively. Now, since T^* is a MAST tree, we can conclude that T_1^* is a MAST tree of T_{v_1} and T'_{w_2}, and that T_2^* is a MAST tree of T_{v_2} and T'_{w_2}. So in this case we have reduced computing $MAST(T_v,T'_w)$ to computing $MAST(T_{v_1},T'_{w_1})$ and $MAST(T_{v_2},T'_{w_2})$ and then taking their union.

The above discussion suggests a straightforward dynamic programming algorithm which involves computing $O(n^2)$ subproblems each of which can be solved in $O(1)$ time (for binary trees).

The running time of the above algorithm is $O(n^2)$ for trees of bounded degree. For general rooted phylogenetic trees the running time is $O(n^{2.5} \log n)$.

4.1 The Maximum Agreement Area Cladogram Algorithm

The difference between the maximum agreement area cladogram and the maximum agreement subset problems is that the former problem takes as input leaf-labeled trees where the mapping between leaves and labels is not one-one. Recall that in the description of the maximum agreement subtree dynamic programming recursion above, p is the *unique* descendant of v such that the homeomorphism mapping T^* to a subtree of T_v maps the root of T^* to p, and q is the *unique* descendant of w such that the homeomorphism mapping T^* to a subtree of T_w maps the root of T^* to q. However, when the map between leaves and labels is not one-one, nodes p and q may not be unique. However, we can remedy this situation by modifying our description thus: in tree T_v, let p be a vertex farthest from v such that the root of T^* is mapped to p, and in T'_w, and let q be a vertex farthest from w such that the root of T^* is mapped to q (note that this modification will not affect the actual algorithm at all, only the proof that the algorithm is correct). The rest of dynamic programming recursion uses only the properties of homeomorphisms, and these properties hold true for homeomorphisms between area cladograms as well. Hence, the maximum agreement subtree algorithm from [SW93] works without change as a maximum agreement area cladogram algorithm.

The Running Time of the MAAC Algorithm. The algorithm is same as the maximum agreement subtree algorithm, and hence the running time of the maximum agreement area cladogram algorithm is $O(n^2)$ for trees of bounded degree and $O(n^{2.5} \log n)$ for trees of unbounded degree.

4.2 Testing Isomorphism Between Two Rooted Area Cladograms

The MAAC distance metric between area cladograms gives us a polynomial-time algorithm for testing isomorphism: we apply the maximum agreement area cladogram algorithm from the previous section to compute the MAAC distance between the two area cladograms, and we conclude that the two cladograms are isomorphic if and only if the distance is zero. The algorithm is adapted from the algorithm for testing rooted tree isomorphism from [AHU74].

The input to the algorithm consists of two rooted area cladograms T_1 and T_2 on n leaves (if the number of leaves is different, then clearly they are not isomorphic). We assume that the leaves are labeled with integers from 1 through n, not all distinct. The algorithm is based on assigning to each node u in the tree, an integer, which we call $index(u)$. For leaves, the index is just their labels. The algorithm is as follows:

1. Compute the *height*, the maximum distance between the root and a leaf, of the two trees. If the heights are not the same, then the trees are not isomorphic, otherwise, let the height be h.

2. Based on the height, assign level numbers to the nodes of the trees. The level number of a node at a distance of d from the root is set to be $h-d$.
3. For each leaf u at level 0, set $index[u]$ to be the leaf-label.
4. Assuming that index has been set for each node at level $i-1$, calculate the indices at level i thus: for each node v at level i, form a *tuple* (an ordered list) consisting of the indices of its children sorted in ascending order. If v is a leaf, then its tuple consists of just its label. Let L_i be the list of tuples of nodes at level i in T_1. Let L_i' be the corresponding list for T_2. Now lexicographically sort L_i and L_i' to obtain S_i and S_i' respectively.
5. If S_i and S_i' are not identical, then declare T_1 and T_2 to be non-isomorphic and quit. Else, assign $index[v]$ for each node v at level i in T_1 thus: $index[v]$ is the *rank* of v's tuple in the sorted list S_i. The ranks start from 1, and all identical tuples receive the same rank. Indices for vertices in T_2 are assigned similarly. The level-i indices can now be used to calculate the indices for level $i+1$.
6. If the roots of T_1 and T_2 are assigned the same index, then the trees are isomorphic, otherwise not.

Proof of Correctness and Running Time: We omit the proof due to space constraints. The running time of the above algorithm for testing isomorphism is $O(n)$, where n is the number of leaves in the input trees (see [AHU74]).

References

[AHU74] A. V. Aho, J. E. Hopcroft, and J. D. Ullman. *The Design and Analysis of Computer Algorithms*. Addison-Wesley Publishing Company, 1974.

[AK97] A. Amir and D. Keselman. Maximum Agreement Subtrees in a Set of Evolutionary Trees: Metrics and Efficient Algorithms. *SIAM Journal of Computing*, 26(6):1656–1669, 1997. A preliminary version of this paper appeared in FOCS '94.

[BL98] J. H. Brown and M. V. Lomolino. *Biogeography*. Sinauer Associates, Sunderland, Massachusetts, second edition, 1998.

[Bro81] D. R. Brooks. Hennig's Parasitological Method: A Proposed Solution. *Systematic Zoology*, 30:229–249, 1981.

[Bun71] P. Buneman. The Recovery of Trees from Measures of Dissimilarity. *Mathematics in the Archaelogical and Historical Sciences*, pages 387–395, 1971.

[CKP03] J. V. Crisci, L. Katinas, and P. Posadas. *Historical Biogeography: An Introduction*. Harvard University Press, Cambridge, Massachusetts, 2003.

[CLW95] M. Crisp, H. P. Linder, and P. Weston. Cladistic Biogeography of Plants in Australia and New Guinea: Congruent Pattern Reveals Two Endemic Tropical Tracts. *Systematic Biology*, 44(4):457–473, 1995.

[Day85] W.H.E. Day. Optimal Algorithms for Comparing Trees with Labeled Leaves. *Journal of Classification*, 2:7–28, 1985.

[EO05] B. C. Emerson and P. Oromi. Diversification of the Forest Beetle Genus Tarphius on the Canary Islands, and the Evolutionary Origins of Island Endemics. *Evolution*, 59(3):586–598, 2005.

[FCPT95a] M. Farach-Colton, T.M. Przytycka, and M. Thorup. On the Agreement of Many Trees. *Information Processing Letters*, 55:297–301, 1995.

[FCPT95b] M. Farach-Colton, T.M. Przytycka, and M. Thorup. The Maximum Agreement Subtree Problem for Binary Trees. 1995. Manuscript.

[FCT94] M. Farach-Colton and M. Thorup. Sparse Dynamic Programming for Evolutionary Tree Comparison. In *Proc. of the 35th Annual Symp. on the Foundations of Computer Science*, pages 770–779, 1994.

[FG85] C.R. Finden and A.D. Gordon. Obtaining Common Pruned Trees. *Journal of Classification*, 2:255–276, 1985.

[GKK94] W. D. Goddard, E. Kubicka, and G. Kubicki. The Agreement Metric for Labeled Binary Trees. *Mathematical Biosciences*, 123:215–226, 1994.

[GvVB02] M. D. Green, M.G.P. van Veller, and D. R. Brooks. Assessing Modes of Speciation: Range Asymmetry and Biogeographical Congruence. *Cladistics*, 18(1):112–124, 2002.

[Jac04a] A. P. Jackson. Cophylogeny of the Ficus Microcosm. *Biological Review*, 79(4):751–768, 2004.

[Jac04b] A. P. Jackson. Phylogeny and Biogeography of the Malagasy and Australasian Rainbowfishes (Teleostei : Melanotaenioidei): Gondwanan Vicariance and Evolution in Freshwater. *Molecular Phylogenetics and Evolution*, 33(3):719–734, 2004.

[JEOH00] C. Juan, B.C. Emerson, P. Oromi, and G. M. Hewitt. Colonization and Diversification: Towards a Phylogeographic Synthesis for the Canary Islands. *Trends in Ecology and Evolution*, 15(3):104–109, 2000.

[NP81] G. Nelson and N. Platnick. *Systematics and Biogeograpy*. Columbia University Press, New York, 1981.

[Pag88] R. D. M. Page. Quantitative Cladistic Biogeograpy: Constructing and Comparing Area Cladograms. *Systematic Zoology*, 37:254–270, 1988.

[Pag90] R. D. M. Page. Temporal Congruence and Cladistic Analysis of Biogeography and Cospeciation. *Systematic Zoology*, 39:205–226, 1990.

[Pag94] R.D.M. Page. Maps Between Trees and Cladistic Analysis of Historical Associations Among Genes. *Systematic Biology*, 43(1):58–77, 1994.

[PC98] R.D.M. Page and M. Charleston. Trees Within Trees: Phylogenies and Historical Associations. *Trends in Ecology and Evolution*, 13(9):356–359, 1998.

[RF81] D.F. Robinson and L.R. Foulds. Comparison of Phylogenetic Trees. *Mathematical Biosciences*, 53:131–147, 1981.

[Ros78] D. E. Rosen. Vicariant Patterns and Historical Explanation in Biogeograpy. *Systematic Zoology*, 27:159–188, 1978.

[SW93] M. Steel and T. Warnow. Kaikoura Tree Theorems: Computing the Maximum Agreement Subtree. *Information Processing Letters*, 48:77–82, 1993.

[vVZK99] M.G.P. van Veller, M. Zandee, and D. J. Kornet. Two Requirements for Obtaining Common Patterns Under Different Assumptions in Vicariance Biogeography. *Cladistics*, 15:393–405, 1999.

On the Complexity of Several Haplotyping Problems*

Rudi Cilibrasi[2,**], Leo van Iersel[1], Steven Kelk[2], and John Tromp[2]

[1] Technische Universiteit Eindhoven (TU/e),
Den Dolech 2, 5612 AX Eindhoven, Netherlands
l.j.j.v.iersel@tue.nl
[2] Centrum voor Wiskunde en Informatica (CWI),
Kruislaan 413, 1098 SJ Amsterdam, Netherlands
Rudi.Cilibrasi@cwi.nl, S.M.Kelk@cwi.nl, John.Tromp@cwi.nl

Abstract. We present several new results pertaining to haplotyping. The first set of results concerns the combinatorial problem of reconstructing haplotypes from incomplete and/or imperfectly sequenced haplotype data. More specifically, we show that an interesting, restricted case of *Minimum Error Correction* (MEC) is NP-hard, question earlier claims about a related problem, and present a polynomial-time algorithm for the ungapped case of *Longest Haplotype Reconstruction* (LHR). Secondly, we present a polynomial time algorithm for the problem of resolving genotype data using as few haplotypes as possible (the *Pure Parsimony Haplotyping Problem*, PPH) where each genotype has at most two ambiguous positions, thus solving an open problem posed by Lancia et al in [15].

1 Introduction

If we abstractly consider the human genome as a string over the nucleotide alphabet $\{A, C, G, T\}$, it is widely known that the genomes of any two humans are more than 99% similar. In other words, it is known that, at most sites along the genome, humans all have the same nucleotide. At certain specific sites along the genome, however, variability is observed across the human population. These sites are known as *Single Nucleotide Polymorphisms* (SNPs) and are formally defined as the sites on the human genome where, across the human population, two or more nucleotides are observed and each such nucleotide occurs in at least 5% of the population. It turns out that these sites, which occur (on average) approximately once per thousand bases, capture the bulk of human genetic variability; the string of nucleotides found at the SNP sites of a human - the *haplotype* of that individual - can thus be thought of as a "fingerprint" for that individual.

* Part of this research has been funded by the Dutch BSIK/BRICKS project.
** Supported in part by the Dutch BSIK/BRICKS project, and by NWO project 612.55.002, and by the IST Programme of the European Community, under the PASCAL Network of Excellence, IST-2002-506778. This publication only reflects the authors' views.

R. Casadio and G. Myers (Eds.): WABI 2005, LNBI 3692, pp. 128–139, 2005.

It is further apparent that, for most SNP sites, only two nucleotides are seen; sites where three or four nucleotides are possible are comparatively rare. Thus, from a combinatorial perspective, a haplotype can be abstractly expressed as a string over the alphabet $\{0,1\}$. Indeed, the biologically-motivated field of SNP and haplotype analysis - which is at the forefront of "real-world" bioinformatics - has spawned an impressively rich and varied assortment of combinatorial problems, which are well described in surveys such as [4] and [8]. In this paper we focus on three such combinatorial problems; the first two are related to the problem of haplotyping a single individual, and the third is related to the problem of explaining the genetic variability of a population using as few haplotypes as possible.

The first two problems are both variants of the *Single Individual Haplotyping Problem* (SIH), introduced in [14]. The SIH problem amounts to determining the haplotype of an individual using (potentially) incomplete and/or imperfect fragments of sequencing data. The situation is further complicated by the fact that, being a *diploid* organism, a human has two versions of each chromosome; one each from the individual's mother and father. Hence, for a given interval of the genome, a human actually has two haplotypes. Thus, the SIH problem can be more accurately described as finding the two haplotypes of an individual given fragments of sequencing data where the fragments potentially have read errors and, crucially, where it is *not* known which of the two chromosomes each fragment was read from. There are four well-known variants of the problem: *Minimum Fragment Removal* (MFR), *Minimum SNP Removal* (MSR), *Minimum Error Correction* (MEC), and *Longest Haplotype Reconstruction* (LHR). In this paper we give results for MEC and LHR and refer the reader to [3] for information about MFR and MSR.

1.1 Minimum Error Correction (MEC)

This is the problem where the input is a matrix M of SNP fragments. Each column of M represents an SNP site and thus each element of the matrix denotes the (binary) choice of nucleotide seen at that SNP location on that fragment. An element of the matrix can thus either be '0', '1' or a *hole*, represented by '-', which denotes lack of knowledge or uncertainty about the nucleotide at that site. We use $M[i,j]$ to refer to the value found at row i, column j of M, and use $M[i]$ to refer to the ith row. We say that two rows r_1, r_2 of the matrix are in *conflict* if there exists a column j such that $M[r_1,j] \neq M[r_2,j]$ and $M[r_1,j], M[r_2,j] \in \{0,1\}$. We say that a matrix is *feasible* if the rows of the matrix can be partitioned into two sets such that all rows within each set are pairwise non-conflicting. The goal with MEC is thus to "correct" (or "flip") as few entries of the input matrix as possible (i.e. convert 0 to 1 or vice-versa) to make the resulting matrix feasible. The motivation behind this is that all rows of the input matrix were sequenced from one haplotype or the other, and that any deviation from that haplotype occurred because of read-errors during sequencing.

In the context of haplotyping, MEC has been discussed - sometimes under a different name - in papers such as [4], [18], [7] and (implicitly) [14]. One question arising from this discussion is how the distribution of holes in the input data affects computational complexity. To explain, let us first define a *gap* (in a string over the alphabet $\{0, 1, -\}$) as a maximal contiguous block of holes that is flanked on both sides by non-hole values. For example, the string `---0010---` has no gaps, `-0--10-111` has two gaps, and `-0-----1--` has one gap.[1] The problem variant *Ungapped-MEC* is where every row of the input matrix is ungapped i.e. all holes appear at the start or end.

In this paper we offer what we believe is the first concrete proof that Ungapped-MEC (and hence the more general MEC) is NP-hard. We do so by reduction from the optimisation version of MAX-CUT. As far as we are aware, other claims of this result are based explicitly or implicitly on results found in [11]; as we discuss in Section 2, we conclude that the results in [11] cannot be used for this purpose. Directly related to this, we define the problem *Binary-MEC*, where the input matrix contains no holes; as far as we know the complexity of this problem is still - intriguingly - open.

1.2 Longest Haplotype Reconstruction (LHR)

In this variant of the SIH problem, the input is again an SNP matrix M with elements drawn from $\{0, 1, -\}$. Recall that the rows of a feasible matrix M can be partitioned into two sets such that all rows within each set are pairwise non-conflicting. Having obtained such a partition, we can reconstruct a haplotype from each set by merging all the rows in that set together. (We define this formally later in Section 3.) With LHR the goal is to remove *rows* such that the resulting matrix is feasible and such that the sum of the lengths of the two resulting haplotypes is maximised. In this paper we show that *Ungapped-LHR* (where ungapped is defined as before) is polynomial-time solvable and give a dynamic programming algorithm for this which runs in time $O(n^2m + n^3)$ for an $n \times m$ input matrix. This improves upon the result of [14] which also showed a polynomial-time algorithm for Ungapped-LHR but under the restricting assumption of non-nested input rows.

1.3 Pure Parsimony Haplotyping Problem (PPH)

As mentioned earlier, there are actually two haplotypes for any given interval of an individual's genome. With current sequencing techniques it is still considered impractical to read the two haplotypes separately; instead, a single string is returned - the *genotype* - which combines the data from the two haplotypes but, in doing so, loses some information. Thus, whereas a haplotype is a string over the $\{0, 1\}$ alphabet, a genotype is a string over the $\{0, 1, 2\}$ alphabet.

[1] The case where each row of the input matrix has at most 1 gap is considered biologically relevant because *double-barreled shotgun sequencing* produces two disjoint intervals of sequencing data.

A '0' (respectively, '1') entry in the genotype means that both chromosomes have a '0' (respectively, '1') at that position. In contrast, a '2' entry means that the two haplotypes *differ* at that location: one has a '0' while the other has a '1' but we don't know which goes where. Thus, a '2'-site of a genotype is called an *ambiguous* position. We say that two haplotypes *resolve* a given genotype if that genotype is the result of combining the two haplotypes in the above manner. For example, the pair of haplotypes 0110 and 0011 resolve the genotype 0212.

It follows that a genotype with $a \geq 1$ ambiguous positions can be resolved in 2^{a-1} ways. Now, suppose we have a population of individuals and we obtain (without errors) the genotype of each individual. The *Pure Parsimony Haplotyping Problem* (PPH) is as follows:- given a set of genotypes, what is the smallest number of haplotypes such that each genotype is resolved by some pair of the haplotypes? In [15] it is shown that PPH is hard (i.e. NP-hard and APX-hard) even in the restricted case where no genotype has more than 3 ambiguous positions. The case of 2 ambiguous positions per genotype is left as an open question in [15]. In this paper we resolve this question by providing a polynomial-time algorithm for this problem that has a running time of $O(mn\log(n) + n^{3/2})$ for n genotypes each of length m.

Since writing the original version of this paper we have learned that, independently, Lancia and Rizzi have come up with a similar result [16] that was submitted for publication at the end of 2004.

2 Minimum Error Correction (MEC)

For a length-m string $X \in \{0,1,-\}^m$, and a length-m string $Y \in \{0,1\}^m$, we define $d(X,Y)$ as being equal to the number of *mismatches* between the strings i.e. positions where X is 0 and Y is 1, or vice-versa; holes do not contribute to the mismatch count. An $n \times m$ SNP matrix M is feasible iff there exist two strings (haplotypes) $H_1, H_2 \in \{0,1\}^m$, such that for all rows $r \in M$, $d(r,H_1) = 0$ or $d(r,H_2) = 0$. A *flip* is where a 0 entry is converted to a 1, or vice-versa. Note that, in our formulation of the problem, we do not allow flipping to or from holes, and the haplotypes H_1 and H_2 may not contain holes.

Problem: *Ungapped-MEC*
Input: An ungapped SNP matrix M
Output: The smallest number of flips needed to make M feasible.

Note that Ungapped-MEC is an optimisation problem, not a decision problem, hence the use of "NP-hard" in the following lemma rather than "NP-complete". A decision version may be obtained by adding a flip upperbound in the range $[0, nm]$.

Lemma 1. *Ungapped-MEC is NP-hard.*

Proof. We give a polynomial-time reduction from the optimisation version of MAX-CUT, which is the problem of computing the size of a maximum cut in a graph.[2] Let $G = (V, E)$ be the input to MAX-CUT, where E is undirected. (Without loss of generality we identify V with the natural numbers $1, 2, ..., |V|$.) We construct an instance M of Ungapped-MEC as follows. M has $2k + |E|$ rows and $2|V|$ columns where $k = 2|E||V|^2$. We use M_0 to refer to the first k rows of M, M_1 to refer to the second k rows of M, and M_G to refer to the remaining $|E|$ rows. The first $k/|V|$ rows of M_0 all have the following pattern: a 0 in the first column, a 0 in the second column, and the rest of the row is holes. The second $k/|V|$ rows of M_0 all have a 0 in the third column, a 0 in the fourth column, and the rest holes; we continue this pattern i.e. each row in the jth block of $k/|V|$ rows in M_0 ($1 \leq j \leq |V|$) has a 0 in column $2j-1$, a 0 in column $2j$, and the rest holes. M_1 is defined identically except that 1s are used instead of 0s. Each row of M_G encodes an edge from E:- for an edge (i, j) (where i is the numerically lower endpoint) we specify that columns $2i-1$ and $2i$ contain 0s, columns $2j-1$ and $2j$ contain 1s, and for all $c \neq i, j$, column $2c-1$ contains 0 and column $2c$ contains 1.

Suppose t is the largest cut possible in G. We claim that:

$$Ungapped\text{-}MEC(M) = |E|(|V| - 2) + 2(|E| - t) \tag{1}$$

From this t (i.e. MAX-CUT(G)) can easily be computed. First, note that the solution to Ungapped-MEC(M) is trivially upperbounded by $|V||E|$. This follows because we could simply flip every 1 entry in M_G to 0; the resulting overall matrix would be feasible because we could just take H_0 as the all-0 string and H_1 as the all-1 string. Now, we say a haplotype H has the *double-entry* property if, for all odd-indexed positions (i.e. columns) j in H, the entry at position j of H is the same as the entry at position $j+1$. We argue that a minimal number of feasibility-inducing flips will *always* lead to two haplotypes H_1, H_2 such that both haplotypes have the double-entry property and, further, H_1 is the bitwise complement of H_2. (We describe such a pair of haplotypes as *partition-encoding*.) This is because, if H_1, H_2 are not partition-encoding, then at least $k/|V| > |V||E|$ (in contrast with zero) entries in M_0 and/or M_1 will have to be flipped, meaning this strategy is doomed to begin with.

Now, for a given partition-encoding pair of haplotypes, it follows that - for each row in M_G - we will have to flip either $|V| - 2$ or $|V|$ entries to reach its nearest haplotype. This is because, irrespective of which haplotype we move a row to, the $|V| - 2$ pairs of columns *not* encoding end-points (for a given row) will always cost 1 flip each to fix. Then either 2 or 0 of the 4 "endpoint-encoding" entries will also need to be flipped; 4 flips will never be necessary because then the row could move to the other haplotype, requiring no extra flips. Ungapped-MEC thus maximises the number of rows which require $|V| - 2$ rather than $|V|$ flips. If we think of H_1 and H_2 as encoding a partition of the vertices of V (i.e. a vertex i is on one side of the partition if H_1 has 1s in columns $2i - 1$ and $2i$, and on the other side if H_2 has 1s in those columns), it follows that each row

[2] The reduction given here can easily be converted into a Karp reduction from the decision version of MAX-CUT to the decision version of Ungapped-MEC.

requiring $|V| - 2$ flips corresponds to a cut-edge in the vertex partition defined by H_1 and H_2. Equation 1 follows. □

Comment - a rediscovered open problem?

The MEC problem, as defined earlier, is technically speaking the *evaluation* variant[3] of the MEC problem. Consider the closely-related *constructive* version:

Problem: *Constructive-MEC*
Input: An SNP matrix M.
Output: For an input matrix M of size $n \times m$, two haplotypes $H_1, H_2 \in \{0,1\}^m$ minimising:

$$D(H_1, H_2) = \sum_{\text{rows } r \in M} \min(d(r, H_1), d(r, H_2)) \tag{2}$$

Owing to space restraints we omit the proof[4] but we can prove that Constructive-MEC is polynomial-time Turing interreducible with its evaluation counterpart, MEC. We mention this correspondence because, when expressed as a constructive problem, it can be seen that MEC is in fact a specific type of *clustering* problem, a topic of intensive study in the literature. More specifically, we are trying to find two representative "median" (or "consensus") strings such that the sum, over all input strings, of the distance between each input string and its nearest median, is minimised. Related to this, let us define a further problem:

Problem: *Binary-Constructive-MEC*
Input: An SNP matrix M that does not contain any holes
Output: As for Constructive-MEC

Our deferred proof of interreducibility between Constructive-MEC and MEC also holds for this restricted version of the problem, proving that Binary-Constructive-MEC is solvable in polynomial time iff Binary-MEC is solvable in polynomial time. This interreducibility is potentially useful because we now argue, in contrast to claims in the existing literature, that the complexity of Binary-MEC / Binary-Constructive-MEC is actually still open.

To elaborate, it is claimed in several papers (such as [1]) that a problem essentially equivalent to Binary-Constructive-MEC is NP-hard. Such claims inevitably refer to the seminal paper *Segmentation Problems* by Kleinberg, Papadimitriou, and Raghavan (KPR), which has appeared in multiple different forms since 1998 (e.g. [11], [12] and [13].) Close examination of the KPR paper(s), and personal communication with the authors [19], has confirmed that the KPR papers actually discuss two superficially similar, but essentially different, problems. One problem is essentially equivalent to Binary-Constructive-MEC, and

---

[3] See [2] for a more detailed explanation of terminology in this area.

[4] The proof will appear in a forthcoming journal version of this paper. Most of the work is reducing the constructive version to the evaluation version; the other direction is trivial and uses only one oracle call.

the other is a more general (and thus, potentially, a more difficult) problem.[5] In the same communication the authors have admitted that they have no proof of hardness for the former problem i.e. the problem that is isomorphic to Binary-Constructive-MEC.

Thus we conclude that the complexity of Binary-Constructive-MEC / Binary-MEC is still open. From an approximation viewpoint the problem has been quite well-studied; the problem has a *Polynomial Time Approximation Scheme* (PTAS) because it is a special form of the *Hamming 2-Median Clustering Problem*, for which a PTAS is demonstrated in [10]. Other approximation results appear in [11], [1], [13], [17] and a heuristic for a similar (but not identical) problem appears in [18]. We also know that, if the number of haplotypes to be found is specified as part of the input (and not fixed as 2), the problem becomes NP-hard; we again defer this proof to a forthcoming, longer version of this paper. Finally, it may also be relevant that the "geometric" version of the problem (where rows of the input matrix are not drawn from $\{0,1\}^m$ but from $\mathbb{R}^m$, and Euclidean distance is used instead of Hamming distance) is also open from a complexity viewpoint [17]. (However, the version using Euclidean-distance-squared *is* known to be NP-hard [5].)

3 Longest Haplotype Reconstruction (LHR)

Suppose an SNP matrix M is feasible. Then we can partition the rows of M into two sets, M_l and M_r, such that the rows within each set are pairwise non-conflicting. (The partition might not be unique.) From M_i ($i \in \{l, r\}$) we can then build a haplotype H_i by combining the rows of M_i as follows: The jth column of H_i is set to 1 if at least one row from M_i has a 1 in column j, is set to 0 if at least one row from M_i has a 0 in column j, and is set to a hole if all rows in M_i have a hole in column j. Note that, in contrast to MEC, this leads to haplotypes that potentially contain holes. For example, suppose one side of the partition contains rows `10--`, `-0--` and `---1`; then the haplotype we get from this is `10-1`. We define the *length* of a haplotype as the number of positions where it does not contain a hole; the haplotype `10-1` thus has length 3, for example. Now, the goal with LHR is to remove *rows* from M to make it feasible but also such that the sum of the lengths of the two resulting haplotypes is maximised. We define the function LHR(M) (which gives a natural number as output) as being the largest value this sum-of-lengths value can take, ranging over all feasibility-inducing row-removals and subsequent partitions.

[5] In this more general problem, rows and haplotypes are viewed as vectors and the distance between a row and a haplotype is their dot product. Further, unlike Binary-Constructive-MEC, this problem allows elements of the input matrix to be drawn arbitrarily from $\mathbb{R}$. This extra degree of freedom - particularly the ability to simultaneously use positive, negative and zero values in the input matrix - is what (when coupled with a dot product distance measure) provides the ability to encode NP-hard problems.

We provide a polynomial-time algorithm for the following variant of LHR:

Problem: *Ungapped-LHR*
Input: An ungapped SNP matrix M
Output: The value LHR(M), as defined above.

The LHR problem for ungapped matrices was proved to be polynomial time solvable by Lancia et. al in [14], but only with the genuine restriction that no fragments are included in other fragments. Our algorithm improves this in the sense that it works for all ungapped input matrices; our algorithm is similar in style to the algorithm that solves MFR in the ungapped case by Bafna et. al. in [3]. The complexity of LHR with gaps is still an open problem. Note that our dynamic-programming algorithm computes Ungapped-LHR(M) but it can easily be adapted to generate the rows that must be removed (and subsequently, the partition that must be made) to achieve this maximum.

Lemma 2. *Ungapped-LHR can be solved in time* $O(n^2m + n^3)$

Proof. Let M be the input to Ungapped-LHR, and assume the matrix has size $n \times m$. For row i define $l(i)$ as the leftmost column that is not a hole and define $r(i)$ as the rightmost column that is not a hole. The rows of M are ordered such that $l(i) \leq l(j)$ if $i < j$. Define the matrix M_i as the matrix consisting of the first i rows of M and two extra rows at the top: row 0 and row -1, both consisting of all holes. Define $OK(i)$ as the set of rows $j < i$ that are not in conflict with row i.

For $h, k \leq i$ and $h, k \geq -1$ and $r(h) \leq r(k)$ define $D[h, k; i]$ as the maximum sum of lengths of two haplotypes such that:-

- each haplotype is a combination of rows from M_i
- each row from M_i can be used to build at most one haplotype (i.e. it cannot be used for both haplotypes)
- row k is one of the rows used to build a haplotype and among such rows maximizes $r(\cdot)$
- row h is one of the rows used to build the other haplotype (than k) and among such rows maximizes $r(\cdot)$

The solution of the problem $LHR(M)$ is given by

$$\max_{h,k | r(h) \leq r(k)} D[h, k; n] \tag{1}$$

We distinguish three different cases in the calculation of the $D[h, k; i]$. The first case is when $h, k < i$. Under these circumstances,

$$D[h, k; i] = D[h, k; i - 1] \tag{2}$$

This is because:-

- If $r(i) > r(k)$: row i cannot be used for the haplotype that row k is used for, because row k has maximal $r(\cdot)$ among all rows that are used for a haplotype
- If $r(i) \leq r(k)$: row i cannot increase the length of the haplotype that row k is used for (because also $l(i) \geq l(k)$)
- the same arguments hold for h

The second case is when $h = i$. In this case:

$$D[i,k;i] = \max_{\substack{j \in OK(i),\ \ j \neq k \\ r(j) \leq r(i)}} D[j,k;i-1] + r(i) - \max\{r(j), l(i)-1\} \tag{3}$$

This results from the following. The definition of $D[i,k;i]$ says that row i has to be used for the other haplotype than k and amongst such rows maximizes $r(\cdot)$. Therefore the maximum sum of lengths is achieved by adding row i to the optimal solution with the restriction that row j is the most-right-ending row, for some j that agrees with i, is not equal to k and ends before i. The term $r(i) - \max\{r(j), l(i)-1\}$ is the increase in length of the haplotype if row i is added.

The last case is when $k = i$:

$$D[h,i;i] = \max_{\substack{j \in OK(i),\ \ j \neq h \\ r(j) \leq r(i)}} \begin{cases} D[j,h;i-1] + r(i) - \max\{r(j), l(i)-1\} & \text{if } r(h) \geq r(j) \\ D[h,j;i-1] + r(i) - \max\{r(j), l(i)-1\} & \text{if } r(h) < r(j) \end{cases} \tag{4}$$

The time for calculating all the $OK(i)$ is $O(n^2m)$. When all the $OK(i)$ are known, it takes $O(n^3)$ time to calculate all the $D[h,k;i]$. This is because we need to calculate $O(n^3)$ values $D[h,k;i]$ $(h,k < i)$ that take $O(1)$ time each and $O(n^2)$ values $D[i,k;i]$ and also $O(n^2)$ values $D[h,i;i]$ that take $O(n)$ time each. This leads to an overall time complexity of $O(n^2m + n^3)$. □

4 The Pure Parsimony Haplotyping Problem (PPH)

We refer the reader to Section 1.3 for definitions, and note once again the similar, independently-discovered result in [16].

Problem: *2-ambiguous Pure Parsimony Haplotyping Problem*
Input: A set G of genotypes such that no genotype has more than 2 ambiguous positions
Output: $PPH(G)$, which is the smallest number of haplotypes that can be used to resolve G.

Lemma 3. *The 2-ambiguous Pure Parsimony Haplotyping Problem can be solved in polynomial-time.*

Proof. We let $n = |G|$ denote the number of genotypes in G and let m denote the length of each genotype in G. We will compute the solution, $PPH(G)$, by reduction to the polynomial-time solvable problem $MaxBIS$, which is the problem of computing the cardinality of the maximum independent set in a bipartite graph. First, some notation. A genotype is *i-ambiguous* if it contains i ambiguous positions. Each genotype in G is thus either 0-ambiguous, 1-ambiguous, or 2-ambiguous. For a 0-ambiguous genotype g, we define h_g as the string g. For a 1-ambiguous genotype g we let $h_{g:0}$ (respectively, $h_{g:1}$) be the haplotype obtained by replacing the ambiguous position in g with 0 (respectively, 1). For a 2-ambiguous genotype g we let $h_{g:i,j}$ - where $i, j \in \{0,1\}$ - be the haplotype obtained by replacing the first (i.e. leftmost) ambiguous position in g with i, and the second ambiguous position with j. A haplotype is said to have even (odd) parity iff it contains an even (odd) number of 1s. Now, observe that there are two ways to resolve a 2-ambiguous genotype g: (1) with haplotypes $h_{g:0,0}$ and $h_{g:1,1}$ and (2) with $h_{g:0,1}$ and $h_{g:1,0}$. Note that - depending on g - one of the ways uses two *even* parity haplotypes, and the other uses two *odd* parity haplotypes.

We build a set H of haplotypes by stepping through the list of genotypes and, for each genotype, adding the 1, 2 or 4 corresponding haplotypes to the set H. (Note that, because H is a set, we discard duplicate haplotypes.) That is, for a 0-ambiguous genotype g add h_g, for a 1-ambiguous genotype g add $h_{g:0}$ and $h_{g:1}$, and for a 2-ambiguous genotype g add $h_{g:0,0}, h_{g:0,1}, h_{g:1,0}$ and $h_{g:1,1}$.

We are now ready to build a bipartite graph $B = (V, E)$ as follows, where V has bipartition $V^+ \cup V^-$. For each $h \in H$ we introduce a vertex, which we also refer to as h; all h with even parity are put into V^+ and all h with odd parity are put into V^-. For each 0-ambiguous genotype $g \in G$ we introduce a set $I_0(g)$ of four vertices and we connect each vertex in $I_0(g)$ to h_g. For each 1-ambiguous genotype $g \in G$ we introduce two sets of vertices $I_1(g,0)$ and $I_1(g,1)$, both containing two vertices. Each vertex in $I_1(g,0)$ is connected to $h_{g:0}$ and each vertex in $I_1(g,1)$ is connected to $h_{g:1}$. Finally, for each 2-ambiguous $g \in G$ we introduce (to V^+ and V^- respectively) two sets of vertices $I_2(g,+)$ and $I_2(g,-)$, each containing 4 vertices. We connect every vertex in $I_2(g,+)$ to every vertex in $I_2(g,-)$, connect every vertex in $I_2(g,+)$ to the two odd parity haplotypes resolving g, and connect every vertex in $I_2(g,-)$ to the two even parity haplotypes resolving g. This completes the construction of B.

A maximum-size independent set (MIS) of B is a largest set of mutually non-adjacent vertices of B. Observe that, in a MIS of B, all the vertices of $I_0(g)$ must be in the MIS, for all 0-ambiguous g. To see this, note firstly that, if at least one vertex of $I_0(g)$ is in the MIS, we should put all of $I_0(g)$ in the MIS. Secondly, suppose all the vertices in $I_0(g)$ are out of the MIS, but h_g is in the MIS. Then we could simply remove h_g from the MIS and add in all the vertices of $I_0(g)$, leading to a larger MIS:- contradiction! By a similar argument we see that, for all 1-ambiguous $g \in G$, all of $I_1(g,0)$ and $I_1(g,1)$ must be in the MIS. Now, consider $I_2(g,+)$ and $I_2(g,-)$, for all 2-ambiguous $g \in G$. We argue that either $I_2(g,+)$ is wholly in the MIS, or $I_2(g,-)$ is wholly in the MIS. Suppose, by way of argument, that there exists a g such that both $I_2(g,+)$ and $I_2(g,-)$ are completely out of the MIS. If we are (wlog) free to add all the vertices in $I_2(g,+)$ to the MIS we have an immediate contradiction. So $I_2(g,+)$ is prevented from being in the MIS by the fact that one or two of the haplotypes to which it is connected are already in the MIS. But we could then build a bigger MIS by removing those (at most) two haplotypes from the MIS and adding the four vertices $I_2(g, +)$; contradiction!

We can think of the presence of an I set in the MIS as denoting that the genotype it represents is resolved using the haplotypes to which it is attached. Hence, every haplotype that is used for at least one resolution will *not* be in the MIS, and unused haplotypes *will* be in the MIS. Hence, a MIS will try and minimise the number of haplotypes used to resolve the given genotypes. Indeed,

$$MaxBIS(B) = 4n + (|H| - PPH(G)) \tag{1}$$

So we can use a polynomial-time algorithm for MaxBIS to compute PPH(G). □

Running time

The above algorithm can be implemented in time $O(mn\log(n) + n^{3/2})$. First we build the graph B. We can without too much trouble build a graph representation of B - that combines adjacency-matrix and adjacency-list features - in $O(mn\log(n))$ time. For each $g \in G$, add its corresponding I set(s) and add the (at most) 4 haplotypes corresponding to g, without eliminating duplicates, and at all times efficiently maintaining adjacency information. Then sort the list of haplotypes and eliminate duplicate haplotypes (by merging their adjacency information into one single haplotype.) It is not too difficult to do this in such a way that, in the final data structure representing the graph, adjacency queries can be answered, and adjacency-lists returned, in $O(1)$ time. This whole graph construction process takes $O(mn\log(n))$ time.

A maximum independent set in a bipartite graph can be constructed from a maximum matching. A maximum matching in B can be found in time $O(n^{3/2})$ because, in our case, $|V| = O(n)$ and $|E| = O(n)$ [9]. Once the maximum matching is found, it needs $O(|E| + |V|)$ time to find a maximum independent set [6]. Thus finding a maximum independent set takes $O(n^{3/2})$ time overall.

References

1. Noga Alon, Benny Sudakov, On Two Segmentation Problems, *Journal of Algorithms* 33, 173-184 (1999)
2. G. Ausiello, P. Crescenzi, G. Gambosi, V. Kann, A. Marchetti-Spaccamela, M. Protasi, Complexity and Approximation - Combinatorial optimization problems and their approximability properties, Springer Verlag (1999)
3. Vineet Bafna, Sorin Istrail, Giuseppe Lancia, Romeo Rizzi, Polynomial and APX-hard cases of the individual haplotyping problem, *Theoretical Computer Science*, (2004)
4. Paola Bonizzoni, Gianluca Della Vedova, Riccardo Dondi, Jing Li, The Haplotyping Problem: An Overview of Computational Models and Solutions, *Journal of Computer Science and Technology* 18(6), 675-688 (November 2003)
5. P. Drineas, A. Frieze, R. Kannan, S. Vempala, V. Vinay, Clustering in large graphs via Singular Value Decomposition, *Journal of Machine Learning* 56, 9-33 (2004)

6. F. Gavril, Testing for equality between maximum matching and minimum node covering, *Information processing letters* 6, 199-202 (1977)
7. Harvey J. Greenberg, William E. Hart, Giuseppe Lancia, Opportunities for Combinatorial Optimisation in Computational Biology, *INFORMS Journal on Computing*, Vol. 16, No. 3, 211-231 (Summer 2004)
8. Bjarni V. Halldorsson, Vineet Bafna, Nathan Edwards, Ross Lippert, Shibu Yooseph, and Sorin Istrail, A Survey of Computational Methods for Determining Haplotypes, *Proceedings of the First RECOMB Satellite on Computational Methods for SNPs and Haplotype Inference*, Springer Lecture Notes in Bioinformatics, LNBI 2983, pp. 26-47 (2003)
9. J.E. Hopcroft, R.M. Karp, An $n^{5/2}$ algorithm for maximum matching in bipartite graphs, *SIAM Journal on Computing* 2, 225-231 (1973)
10. Yishan Jiao, Jingyi Xu, Ming Li, On the k-Closest Substring and k-Consensus Pattern Problems, *Combinatorial Pattern Matching: 15th Annual Symposium* (CPM 2004) 130-144
11. Jon Kleinberg, Christos Papadimitriou, Prabhakar Raghavan, Segmentation Problems, *Proceedings of STOC 1998*, 473-482 (1998)
12. Jon Kleinberg, Christos Papadimitriou, Prabhakar Raghavan, A Microeconomic View of Data Mining, *Data Mining and Knowledge Discovery* 2, 311-324 (1998)
13. Jon Kleinberg, Christos Papadimitriou, Prabhakar Raghavan, Segmentation Problems, *Journal of the ACM* 51(2), 263-280 (March 2004) Note: this paper is somewhat different to the 1998 version.
14. Giuseppe Lancia, Vineet Bafna, Sorin Istrail, Ross Lippert, and Russel Schwartz, SNPs Problems, Complexity and Algorithms, *Proceedings of the 9th Annual European Symposium on Algorithms*, 182-193 (2001)
15. Giuseppe Lancia, Maria Christina Pinotti, Romeo Rizzi, Haplotyping Populations by Pure Parsimony: Complexity of Exact and Approximation Algorithms, *INFORMS Journal on Computing*, Vol. 16, No.4, 348-359 (Fall 2004)
16. Giuseppe Lancia, Romeo Rizzi, A polynomial solution to a special case of the parsimony haplotyping problem, to appear in *Operations Research Letters*
17. Rafail Ostrovsky and Yuval Rabani, Polynomial-Time Approximation Schemes for Geometric Min-Sum Median Clustering, *Journal of the ACM* 49(2), 139-156 (March 2002)
18. Alessandro Panconesi and Mauro Sozio, Fast Hare: A Fast Heuristic for Single Individual SNP Haplotype Reconstruction, *Proceedings of 4th Workshop on Algorithms in Bioinformatics* (WABI 2004), LNCS Springer-Verlag, 266-277
19. Personal communication with Christos H. Papadimitriou, June 2005
20. Romeo Rizzi, Vineet Bafna, Sorin Istrail, Giuseppe Lancia: Practical Algorithms and Fixed-Parameter Tractability for the Single Individual SNP Haplotyping Problem, *2nd Workshop on Algorithms in Bioinformatics* (WABI 2002) 29-43

A Hidden Markov Technique for Haplotype Reconstruction

Pasi Rastas, Mikko Koivisto, Heikki Mannila, and Esko Ukkonen

Department of Computer Science & HIIT Basic Research Unit,
P.O. Box 68,(Gustaf Hällströmin katu 2b), FIN-00014, University of Helsinki, Finland
firstname.lastname@cs.helsinki.fi

Abstract. We give a new algorithm for the genotype phasing problem. Our solution is based on a hidden Markov model for haplotypes. The model has a uniform structure, unlike most solutions proposed so far that model recombinations using haplotype blocks. In our model, the haplotypes can be seen as a result of iterated recombinations applied on a few founder haplotypes. We find maximum likelihood model of this type by using the EM algorithm. We show how to solve the subtleties of the EM algorithm that arise when genotypes are generated using a haplotype model. We compare our method to the well-known currently available algorithms (PHASE, HAP, GERBIL) using some standard and new datasets. Our algorithm is relatively fast and gives results that are always best or second best among the methods compared.

1 Introduction

The DNA differences between individuals of the same species are typically on single nucleotide locations in which more than only one nucleotide (allele) occurs in the population. Such differences, due to point mutations, and their sites are called *single nucleotide polymorphisms* (SNPs). SNPs can be used as genetic markers that can be utilized, for example, in finding disease causing mutations. For a diploid species, when an SNP is *typed* (observed) for an individual, the following problem arises: There are two near-identical copies of each chromosome of a diploid organism, but the common techniques for SNP typing do not provide the allele information separately for each of the two copies. Instead, they just give genotype information, i.e., for each SNP an unordered pair of allele readings is found, one from each copy. The alleles coming from the *same* chromosome copy are called a haplotype, while a genotype combines alleles from the two copies. So a genotype {A, C}, {T, T}, {G, T} could result from two haplotype pairs: (ATG, CTT) and (ATT, CTG).

A genotype with two identical alleles in a site is called *homozygote*, while a genotype with two different alleles is called *heterozygote* in that site. Given a set of genotypes, the problem of finding the corresponding two haplotypes for each genotype is called *phasing* or *resolving* the genotypes. Resolving is done simultaneously for all genotypes, based on some assumptions on how the haplotypes have evolved. The first approach to resolve haplotypes was Clark's method

R. Casadio and G. Myers (Eds.): WABI 2005, LNBI 3692, pp. 140–151, 2005.

[1] based on a greedy resolution rule. Clark's method is sometimes referred to as parsimony-based, but pure parsimony was investigated later in [2]. In pure parsimony one asks for finding a smallest set of haplotypes able to resolve all the genotypes. Different probabilistic approaches, still without recombination, have been proposed by e.g. [3,4,5,6]. Yet another combinatorial method was proposed by Gusfield [7], aiming at finding a set of resolving haplotypes that admits a *perfect phylogeny.* Gusfield's method works on genotype blocks within which no recombination is allowed; the block structure has to be uncovered separately. Greenspan and Geiger [8] were able to combine block finding and haplotype resolution by using a Bayesian network model. Very recently, Kimmel and Shamir [9,10] gave another such method, with improved phasing results.

In this paper we describe an approach to the phasing problem based on looking at the haplotypes as a result of recombinations applied on some small number of underlying founder haplotypes. This can be formalized as a simple hidden Markov model. The model has transitions along each founder and between the founders. A haplotype is generated along the transition paths: at each state of the model some allele is emitted, according to the emission probability distribution of the state. Transitions are taken according to the associated distributions, the transitions between different founders (i.e., transitions with low probability) indicating recombination events.

To solve the phasing problem for a given set of genotypes, we learn a maximum likelihood hidden Markov model from the genotype data, and then for each genotype in the data we find a resolving pair of haplotypes that has the highest probability in this model. In practice we use the EM algorithm for estimating the parameters of the model and the Viterbi algorithm for finding the resolving haplotype pairs. We need to modify the standard versions of these algorithms [11], as the data does not contain haplotypes but unphased genotypes.

We have tested the method on some real datasets and compared its performance to the state-of-art phasing softwares PHASE [5] (version 2.1.1), HAP [12] (version 3.0), and GERBIL [10]. A prototype implementation of our method, called HIT (a Haplotype Inference Technique), gives results that are always best or second best among the methods compared, when the phasing accuracy is measured by using the switch distance [13]. PHASE is the strongest competitor but it is clearly slower than our method.

2 A Hidden Markov Model for Recombinant Haplotypes

We consider m SNP markers from the same chromosome, numbered $1, \ldots, m$ from left to right in the physical order of appearance along the chromosome. Let A_j be the set of possible alleles (values) of marker j. Then a *haplotype* is a sequence in $A_1 \times \ldots \times A_m$. A *genotype* is an unphased haplotype pair and can be defined as a sequence in $A'_1 \times \ldots \times A'_m$, where each $A'_j = A_j \times A_j$. A genotype g is *homozygous* at marker j, if $g_j = (x, y)$ and $x = y$, and *heterozygous* if $x \neq y$. We use the encoding $A_j = \{1, 2\}$ where 1 and 2 refer, respectively, to the most frequent and the second frequent allele of the SNP j.

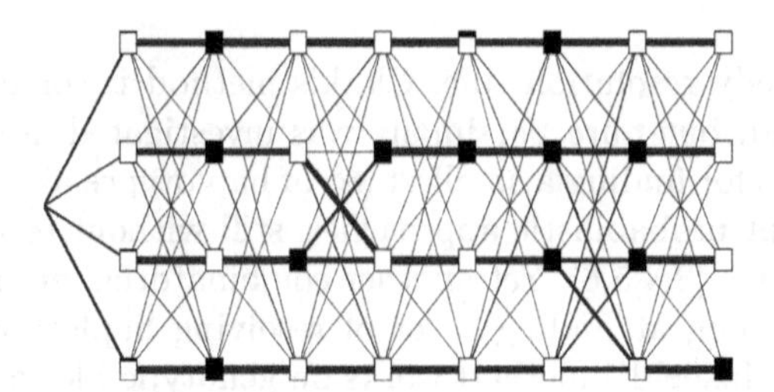

Fig. 1. An example HMM for $m = 8$ markers and $K = 4$ founders. The states are represented as boxes, the jth column of four states corresponding to the jth marker. The black area within each state encodes the emission probability of allele 1 from that state (each marker has only two alleles). The thickness of each transition line encodes the corresponding transition probability.

Our hidden Markov model (HMM) model is a pair $M = (S, \theta)$ where S is the set of states and $\theta = (\tau, \varepsilon)$ consists of the state transition probabilities τ, and the allele emission probabilities ε. The set of states $S = \{s_0\} \cup S_1 \cup \ldots \cup S_m$ consists of disjoint sets S_j, the states at marker j. The transition probabilities $\tau(s_{j-1}, s_j)$ are defined for all $s_{j-1} \in S_{j-1}$ and $s_j \in S_j$, i.e., only transitions from states in S_{j-1} to states in S_j are allowed for all $j = 1, \ldots, m$. The transition probabilities from each fixed s_j form a probability distribution, i.e., their sum equals 1. Each state $s_j \in S_j$ has a probability distribution emitting the alleles in A_j, i.e., probability $\varepsilon(s_j, a)$ of emitting $a \in A_i$. We restrict our consideration to the case that all sets S_j contain the same number K of states. The parameter K, called the *number of the founders* of M and the number m of the markers determine the topology of the HMM. The *initial state* s_0 is a dummy state from which the HMM does not emit any letter. Any path from the dummy state to a state in S_m generates a haplotype in $A_1 \times \ldots \times A_m$, with a probability determined as the product of the transition and emission probabilities along the path.

Our HMM can also handle missing data. We assume that the unobserved values are missing at random, i.e., the fact that a value is unobserved provides no information about the underlying allele; if a data point is missing, the probability of emitting is considered to be 1.

The connection to the idea of the founder sequences is as follows: The current haplotype sequences are seen as results of iterated recombinations on the haplotypes of some ancient founder population whose offspring the observed population is. The current sequences should therefore be built of fragments of the founder sequences, and some such preserved fragments should be seen in several current sequences. Our model M represents the current sequences, based on K founders. A high transition probability $\tau(s_{j-1}, s_j)$ suggests that states s_{j-1} and s_j refer to the same haplotype, i.e., there is a conserved piece of some founder. A low transition probability suggests a cross-over (recombination) between the two states.

A HMM with similar topology appears in [14,15]. Our HMM can also be seen as a probabilistic generalization of the combinatorial approach of [16] to parse haplotypes with respect An example of our model is given in Figure 1.

3 HMM Estimation and Haplotype Reconstruction

In this section we show how, given a set of unphased genotypes, the popular expectation-maximization algorithm can be efficiently applied to maximum likelihood estimation of the parameters of the hidden Markov model. We also show how the estimated HMM can be used for haplotype reconstruction.

3.1 The EM Algorithm for Maximum Likelihood Estimation

We use the maximum likelihood principle to fit our hidden Markov model to the observed genotype data G we want to phase. That is, for a fixed number of founders, we search for the parameters $\theta = (\tau, \varepsilon)$ so as to maximize the likelihood $P(G \mid \theta)$. This estimation problem is known to be hard in general HMMs, and this seems to be the case also in our application. Therefore we resort to the commonly adopted family of expectation-maximization (EM) algorithms, which are guaranteed to converge to a *local* optimum [17].

The generic EM algorithm approaches an intractable optimization problem by completing the original data with auxiliary hidden data. Then the expected log-likelihood of the complete data – where the expectation is with respect to the distribution of the hidden data given the current parameter values – is maximized in an iterative fashion. Usually the choice of the hidden data is natural and direct from the problem. For the standard HMMs the hidden data contains the unobserved hidden states.

In our case it is natural to treat the hidden state sequences, two per genotype, as the hidden data. This is, in essence, the choice that has been made in a number of related applications of EM to the haplotype reconstruction problem; e.g., [8,9,10]. While this approach works nicely when a state is deterministically related to an allele, computational problems will arise as soon as emission parameters are included in the model [9]. In such a case Kimmel and Shamir [9,10] use a (multivariate) numerical maximization routine within each EM iteration.

We propose an alternative instantiation of the EM algorithm that yields efficient closed-form expressions for the maximizing parameter values within each EM iteration. The idea is simple: in the hidden data we include not only the hidden states but also indicators which for any pair of states and the corresponding observed pair of alleles determine which one of the two states emitted the first allele in the pair, the second allele being emitted by the other state. We next provide some technical details.

3.2 Hidden Data and the Maximization Step

Let $G = \{g_1, \ldots, g_n\}$ be a set of n genotypes over m markers. We suppose the topology (the state space S) of our HMM $M = (S, \theta)$ is fixed and we wish to find parameter values $\theta = (\tau, \epsilon)$ that maximize the probability of the genotype data, $P(G \mid \theta)$.

In this setting, the EM algorithm is as follows. Starting from some initial values $\theta^{(0)}$ the algorithm iteratively improves the current values $\theta^{(r)}$ by setting

$$\theta^{(r+1)} := \arg\max_{\theta} \sum_{Z} P(Z \mid G, \theta^{(r)}) \ln P(G, Z \mid \theta), \tag{1}$$

where Z runs through a chosen set of additional (hidden) data. In words, the new parameter values are obtained by maximizing the expected log-likelihood of the complete data. For a large enough r the increment in the likelihood becomes negligible and the algorithm terminates.

We choose the hidden data Z such that the complete likelihood $P(G, Z \mid \theta)$ factorizes into a product of individual transition and emission probabilities, as described below. This is the key to obtain a computationally efficient evaluation of (1). Recall that our HMM $M = (S, \theta)$ defines a probability distribution over singleton haplotypes. A genotype is obtained as a pair of two independent haplotypes, each generated by M along a path through some m states of M. From this generative model we extract the hidden data Z as the the combination of (a) the two state sequences per observed genotype and (b) the alleles emitted from the states.

The paths are given by an $n \times m \times 2$ matrix $T = (t_{ijk})$ of states of M. The entry $t_{ijk} \in S_j$ gives the state from which the jth allele for the first ($k = 1$) or the second ($k = 2$) haplotype for building g_i is to be emitted. The emitted allele from the possible alternatives that are consistent with g_i is indicated by an $n \times m \times 2$ matrix $U = (u_{ijk})$. The entries of U are selector variables that take values in $\{1, 2\}$. Recall that g_i consists of observed genotypes $g_{i1}, \ldots, g_{im}$ over the m markers, each genotype being a pair $g_{ij} = (g_{ij1}, g_{ij2})$ of alleles; note that we do not know which of the two alleles comes from which of the two underlying haplotypes. Here we only have arbitrarily fixed the order of the two observations. Element u_{ijk} of U specifies the jth allele of the first ($k = 1$) or of the second ($k = 2$) haplotype for building g_i: if $u_{ijk} = 1$ then the allele is g_{ij1} and if $u_{ijk} = 2$ then the allele is g_{ij2}. Both alleles must always be used, so we require that $\{u_{ij1}, u_{ij2}\} = \{1, 2\}$.

The point in introducing the hidden data $Z = (T, U)$ is that the complete likelihood factorizes into

$$P(G, T, U \mid \theta) = \left(\frac{1}{2}\right)^n \prod_{i=1}^{n} \prod_{j=1}^{m} \prod_{k=1,2} \tau(t_{i(j-1)k}, t_{ijk}) \varepsilon(t_{ijk}, g_{iju_{ijk}}) .$$

Here the coefficient $(1/2)^n$ appears, since all the 2^n values for U are a priori equally likely (independently of θ). Thus, the expected log-likelihood is

$$\sum_{T,U} P(T, U \mid G, \theta^{(r)}) \ln P(G, T, U \mid \theta) = \sum_{j=1}^{m} A_j(\tau) + \sum_{j=1}^{m} B_j(\varepsilon) - n \ln 2 ,$$

where

$$A_j(\tau) = \sum_{i=1}^{n} \sum_{k=1,2} \sum_{T,U} P(T, U \mid G, \theta^{(r)}) \ln \tau(t_{i(j-1)k}, t_{ijk}) ,$$

$$B_j(\varepsilon) = \sum_{i=1}^{n} \sum_{k=1,2} \sum_{T,U} P(T, U \,|\, G, \theta^{(r)}) \ln \varepsilon(t_{ij}, g_{iju_{ijk}}) \,.$$

Furthermore, each A_j only depends on the transition probability parameters for transitions from a state in S_{j-1} to a state in S_j. Similarly B_j only depends on the emission probability parameters for states in S_j. Thus, the maximizing parameter values can be found separately for each A_j and B_j.

Standard techniques of constrained optimization (e.g., the general Lagrange multiplier method [18] or the more special Kullback–Leibler divergence minimization approach [17]) now apply. For the transition probabilities $\tau(a, b)$, with $a \in S_{j-1}, b \in S_j$, we obtain the update equation

$$\tau^{(r+1)}(a, b) = c \sum_{i=1}^{n} \sum_{k=1,2} P(t_{i(j-1)k} = a, t_{ijk} = b \,|\, G, \theta^{(r)}) \,, \tag{2}$$

where c is the normalization constant of the distribution $\tau^{(r+1)}(a, \cdot)$. That is, $\tau^{(r+1)}(a, b)$ is proportional to the expected number of transitions from a to b. Note that the hidden data U plays no role in this expression. Similarly, for the emission probabilities $\varepsilon(b, y)$, with $b \in S_j, y \in A_j$, we obtain

$$\varepsilon^{(r+1)}(b, y) = c \sum_{i=1}^{n} \sum_{k=1,2} P(t_{ijk} = b, g_{iju_{ijk}} = y \,|\, G, \theta^{(r)}) \,, \tag{3}$$

where c is the normalization constant of the distribution $\varepsilon^{(r+1)}(b, \cdot)$. That is, $\varepsilon^{(r+1)}(b, y)$ is proportional to the expected number of emissions from b to y. Note that the variable u_{ijk} is free meaning that the expectation is over both its possible values.

3.3 Computation of the Maximization Step

We next show how the well-known forward–backward algorithm of hidden Markov Models [11] can be adapted to evaluation of the update formulas (2) and (3).

Let a_j and b_j be states in S_j. For a genotype $g_i \in G$, let $L(a_j, b_j)$ denote the (left or backward) probability of emitting the initial segment $g_{i1} \ldots g_{i(j-1)}$ and ending at (a_j, b_j) along the pairs of paths of M that start from s_0. It can be shown that

$$\begin{aligned} L(a_0, b_0) &= 1 \qquad \text{and} \\ L(a_{j+1}, b_{j+1}) &= \sum_{a_j, b_j} P(g_{ij} \,|\, a_j, b_j, \varepsilon) L(a_j, b_j) \tau(a_j, a_{j+1}) \tau(b_j, b_{j+1}) \end{aligned} \tag{4}$$

where

$$P(g_{ij} \,|\, a_j, b_j, \varepsilon) = \frac{1}{2} \varepsilon(a_j, g_{ij1}) \varepsilon(b_j, g_{ij2}) + \frac{1}{2} \varepsilon(a_j, g_{ij2}) \varepsilon(b_j, g_{ij1}) \,.$$

(Recall that here we treat g_i as an ordered pair, though the ordering of the alleles is arbitrary.) Then the probability of the genotype g_i is obtained as $P(g_i \mid \theta) = \sum_{a_m,b_m} L(a_m, b_m)P(g_{im} \mid a_m, b_m, \varepsilon)$ and the probability of the entire data set is $P(G|\theta) = \prod_{g_i \in G} P(g_i \mid \theta)$. Note that for each g_i we have its own $L(\cdot, \cdot)$.

Direct evaluation of (4) would use $O(|G| \sum_j |S_j|^4) = O(nmK^4)$ time in total. By noting that

$$L(a_{j+1}, b_{j+1}) = \sum_{a_j} \tau(a_j, a_{j+1}) \sum_{b_j} L(a_j, b_j)P(g_{ij} \mid a_j, b_j, \varepsilon)\tau(b_j, b_{j+1})$$

and by storing the sum $\sum_{b_j} L(a_j, b_j)\tau(b_j, b_{j+1})$ for each a_j and b_{j+1} the running time reduces to $O(nmK^3)$. The space requirement is $O(mK^2)$.

We call $L(\cdot, \cdot)$ the forward (or left) table. Similarly, we define the backward (or right) table $R(\cdot, \cdot)$. For a genotype $g_i \in G$, let $L(a_j, b_j)$ denote the probability of emitting the end segment $g_{i(j+1)} \cdots g_{im}$ along the pairs of paths of M that visit (a_j, b_j).

We are now ready to show how formulas (2) and (3) can be evaluated. We consider the latter formula; the former is handled similarly. First notice that it is sufficient to consider evaluation of

$$P(t_{ijk} = b, g_{iju_{ijk}} = y \mid G, \theta^{(r)}) = P(t_{ijk} = b, g_{iju_{ijk}} = y, g_i \mid \theta^{(r)}) \big/ P(g_i \mid \theta^{(r)}) .$$

We already described a way to compute the denominator. The numerator can be written as

$$\sum_{a_j} \sum_{u_{ijk}=1,2} I(g_{iju_{ijk}} = y)\frac{1}{2}L(a_j, b)\varepsilon(a_j, g_{iju_{ijk}})\varepsilon(b, g_{ij(3-u_{ijk})})R(a_j, b) ,$$

where $I(\cdot)$ is the 0, 1-valued indicator function. Note that both u_{ijk} and $3 - u_{ijk}$ take values in $\{1, 2\}$. For update (3) a similar forward–backward expression is found. Thus, the total time complexity of an EM iteration is the above given $O(nmK^3)$.

3.4 Initialization and Model Training

As the EM algorithm is guaranteed to find only a local optimum, it is important to find a good initial configuration of the model parameters. Our initialization routine greedily finds a promising region in the parameter space. It consists of three steps.

First, we fix the transition probabilities and emission probabilities without looking at the data, as follows. Let $s_{j1}, \ldots, s_{jK}$ be the states in S_j. For the first transition we set $\tau(s_0, s_{1l}) = 1/K$ for $l = 1, \ldots, K$. Then for each $j = 1, \ldots, m$, we set $\tau(s_{(j-1)l}, s_{1l'})$ to $1 - \rho$, if $l = l'$, and to $\rho/(K - 1)$ otherwise. The emission probabilities for $s_j \in S_j$ are initialized by setting $\varepsilon(s_j, b) = 1 - \nu$ for a selected major allele b specific to s_j, and $\varepsilon(s_j, a) = \nu/(|A_j| - 1)$ for the other alleles $a \neq b$.

Second, we select the major alleles in a greedy manner based on the observed data. We traverse the sets S_j from left to right and assign to the states in S_j the major alleles that locally maximize the likelihood of the initial segments of G up to marker j. This is done by simply trying all $|A_j|^K$ possible choices. Using dynamic programming the pass takes time $O(nmK^3 2^K)$ for SNP markers. We then make another pass from left to right and again choose the locally optimal major alleles but now in the context of the current solution on both sides of S_j.

Finally, the probability distributions are perturbed a bit by multiplying each parameter value by e^X, where X is drawn uniformly from $[-\eta, \eta]$, independently for each parameter, and η is a noise parameter. The perturbed distributions are obtained by normalizing the perturbed values. The constants ρ, ν, and η are specified by the user. In our tests, reported in Section 4, we used $\rho = 0.1$, $\nu = 0.01$, and $\eta = 0.8$.

Starting from the perturbed initial model, we then apply the EM algorithm to find a maximum likelihood HMM for the genotype data G. In practice, we repeat this training scheme several times, and then pick the highest likelihood HMM as the final model from which the haplotypes for G are read, as will be described in Section 3.5.

Another parameter to be fixed is the number of founders, K. Our experiments show that too small a K gives poor results, but as long as K is sufficiently large (for our test data typically K should be at least 5) varying K has a rather small effect on the quality of haplotyping result.

3.5 Reading the Haplotypes from a HMM

We reconstruct from a trained HMM $M = (S, \theta)$ the haplotypes of each $g \in G$ as follows. First we find for g the Viterbi path from M, that is, a pair (p, p') of paths through M such that emitting g from (p, p') has the highest probability among all path pairs (q, q'), i.e.,

$$P(g, p, p' \,|\, \theta) = \max_{(q,q')} P(g, q, q' \,|\, \theta) \,.$$

This can be done by a variant of (4) followed by standard trace-back. Then generate from p a haplotype h and from p' a haplotype h' such that they together give genotype g and $P(h \mid p, \theta)P(h' \mid p', \theta)$ is largest possible. This is simple local maximization at heterozygous markers of g. Haplotypes $\{h, h'\}$ are the reconstructed haplotypes for g according to our method.

4 Test Results

We have implemented the presented phasing method in a prototype program HIT (Haplotype Inference Technique). In this section, we report the results we have obtained on a few real datasets. We compare the performance of HIT against HAP version 3.0 [12], PHASE version 2.1.1[5], and GERBIL [10].

4.1 Datasets

We tested our method on five real datasets. Daly's et al. [19] commonly used benchmark dataset is a sample from a European-derived population and spans a 500-kb region on human chromosome 5q31 which contains a genetic risk factor for Crohn disease. From that area there are genotypes for 103 SNP markers, collected from 129 trios (of mother, father, and child). The trios were used to infer the true haplotypes for the 129 genotypes of the children.

The second dataset is a fragment of the data provided recently by Hinds et al. [20]. This data consists of 71 haplotype pairs over 1,586,383 SNPs that cover the entire human genome. We took the haplotypes for the SNPs 1,000–1,199 of chromosome 2. We notice that these haplotypes were inferred from genotypes with some version of HAP [12].

The rest three datasets are genotype samples over 68 SNP markers from three datasets from Finland [21]. We call these datasets Population1 (32 haplotypes), Population2 (108 haplotypes), and Population3 (108 haplotypes).

The latter four datasets were available in a haplotyped form. For our tests we constructed the input genotypes simply by merging the two haplotypes of each individual.

4.2 Switch Distance

We measure the quality of haplotyping results using the commonly adopted switch distance [13]. Switch distance between two pairs of haplotypes $\{h, h'\}$ and $\{f, f'\}$ is the minimum number of phase shifts needed to turn $\{h, h'\}$ into $\{f, f'\}$. For example if the true haplotypes for a genotype are $\{111111, 222222\}$, then the switch distance is 1 to $\{111222, 222111\}$ and 5 to $\{121212, 212121\}$.

Unfortunately, the basic switch distance is undefined when no phase shifts can transform a haplotype pair into another pair. This may happen when the data has missing values or genotyping errors. For example, suppose the observed genotype is $\{1,2\}\{1,2\}\{-,-\}\{1,2\}\{1,2\}$, where a "$-$" stands for a missing allele. Then it is possible that our model gives haplotypes $\{112111, 222222\}$, thus imputing the missing values. However, if the underlying true pair of haplotypes is $\{111111, 222222\}$, the distance between these haplotype pairs is not defined.

In such a situation one needs a generalized switch distance. Define $errors(\{h, h'\}, \{f, f'\})$ as the minimum number of allele substitutions to $\{f, f'\}$ that are needed to make the switch distance defined, and let J be the markers where no changes are needed. Then our generalized switch distance is defined as $sd'(\{h, h'\}, \{f, f'\}) = errors(\{h, h'\}, \{f, f'\}) + sd_J((\{h, h'\}, \{f, f'\})$ where sd_J is switch distance restricted on J.

Another possibility, used by some authors [13,22], is just to ignore inconsistent markers and report the basic switch distance on the remaining markers; denote this distance by sd''. In our tests, we needed sd' and sd'' only for the Daly et al. dataset [19].

The relative versions of these distances are obtained by dividing by the total number of heterozygote sites of the genotypes minus the number of genotypes.

4.3 Comparison Results

Figure 2 shows how the performance of HIT depends on the number of founders. We see that increasing the number of founders consistently increases the good-

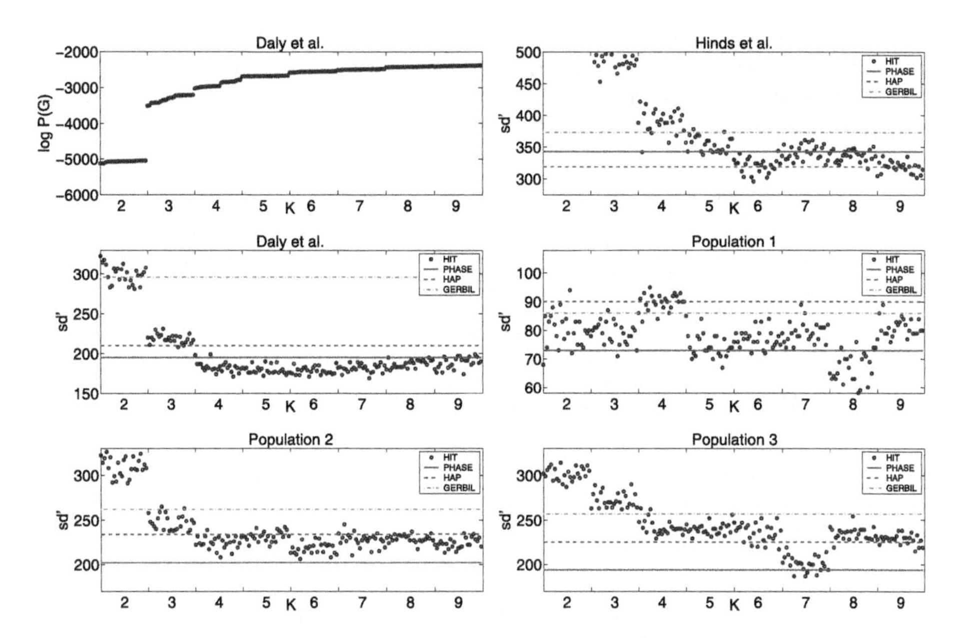

Fig. 2. The phasing accuracy (vertical axis) as a function of the number of founders (horizontal axis) for five real data sets. Shown the achieved total switch distance for 25 random restarts of HIT (in increasing order of likelihood), for 2 to 9 founders. For the Daly et al. data also shown the growth of likelihood (top left); for the other datasets the curves behave similarly (not shown). The results for PHASE [5,22], GERBIL [9,10], and HAP [12], shown as vertical lines, were obtained with their default parameter values.

ness of fit to the data, as expected. However, overfitting does not seem to impede the performance of HIT in phasing. For example, for the Daly et al. data HIT gives the best result consistently for $K \geq 4$.

The effect of starting the EM algorithm from slightly different initial settings is also shown in Figure 2, indicating a fairly robust behaviour. We note that the correlation of the achieved data likelihood and switch distance is surprisingly small. Thus the plain likelihood is perhaps not the best criterion for choosing the model for haplotyping.

Table 1 summarizes the phasing accuracy of HIT when we set the number of founders K to 7, performed 25 restarts of the EM algorithm, and used the highest likelihood HMM for haplotyping. We note that in the Daly et al. dataset the handling of the missing data (sd' or sd'') has a clear effect on the results, yet the relative differences are about the same for both measures. The fact that GERBIL treats an allele pair where one allele is missing as completely missing data, explains its relatively poor performance w.r.t. sd'' on the Daly et al. dataset. We note that HIT always gives the best or second best result.

Table 2 displays the running times of the compared methods. Clearly, GERBIL and HAP are very fast, whereas PHASE becomes rather slow for the largest

Table 1. Phasing accuracy of HIT ($K = 7$ founders), PHASE [5,22], GERBIL [9,10], and HAP [12] on five real data sets, measured using switch distance. For the Daly et al. dataset the first and the second line show switch distance sd'' and sd', respectively; for the other cases all variants coincide. The relative distances are given in parentheses

Dataset	HIT	PHASE	GERBIL	HAP
Daly et al.	**80 (0.021)**	*86 (0.023)*	*86 (0.023)*	89 (0.024)
Daly et al.	**185 (0.049)**	*195 (0.052)*	296 (0.079)	210 (0.056)
Hinds et al.	*329 (0.093)*	343 (0.097)	373 (0.11)	**319 (0.090)**
Population1	*82 (0.24)*	**73 (0.21)**	86 (0.25)	90 (0.26)
Population2	*219 (0.17)*	**202 (0.15)**	262 (0.20)	234 (0.18)
Population3	**194 (0.16)**	**194 (0.16)**	257 (0.22)	225 (0.19)

Table 2. The running time in seconds for HIT ($K = 7$ founders, median over 25 EM restarts), PHASE [5,22], GERBIL [9,10], and HAP [12] on five real data sets. HAP was run on its own server, the other programs on a Pentium IV 3.0 GHz with 1 GB of RAM.

Dataset	HIT	PHASE	GERBIL	HAP
Daly et al.	126	9290	45	25
Hinds et al.	88	71100	52	29
Population1	9	773	19	2
Population2	28	5180	89	7
Population3	29	4520	10	7

datasets. The speed of HIT (Java implementation) per EM restart is comparable to the speed of GERBIL and HAP, but slower when tens of restarts are used. Yet, HIT scales nicely to large datasets, opposite to PHASE.

References

1. Clark, A.G.: Inference of haplotypes from PCR-amplified samples of dipoid populations. Molecular Biology and Evolution **7** (1990) 111–122
2. Gusfield, D.: Haplotype inference by pure parsimony. Technical Report CSE-2003-2, Department of Computer Science, University of California (2003)
3. Excoffier, L., Slatkin, M.: Maximum-likelihood estimation of molecular haplotype frequencies in a diploid population. Molecular Biology and Evolution **12** (1995) 921–927
4. Long, J.C., Williams, R.C., Urbanek, M.: An E-M algorithm and testing strategy for multiple-locus haplotypes. American Journal of Human genetics **56** (1995) 799–810
5. Stephens, M., Smith, N., Donnelly, P.: A new statistical method for haplotype reconstruction from population data. American Journal of Human Genetics **68** (2001) 978–989
6. Niu, T., Qin, Z., Xu, X., Liu, J.: Bayesian haplotype inference for multiple linked single nucleotide polymorphisms. American Journal of Human Genetics **70** (2002) 157–169

7. Gusfield, D.: Haplotyping as perfect phylogeny: conceptual framework and efficient solutions. In: Research in Computational Molecular Biology (RECOMB '02), ACM Press (2002) 166–175
8. Greenspan, G., Geiger, D.: Model-based inference of haplotype block variation. In: Research in Computational Molecular Biology (RECOMB '03), ACM Press (2003) 131–137
9. Kimmel, G., Shamir, R.: Maximum likelihood resolution of multi-block genotypes. In: Research in Computational Molecular Biology (RECOMB '04), ACM Press (2004) 2–9
10. Kimmel, G., Shamir, R.: Genotype resolution and block identification using likelihood. Proceeding of the National Academy of Sciences of the United States of America (PNAS) **102** (2005) 158–162
11. Rabiner, L.R.: A tutorial on hidden Markov models and selected applications in speech recognition. Proceedings of the IEEE **77** (1989) 257–285
12. Halperin, E., Eskin, E.: Haplotype reconstruction from genotype data using imperfect phylogeny. Bioinformatics **20** (2004) 104–113
13. Lin, S., Cutler, D.J., Zwick, M.E., Chakravarti, A.: Haplotype inference in random population samples. American Journal of Human Genetics **71** (2002) 1129–37
14. Schwartz, R., Clark, A.G., Istrail, S.: Methods for inferring block-wise ancestral history from haploid sequences. In: Workshop on Algorithms in Bioinformatics (WABI '02), Springer (2002) 44–59
15. Jojic, N., Jojic, V., Heckerman, D.: Joint discovery of haplotype blocks and complex trait associations from snp sequences. In: Proceedings of the 20th conference on Uncertainty in artificial intelligence (UAI '04), AUAI Press (2004) 286–292
16. Ukkonen, E.: Finding founder sequences from a set of recombinants. In: Algorithms in Bioinformatics (WABI '02), Springer (2002) 277–286
17. McLachlan, G.J., Krishnan, T.: The EM Algorithm and Extensions. John Wiley and Sons (1996)
18. Bertsekas, D.P.: Constrained Optimization and Lagrange Multiplier Methods. Academic Press, New York (1982)
19. Daly, M.J., Rioux, J.D., Schaffner, S.F., et al.: High-resolution haplotype structure in the human genome. Nature Genetics **29** (2001) 229–232
20. Hinds, D.A., Stuve, L.L., Nilsen, G.B., et al.: Whole-genome patterns of common dna variation in three human populations. Science **307** (2005) 1072–1079
21. Koivisto, M., Perola, M., Varilo, T., et al.: An MDL method for finding haplotype blocks and for estimating the strength of haplotype block boundaries. In: Pacific Symposium on Biocomputing (PSB '03), World Scientific (2003) 502–513
22. Stephens, M., Scheet, P.: Accounting for decay of linkage disequilibrium in haplotype inference and missing-data imputation. Americal Journal of Human Genetics **76** (2005) 449–462

Algorithms for Imperfect Phylogeny Haplotyping (IPPH) with a Single Homoplasy or Recombination Event

Yun S. Song, Yufeng Wu, and Dan Gusfield

Department of Computer Science,
University of California, Davis, CA 95616, USA
yssong@cs.ucdavis.edu, wuyu@cs.ucdavis.edu,
gusfield@cs.ucdavis.edu

Abstract. The haplotype inference (HI) problem is the problem of inferring $2n$ haplotype pairs from n observed genotype vectors. This is a key problem that arises in studying genetic variation in populations, for example in the ongoing HapMap project [5]. In order to have a hope of finding the haplotypes that actually generated the observed genotypes, we must use some (implicit or explicit) genetic model of the evolution of the underlying haplotypes. The Perfect Phylogeny Haplotyping (PPH) model was introduced in 2002 [9] to reflect the "neutral coalescent" or "perfect phylogeny" model of haplotype evolution. The PPH problem (which can be solved in polynomial time) is to determine whether there is an HI solution where the inferred haplotypes can be derived on a perfect phylogeny (tree).

Since the introduction of the PPH model, several extensions and modifications of the PPH model have been examined. The most important modification, to model biological reality better, is to allow a *limited* number of biological events that violate the perfect phylogeny model. This was accomplished implicitly in [7,12] with the inclusion of several heuristics into an algorithm for the PPH problem [8]. Those heuristics are invoked when the genotype data cannot be explained with haplotypes that fit the perfect phylogeny model. In this paper, we address the issue *explicitly*, by allowing one recombination or homoplasy event in the model of haplotype evolution. We formalize the problems and provide a polynomial time solution for one problem, using an additional, empirically-supported assumption. We present a related framework for the second problem which gives a practical algorithm. We believe the second problem can be solved in polynomial time.

1 Introduction

In diploid organisms (such as humans) there are two (not completely identical) "copies" of each chromosome, and hence of each region of interest. A description

* Research partially supported by grant EIA-0220154 from the National Science Foundation.

R. Casadio and G. Myers (Eds.): WABI 2005, LNBI 3692, pp. 152–164, 2005.

of the data from a single copy is called a *haplotype*, while a description of the conflated (mixed) data on the two copies is called a *genotype*. In complex diseases (those affected by more than a single gene) it is often much more informative to have haplotype data (identifying a set of gene alleles inherited together) than to have only genotype data.

Today, the underlying data that forms a haplotype is usually a vector of values of m *single nucleotide polymorphisms (SNP's)*. A SNP is a single nucleotide site where exactly two (of four) different nucleotides occur in a large percentage of the population. Genotype data is represented as an n by m 0-1-2 (ternary) matrix G. Each row is a genotype. A pair of binary vectors of length m (haplotypes) *generate* a row i of G if for every position c both entries in the haplotypes are 0 (or 1) if and only if $G(i, c)$ is 0 (or 1) respectively, and exactly one entry is 1 and one is 0 if and only if $G(i, c) = 2$. The international Haplotype Map Project [5] is focused on determining the common SNP haplotypes in several diverse human populations.

Given an input set of n genotype vectors of length m, the *Haplotype Inference (HI) Problem* is to find a set of n pairs of binary vectors (with values 0 and 1), one pair for each genotype vector, such that each genotype vector is generated by the associated pair of haplotypes. The ultimate goal is to computationally infer the true haplotype pairs that generated the genotypes. This would be impossible without the implicit or explicit use of some genetic model to guide the algorithm in constructing a solution. A powerful genetic model that has been used in the HI problem is the population-genetic concept of a *coalescent* [14,21].

The coalescent model of SNP haplotype evolution implies that the evolutionary history of $2n$ haplotypes, one from each of $2n$ individuals, can be displayed as a rooted tree T with $2n$ leaves, where some ancestral sequence labels the root of the tree, and where each of the m sites labels *exactly* one edge of the tree. A label i on an edge indicates the (unique) point in history where a mutation at site i occurred. Sequences evolve down the tree, starting from the ancestral sequence, changing along a branch $e = (u, v)$ by changing the state of any site that labels edge e. The state changes from what it is at u to the opposite state, recorded at v. The tree "generates" the resulting sequences that appear at its leaves. In more computer science terminology, the coalescent model says that $2n$ haplotype (binary) sequences that appear at the leaves of T are generated on a *perfect phylogeny*. See [9] for further explanation and justification of the perfect phylogeny haplotype model.

Generally, most solutions to the HI problem will not fit a perfect phylogeny, and this leads to **The Perfect Phylogeny Haplotyping (PPH) Problem**: Given an n by m matrix M that holds n genotypes from m sites, find n pairs of haplotypes that generate M on a perfect phylogeny.

It is the requirement that the haplotypes fit a perfect phylogeny, and the fact that most solutions to the HI problem will not, that enforce the coalescent model of haplotype evolution, and make it plausible that a solution to the PPH problem (when there is one) is biologically meaningful.

There are several polynomial-time solutions to the PPH problem [2,4,8,9] and a linear-time algorithm [6]. An additional empirical result that will be exploited in this paper is that when there is a PPH solution, it is highly likely that there is only a single, unique, PPH solution [3,4]. The frequency of uniqueness increases with the number of genotypes, and a surprisingly small number of genotypes is needed before the solution is highly likely to be unique.

Since the introduction of the PPH model, a central goal has been to extend the range of applicability of model by incorporating more biological complexity, yet preserving polynomial-time solvability. Biologically, the most important extension is to allow a limited amount of recombination or homoplasy (recurrent or back mutation) in the model of haplotype evolution. Homoplasy allows a site to mutate more than once, and hence allows a site to label more than one edge in the tree. As before, if a site i labels the directed edge (u, v), then the state of site i at u mutates to the opposite state, which is the state of i at v. If the two occurrences of site i are on the same path from the root of T, then the second occurrence is a "back mutation", otherwise it is a "recurrent mutation".

An *H-1 Phylogenetic Tree* T is derived from a perfect phylogeny by allowing exactly one site to label two distinct edges of T. An H-1 phylogenetic tree generates M if the sequences in M label the leaves of T.

A single-crossover recombination between two equal-length sequences P and S creates a third "recombinant" sequence of the same length consisting of a prefix of P followed by a suffix of S. The point where the recombinant sequence changes from P to S is called the "crossover" or "break" point. Recombination occurs during meiosis (and in other contexts) and is a primary mechanism creating genomic diversity in a population.

An *R-1 Phylogenetic Network* N is derived from a perfect phylogeny by allowing one recombination event, represented at a node v of N by two edges entering v, one from the node labeled by sequence P, and one labeled by sequence S. The crossover point is also noted at v. a recombination node (e.g., to have two incoming edges), where a single-crossover recombination occurs. An R-1 phylogenetic network generates M if the sequences of M label the terminal nodes (nodes with no descendants) of N. An R-1 phylogenetic network can also be described as a galled-tree with exactly one gall [11].

Given an input set of genotypes M, we define two problems.

1. The H-1 Imperfect Phylogeny Haplotyping (IPPH) Problem: Find an H-1 Phylogenetic Tree generating haplotypes that solve the HI problem for M.
2. The R-1 Imperfect Phylogeny Haplotyping (IPPH) Problem: Find an R-1 phylogenetic Network generating haplotypes that solve the HI problem for M.

In this paper, we develop algorithms for both problems. Both solutions first solve a PPH problem for a subset of the data, and we will assume (following the observations in [3,4]) that those solutions are unique. Given that assumption, our solution to the H-1 IPPH problem runs in polynomial time. We have implemented our H-1 IPPH algorithm in C++ and evaluated its performance using simulated data. As we elaborate later, our study shows that our method is both practical and highly accurate. We are currently working on extending our method to IPPH with multiple homoplasy events. Our present solution for the R-1 IPPH problem takes exponential time, but a polynomial-time algorithm for that approach looks promising, and we believe a related, more complex, method runs in polynomial time.

In what follows, we use M to denote an $n \times m$ genotype matrix. Its haplotype matrix, of size $2n \times m$, is denoted by M'. Following [2], we say that two rows in M' are *mates* if they come from a single row in M.

2 IPPH with a Single Homoplasy Event

In this section, we construct an IPPH method that allows for exactly one back or recurrent mutation. Suppose that a genotype matrix M does not admit a PPH solution. If that is due to a single homoplasy event at a particular site, then removing the column in M corresponding to that site will render the remaining data M_r compatible with a perfect phylogeny, and therefore there will be a PPH solution for M_r. In our work, we consider partitioning M column-wise into two parts, one (denoted M_s) containing exactly one column of M and the other (denoted M_r) the rest. We denote this partitioning by $M \longmapsto M_r \oplus M_s$. Our algorithm proceeds by trying all such partitions of M and checking whether M_r admits a PPH solution (i.e. a perfect phylogeny T), and, if so, whether the single column in M_s can be explained by a single homoplasy event in T, possibly after some refinement. If these conditions are satisfied by at least one partition of M, then we say that there exists a solution to IPPH with a single homoplasy event.

2.1 H-1 IPPH When There Exists a Unique PPH Solution for M_r

In what follows, we describe our main ideas through an explicit example. Consider the genotype matrix M shown on the left hand side of Figure 1. The only partition of that genotype matrix that leads to a PPH solution for M_r is the one shown on the right hand side of Figure 1. In fact, that M_r admits a unique PPH solution M'_r, shown on the left hand side of Figure 2; a PPH solution M'_s for M_s is shown there as well. The question that remains is whether we can appropriately combine M'_r with M'_s to create an H-1 IPPH solution M' for the entire genotype matrix M; i.e., for each $1 \leq i \leq n$, we want to ask whether we

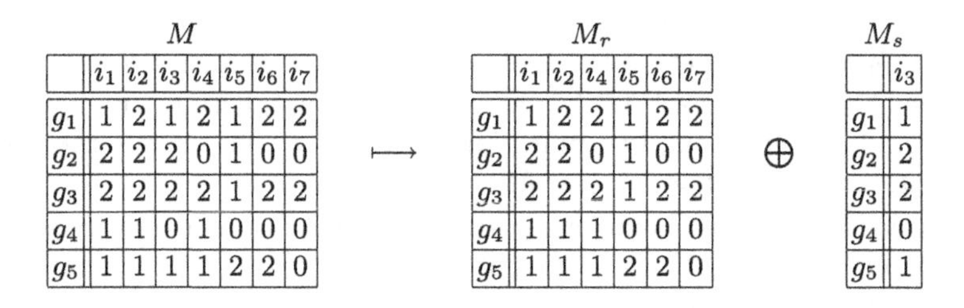

M

	i_1	i_2	i_3	i_4	i_5	i_6	i_7
g_1	1	2	1	2	1	2	2
g_2	2	2	2	0	1	0	0
g_3	2	2	2	2	1	2	2
g_4	1	1	0	1	0	0	0
g_5	1	1	1	1	2	2	0

$\longmapsto$

M_r

	i_1	i_2	i_4	i_5	i_6	i_7
g_1	1	2	2	1	2	2
g_2	2	2	0	1	0	0
g_3	2	2	2	1	2	2
g_4	1	1	1	0	0	0
g_5	1	1	1	2	2	0

$\oplus$

M_s

	i_3
g_1	1
g_2	2
g_3	2
g_4	0
g_5	1

Fig. 1. Partition of M into M_r and M_s, where M_s contains column i_3

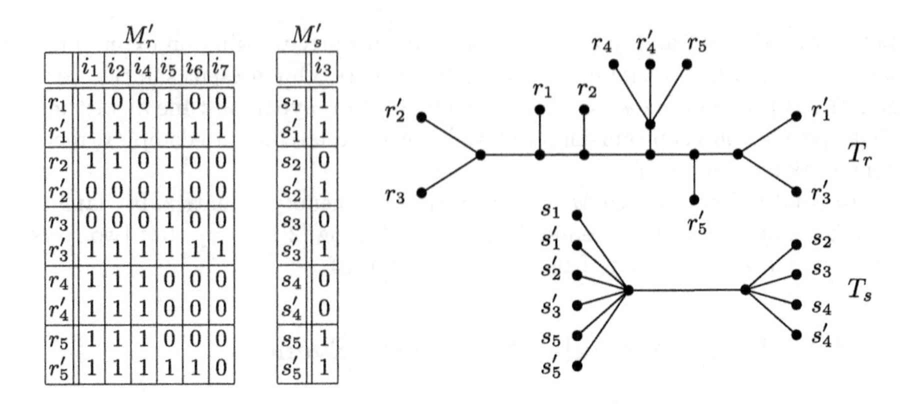

M'_r

	i_1	i_2	i_4	i_5	i_6	i_7
r_1	1	0	0	1	0	0
r'_1	1	1	1	1	1	1
r_2	1	1	0	1	0	0
r'_2	0	0	0	1	0	0
r_3	0	0	0	1	0	0
r'_3	1	1	1	1	1	1
r_4	1	1	1	0	0	0
r'_4	1	1	1	0	0	0
r_5	1	1	1	0	0	0
r'_5	1	1	1	1	1	0

M'_s

	i_3
s_1	1
s'_1	1
s_2	0
s'_2	1
s_3	0
s'_3	1
s_4	0
s'_4	0
s_5	1
s'_5	1

Fig. 2. Separate PPH solutions and perfect phylogenies for M_r and M_s in Figure 1

can appropriately order the rows s_i, s'_i in M'_s with respect to the rows r_i, r'_i in M'_r, such that the combined matrix is an H-1 IPPH solution for M. For each $1 \leq i \leq n$, row r_i can get paired with either s_i or s'_i. Therefore, if r_i and r'_i are distinct, and so are s_i and s'_i, for all $1 \leq i \leq n$, then there are 2^n possible ways of pairing the rows. Hence, checking whether there exists an H-1 IPPH solution for each way of pairing would be impractical when n is large. The approach we take is to work not with haplotype matrices directly but with perfect phylogenies. The problem of finding an H-1 IPPH solution therefore translates to a graph theoretical problem.

Returning to our example, consider the perfect phylogenies shown on the right hand side of Figure 2. Trees T_r and T_s correspond to the PPH solutions M'_r and M'_s, respectively. To create an H-1 IPPH solution for the entire genotype matrix M, we need to combine the information contained in T_r with that in T_s, but, before we can do that, we first need to identify the leaf labels r_i, r'_i in T_r with the leaf labels s_i, s'_i in T_s. There are $O(2^n)$ ways to do this in general. But, an important observation allows us to avoid considering all $O(2^n)$ pairings of leaf labels explicitly. Because we do not know *a priori* how the leaf labels r_i, r'_i in T_r should be paired up with the leaf labels s_i, s'_i in T_s, we can actually use that freedom to set r_i, r'_i, s_i, s'_i equal to a new label, say x_i, and study the re-labeled trees $\widetilde{T}_r$ and $\widetilde{T}_s$ to see whether there exists an H-1 IPPH solution; an H-1 IPPH solution exists if having two mutation events in $\widetilde{T}_r$ can induce the same bipartition of the *multiset* $\{x_1, x_1, x_2, x_2, \ldots, x_n, x_n\}$ as that captured by the tree topology of $\widetilde{T}_s$. Once the location of the two mutation events in $\widetilde{T}_r$ is determined, we can go back to T_r and determine the phase of the entire data M; i.e., the location of the two mutation events in T_r determines the order of s_i, s'_i in M_s with respect to r_i, r'_i in M_r.

Equivalently, if $\widetilde{T}_r$ is a binary tree, an H-1 IPPH solution exists if there exist two edges e_1 and e_2 in $\widetilde{T}_r$ that are not incident with a common vertex, such that removing those edges partitions $\widetilde{T}_r$ into three subtrees, of which two are non-adjacent, with the following properties:

(i) The multiset union of leaf labels for the two non-adjacent subtrees—i.e., those subtrees not joined by e_1 or e_2 in $\widetilde{T}_r$—is equal to the multiset of leaf labels on one side of the unique interior edge e_I in $\widetilde{T}_s$, and

(ii) the multiset of leaf labels for the remaining subtree is equal to that on the other side of e_I in $\widetilde{T}_s$.

Whether there exists such a pair of edges can easily be answered in polynomial time. If the answer is affirmative, then the phasing of the single column in M_s can be determined, up to exchange of 0s with 1s in the entire column, by looking at the topology of T_r and the position of the two mutation events in T_r.

If $\widetilde{T}_r$ is not binary, things are more complicated. For ease of discussion, we introduce the following definition (see [17] for graph theoretical terminology):

Definition 1 (Cut-subtree). *By a cut-subtree τ of a leaf-labeled unrooted tree T, we mean a subtree of T that can be obtained by removing an edge e in T, or by first refining a vertex of degree ≥ 4 and then removing the newly created edge e. The remaining part of T, after deleting any degree-2 vertex created by removing e, is denoted $T \setminus \tau$.*

If $\widetilde{T}_r$ is not binary, we need to ask whether there exist two disjoint *cut-subtrees* τ_1 and τ_2 of $\widetilde{T}_r$, such that

(i′) the multiset union of leaf labels for τ_1 and τ_2 is equal to that for one side of e_I in $\widetilde{T}_s$, and

(ii′) the remaining part of T_r has a label multiset equal to that for the other side of e_I in $\widetilde{T}_s$.

A polynomial-time algorithm for finding such cut-subtrees, if they exist, is described in Section 2.2. For now, we return to the simple example in Figure 2. It

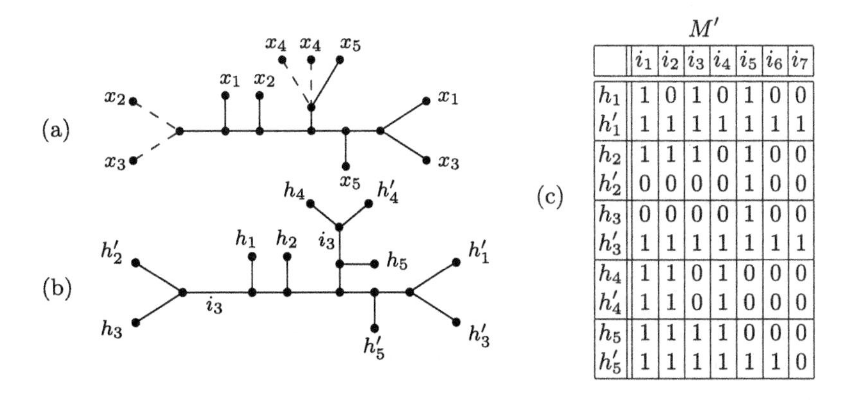

	i_1	i_2	i_3	i_4	i_5	i_6	i_7
h_1	1	0	1	0	1	0	0
h'_1	1	1	1	1	1	1	1
h_2	1	1	1	0	1	0	0
h'_2	0	0	0	0	1	0	0
h_3	0	0	0	0	1	0	0
h'_3	1	1	1	1	1	1	1
h_4	1	1	0	1	0	0	0
h'_4	1	1	0	1	0	0	0
h_5	1	1	1	1	0	0	0
h'_5	1	1	1	1	1	1	0

Fig. 3. (a) Cut-subtrees, denoted by dashed lines, in $\widetilde{T}_r$ that satisfy properties (i′) and (ii′). (b) Imperfect phylogenetic tree with two mutations for column i_3. The two edges on which mutations for column i_3 occur are labeled i_3. (c) The corresponding H-1 IPPH solution for the entire genotype matrix M.

is easy to see that there does not exist a pair of edges in $\widetilde{T}_r$ such that properties (i) and (ii) shown above are both satisfied. However, there exist two cut-subtrees, shown in Figure 3a, that satisfy properties (i′) and (ii′). Having identified the appropriate cut-subtrees, we can now go back to the original tree T_r and determine the phase of M_s. An imperfect phylogenetic tree with two mutations for column i_3 and the corresponding H-1 IPPH solution are shown in Figures 3b and 3c, respectively, where haplotype mates are now labeled h_i and h'_i. Note that the H-1 IPPH solution is unique. In terms of pairing the rows in M'_r and M'_s, the H-1 IPPH solution corresponds to pairing s_2 with r'_2 (and hence s'_2 with r_2) and s_3 with r_3 (and hence s'_3 with r'_3). In this example, the imperfect phylogeny for M' is binary. In general, an imperfect phylogeny for M' may still contain vertices of degree greater than 3.

2.2 Algorithm for Finding Appropriate Cut-Subtrees in Non-binary Trees for a Single Homoplasy Event

Suppose that $\widetilde{T}_r$ is non-binary. In what follows, the reader should refer to Figure 4 for illustration of notation. If we remove two existing edges E_1 and E_2 from $\widetilde{T}_r$, then that defines two non-adjacent subtrees T_1 and T_2. Our goal is to check whether we can choose a set of edges from $e_1, \ldots, e_p$ to create a cut-subtree τ_1 and, similarly, choose a set of edges from $e_{p+1}, \ldots, e_q$ to create a cut-subtree τ_2, such that properties (i′) and (ii′) in Section 2.1 are satisfied. Below we provide a polynomial-time algorithm for finding all possible such pairs of disjoint cut-subtrees, when they exist.

Our algorithm is based on coloring a graph G whose q vertices $v_1, \ldots, v_q$ are in one-to-one correspondence with $e_1, \ldots, e_q$; v_i in G is related to e_i in $\widetilde{T}_r$. Edges in G will be defined shortly. A coloring procedure may terminate before reaching the end, if inconsistency is encountered; i.e., if a vertex is assigned more than one color. If G is colored consistently, the final coloring of the vertices in G determines which of $e_1, \ldots, e_q$ should be chosen to construct the desired cut-subtrees τ_1 and τ_2. In our convention, if vertex v_i is colored "red," then it means we should "take" e_i. If it is colored "black," then we should "not take" e_i.

We use $\mathcal{L}(T)$ to denote the set of leaf labels in T. More generally, $\mathcal{L}(T)$ is a multiset when we consider trees with duplicate labels. Let $X|Y$ be the bipartition of $\mathcal{L}(\widetilde{T}_s)$ defined by the single interior edge in $\widetilde{T}_s$, and recall that $\widetilde{T}_r$ and $\widetilde{T}_s$ carry duplicate leaf labels. Our algorithm is as follows.

1. Choose a pair of existing edges E_1, E_2 in $\widetilde{T}_r$ that has not been tried so far. If no such choice remains, terminate the algorithm.
2. Remove E_1, E_2 to partition $\widetilde{T}_r$ into three subtrees. Let T_1, T_2 be the two non-adjacent subtrees as depicted in Figure 4. If $\mathcal{L}(T_1) \cup \mathcal{L}(T_2)$ contains either X or Y, or both, create q vertices $v_1, \ldots, v_q$, and go to next step. If not, go back to step 1; no solution is possible for the current choice of E_1, E_2.
3. For each $Z \in \{X, Y\}$ satisfying $Z \subset \mathcal{L}(T_1) \cup \mathcal{L}(T_2)$, check whether the following conditions hold:

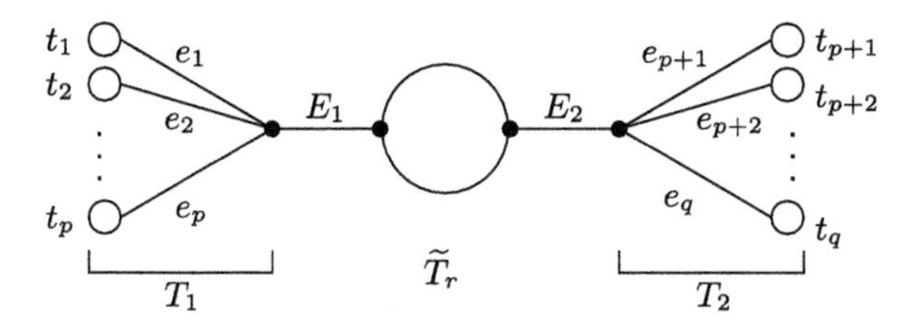

Fig. 4. Schematic depiction of $\widetilde{T}_r$. Here, t_i denote subtrees and the big circle in the center schematically represents the rest of $\widetilde{T}_r$.

   (a) If x appears only once in Z, then there does not exist a t_k in T_1 or T_2 such that $\mathcal{L}(t_k)$ contains two xs.
   (b) For every x that appears exactly once in Z, create an edge between v_i and v_j if $x \in \mathcal{L}(t_i)$ and $x \in \mathcal{L}(t_j)$. Let G_Z denote the resulting graph. Then, every non-trivial connected component of G_Z is bipartite[1].

   If no Z satisfies the above conditions, go back to step 1; there is no solution for the current choice of E_1, E_2. Otherwise, pass G_Z to the next step.
4. For each G_Z passed, check whether it is possible to color the vertices in G_Z as describe below without encountering inconsistency.
   (a) If $x \notin Z$, find all t_k such that $x \in \mathcal{L}(t_k)$ and color v_k black.
   (b) If x occurs twice in Z, find all t_k such that $x \in \mathcal{L}(t_k)$ and color v_k red.
   (c) If x occurs only once in Z and there exists exactly one t_k such that $\mathcal{L}(t_k)$ contains a single x, color v_k red.

   If no G_Z admits consistent coloring, go back to step 1. Otherwise, pass consistently colored G_Z to the next step.
5. For each non-trivial connected component of G_Z, see whether any of the vertices has been colored. If so, color the remaining uncolored vertices, if there are any, in that connected component to respect the bipartite structure (i.e. red on one side and black on the other). If this is not possible, return to step 1.
6. If there are k totally-uncolored non-trivial connected components of G_Z, then there are 2^k ways to color them consistently, and hence there are 2^k solutions for the current choice of E_1, E_2. Go back to step 1.

The above algorithm can be implemented to run in $O(n^4)$ time, where $2n$ is the number of leaves.

2.3 Empirical Results

We implemented our H-1 IPPH algorithm in C++ and compared its performance on simulated data with that of PHASE [20]. The input datasets were generated as follows. We first used Hudson's program MS [15] to generate homoplasy-free haplotype datasets satisfying the 5% rule[2] described in [4]. To introduce

[1] If a non-trivial connected component is not bipartite, then it has an odd-length cycle, which leads to inconsistent coloring.

[2] The 5% rule is a biologically relevant restriction that every column in the haplotype matrix has minor allele frequency $\geq 5\%$.

Table 1. Comparison of our H-1 IPPH method with PHASE for genotype matrices of size $n \times m$. Shown here are average accuracy measures and average running time (on a 2.8 GHz Pentium PC) per dataset, based on 100 datasets of each size. Our method seems comparable to PHASE in accuracy, while being significantly faster than PHASE.

	Our H-1 IPPH method				PHASE			
	50×50	100×50	50×100	100×100	50×50	100×50	50×100	100×100
Standard error	0.005	0.005	0.006	0.003	0.006	0.002	0.007	0.001
Switch accuracy	0.999	0.999	1.000	1.000	0.998	0.999	0.999	1.000
% of misphased 2s	0.03%	0.03%	0.02%	0.01%	0.07%	0.02%	0.03%	0.01%
Running time	0.22s	0.41s	1.52s	3.09s	14.2s	27.7s	43.6s	85.8s

Table 2. Performance of our H-1 IPPH method in more detail. The number of datasets shown is out of 100.

	50×50	100×50	50×100	100×100
# of datasets admitting PPH solutions	20	19	16	15
# of datasets admitting H-1 IPPH solutions (with a unique PPH solution for M_r)	80 (80)	81 (81)	84 (84)	85 (85)
Frequency of correctly identifying the homoplasy column when M admits no PPH solution	95%	98%	96%	98%

a homoplasy event, we randomly chose two distinct edges (on which mutations occur) in the underlying rooted genealogical tree of each haplotype dataset, with the probability of choosing an edge being proportional to its length. This process was repeated for each dataset until a homoplasy site satisfying the 5% rule got generated. We then randomly inserted the so obtained homoplasy column into the original haplotype matrix. Finally, a genotype matrix was created by pairing row $2i$ with row $2i - 1$ in the modified haplotype matrix.

Let G_M denote the set of genotypes in dataset M with more than one heterozygous site. We used the following three measures of haplotype reconstruction accuracy: (a) The *standard error* [20] is the ratio of the number of genotypes in G_M whose haplotypes are incorrectly inferred to the total number of genotypes in G_M. (b) For a genotype g in G_M, the *switch accuracy* [16] of its inferred haplotypes is defined as $(h - w - 1)/(h - 1)$, where h is the number of heterozygous sites in g and w is the number of switches between neighboring heterozygous sites needed to transform the inferred haplotypes to the true haplotypes. The switch accuracy averaged over the genotypes in G_M defines the switch accuracy of the entire inferred haplotype matrix M'. (c) The last measure is the percentage of misphased 2s with respect to the total number of 2s in G_M.

Our simulation results are summarized in Table 1. For each size $n \times m$, we used 100 simulated genotype matrices. As the table shows, our method is comparable to PHASE in terms of accuracy, while being tens of times faster than PHASE. As shown in Table 2, for each combination of n and m used, more than 80 out of 100 datasets did not admit PPH solutions. Every such a dataset had an

H-1 IPPH solution with a *unique* PPH solution for the genotype matrix M_r, in agreement with our assumption. Also, note that, for datasets with no PPH solutions, our H-1 IPPH method correctly identified the homoplasy column (a feature that PHASE does not have) with very high accuracy. This study shows that our method is both practical and highly accurate.

3 IPPH with a Single Recombination Event

The case with exactly one single-crossover recombination event is similar in spirit to the case of a single homoplasy event. Suppose that an $n \times m$ genotype matrix M does not admit a PPH solution. If that is due to a single recombination event with a breakpoint[3] b somewhere between 1 and m, then the part to the left of b and that to the right of b should each admit a PPH solution. In our approach, we choose a recombination breakpoint b somewhere between 1 and m, and consider partitioning M column-wise into two parts, one containing the columns to the left of b and the other the columns to the right of b. We denote a partitioning by $M \longmapsto M_L \oplus M_R$, with L (resp. R) denoting left (resp. right). Our algorithm proceeds by trying all such partitions of M and checking whether each of M_L and M_R admits a PPH solution, and, if so, whether the PPH solutions from the two parts can be combined in a way consistent with there being a single recombination event, i.e. a galled tree with one gall [11,10]. If these conditions are satisfied by at least one partition of M, then we say that there exists a solution to IPPH with a single recombination event.

3.1 R-1 IPPH When There Exist Unique PPH Solutions for Each Side

Given PPH solutions M'_L and M'_R for M_L and M_R, respectively, the main question that we need to ask is whether we can appropriately pair up the mates L_i, L'_i in M'_L with the mates R_i, R'_i in M'_R, such that there is a galled tree with one gall for the combined data. If there are n genotypes in M, then in the worst case there are 2^n ways of doing the pairing. Similar to what we discussed in Section 2.1 for the case of a single homoplasy event, we propose to solve this problem in the following way: First, for all $1 \leq i \leq n$, we set $L_i = L'_i = R_i = R'_i = x_i$, where x_i is a new leaf label. Then, we work with perfect phylogenies, not with haplotype matrices, as described below.

Let $\widetilde{T}_L$ and $\widetilde{T}_R$ denote the re-labeled perfect phylogenies corresponding to M'_L and M'_R, respectively. Then, what properties of $\widetilde{T}_L$ and $\widetilde{T}_R$ would imply that there exists an R-1 IPPH solution? In a galled tree with exactly one gall, whose recombination breakpoint is denoted b, the rooted tree τ_L to the left of b and the rooted tree τ_R to the right of b are closely related. More precisely, there exists a single tree rearrangement operation, called the subtree-prune-and-regraft (SPR) operation, that one can perform on τ_L to transform it into τ_R, or vice versa [1,13,18,19]. This implies that $\widetilde{T}_L$ and $\widetilde{T}_R$ cannot be two arbitrary

[3] A breakpoint occurs between two sites.

trees for there to be an R-1 IPPH solution. Rather, they, too, should be related by an SPR-like operation. There is an important point that we should highlight here. For SPR operations to be biologically meaningful, certain restrictions must be imposed to avoid possible contradictions [18,19]. For example, time-ordering of certain associated biological events must be obeyed. The reader should be warned that, as there exists no fixed sense of time direction in unrooted trees, performing SPR operations on them and drawing conclusions about evolutionary histories, of which time is an essential component, may not be the right thing to do. However, because we are working with the case of a single recombination event, which involves a single SPR operation, it is always possible to root the two unrooted trees involved to construct a consistent evolutionary history. To recapitulate, there exists an R-1 IPPH solution if $\widetilde{T}_L$ and $\widetilde{T}_R$ are related by a single SPR operation.

Determining whether two binary trees $\widetilde{T}_L$ and $\widetilde{T}_R$ are exactly one SPR operation away is equivalent to checking whether there exists a common cut-subtree t in $\widetilde{T}_L$ and $\widetilde{T}_R$ such that $\widetilde{T}_L \setminus t$ is identical to $\widetilde{T}_R \setminus t$. It is straightforward to do this in polynomial time. It is important to note that, in general, perfect phylogenies that we need to compare may not be binary. When either one or both of $\widetilde{T}_L$ and $\widetilde{T}_R$ are non-binary, to determine whether they are exactly one SPR operation way, we need to check whether there exist a cut-subtree t_L of $\widetilde{T}_L$ and a compatible[4] cut-subtree t_R of $\widetilde{T}_R$, such that $\widetilde{T}_L \setminus t_L$ is compatible with $\widetilde{T}_R \setminus t_R$. A brute-force way of checking the 1-SPR condition is as follows. As shown in Figure 4, remove two edges E_1 and E_2 from $\widetilde{T}_L$ to obtain two subtrees T_1 and T_2. We need to find two cut-subtrees $\tau_1 \subseteq T_1$ and $\tau_2 \subseteq T_2$ so that we can prune τ_1 and regraft it next to τ_2, or vice versa, and check whether the resulting tree $\widetilde{T}'_L$ is compatible with $\widetilde{T}_R$. To do this, simply enumerate all possible ways of generating cut-subtrees τ_1 and τ_2. After pruning and regrafting, test for compatibility. This simple method is feasible when there are not too many unrefined vertices of large degree. We believe that there is a polynomial-time algorithm for checking the 1-SPR condition, as well as a related, more complex, method for solving the R-1 IPPH problem that runs in polynomial time.

3.2 An Example

Consider the genotype matrix shown on the left hand side of Figure 5. For every partition $M \mapsto M_L \oplus M_R$, we first need to check whether there exist PPH solutions for M_L and M_R. For the particular partition shown on the right hand side of Figure 5, there exists a unique PPH solution for each of M_L and M_R. These PPH solutions M'_L and M'_R, and the corresponding the perfect phylogenies T_L and T_R, respectively, are shown in Figure 6. After redefining $L_i = L'_i = R_i = R'_i = x_i$, we obtain the trees $\widetilde{T}_L$ and $\widetilde{T}_R$ shown in Figure 7. Now, note that, $\widetilde{T}_L$ and $\widetilde{T}_R$ contain a common cut-subtree t such that $\widetilde{T}_L \setminus t$ is identical to $\widetilde{T}_R \setminus t$. Hence, $\widetilde{T}_L$ and $\widetilde{T}_R$ can be related by a single SPR operation in which t gets

[4] Two trees T_1, T_2 are said to be compatible if there exists a third tree that is a refinement of both T_1 and T_2.

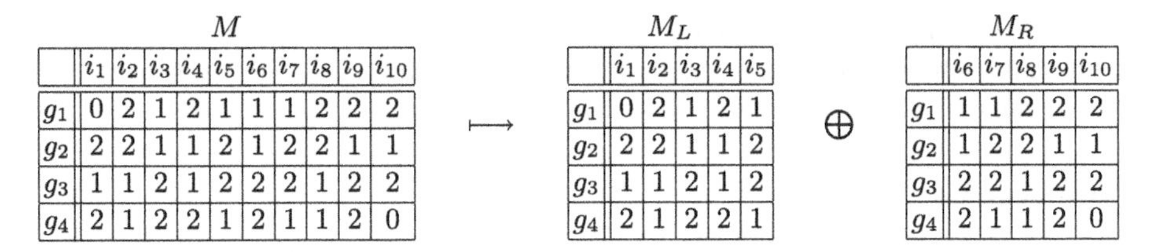

M

	i_1	i_2	i_3	i_4	i_5	i_6	i_7	i_8	i_9	i_{10}
g_1	0	2	1	2	1	1	1	2	2	2
g_2	2	2	1	1	2	1	2	2	1	1
g_3	1	1	2	1	2	2	2	1	2	2
g_4	2	1	2	2	1	2	1	1	2	0

$\longmapsto$

M_L

	i_1	i_2	i_3	i_4	i_5
g_1	0	2	1	2	1
g_2	2	2	1	1	2
g_3	1	1	2	1	2
g_4	2	1	2	2	1

$\oplus$

M_R

	i_6	i_7	i_8	i_9	i_{10}
g_1	1	1	2	2	2
g_2	1	2	2	1	1
g_3	2	2	1	2	2
g_4	2	1	1	2	0

Fig. 5. A partition of M into M_L and M_R

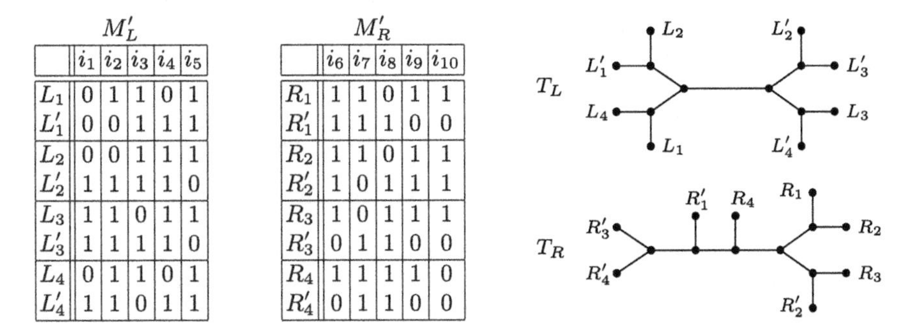

M'_L

	i_1	i_2	i_3	i_4	i_5
L_1	0	1	1	0	1
L'_1	0	0	1	1	1
L_2	0	0	1	1	1
L'_2	1	1	1	1	0
L_3	1	1	0	1	1
L'_3	1	1	1	1	0
L_4	0	1	1	0	1
L'_4	1	1	0	1	1

M'_R

	i_6	i_7	i_8	i_9	i_{10}
R_1	1	1	0	1	1
R'_1	1	1	1	0	0
R_2	1	1	0	1	1
R'_2	1	0	1	1	1
R_3	1	0	1	1	1
R'_3	0	1	1	0	0
R_4	1	1	1	1	0
R'_4	0	1	1	0	0

Fig. 6. Separate PPH solutions for M_L and M_R in Figure 5, and their corresponding perfect phylogenies

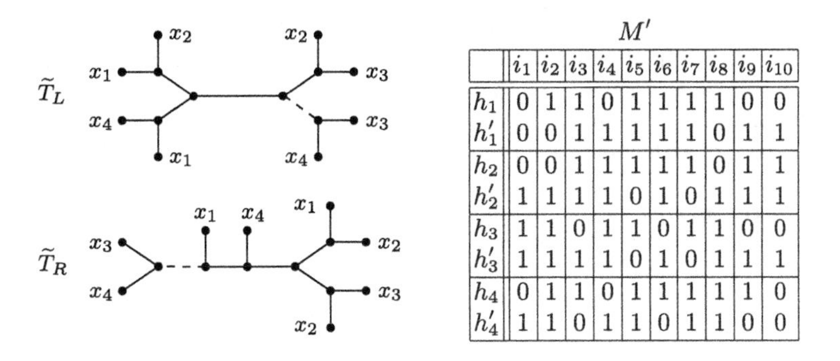

M'

	i_1	i_2	i_3	i_4	i_5	i_6	i_7	i_8	i_9	i_{10}
h_1	0	1	1	0	1	1	1	1	0	0
h'_1	0	0	1	1	1	1	1	0	1	1
h_2	0	0	1	1	1	1	1	0	1	1
h'_2	1	1	1	1	0	1	0	1	1	1
h_3	1	1	0	1	1	0	1	1	0	0
h'_3	1	1	1	1	0	1	0	1	1	1
h_4	0	1	1	0	1	1	1	1	1	0
h'_4	1	1	0	1	1	0	1	1	0	0

Fig. 7. Re-labeled perfect phylogenies and an R-1 IPPH solution for the entire genotype matrix M in Figure 5. If the edges denoted by dashed lines are removed, then a common 2-leaved subtree t labeled by x_3 and x_4 gets pruned, and $\widetilde{T}_L \setminus t$ and $\widetilde{T}_R \setminus t$ become identical.

pruned and regrafted to transform $\widetilde{T}_L$ into $\widetilde{T}_R$, or vice versa. Going back to T_L and T_R with original leaf labels, it is then possible to conclude that L_1 should get paired with R'_1, L_2 with R_2, L_3 with R'_3, and L_4 with R_4. The R-1 IPPH solution just described is shown on the right hand side of Figure 7, where h_i and h'_i denote haplotype mates for genotype g_i.

The partition shown in Figure 5 led to two binary trees. Other partitions do not have this nice property, however. For instance, the partition that divides M between columns 4 and 5 leads to a non-binary tree $\widetilde{T}_L$ for the left part and a binary tree $\widetilde{T}_R$ for the right part. There are several other possible partitions such that each part admits a PPH solution; but, in fact, all such partitions lead to the same R-1 IPPH solution shown in Figure 7.

References

1. B. L. Allen and M. Steel. Subtree transfer operations and their induced metrics on evolutionary trees. *Ann. Combin.*, 5:1–13, 2001.
2. V. Bafna, D. Gusfield, G. Lancia, and S. Yooseph. Haplotyping as perfect phylogeny: A direct approach. *J. Comput. Biol.*, 10:323–340, 2003.
3. T. Barzuza, J.S. Beckman, R. Shamir, and I. Pe'er. Computational problems in perfect phylogeny haplotyping: XOR genotypes and tag SNPs. In *Proc. of CPM*, pages 14–31, 2004.
4. R.H. Chung and D. Gusfield. Empirical exploration of perfect phylogeny haplotyping and haplotypers. In *Proc. of COCOON*, pages 5–19, 2003.
5. International HapMap Consortium. The HapMap project. *Nature*, 426:789–796, 2003.
6. Z. Ding, V. Filkov, and D. Gusfield. A linear-time algorithm for the perfect phylogeny haplotyping problem. In *Proc. of RECOMB*, pages 585–600, 2005.
7. E. Eskin, E. Halperin, and R. Karp. Large scale reconstruction of haplotypes from genotype data. In *Proc. of RECOMB*, pages 104–113, 2003.
8. E. Eskin, E. Halperin, and R.M. Karp. Efficient reconstruction of haplotype structure via perfect phylogeny. *J. Bioinf. Comput. Biol.*, 1:1–20, 2003.
9. D. Gusfield. Haplotyping as perfect phylogeny: Conceptual framework and efficient solutions (Extended Abstract). In *Proc. of RECOMB*, pages 166–175, 2002.
10. D. Gusfield. Optimal, efficient reconstruction of Root-Unknown phylogenetic networks with constrained recombination. *J. Comput. Sys. Sci.*, 70:381–398, 2005.
11. D. Gusfield, S. Eddhu, and C. Langley. Optimal, efficient reconstruction of phylogenetic networks with constrained recombination. *J. Bioinf. Comput. Biol.*, 2(1):173–213, 2004.
12. E. Halperin and E. Eskin. Haplotype reconstruction from genotype data using Imperfect Phylogeny. *Bioinformatics*, 20:1842–1849, 2004.
13. J. Hein. Reconstructing evolution of sequences subject to recombination using parsimony. *Math. Biosci.*, 98:185–200, 1990.
14. R. Hudson. Gene genealogies and the coalescent process. *Oxford Survey of Evolutionary Biology*, 7:1–44, 1990.
15. R. Hudson. Generating samples under the Wright-Fisher neutral model of genetic variation. *Bioinformatics*, 18:337–338, 2002.
16. S. Lin, D.J. Cutler, M.E. Zwick, and A. Chakravarti. Haplotype inference in random population samples. *Am. J. Hum. Genet.*, 71:1129-1137, 2002.
17. C. Semple and M. Steel. *Phylogenetics*. Oxford University Press, UK, 2003.
18. Y. S. Song. On the combinatorics of rooted binary phylogenetic trees. *Ann. Combin.*, 7:365–379, 2003.
19. Y. S. Song and J. Hein. Constructing minimal ancestral recombination graphs. *J. Comput. Biol.*, 12:147–169, 2005.
20. M. Stephens, N.J. Smith, and P. Donnelly. A new statistical method for haplotype reconstruction from population data. *Am. J. Hum. Genet.* 68:978–989, 2001.
21. S. Tavaré. Calibrating the clock: Using stochastic processes to measure the rate of evolution. In E. Lander and M. Waterman, editors, *Calculating the Secrets of Life*. National Academy Press, 1995.

A Faster Algorithm for Detecting Network Motifs

Sebastian Wernicke*

Institut für Informatik, Friedrich-Schiller-Universität Jena,
Ernst-Abbe-Platz 2, D-07743 Jena, Germany
wernicke@minet.uni-jena.de

Abstract. Motifs in a network are small connected subnetworks that occur in significantly higher frequencies than in random networks. They have recently gathered much attention as a useful concept to uncover structural design principles of complex networks. Kashtan et al. [*Bioinformatics*, 2004] proposed a sampling algorithm for efficiently performing the computationally challenging task of detecting network motifs. However, among other drawbacks, this algorithm suffers from sampling bias and is only efficient when the motifs are small (3 or 4 nodes). Based on a detailed analysis of the previous algorithm, we present a new algorithm for network motif detection which overcomes these drawbacks. Experiments on a testbed of biological networks show our algorithm to be orders of magnitude faster than previous approaches. This allows for the detection of larger motifs in bigger networks than was previously possible, facilitating deeper insight into the field.

1 Introduction

Motivation. Based on the idea that "evolution preserves modules that define specific [...] functions" [20], Milo et al. [14,15] propose to uncover the structural design principles of biological networks[1] by detecting small subnetworks which occur in significantly higher frequencies than in random networks. These "topological modules" [20] are called *network motifs.*[2]

Some excitement has surrounded the network motif approach with the original paper by Milo et al. [15] being cited well over 40 times in some major scientific journals as of June 2005. The analysis of network motifs has led to interesting results (of which we only name a few here), e.g., in the areas of protein-protein interaction prediction [1] and hierarchical network decomposition [7]. The transcriptional network of *Escherichia Coli* displays motifs to which specific functionalities such as the generation of temporal expression programs or the response to

* Supported by Deutsche Telekom Stiftung and Studienstiftung des deutschen Volkes.

1 We use the terms "network" and "node" for fields outside mathematics and computer science. The terms "graph" and "vertex" are used for discussing algorithmic aspects.

2 Note that the term "network motif" has been used in other contexts as well and, e.g., may also refer to a common subnetwork in a set of given networks [17] or to any small labeled subnetwork (without considering connectivity or isomorphy) [5].

R. Casadio and G. Myers (Eds.): WABI 2005, LNBI 3692, pp. 165–177, 2005.

fluctuating external signals can be attributed [15,18], suggesting that network motifs play key information processing roles in this type of network [9]. The same motifs as in the transcriptional interaction network of *E. Coli* were also identified for the yeast *Saccharomyces Cerevisiae*, possibly hinting that common network function implies the sharing of common motifs [11].

To put motif research in proper perspective, it should be noted that it has also been met with some criticism. Artzy-Randrup et al. [2] found that certain random network models lead to a display of motifs although there is no explicit selection mechanism for local structures (Milo et al. answer this criticism in [13]). Vázquez et al. [19] demonstrated that global network features such as the clustering coefficient also influence local features such as the abundance of certain subgraphs.

Previous Work. Much work related to network motifs has been spent on interpreting and applying the general concept, but considerably less on the involved algorithmics. Finding network motifs consists of three subtasks:

1. Find which subgraphs occur in the input graph (and in which number).
2. Determine which of these subgraphs are topologically equivalent (i.e., isomorphic) and group them into subgraph classes accordingly.
3. Determine which subgraph classes are displayed at a much higher frequency than in random graphs (under a specified random graph model).

Performing the first subtask by explicitly enumerating all subgraphs of a certain size can be time consuming due to their potentially large number even in small, sparse networks. For this reason, Kashtan et al. [9] propose an algorithm that estimates subgraph occurrences from a randomly sampled set of subgraphs. We discuss this algorithm in full detail in Section 2, mentioning only in passing here that it provides only biased sampling. This leads to considerable drawbacks such as an inconsistent sampling quality and the need for a computationally expensive bias correction. Besides [9], we are only aware of the work by Duke et al. [6] on approximating the number of size-k subgraphs in a given graph. Their algorithm, however, has no practical relevance since the input graph has to be astronomically large (as compared to k) in order to ensure a reasonable quality of approximation.

Much work has already been done concerning the second subtask and we rely on the *nauty* algorithm [12] for performing it in practice.

As to the third subtask, the standard approach for determining subgraph significance so far has been to explicitly generate an ensemble of random graphs under a given random graph model. One popular of these random graph models—which we also focus on in this work—is that of random graphs which preserve the degree sequence of the original graph. (Alternative choices, e.g., additionally preserve the number of bidirectional edges.) While there has been some research concerning the properties of graphs with prescribed degree sequence (such as the average path length [16]), the problem of subgraph distribution within such graphs has only been studied for directed sparse random graphs with *expected* degree sequences [8].

Contribution and Structure of this Work. We give significant improvements for the first and third subtask of motif detection. Based on a comprehensive analysis of the drawbacks encountered when using the subgraph sampling approach proposed by Kashtan et al. [9], Section 2 presents a new algorithm for subgraph sampling which does not suffer from these drawbacks (and has some additional useful features). While this comes at the price of only being able to control the *expected* number of samples, our proposed algorithm is much easier to implement and experiments in Section 4 reveal it to be orders of magnitude faster than the algorithm of Kashtan et al. As to the task of determining subgraph significance, Section 3 proposes a new approach that does not require the explicit generation of random graphs with a prescribed degree sequence. This approach leads to a faster algortithm that is moreover able to focus on determining the significance of specific subgraphs (which is not possible with previous approaches).

The proposed new algorithms have been implemented in C++, the source code is freely available online at http://www.minet.uni-jena.de/~wernicke/motifs/. We show in Section 4 that in a testbed of biological networks, our algorithm detects network motifs significantly faster than the implementation of Kashtan et al. This enables the analysis of larger networks and more complex motifs than previously possible.

2 A Faster Algorithm for Subgraph Sampling

Introduction. The algorithm for subgraph sampling suggested by Kashtan et al. [9] is based on the idea that we start by selecting a random edge in the input graph and then randomly extend this subgraph until we obtain a connected subgraph with the desired number of vertices. Subsection 2.1 discusses this approach and its main drawbacks. We present a new approach to subgraph sampling (which is based on randomized enumeration) in Subsection 2.2. Note that, due to lack of space, we omit the proofs of the theorems and lemmas presented in this section.

Notation. Basic familiarity with graph-theoretic terminology is assumed. Given a graph $G = (V, E)$ (which can be directed), we let $n := |V|$ and assume that all vertices in V are uniquely labeled by the integers $1, \ldots, n$. We write "$u > v$" to abbreviate "label(u) $>$ label(v)." For a set $V' \subseteq V$ of vertices, its neighborhood $N(V')$ is the set of all vertices from $V \setminus V'$ which are adjacent to at least one vertex in V'.

A connected subgraph that is induced by a vertex set of cardinality k is called *size-k subgraph.* For a given integer k, the set of all size-k subgraphs in G can be partitioned into sets $\mathcal{S}_k^i(G)$ called *subgraph classes* where two size-k subgraphs belong to the same subgraph class if and only if they are isomorphic. The *concentration* $\mathcal{C}_k^i(G)$ of a subgraph class $\mathcal{S}_k^i(G)$ is defined as $\mathcal{C}_k^i(G) := |\mathcal{S}_k^i(G)| \cdot (\sum_j |\mathcal{S}_k^j(G)|)^{-1}$. For a graph G, an integer k, and a set $\mathcal{R}$ of size-k subgraphs that were randomly sampled in G by an algorithm $\mathcal{A}$, a mapping $\hat{\mathcal{C}}^i_k : (\mathcal{R}, G) \to [0,1]$ is called an *estimator* for $\mathcal{C}^i_k(G)$. We say that $\hat{\mathcal{C}}^i_k(\mathcal{R}, G)$ is *unbiased* (with respect to $\mathcal{A}$) if the expected value of $\hat{\mathcal{C}}^i_k(\mathcal{R}, G)$ equals $\mathcal{C}^i_k(G)$ and *biased* otherwise.

2.1 The Previous Approach: Edge Sampling

For a given graph $G = (V, E)$ and an integer $k \geq 3$, Kashtan et al. [9] suggest to sample a random subgraph by starting with a randomly chosen edge and then adding neighboring vertices until a subgraph of the desired size k is obtained:

Algorithm: Edge Sampling(G, k) (ESA)
Input: A graph $G = (V, E)$ and an integer $2 \leq k \leq |V|$.
Output: Vertices of a randomly chosen size-k subgraph in G.

```
01  {u, v} ← random edge from E
02  V' ← {u, v}
03  while |V'| ≠ k do
04      {u, v} ← random edge from V' × N(V')
05      V' ← V' ∪ {u} ∪ {v}
06  return V'
```

As already noted in [9], ESA has a bias for sampling certain subgraphs more often than others. Figure 1 shows a concrete example we have constructed to illustrate this. The total number of connected size-3 subgraphs both in G_1 and G_2

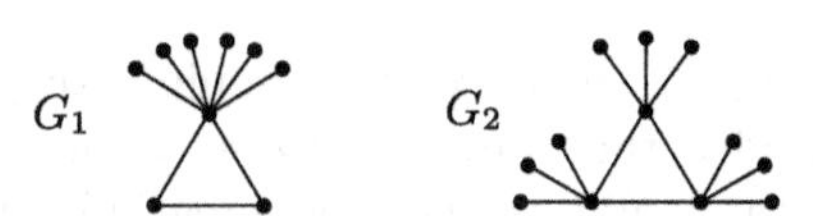

Fig. 1. Graphs G_1 and G_2 have an equal number of (connected) size-3 subgraphs. The subgraph $\triangle$ occurs exactly once in each of them. As outlined in the text, ESA oversamples the subgraph $\triangle$ in both G_1 and G_2. The oversampling is worse for G_1.

is 28. Since the subgraph $\triangle$ occurs exactly once each in G_1 and G_2, we should expect that ESA samples $\triangle$ with probability $\frac{1}{28}$ within both graphs. However, $\Pr[\text{ESA samples } \triangle \text{ in } G_1] = \frac{1}{9} \cdot 1 + \frac{2}{9} \cdot \frac{2}{8} = \frac{1}{6}$ and $\Pr[\text{ESA samples } \triangle \text{ in } G_2] = \frac{3}{12} \cdot \frac{2}{8} = \frac{1}{16}$. This illustrates some crucial problems of ESA: The subgraph $\triangle$ is oversampled and—as a direct consequence—the only other occurring size-3 subgraph $\vee$ is undersampled. The oversampling of $\triangle$ is worse for G_1 than it is for G_2 and it is possible to show (using an adaption of the above example) that the magnitude of the oversampling cannot be estimated simply from the number of edges neighboring the oversampled subgraph. Given a set $\mathcal{R}$ of size-k subgraphs that were randomly sampled using ESA, the demonstrated bias can be overcome by using the following (unbiased) estimator [9]:

$$\hat{\mathcal{C}}^i_k(\mathcal{R}, G) := \frac{\sum_{\{G' \in \mathcal{R} \mid G' \in \mathcal{S}^i_k(G)\}} (\Pr[G' \text{ is sampled by ESA}])^{-1}}{\sum_{G' \in \mathcal{R}} (\Pr[G' \text{ is sampled by ESA}])^{-1}} . \tag{1}$$

The main idea here is that each subgraph is (ex post facto) scored inversely proportional to the probability that ESA samples it. While it is possible to correctly estimate $C_k^i(G)$ in this way, several disadvantages remain:

- The bias itself remains. E.g., subgraphs which appear in low concentration and are at the same time undersampled by ESA are hardly ever found.[3]
- Computing (1) is expensive since the calculation of *each single* probability can require as much as $\mathcal{O}(k^k)$ time [9].
- We have no estimate as to what *fraction* of subgraphs has been sampled.
- ESA can sample the same subgraph multiple times.

In the next subsection we suggest a new approach to subgraph sampling that overcomes these problems.

2.2 The New Approach: Randomized Enumeration

The idea here is to start with an algorithm that efficiently enumerates all size-k subgraphs. This algorithm is then modified to randomly "skip" some of these subgraphs during its execution, yielding an unbiased subgraph sampling algorithm.

Enumerating all size-k subgraphs. Given a graph $G = (V, E)$, the following algorithm enumerates all of its size-k subgraphs (with $N_{excl}(v, V') := N(\{v\}) \setminus N(V')$ being the *exclusive neighborhood of* v with respect to $V' \subseteq V$):

Algorithm: ENUMERATESUBGRAPHS(G, k) (ESU)
Input: A graph $G = (V, E)$ and an integer $1 \leq k \leq |V|$.
Output: All size-k subgraphs in G.

01 **for** each vertex $v \in V$ **do**
02 $\quad V_{Extension} \leftarrow \{u \in N(\{v\}) \mid u > v\}$
03 $\quad$ **call** EXTENDSUBGRAPH$(\{v\}, V_{Extension}, v)$
04 **endfor**

EXTENDSUBGRAPH$(V_{Subgraph}, V_{Extension}, v)$
E1 **if** $|V_{Subgraph}| = k$ **then output** $G[V_{Subgraph}]$ and **return**
E2 **while** $V_{Extension} \neq \emptyset$ **do**
E3 $\quad$ Remove an arbitrarily chosen vertex w from $V_{Extension}$
E4 $\quad V'_{Extension} \leftarrow V_{Extension} \cup \{u \in N_{excl}(w, V_{Subgraph}) \mid u > v\}$
E5 $\quad$ **call** EXTENDSUBGRAPH$(V_{Subgraph} \cup \{w\}, V'_{Extension}, v)$
E6 **return**

The basic idea of ESU is that—starting with a vertex v from the input graph—we add only those vertices to the $V_{Extension}$ set that have two properties: Their

[3] Kashtan et al. [9] observe that ESA can accurately estimate the concentration of $\mathcal{S}_k^i(G)$ with less than $(C_k^i(G))^{-1}$ samples for subgraphs which are oversampled. In return however, other subgraphs might be missed completely for far more than $(C_k^i(G))^{-1}$ samples and would consistently be overlooked as motif candidates.

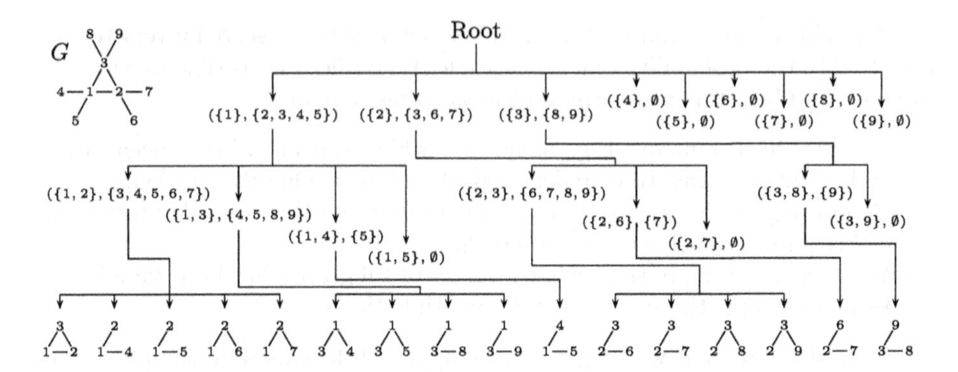

Fig. 2. The above ESU-tree corresponds to calling ENUMERATESUBGRAPHS$(G, 3)$. The tree has 16 leafs which correspond to the 16 size-3 subgraphs in G.

label must be larger than that of v and they must not be neighbor to a vertex in $V_{Subgraph}$ (other than the newly added vertex w). Some more insight into the structure of ESU can be gained by the following visualization.

Definition 1. *With a call to* ENUMERATESUBGRAPHS(G, k), *we associate a tree-graph called* ESU-tree *which represents the recursive function calls. The root at depth 0 represents the call of* ENUMERATESUBGRAPHS(G, k). *Each call of* EXTENDSUBGRAPH$(V_{Subgraph}, V_{Extension}, v)$ *is represented by an edge from the vertex representing the caller function to a vertex representing the callee. The callee vertex is labeled* $(V_{Subgraph}, V_{Extension})$ *and located at depth* $|V_{Subgraph}|$.

The structure of the tree is illustrated in an example in Figure 2. Omitting the proof here, it is also the basis to establish the correctness of the ESU algorithm.

Theorem 2. *Given a graph G and $k \geq 2$,* ESU *enumerates all size-k subgraphs in G (each size-k subgraph is output exactly once).* □

The tree structure to represent ESU exposes some useful properties. E.g., using a technique by Knuth [10], we can randomly explore paths in the tree in order to quickly estimate the total number of size-k subgraphs in the input graph. Probably the most important feature of the ESU-tree, however, is that we can use it to efficiently sample subgraphs uniform at random (i.e., without bias).

Uniformly sampling size-k subgraphs. The ESU algorithm completely traverses its corresponding ESU-tree. Where complete traversal is too time-expensive, we can explore only parts of the ESU-tree such that each leaf is reached with equal probability. For this purpose, a probability $0 < p_d \leq 1$ is introduced for each depth $1 \leq d \leq k$ in the tree. With p_d, we determine for each child vertex at depth d whether we traverse the subtree rooted at it. This is implemented by replacing line *03* of the ESU algorithm with "With probability p_1, call EXTENDSUBGRAPH(...)" and line *E5* with "With probability p_d, call EXTENDSUB-

GRAPH(...)" (where $d := |V_{Subgraph}|+1$).[4] We call this new algorithm RAND-ESU. (To simplify the discussion, we will also use this name when all p_d are set to 1, in which case RAND-ESU is equivalent to ESU.) RAND-ESU visits each leaf of the ESU-tree with equal probability and hence estimating subgraph concentrations from its output is straightforward (the proofs are omitted).

Lemma 3. *RAND-ESU visits each leaf in the ESU-tree with probability* $\prod_d p_d$. □

Theorem 4. *Given a graph G, an integer k, and $0 < p_d \leq 1$ for $1 \leq d \leq k$. Let $\mathcal{R}$ be a set of size-k subgraphs obtained by running RAND-ESU on G using the probabilities p_d. Then, $\hat{C}_k^i(\mathcal{R}, G) := |\{G' \in \mathcal{R} \mid G' \in \mathcal{S}_k^i(G)\}| \;/\; |\mathcal{R}|$ is an unbiased estimator for $C_k^i(G)$.* □

It remains to discuss how the values p_d should be chosen. If we wish to sample an expected fraction $0 < q < 1$ of all size-k subgraphs using RAND-ESU, we have to ensure that $\prod_{1 \leq d \leq k} p_d = q$ (we omit a rigorous proof of this here). However, this still leaves us to choose the individual values, i.e., do we uniformly set every p_d equal to $\sqrt[k]{q}$ or are there better choices? Some observations are:

- Choosing whether or not to explore a subtree whose root is close to the root of the ESU-tree generally has a higher influence on the total number of explored leafs than for a subtree whose root is farther from it.
- The parameters p_d influence the distribution of the sampling, i.e., if p_d is small for small d, some local neighborhoods in the input graph are likely not to be explored at all while others will be explored extensively.
- The running time is influenced from an amortized point of view: If the p_d values are large for small values of d (and hence small for larger d), much of the ESU-tree is explored but only comparably few leafs are reached.

As a general rule from these observations, the parameters p_d should be larger for small d and become smaller as d increases—as long as the sacrifice made with respect to the amortized running time per sample is acceptable. This ensures a lower variance for the number of samples and the exploration of many different regions in the input graph.

Concluding this section, while RAND-ESU—as compared to ESA—requires a choice of sampling parameters and only allows for controlling the expected number of samples, it has a lot to offer in return. Most importantly it is unbiased, which rules out the respective disadvantages of ESA. Also, it is much faster (see Section 4) and easier to implement since we do not require any bias-correcting parts. Contrary to ESA, our new algorithm never samples more subgraphs than the input graph contains and results become exact as the number of samples reaches the total number of size-k subgraphs in the input graph.

[4] In order to reduce the sampling variance, the following more sophisticated method may be used: For a tree vertex at depth d with x children, randomly choose x' of the x children (where $x' = \lfloor x \cdot p_d \rfloor$ with probability $1 - (x \cdot p_d - \lfloor x \cdot p_d \rfloor)$ and $x' = \lceil x \cdot p_d \rceil$ otherwise) and explore exactly these. It can be shown that this does not change the probability of a leaf being explored.

3 Direct Calculation of Motif Significance

The Previous Approach. As already mentioned in the introduction, motif detection includes the subtask of determining subgraph significance. In this work, we consider the case where the significance of a subgraph is determined by comparing its concentration in the given graph G to its mean concentration $\langle C_k^i(G)\rangle$ in random graphs with the same degree sequence [15,9]. It is suggested in [15,9] to estimate $\langle C_k^i(G)\rangle$ by generating a large ensemble of random graphs (typically at least 1000) with the same degree sequence as the original graph and then sampling subgraphs in these random graphs. The random graphs are generated from the original graph by randomly switching edges between vertices, which requires a lot of switching operations while at the same time it is never certain when proper randomization has been reached. Also with this method, we are likely to spend lots of excess computational efforts estimating the concentrations of subgraph classes we are not interested in.[5] In this section we propose an algorithm for determining subgraph significance without the need to explicitly generate random graphs (assuming the background model of random graphs with the same degree sequence). We also gain the ability to focus our estimation of significance on specific subgraphs.

Direct Calculation of Subgraph Significance. Milo et al. observe that the total number of size-k subgraphs within an ensemble of large graphs with the same degree sequence does not vary much (see supplementary online material to [14] for details). This allows us to estimate $\langle C_k^i(G)\rangle$ by

$$\langle C_k^i(G)\rangle \approx \langle \hat{C}_k^i(G)\rangle := \frac{\sum_{G' \in \textsc{DegSeq}(G)} |S_k^i(G')|}{\sum_{G' \in \textsc{DegSeq}(G)} \sum_i |S_k^i(G')|} \tag{2}$$

where $\textsc{DegSeq}(G)$ is the set of all graphs G' that have the same degree sequence as G. Since all graphs G' can be viewed as graphs over the same set of vertices (because they differ only in their edge sets), Equation (2) can also be written as

$$\langle \hat{C}_k^i(G)\rangle = \frac{\sum_{\{v_1,\dots,v_k\}\subseteq V} |\{G' \in \textsc{DegSeq}(G) \mid G'[\{v_1,\dots,v_k\}] \in S_k^i\}|}{\sum_{\{v_1,\dots,v_k\}\subseteq V} |\{G' \in \textsc{DegSeq}(G) \mid G'[\{v_1,\dots,v_k\}] \text{ connected}\}|}. \tag{3}$$

Both the nominator and denominator of this equation can be estimated in a Monte Carlo approach—i.e., by randomly sampling size-k subsets of the vertices in the input graph—as long as we are able to perform the following calculation: Given G and $\{v_1,\dots,v_k\}$, find $|\{G' \in \textsc{DegSeq}(G) \mid G'[\{v_1,\dots,v_k\}] \in S_k^i\}|$. As it turns out, this number is indeed possible to calculate using two theorems (one for undirected graphs and one for the directed case) due to Bender and Canfield [3,4]. Without going into technical details here, these theorems allow us

[5] This is especially important for sparse networks where a randomly sampled subgraph is likely to be a tree. Trees, however, are often considered to be uninteresting motifs [5].

Table 1. Number of size-k subgraphs and the number of respective subgraph classes that occur in our test instances for $3 \leq k \leq 6$. (All instances are directed graphs).

	nodes	edges	subgraphs				subgraph classes			
			size-3	size-4	size-5	size-6	size-3	size-4	size-5	size-6
COLI	423	519	5 206	83 893	1 433 502	22 532 584	4	17	83	390
YEAST	688	1 079	13 150	183 174	2 508 149	32 883 898	7	33	173	888
ELEGANS	306	2 345	47 322	1 394 259	43 256 069	1 309 307 357	13	197	7 071	286 375
YTHAN	135	597	9 487	169 733	2 908 118	45 889 039	8	57	629	9 339

to calculate for a given degree sequence how many graphs there are which realize exactly this degree sequence under the constraint that a certain subgraph is fixed. Given a subgraph class $\mathcal{S}_k^i$ and k vertices $\{v_1, \ldots, v_k\}$, we can thus consider all (at most $k!$) ways in which $\{v_1, \ldots, v_k\}$ can induce a subgraph from $\mathcal{S}_k^i$ and hence estimate the nominator in Equation (3).

An analogous approach (considering all ways in which the given vertices can be connected) can be used to estimate the denominator in Equation (3). Omitting the details here, it is possible to show that this does not require the explicit consideration of every connected size-k subgraph but only of k^{k-2} subgraphs for undirected graphs and $2 \cdot (2k)^{k-2}$ in the directed case. At first glance it might seem as if this is prohibitively expensive to calculate, but for two reasons it actually promises a gain in efficiency: Firstly, the denominator in Equation (3) is the same for all subgraph classes and hence has to be calculated only once. Secondly, the number of occurring subgraph classes is often far less than the total number of subgraphs (see Table 1). Experiments which are discussed in the next section confirm this expected performance gain.

4 Experimental Studies

Method and Results. We have implemented our algorithms from Sections 2 and 3 in C++. The source code is freely obtainable online at http://www.minet.uni-jena.de/~wernicke/motifs/. As a comparison, we used the *mfinder* 1.1 tool[6] by Kashtan et al. which implements the ESA algorithm. All tests were performed on an AMD Athlon 64 3400+ with 2.4 GHz, 512 KB cache, and 1 GB main memory running under the Debian GNU/Linux 3.1 operating system. Sources were compiled with the GNU gcc/g++ 3.4.3 compiler using the option "-O3."
The network instances for testing the algorithms were up-to-date versions of the motif detection testbed used by Kashtan et al. [9]. The testbed consists of the instances COLI (transcriptional network of *Escherichia Coli* [18]), YEAST (transcriptional network of *Saccharomyces Cerevisiae* [15]), ELEGANS (neuronal network of *Caenorhabditis Elegans* [9]), and YTHAN (food web of the Ythan estuary [21]). Some properties of these networks are summarized in Table 1. The algorithms were compared both for their speed and quality; results and some details as to the experimental setting are shown in Figure 3 and Table 2.

[6] Source at http://www.weizmann.ac.il/mcb/UriAlon/groupNetworkMotifSW.html

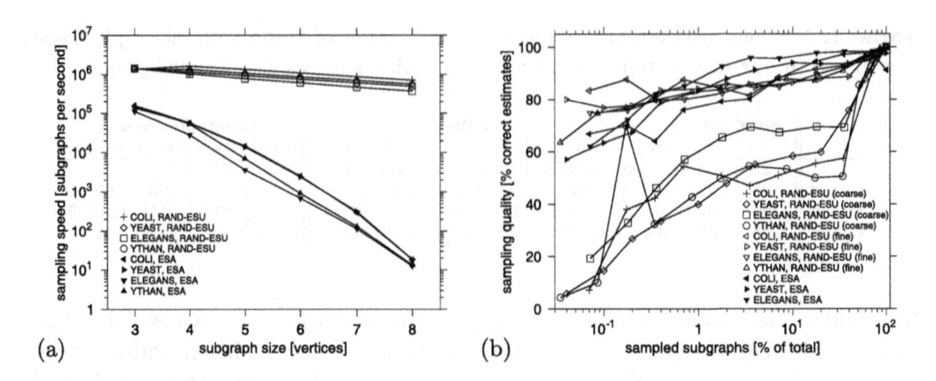

Fig. 3. (a) Sampling speed for different subgraph sizes on a semi-log scale (time measurement does not include the grouping of sampled subgraphs into classes). For ESU, the curve shows the mean speed for three different settings of $(p_1, \ldots, p_k)$ that lead to sampling of an expected 10% of all subgraphs: $(1, \ldots, 1, .316, .316)$, $(1, \ldots, 1, .5, .2)$, and $(1, \ldots, 1, .1)$. (The speed of the deterministic ESU algorithm—not shown here—is slightly faster than that of RAND-ESU.) **(b)** Sampling quality for size-5 subgraphs (size-4 for YTHAN) versus the percentage p of sampled subgraphs (semi-log scale). We define the sampling quality as the percentage of subgraph classes $\mathcal{S}_i^k$ for which C_i^k is estimated with at most 20% relative error (considering only those subgraph classes for a given p that we would expect to sample at least 10 times on average). RAND-ESU was run with two different settings of the p_d values we refer to as "coarse" $(1, \ldots, 1, \sqrt{p}, \sqrt{p})$ and "fine" $(1, \ldots, 1, p)$. For our YTHAN instance, the *mfinder* tool reproducibly failed to report results for more than 100 samples, hence this curve is not shown.

Discussion. Most notable in Figure 3a, RAND-ESU is much faster than the ESA sampling in *mfinder*. This amounts to several orders of magnitude for larger subgraphs ($k \geq 5$). For small sampling quantities, the "coarse" variant of RAND-ESU proved to be faster than the "fine" variant (not explicitly shown in Figure 3a). However, Figure 3b shows that the resulting sampling quality from using "coarse" settings for the p_d values is relatively low when compared to that of ESA. The qualities are roughly equal for the "fine" variant with ESA having a slight advantage for sampling sizes above 1% and close to 100%. (Note that for 100%, RAND-ESU is equivalent to ESU and the results are exact.) Two things are to be noted in this respect, though: Firstly, RAND-ESU is much faster and can, e.g., fully enumerate all size-5 subgraphs in roughly the same time that ESA needs to sample 1% of them. Secondly, the sampling quality of the "fine" variant appears to be more consistent for different networks, e.g., in some percentage ranges ESA has a very good sampling quality for ELEGANS and a comparably fair one for COLI whereas the "fine" RAND-ESU remains much more consistent here. Also note that—contrary to ESA—statistical estimates about the achieved sampling quality can be made with RAND-ESU because of its unbiasedness and the ability to estimate the total number of subgraphs (especially with the "fine" variant where individual samples are fully independent of each other).

Table 2. For directed size-3 subgraphs, the table shows the approximate subgraph concentrations in random graphs based on the methods discussed in Section 3. For estimating $\langle C_k^i(G)\rangle$, 10 000 random graphs were generated. The $\langle \hat{C}_k^i(G)\rangle$ values are based on 100 000 samples. Compared to the $\langle C_k^i(G)\rangle$ values, their calculation was a few hundred times faster on our machine. For most subgraphs with $\langle C_k^i(G)\rangle > 10^{-5}$, $\langle \hat{C}_k^i(G)\rangle$ appears to be a good approximation for $\langle C_k^i(G)\rangle$ (since often here, the ratio $\langle C_k^i(G)\rangle / \langle \hat{C}_k^i(G)\rangle$ is close to one).

COLI $\langle C_k^i\rangle$	9.1e-1	3.7e-2	1.9e-4	5.0e-2	1.4e-3	2.1e-6	7.6e-8	3.4e-7	2.9e-6	2.9e-5	8.0e-7	–	–
$\langle \hat{C}_k^i\rangle$	9.0e-1	4.2e-2	2.6e-4	5.5e-2	1.4e-3	2.1e-6	1.3e-7	8.7e-8	2.3e-6	4.4e-5	1.1e-7	8e-12	6e-15
$\langle C_k^i\rangle/\langle \hat{C}_k^i\rangle$	1.0	0.9	0.7	0.9	1.0	1.0	0.6	3.9	1.3	0.7	7.4	–	–
YEAST $\langle C_k^i\rangle$	9.1e-1	3.7e-2	1.8e-4	5.0e-2	1.4e-3	9.5e-7	–	2.6e-7	2.3e-6	2.9e-5	3.4e-7	–	–
$\langle \hat{C}_k^i\rangle$	8.9e-1	3.0e-2	1.2e-4	7.6e-2	1.2e-3	1.5e-6	2.8e-8	4.4e-8	5.4e-7	1.0e-5	1.0e-7	1e-14	1e-15
$\langle C_k^i\rangle/\langle \hat{C}_k^i\rangle$	1.0	1.2	1.5	0.6	1.2	0.7	–	6.1	4.3	2.9	3.3	–	–
ELEG. $\langle C_k^i\rangle$	2.0e-1	3.3e-1	2.7e-2	3.7e-1	3.3e-2	1.7e-3	1.5e-3	2.0e-3	4.4e-3	2.9e-2	1.4e-3	3.8e-4	1.5e-5
$\langle \hat{C}_k^i\rangle$	2.0e-1	3.3e-1	2.9e-2	3.6e-1	3.6e-2	2.0e-3	1.9e-3	2.3e-3	4.7e-3	3.0e-2	1.5e-3	4.0e-4	1.5e-5
$\langle C_k^i\rangle/\langle \hat{C}_k^i\rangle$	1.0	1.0	0.9	1.0	0.9	0.9	0.8	0.9	0.9	1.0	0.9	0.9	1.0
YTHAN $\langle C_k^i\rangle$	4.1e-1	2.3e-1	3.3e-2	2.2e-1	5.1e-2	3.0e-3	2.7e-3	2.8e-3	2.0e-3	3.6e-2	5.3e-3	1.1e-3	5.8e-5
$\langle \hat{C}_k^i\rangle$	3.7e-1	2.4e-1	3.9e-2	2.2e-1	5.6e-2	3.5e-3	4.8e-3	5.0e-3	3.0e-3	5.2e-2	8.1e-3	2.7e-3	7.5e-4
$\langle C_k^i\rangle/\langle \hat{C}_k^i\rangle$	1.1	1.0	0.9	1.0	0.9	0.8	0.6	0.6	0.6	0.7	0.6	0.4	0.1

As to the estimation of subgraph significance, Table 2 shows that for most subgraphs with $\langle C_k^i(G)\rangle > 10^{-5}$, $\langle \hat{C}_k^i(G)\rangle$ is a good approximation in our experimental setting. Further research should investigate the few exceptions, which might hint that for some subgraphs a larger number of samples is needed. Given that direct calculation with our tool was much faster than the explicit generation of random networks, further investigation in this respect appears to be worthwhile. Also, note that with our new approach, the frequency of some subgraphs could be estimated for which the explicit generation of subgraphs did not give any results due to an extremely low average concentration in the explicitly generated random graphs.

5 Conclusion

Based on a detailed analysis of previous approaches we have presented new algorithmic techniques which allow for a faster detection of network motifs and offer useful additional features such as unbiased subgraph sampling and a specifically targeted detection of subgraph significance. This enables motif detection for larger motifs and larger networks than was previously possible and hopefully facilitates future research in the field.

Further research could improve the presented sampling technique, e.g., by examining how the labeling of the vertices in the input graph affects the sampling quality or seeing if RAND-ESU can be tweaked to selectively sample "interesting" parts of the input graph. For subgraph significance, we have shown that a direct calculation scheme may serve as a fast and accurate alternative to the explicit generation of random networks. It would be interesting to further explore this path by extending the scheme to classes of random background models other than those that solely preserve the degree sequence.

Acknowledgments. The author is grateful to Jens Gramm (Tübingen), Falk Hüffner (Jena), and Rolf Niedermeier (Jena) for helpful discussions and comments and to an anonymous referee of WABI 2005 for some insightful remarks on this work.

References

1. I. Albert and R. Albert. Conserved network motifs allow protein-protein interaction prediction. *Bioinformatics*, 20(18):3346–3352, 2004.
2. Y. Artzy-Randrup, S. J. Fleishman, N. Ben-Tal, and L. Stone. Comment on "network motifs: Simple building blocks of complex networks" and "superfamilies of designed and evolved networks". *Science*, 305:1007c, 2004.
3. E. A. Bender. The asymptotic number of non-negative matrices with given row and column sums. *Disc. Appl. Math.*, 10:217–223, 1974.
4. E. A. Bender and E. R. Canfield. The asymptotic number of labeled graphs with given degree sequences. *J. Comb. Theor. A*, 24:296–307, 1978.
5. J. Berg and M. Lässig. Local graph alignment and motif search in biological networks. *PNAS*, 101(41):14689–14694, 2004.
6. R. A. Duke, H. Lefmann, and V. Rödl. A fast approximation algorithm for computing the frequencies of subgraphs in a given graph. *SIAM J. Comp.*, 24(3):598–620, 1995.
7. S. Itzkovitz, R. Levitt, N. Kashtan, et al. Coarse-graining and self-dissimilarity of complex networks. *Phys. Rev. E*, 71:016127, 2005.
8. S. Itzkovitz, R. Milo, N. Kashtan, et al. Subgraphs in random networks. *Phys. Rev. E*, 68(026127), 2003.
9. N. Kashtan, S. Itzkovitz, R. Milo, and U. Alon. Efficient sampling algorithm for estimating subgraph concentrations and detecting network motifs. *Bioinformatics*, 20(11):1746–1758, 2004.
10. D. E. Knuth. Estimating the efficiency of backtrack programs. In *Selected papers on Analysis of Algorithms*. Stanford Junior University, Palo Alto, 2000.
11. T. I. Lee, N. J. Rinaldi, F. Robert, et al. Transcriptional regulatory networks in Saccharomyces Cerevisiae. *Science*, 298:799–804, 2002.
12. B. D. McKay. Practical graph isomorphism. *Congr. Numer.*, 30:45–87, 1981.
13. R. Milo, S. Itzkovitz, N. Kashtan, et al. Response to comment on "network motifs: Simple building blocks of complex networks" and "superfamilies of designed and evolved networks". *Science*, 305:1007d, 2004.
14. R. Milo, S. Itzkovitz, N. Kashtan, et al. Superfamilies of designed and evolved networks. *Science*, 303(5663):1538–1542, 2004.
15. R. Milo, S. S. Shen-Orr, S. Itzkovitz, et al. Network motifs: Simple building blocks of complex networks. *Science*, 298(5594):824–827, 2002.
16. M. E. J. Newman, S. H. Strogatz, and D. J. Watts. Random graphs with arbitrary degree distributions and their applications. *Phys. Rev. E*, 64:026118, 2001.
17. S. Ott, A. Hansen, S. Kim, and S. Miyano. Superiority of network motifs over optimal networks and an application to the revelation of gene network evolution. *Bioinformatics*, 21(2):227–238, 2005.

18. S. S. Shen-Orr, R. Milo, S. Mangan, and U. Alon. Network motifs in the transcriptional regulation network of Escherichia Coli. *Nature Gen.*, 31(1):64–68, 2002.
19. A. Vázquez, R. Dobrin, D. Sergi, et al. The topological relationship between the large-scale attributes and local interaction patterns of complex networks. *PNAS*, 101(52):17940–17945, 2004.
20. A. Vespignani. Evolution thinks modular. *Nature Gen.*, 35(2):118–119, 2003.
21. R. J. Williams and N. D. Martinez. Simple rules yield complex food webs. *Nature*, 404:180–183, 2000.

Reaction Motifs in Metabolic Networks

Vincent Lacroix[1,2,*], Cristina G. Fernandes[3], and Marie-France Sagot[1,2,4]

[1] Équipe BAOBAB, Laboratoire de Biométrie et Biologie Évolutive, Université Lyon I, France
[2] Projet Helix, INRIA Rhône-Alpes, France
lacroix@biomserv.univ-lyon1.fr
[3] Instituto de Matemática e Estatística, Universidade de São Paulo, Brazil
[4] Department of Computer Science, King's College London, England

Abstract. The classic view of metabolism as a collection of metabolic pathways is being questioned with the currently available possibility of studying whole networks. Novel ways of decomposing the network into modules and motifs that could be considered as the building blocks of a network are being suggested. In this work, we introduce a new definition of motif in the context of metabolic networks. Unlike in previous works on (other) biochemical networks, this definition is not based only on topological features. We propose instead to use an alternative definition based on the functional nature of the components that form the motif. After introducing a formal framework motivated by biological considerations, we present complexity results on the problem of searching for all occurrences of a reaction motif in a network, and introduce an algorithm that is fast in practice in most situations. We then show an initial application to the study of pathway evolution.

1 Introduction

Network biology is a general term for an emerging field that concerns the study of interactions between biological elements [2]. The term *molecular interaction networks* may designate several types of networks depending on the kind of molecules involved. Classically, one distinguishes between gene regulatory networks, signal transduction networks and metabolic networks. Protein-protein interaction networks represent yet another type of network, but this term is rather linked to the techniques (such as Yeast-2-hybrid) used to produce the data and covers possibly several biological processes (including, for example, the formation of complexes and phosphorylation cascades) [16].

One of the declared objectives of network biology (or systems biology in general) is whole cell simulation [9]. However, dynamic simulation requires knowledge on reaction mechanisms such as the kinetic parameters describing a Michaelis-Menten equation. Besides the fact that such knowledge is often unavailable or unreliable, the study of the static set of reactions that constitute metabolism is equally important, both as a first step towards introducing dynamics, and in itself. Indeed, such static set represents not what is happening at a given time in a given cell but instead the capabilities of the cell, including capabilities the cell does not use. A careful analysis of this set of reactions for a given organism, alone or in comparison with the set of other organisms, may also help to arrive at a better understanding on how metabolism evolves. It is this

* Corresponding author.

R. Casadio and G. Myers (Eds.): WABI 2005, LNBI 3692, pp. 178–191, 2005.

set we propose to study in this paper. More precisely, in the following sections, the term "metabolism" should be understood as the static set of reactions involved in the synthesis and degradation of small molecules. Regulation information is not taken into consideration for now. It may be added in a later step, as the "software" running on the "hardware" of a metabolic network [15].

A major issue concerning the study of biochemical networks is the problem of their organisation. Several attempts have been made to decompose complex networks into parts. These "parts" have been called modules or motifs, but no definition of such terms seems to be completely satisfying.

Modules have first been mentioned by Hartwell *et al.* [6] who outline the general features a module should have but provide no clear definition for it. In the context of metabolic networks, a natural definition of modules could be based on the partition of a metabolic network into the metabolic pathways one can find in databases: modules would thus be the pathways as those have been established. The advantage of this partition, and thus of modules representing pathways, is that it reflects the way metabolism has been discovered experimentally (starting from key metabolites and studying the ability of an organism to synthesize or degrade them). The drawback is that it is not based on objective criteria and therefore is not universal (indeed, the number of metabolic pathways and the frontiers between them vary from one database to the other).

Several attempts to give systematic and practical definitions have been made using graph formalisms [14,10,5] and constraint-based approaches [11]. Graph-based methods range from a simple study of the local connectivity of metabolites in the network [14] to the maximisation of a criterion expressing modularity (number of links within modules) [5]. The only information used in these methods is the topology of the network. In the case of constraint-based approaches, the idea is quite different. First, a decomposition of the network into functional sets of reactions is performed (by analysis of the stoichiometric matrix [12]) and then modules are defined from the analysis of these functional states. The result is not a partition in the sense that all reactions might not be covered and a single reaction might belong to several modules.

Unlike the definition of module, the notion of motif has not been studied in the context of metabolic networks. In general, depending on what definition is adopted for modules and motifs, there is no clear limit between the two notions besides the difference in size. In the context of regulatory networks, motifs have been defined as small, repeated and perhaps evolutionary conserved subnetworks. In contrast with modules, motifs do not function in isolation. Furthermore, they may be nested and overlapping [22]. This definition refers to general features that regulatory motifs are believed to share but it provides no practical way to find them. A more practical definition has been proposed, still in the context of gene regulatory networks (and other types of non-biological networks such as the web or social networks). These are "network motifs" and represent patterns of interconnections that recur in many different parts of a network at frequencies much higher than those found in randomized networks [17]. This definition is purely topological and disregards the nature of the components in a motif. It assumes that the local topology of the network is sufficient to model function (which is understood here as the dynamic behaviour of the motif). This assumption seems acceptable when studying the topology of the internet and may also hold when analysing gene regulatory networks, but it appears not adapted to metabolic networks. In a static context, a topological definition of motif seems indeed inappropriate as similar topologies can give rise to very different functions.

In the definition of motif we introduce, the components of the network play the central part and the topology can be added as a further constraint only. This is the main biological contribution of this paper.

Its main algorithmical contribution comes from the fact that the definition of motif we adopt leads to new questions. Indeed, if searching for "purely" topological motifs may be formally modelled as a subgraph isomorphism problem, this no longer applies when searching for motifs where the features describing the components are the important elements and topology is initially indifferent (connectivity only is taken into account). Observe that the problem we address is different from pathway alignment because we wish to go beyond the notion of pathway in order to study the network as a whole. Moreover, in [19] and [13], the pathways are modelled as, respectively, chains and trees to simplify the problem. This simplification may seem reasonable in the case of a pathway alignment, it is no longer so in the case of general networks.

The paper addresses complexity issues related to this new definition of a graph motif, providing hardness results on the problem, and then presents an exact algorithm that is fast in practice for searching for such motifs in networks representing the whole metabolism of an organism. The paper ends with an initial application of the algorithm to the formulation of hypotheses on the evolution of pathways.

2 Preliminaries

2.1 Data

The metabolic network analysed in this work was obtained from the PATHWAY database from KEGG [8]. Data describing reactions, compounds and enzymes were downloaded and stored locally using a relational database management system (postgreSQL). The KEGG database contains metabolic data concerning 209 sequenced organisms. The network we built from such data is therefore a consensus of our current knowledge on the metabolisms of all those organisms. As a consequence, sequences of reactions present in the network may have been observed in no organism. To avoid this configuration, one can "filter" the consensus network by an organism of interest, keeping only in the dataset reactions catalysed by enzymes the organism is considered to be able to synthetize. We adopt a different strategy by choosing to perform our motif search on the consensus network and to possibly filter the results in a second step, allowing for easier comparative analysis between organisms.

Moreover, we use an additional information present in KEGG: the notion of primary/secondary metabolites. Indeed, in the KEGG reference pathway diagrams (maps), only primary metabolites are represented and connect reactions together, whereas secondary metabolites are not drawn (even though they participate in the reaction). A typical example of a secondary metabolite is the ATP molecule in an ATP-consuming reaction. (Observe that, unlike the notion of ubiquitous compound [14], the notion of primary/secondary metabolite is relative to a reaction.) Keeping all metabolites in the network leads to the creation of artefactual links between reactions and the bias introduced can lead to inaccurate results such as considering metabolic networks as small-world networks as shown in [3]. Withdrawing secondary metabolites may not be the best strategy to adopt, but it represents a simple way of avoiding this bias.

2.2 Graph Models

Several formal models have been in use to study metabolic networks. The choice of a formal model seems to depend mainly on the nature of the hypotheses one wishes to test (qualitative or quantitative, static or dynamic) and on the size of the network under study. Differential equations seem well adapted to study the dynamic aspects of very small networks whereas graphs enable the static study of very large networks.

Between these two ends of the spectrum, semi-quantitative models have been proposed. For example, Petri nets allow for the simulation and dynamical analysis of small networks [21], while constraint-based models provide a mathematical framework enabling to decompose the network into functional states starting only from information on stoichiometry and making the assumption that the network is at steady-state [12].

As our goal is to deal with large networks and work with the least possible *a priori*, graph models seem appropriate. In previous genome-scale studies [7], graphs have been used mainly for topological analyses regardless of the nature of their components (reactions, compounds and enzymes). We propose to enrich the graph models and take into consideration some of the features of such components.

Formally, a graph G is defined as a pair (V, E), with V a set of *vertices* and $E \subseteq V \times V$ a set of *edges*. The edges represent the relations between the vertices and may be directed or undirected. The vertices and edges of the graph can be labelled.

The most intuitive graph representation of a metabolic network is provided by a bipartite graph. A bipartite graph has two types of vertices which in the context of metabolic networks represent, respectively, reactions and chemical compounds. The compound graph is a compact version of the bipartite graph where only compound vertices are kept and information on the reactions is stored as edge labels. The reaction graph is the symmetric representation of a compound graph (*i.e.*, reaction vertices are kept and information on the compounds is stored as edge labels). Directed versions of these graphs can be drawn expressing the irreversibility of some reactions. The information concerning the reversibility of reactions is generally not well-known. Indeed, contradictions may be found within a same database. We therefore consider this information as uncertain and, in an initial step, assume that all reactions are reversible. This apparently strong hypothesis seems preferable than considering a reaction as irreversible when it actually is reversible (leading to a loss of information).

In the following sections, we denote by C a finite set of labels, which we refer as *colours*, that correspond to reaction labels. Also, we assume the graph $G = (V, E)$ is undirected and that we are given, for each vertex, a set of colours from C. Reversibility and edge labels will not be used. If needed, one can use them in a later step.

2.3 Motif Definition

We define a motif using the nature of the components it contains.

Definition 1. *A* motif *is a multiset of elements from the set C of colours.*

As mentioned earlier, we choose in this definition not to introduce any constraint on the order of the reactions nor on topology. This choice is motivated by the wish to explore the network with the least possible *a priori* information on what we are searching for. Topology and order of the reactions can be used later as further constraints. The advantage of this strategy is that the impact of each additional constraint can then be measured.

2.4 Occurrence Definition

Intuitively, an occurrence is a connected set of vertices labelled by the colours of the motif. For a precise definition, let R be a set of vertices of G and let M be a motif of the same size as R. Let $H(R, M)$ denote the bipartite graph whose set of vertices is $R \cup M$ and where there is an edge between a vertex v of R and a vertex c of M if and only if v has c as one of its colours.

Definition 2. Definition of an exact occurrence of a motif
An exact occurrence *of a motif M is a set R of vertices of G such that $H(R, M)$ has a perfect matching and R induces a connected subgraph of G.*

If one is strict on the relation of similarity between colours (colours are considered the same only if they are identical), the risk is to find a single occurrence, or none, of any given motif in the network [3]. Moreover, since studying the evolution of what the graph G represents is one of our main objectives, it seems relevant to allow for flexibility in the search for occurrences of a motif.

With this in mind, we introduce a function S (detailed later) that assigns, to each pair c_i, c_j in $C \times C$, a score which measures the similarity between c_i and c_j. Two colours are considered similar if this score is superior to a threshold s. We then adapt our definition of exact occurrence by modifying $H(R, M)$ in the following way. There will be an edge between a vertex v in R and a colour c in M if and only if there exists a colour c' of v such that the value of $S(c', c) \geq s$. Further, we generalise this to the case where the threshold s is different for every element c in M. The latter is motivated by the idea that some elements in the motif we are searching for may be more crucial than others. Observe that these considerations are independent of the definition of S that is discussed in the next section.

Another type of flexibility can then be added, that allows for gaps in the occurrences. By this we mean, roughly, allowing the occurrence to have more vertices just to achieve the connectivity requirement. These extra vertices are not matched to the elements of the motif. Two types of control on the number of gaps are considered: local and global. Intuitively, a local gap control policy bounds the maximum number of consecutive gaps allowed between a pair of matched vertices of R. A global control policy bounds the total number of gaps in an occurrence.

This leads to the following definition of an approximate occurrence of a motif, where we denote by G_R the subgraph of G induced by a set R of vertices of G.

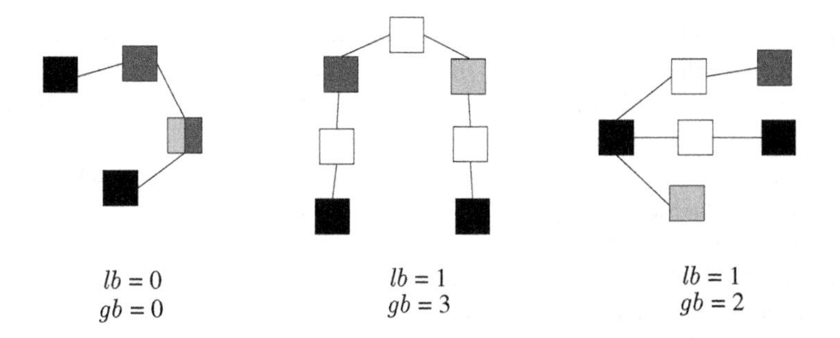

Fig. 1. Subgraphs induced by occurrences for the motif {black, black, dark grey, light grey}.

Definition 3. Definition of an approximate occurrence of a motif

Let lb and gb be the local and global gap control bounds and let M be a motif. For each c in M, let s_c be a number. An approximate *occurrence of M (with respect to lb, gb and the thresholds s_c) is any minimal set R of vertices of G that has a subset R' that satisfies the following conditions:*

1. *the bipartite graph $H(M \cup R', E_H)$ with $E_H = \{\{c, v\} \in M \times R' |$ there exists a colour c' of v such that $S(c', c) \geq s_c\}$ contains a perfect matching;*
2. *for each subset B of R' such that $B \neq \emptyset$ and $R' \setminus B \neq \emptyset$, the length of a shortest path in G_R between B and $R' \setminus B$ is at most lb;*
3. *$|R| - |R'| \leq gb$.*

The minimality requirement on the set R avoids uninteresting approximate occurrences that are simple copies of other occurrences with extra vertices connected to them.

Observe that when no gaps are allowed then $R = R'$ and condition 2 means simply that G_R is connected. An example is given in Figure 1.

2.5 Reaction Similarity

We now discuss function S for the problem of metabolic networks and reaction motifs in such networks. Various functions of different nature may be used. We present here two possible ways to define S.

The first one is based on alignment. Indeed, in order to compare reactions, which is what function S is used for, one can compare the enzymes that catalyse these reactions by performing an alignment of their sequences (or structures). An element of C would then be a protein sequence (or structure). The function S assigns a sequence (or structure) alignment score and s is a user-defined threshold that has to be met to consider the sequences (structures) similar. In the case of whole networks, sequences are preferable since many structures are not known.

The second example is the one we adopt in this paper. It is based on a hierarchical classification of enzymes developed by the International Union of Biochemistry and Molecular Biology (IUBMB) [1]. It consists in assigning to each enzyme a code with 4 numbers expressing the chemistry of the reaction it catalyses. This code is known as the enzyme's EC number (for Enzyme Commission Number). The first number of the EC number can take values in $[1..6]$, each number symbolizing the 6 broad classes of enzymatic activity. (1. Oxidoreductase, 2. Transferase, 3. Hydrolase, 4. Lyase, 5. Isomerase, 6. Ligase.) Then each of the three remaining numbers of the EC number provides additional levels of detail. For example, the EC number 1.1.1.1 refers to an oxidoreductase (1) with CH-OH as donor group and NAD+ as acceptor group.

An element of C is in this case an EC number. The function S then assigns a similarity score between two EC numbers that corresponds to the index of the deepest level down to which they remain identical. For example, $S(1.1.1.2, 1.1.1.3) = 3$. Two EC numbers are considered similar if their similarity score is above a user-defined cut-off value s in $[0..4]$. The advantage of this definition of similarity between colours, *i.e.*, reactions, is that it is more directly linked to the notion of function. Reactions compared with this measure are likely to be functionally related (and possibly evolutionarily related also).

3 Algorithmics

3.1 Hardness Results

The formal problem we address is the following:

Search Problem. Given a motif M and a labelled undirected graph G, find all occurrences of M in G.

As mentioned earlier, this problem is different from subgraph isomorphism because the topology is not specified for the motif.

For this problem, we may assume the graph is connected and all vertices have colours that appear in the motif. Otherwise, we preprocess the graph throwing away all the vertices having no colour appearing in the motif and solve the problem in each component of the resulting graph.

A natural variant of the Search Problem consists in, given a motif and a labelled graph, deciding whether the motif occurs in the graph or not. As before, we may assume the graph is connected, all vertices are labelled with colours and all colours appear in the motif. It is easy to see this decision version of the Search Problem is in NP. We show next that it is NP-complete even if G is a tree, which implies that the Search Problem is NP-complete for trees. For the following proof, we consider the version where no gaps are allowed.

NP-Complete for Trees. We have the following proposition.

Proposition 1. *The Search Problem is NP-complete even if G is a tree.*

Proof. We present a reduction from EXACT COVER BY 3-SETS (X3C):

INSTANCE: Set X with $|X| = 3q$ and a collection $\mathcal{C}$ of 3-element subsets of X.

QUESTION: Does $\mathcal{C}$ contain an exact cover for X, *i.e.*, a subcollection $\mathcal{C}' \subseteq \mathcal{C}$ such that every element of X occurs in exactly one member of $\mathcal{C}'$?

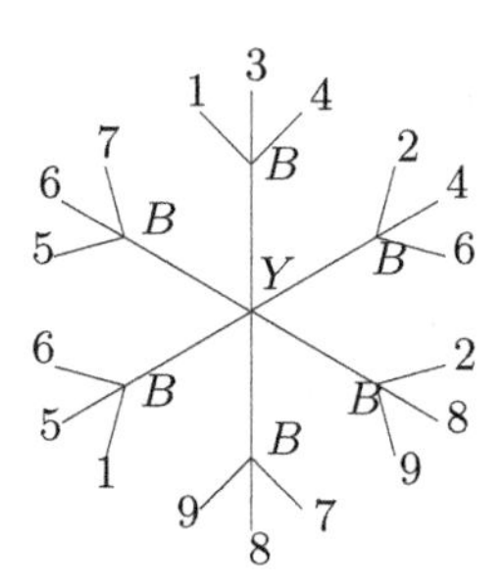

Fig. 2. Tree T and its labels for $X = \{1, \ldots, 9\}$ and $\mathcal{C} = \{\{1,3,4\}, \{2,4,6\}, \{2,8,9\}, \{7,8,9\}, \{1,5,6\}, \{5,6,7\}\}$. The motif M in this case is $\{Y, B, B, B, 1, \ldots, 9\}$.

Let $X = \{1, \ldots, 3q\}$ and $\mathcal{C} = \{C_1, \ldots, C_n\}$ be an instance of X3C. The instance for the decision version of the Search Problem consists of a motif $M = \{Y, B, \ldots, B, 1, \ldots, 3q\}$, where B appears q times in M, and a tree T as follows. (See Figure 2 for an example.) There are four vertices in T for each i, $1 \leq i \leq n$, three of them are leaves in T, each one labelled by one of the elements of C_i. The fourth vertex, named r_i, is adjacent to the three leaves and has colour B. Besides these $4n$ vertices, there is only one more vertex in T, which is labelled Y and is adjacent to each r_i. This completes the description of the instance. Clearly it has size polynomial in the size of X and $\mathcal{C}$.

To complete the reduction, we need to argue that the motif M occurs in T if and only if there is a subcollection $\mathcal{C}'$ of $\mathcal{C}$ such that each element of X occurs exacly in one member of $\mathcal{C}'$.

Suppose there is such a $\mathcal{C}'$. Clearly $|\mathcal{C}'| = q$. Let R be the set of vertices of T consisting of the vertex labelled Y and the four vertices of each C in $\mathcal{C}'$. The subgraph of T induced by R is connected. Also, in R, there is a vertex labelled Y, q vertices labelled B (one for each C in $\mathcal{C}'$) and one labelled by each element in X (because of the property of $\mathcal{C}'$). That is, R is an occurrence of M in T.

Now, suppose there is an occurrence of M in T, that is, there is a set R of $1 + 4q$ vertices of T that induces a connected subgraph of T and has a vertex labelled by each of the colours in M. Let $\mathcal{C}'$ consist of the sets C_i in $\mathcal{C}$ whose vertex r_i in T is in R. Let us prove that each element of X appears in exactly one of the sets in $\mathcal{C}'$. First, note that the vertex labelled Y is necessarily in R, because it is the only one labelled Y and there is a Y in M. Then, as R induces a connected graph, a leaf from a set C_i is in R if and only if r_i is also in R. But R must contain exactly q vertices labelled B. Consequently, $|\mathcal{C}'| = q$ and, as R must contain $1 + 4q$ vertices, all three leaves of each C in $\mathcal{C}'$ must be in R, and these are all vertices in R. As R must contain a vertex labelled after each element in X, there must be exactly one set in $\mathcal{C}'$ containing each element in X. □

Fixed Parameter Tractability. This problem is fixed-parameter tractable with parameter k. Indeed, a naive fixed-parameter algorithm consists in generating all possible topologies for the input motif M, and then searching for each topology by using a subtree isomorphism algorithm. Since it is enough to generate all possible tree topologies for M, the number of topologies to consider depends (exponentially) on k only, and subtree isomorphism is polynomial in the size of both the motif M and the tree T where

M is sought. This reasoning is not valid anymore when the motif must be searched in a general graph G as subgraph isomorphism is NP-complete even when the motif is a tree [4].

General Complexity Results. Table 1 summarizes the complexity of the Search Problem for various types of motifs and graphs. As mentioned, it is enough to consider that our motifs are trees (or paths). This is because topology is indifferent (only connectivity matters).

By *fixed* in the Table, we mean that the colours of the vertices in a path (respectively tree) are fixed, otherwise (*i.e.* path/tree *not fixed*) we mean that we are searching for a path (respectively tree) with the given vertex colours but do not care in what order they appear, provided they all appear.

Motifs that are paths are already hard problems for general graphs G. This can be shown by a reduction from the Hamiltonian path problem.

Table 1. Complexity results for the motif Search Problem

MOTIF \ TYPE OF GRAPH		path	tree	graph
path	fixed	polynomial	polynomial	NP-complete
	not fixed	polynomial	polynomial	NP-complete
tree	fixed	—	polynomial	NP-complete
	not fixed	—	NP-complete, FPT in k	NP-complete

Since the instances we have to consider in the case of metabolic networks are relatively small (3184 vertices and 35284 edges for the network built from the KEGG Pathway database), it is possible to solve the problem exactly, provided some efficient pruning is applied. This is described in the next section.

3.2 Exact Algorithm

Version with no Gaps. We now present an exact algorithm which solves the Search Problem. We first explain it for the simple case where the gap parameters lb and gb are set to 0 and then we show how it can be extended to the general case.

Let M be the motif we want to search for. A very naive algorithm would consist in systematically testing all sets R of k vertices as candidates for being an occurrence, where $k = |M|$. For R to be considered an occurrence of M, the subgraph induced by R must be connected and there must be a perfect matching in the bipartite graph $H(R, M)$ that has an edge between $r \in R$ and $c \in M$ if and only if c is similar to one of the colours at vertex r. The search space of all combinations of k vertices among the n vertices in G is huge. We therefore show two major pruning ideas arising from the two conditions that R has to fulfill to be validated as an occurrence of M.

The connectivity condition can be checked by using a standard method for graph traversal, such as breadth first search (BFS). In our case, a BFS mixed with a backtracking strategy is performed starting from each vertex in the graph. At each step of the search, a subset of the vertices in the BFS queue is marked as part of the candidate set R. The queue, at each step, contains only marked vertices and neighbours in G of marked vertices. Also, there is a pointer p to the last marked vertex in the queue. At each step, there are two cases to be analysed: either there are k vertices marked or not. If there are k vertices marked, we have a candidate set R at hand. We submit R to the test of the colouring condition, described below, and we backtrack to find the next candidate set. If there are less than k vertices marked, then there are two possible cases to be analysed: either p is pointing to the last vertex in the queue or not. If p is not pointing to the last vertex in the queue, we move p one position ahead in the queue, mark the next vertex and queue its neighbours that are not in the queue already (checking the latter can be done in constant time by adding a flag to each vertex in the original graph). Then we repeat, that is, start a new step. If, on the other hand, p is pointing to the last vertex in the queue, then we backtrack. The backtracking consists of unmarking the vertex pointed to by p, unqueueing its neighbours that were added when it was marked, moving p to the previous marked vertex in the queue and starting a new step. (If no such vertex exists, the search is finished.) Next we describe the test of the colouring condition.

Given a candidate set R, one can verify the colouring condition by building the graph H and checking whether it has a perfect matching or not. In fact, we can apply a variation of this checking to a partial set R, that is, we can, while constructing a candidate set R, be checking whether the corresponding graph H has or not a complete matching. The latter is a matching that completely covers the partial candidate set R. If there is no such matching, we can move the search ahead to the next candidate set. This verification can be done in constant time using additional data structures that are a constant time the size of the motif.

Extra optimisations can also be added. For instance, instead of using every vertex as a seed for the BFS, we can use only a subset of the vertices: those coloured by one of the colours from the motif, preferably the less frequent in the graph.

Allowing for Gaps. Allowing for local but not global gaps (*i.e.*, setting $lb > 0$ and $gb = \infty$) can easily be done by performing the $lb-$transitive closure of the initial graph G and applying the same algorithm as before to the graph with augmented edge set. The $p-$transitive closure of a graph G for p a positive integer is the graph obtained from G by adding an edge between any two vertices u and v such that the length l of a shortest path from u to v in the original graph satisfies $1 < l \leq p$. The $p-$transitive closure can be done at the beginning of the algorithm or on the fly. In the latter case, when a next vertex is added to the queue, instead of queueing its neighbours only, all vertices at distance at most p from it are queued (if they are not already in the queue) where by distance between any two vertices we mean the number of vertices other than these two in a shortest path between them.

Allowing for global gaps as well as local ones is more tricky. The reason is that an unmarked vertex can be put in the queue because of many different marked vertices. When backtracking in the queue at any step in the algorithm, unmarked vertices that have been queued only because of the marked vertex v that is being dequeued can be safely eliminated from the queue. Unmarked vertices $\{v_i\}$ that were queued because of the vertex being dequeued *and* of at least one other marked vertex will remain (somewhere) in the queue. Therefore, in order to correctly account for the global number of gaps introduced so far in the current occurrence, one must consider all the remaining marked vertices that implied the queuing of $\{v_i\}$. Extra information must be kept to locate in constant time the unmarked vertices $\{v_i\}$ and to update the global count of gaps. This information can be kept in a balanced tree of size proportional to $k = |M|$ associated with each queued unmarked vertex u'. Each node in the tree corresponds to a marked vertex u that could have led to the queuing of u' and is labelled by the distance from u' to u (this distance is at most lb). Keeping, updating and using the extra information adds a multiplicative term in $O(k \log k)$ to the time complexity of the algorithm, which seems reasonable.

On average, searching for all occurrences of a motif of size 4 with no gaps and threshold $s = 3$ takes 8 microseconds of CPU time on a Pentium 4 (CPU 1.70 GHz) with 512 Mb of memory.

4 Application

The approach we propose, and have described in the previous sections, should enable both to generate some hypotheses on the evolution of metabolic pathways, and to analyse global features of the whole network.

We start by presenting a case study motivated by trying to understand how metabolic pathways evolve. We do not directly answer this question, which is complex and would be out of the scope of this paper. Instead, we give a first example of the type of evolutionary question people have been asking already and have addressed in different, often semi-manual ways in the past [20], and that the algorithm we propose in this paper might help treat in a more systematic fashion.

As in [20], one is often interested in a specific pathway, and, for instance, in finding whether this pathway can be considered similar to other pathways in the whole metabolic network thus suggesting a common evolutionary history. The metabolic pathway we chose as example is valine biosynthesis. Focusing on the last five steps of the pathway, we derived a motif $M = \{1.1.1.86, 1.1.1.86, 4.2.1.9, 2.6.1.42, 6.1.1.9\}$ and performed the search for this motif using initially a cut-off value s of 4 for the similarity score between two EC numbers (that is, between two reaction labels). With this cut-off value, the motif was found to occur only once. (see Figure 3).

From this strictly defined motif, we then relaxed constraints by first lowering the cut-off value s from 4 to 3 and then setting the gap parameters to 1 (motif denoted by M'). Additional occurrences were found. Three of them particularly drew our attention (see Figure 3).

The first one corresponds to the five last steps of the isoleucine biosynthesis. The second one corresponds to the five last steps of the leucine biosynthesis. Together, they suggest a common evolutionary history for the biosynthesis pathways of valine, leucine and isoleucine.

An interesting point concerning the second occurrence is the fact that the order of the reactions is not the same as in the other pathways. This occurrence would not have been found if we had used a definition of motif where the order was specified.

Finally, the third occurrence that drew our attention was formed by reactions from both the biosynthesis of valine and a distinct metabolic pathway, namely the biosynthesis of Panthotenate and CoA. This latter case illustrates a limit of our current general

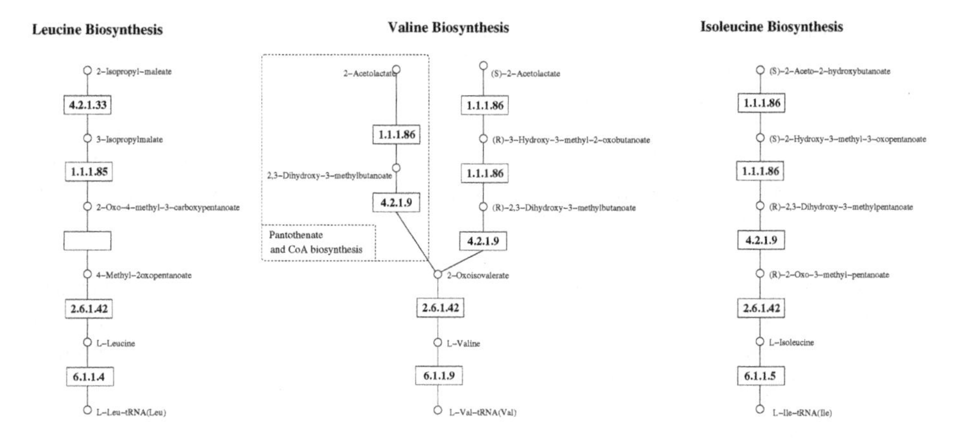

Fig. 3. Bipartite representation of the results obtained when searching for the following motif : $M' = \{1.1.1, 4.2.1, 2.6.1.42, 6.1.1\}$ with local and global gap bounds set to 1. The empty box in the leucine biosynthesis represents a spontaneous reaction.

way of thinking about metabolism: frontiers between metabolic pathways as defined in databases are not tight. If we had taken such frontiers into account, we would not have found this occurrence that overlaps two different pathways. Yet such occurrence can be given a biological meaning: it can be seen as a putative alternative path for the biosynthesis of valine.

To complement this analysis, one should add that the results presented in this section hold for 125 organisms in KEGG among which *S. cerevisiae* and *E. coli*.

Intrigued by the potential importance of inter-pathway occurrences, we computed their proportion in the general case of a randomly chosen motif. By systematically testing all motifs of size 3 and 4 (with cut-off values set to 3), we found that, on average, a motif of size 3 (respectively 4) has 74% (respectively 92%) of its occurrences that are inter-pathway occurrences. All inter-pathway occurrences may not represent biologically meaningful chemical paths but the proportions above suggest that a lot of information may be lost when studying pathways and not networks.

5 Conclusion

In this paper, we presented a novel definition of motif, called a "reaction motif", in the context of metabolic networks. Unlike previous works, the definition of motif is focused on reaction labels while the topology is not specified. Such novel definition raises original algorithmic issues of which we discuss the complexity in the case of the problem of searching for such motifs in a network. To demonstrate the utility of our definition, we show an example of application to the comparative analysis of different amino-acid biosynthesis pathways. This work represents a first step in the process of exploring the building blocks of metabolic networks. It seems promising in the sense that, with a simple definition of motif, biologically meaningful results are found.

We are currently working on an enriched definition of motif that will take into account information on input and output compounds. The current definition already enables to discover regularities in the network. Enriched definitions should enable to test more precise hypotheses.

In this paper, we used a particular formalism for analysing a metabolic network through the identification of motifs. Other formalisms have been employed or could be considered. As J. Stelling indicated in his review of 2004 [18], each formalism gives a different perspective and confronting them seems to be a promising way of getting at a deeper understanding of such complex networks.

Acknowledgements. The authors would like to thank Anne Morgat, Alain Viari and Eric Tannier for very fruitful discussions. The work presented in this paper was funded in part by the ACI Nouvelles Interfaces des Mathématiques (project π-*vert*) of the French Ministry of Research, and by the ARC (project *IBN*) from the INRIA.

References

1. *Recommendations of the Nomenclature Commitee of the International Union of Biochemistry and Molecular Biology on the Nomenclature and Classificationof Enzymes.* Oxford University Press, 1992.
2. E. Alm and A. Arkin. Biological networks. *Current opinion in Structural Biology*, 13:193–202, 2003.
3. M. Arita. The metabolic world of *escherichia coli* is not small. *PNAS*, 101(6):1543–1547, 2004.
4. M. R. Garey and D. S. Johnson. *Computers and Intractability. A Guide to the Theory of NP-Completeness.* Freeman, 1979.
5. R. Guimerà and LA. Nunes Amaral. Functional cartography of complex metabolic networks. *Nature*, 433(7028):895–900, 2005.
6. L. Hartwell, J. Hopfield, A. Leibler, and A. Murray. From molecular to modular cell biology. *Nature*, 402:c47–c52, 1999.
7. H. Jeong, B. Tombor, R. Albert, Z.N. Oltvai, and AL. Barabasi. The large-scale organization of metabolic networks. *Nature*, 407:651–654, 2000.
8. M. Kanehisa, S. Goto, S. Kawashima, Y. Okuno, and M. Hattori. The KEGG resource for deciphering the genome. *Nucleic Acids Research*, 32:277–280, 2004.
9. H. Kitano. Systems biology: A brief overview. *Science*, 295:1662–1664, 2002.
10. HW. Ma, XM. Zhao, YJ. Yuan, and AP Zeng. Decomposition of metabolic network into functional modules based on the global connectivity structure of reaction graph. *Bioinformatics*, 20(12):1870–1876, 2004.
11. JA. Papin, JL. Reed, and BO. Palsson. Hierarchical thinking in network biology: the unbiased modularization of biochemical networks. *Trends Biochem Sci.*, 29(12):641–7, 2004.
12. JA. Papin, J. Stelling, ND. Price, S. Klamt, S. Schuster, and BO. Palsson. Comparison of network-based pathway analysis methods. *Trends Biotechnol.*, 22(8):400–5, 2004.
13. RY. Pinter, O. Rokhlenko, D. Tsur, and M. Ziv-Ukelson. Approximate labelled subtree homeomorphism. In *Proceedings of the 15th Annual Symposium on Combinatorial Pattern Matching (CPM)*, volume 3109 of *LNCS*, pages 59–73, 2004.
14. S. Schuster, T. Pfeiffer, F. Moldenhauer, I. Koch, and T. Dandekar. Exploring the pathway structure of metabolism: decomposition into subnetworks and application to *Mycoplasma pneumoniae*. *Bioinformatics*, 18(2):351–361, 2002.
15. D. Segrè. The regulatory software of cellular metabolism. *Trends Biotechnol.*, 22(6):261–5, 2004.

16. P. Shannon, A. Markiel, O. Ozier, NS. Baliga, JT. Wang, D. Ramage, N. Amin, B. Schwikowski, and T. Ideker. Cytoscape: A software environment for integrated models of biomolecular interaction networks. *Genome Res.*, 13(11):2498–504, 2003.
17. S. Shen-Orr, R. Milo, S. Mangan, and U. Alon. Network motifs in the transcriptional regulation network of *escherichia coli. Nat. Genet.*, 31(1):64–8, 2002.
18. J. Stelling. Mathematical models in microbial systems biology. *Curr Opin Microbiol.*, 7(5):513–8, 2004.
19. Y. Tohsato, H. Matsuda, and A. Hashimoto. A multiple alignment algorithm for metabolic pathway analysis using enzyme hierarchy. In *Proc. Int. Conf. Intell. Syst. Mol. Biol.*, pages 376–383, 2000.
20. A. M. Velasco, J. I. Leguina, and A. Lazcano. Molecular evolution of the lysine biosynthetic pathways. *J. Mol. Evol.*, 55:445–459, 2002.
21. K. Voss, M. Heiner, and I. Koch. Steady state analysis of metabolic pathways using Petri nets. *In Silico Biol.*, 3(3):367–387, 2003.
22. D. Wolf and A. Arkin. Motifs, modules and games in bacteria. *Curr. Opin. Microbiol.*, 6(2):125–134, 2003.

Reconstructing Metabolic Networks Using Interval Analysis

Warwick Tucker[1] and Vincent Moulton[2]

[1] Department of Mathematics, Uppsala University, Box 480, Uppsala, Sweden
warwick@math.uu.se
http://www.math.uu.se/~warwick
[2] School of Computing Sciences, University of East Anglia, Norwich, NR4 7TJ, UK
Vincent.Moulton@cmp.uea.ac.uk
http://www.cmp.uea.ac.uk

Abstract. Recently, there has been growing interest in the modelling and simulation of biological systems. Such systems are often modelled in terms of coupled ordinary differential equations that involve parameters whose (often unknown) values correspond to certain fundamental properties of the system. For example, in metabolic modelling, concentrations of metabolites can be described by such equations, where parameters correspond to the kinetic rates of the underlying chemical reactions. Within this framework, the increasing availability of time series data opens up the attractive possibility of reconstructing approximate parameter values, thus enabling the *in silico* exploration of the behaviour of complex dynamical systems. The parameter reconstruction problem, however, is very challenging – a fact that has resulted in a plethora of heuristics methods designed to fit parameters to the given data.

In this paper we propose a completely deterministic method for parameter reconstruction that is based on interval analysis. We illustrate its utility by applying it to reconstruct metabolic networks using S-systems. Our method not only estimates the parameters very precisely, it also determines the appropriate network topologies. A major strength of the proposed method is that it proves that large portions of parameter space can be disregarded, thereby avoiding spurious solutions.

1 Introduction

A well-known and difficult problem in metabolic modeling is that of *parameter reconstruction*. A metabolic model is often given in terms of a system of ordinary differential equations $\dot{x} = f(x;p)$, where the right-hand side (the vector field) depends on a (multi-dimensional) parameter p. The problem is then to search for a particular $p^\star$ within a parameter space $\mathbb{P}$ such that the solutions of the system $\dot{x} = f(x;p^\star)$ match a given data set, in some pre-specified manner. Typically, the data set is a time series, that is, samples taken along one or several trajectories of the target system.

R. Casadio and G. Myers (Eds.): WABI 2005, LNBI 3692, pp. 192–203, 2005.

As in many other settings, parameter reconstruction in metabolic modelling is often recast as a global optimization problem. Due to the high dimensionality of the problem, however, straight-forward optimization strategies rarely produce accurate parameter values. Today, most methods used for parameter reconstruction are thus based on heuristic algorithms such as, for example, machine learning, genetic algorithms and PL-models – see [9] for a recent overview. In this paper, we describe a very general and completely deterministic approach to solving the parameter reconstruction problem, which is based on interval analysis. This allows us to examine entire sets of parameters, and thus to exhaust the global search within a finite number of steps.

Although our new approach is very general, we will focus on a particular class of differential equations commonly used to model biochemical networks, known as *S-systems* [17]. These have been extensively studied (see e.g. [5,3,10,7,18]), and have the appealing feature that the underlying metabolic network topology can be estimated along with the other parameters. In addition, several methods have been recently described for parameter reconstruction in S-systems (e.g. [10,18,16]).

2 Methods

2.1 Component-Wise Reconstruction via Slopes

Suppose that we are given a d-dimensional system of ordinary differential equations $\dot{x} = f(x;p)$, sampled at N distinct times (excluding the initial point, which is assumed to be known at time t_0), producing the data set $\{x(t_j)\}_{j=0}^{N}$, where each sample $x(t_j) = (x_1(t_j), \ldots, x_d(t_j))$ has d components. Rather than attempting to reconstruct parameters by solving the entire system $\dot{x} = f(x;p)$, it can be more helpful to obtain more detailed information localized at individual sample points. One way to do this is to use the samples to reconstruct the trajectories (e.g. via piece-wise splines) with some degree of smoothness. This enables the computation of an approximation of the vector field at each sample point:

$$s_{i,j} \approx f_i(x(t_j); p^\star), \qquad i = 1, \ldots, d; \quad j = 0, \ldots, N.$$

The number $s_{i,j}$ corresponds to the slope of the trajectory's i:th component at time t_j, see Figure 1.

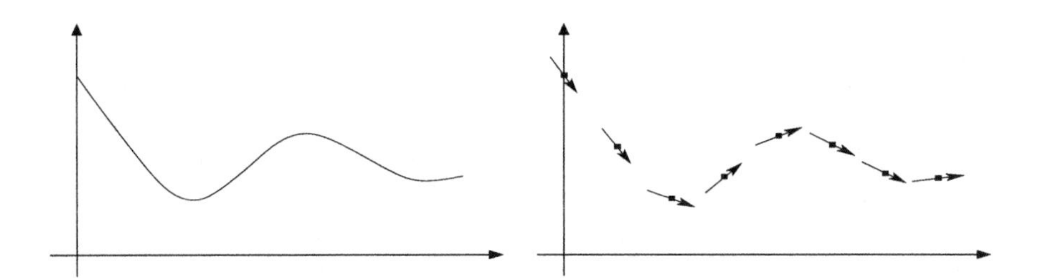

Fig. 1. (a) One component of a trajectory. (b) Sample data with slopes.

Equipped with this *enhanced* sample data, we can try to locate a point in $\mathbb{P}$ that minimizes the *defect*:

$$\Delta(p) = \sum_{i=1}^{d} \Delta_i(p) \stackrel{\text{def}}{=} \sum_{i=1}^{d} \sum_{j=0}^{N} \|f_i(x(t_j); p) - s_{i,j}\|,$$

for some convenient norm $\|\cdot\|$. Using the defect as a measurement of quality is not new, see e.g. [6] or, in the context of S-systems, [18]. The major advantage of this approach is that the system *decouples*, i.e., the computation of each $\Delta_i(p)$ depends only on a fraction of the total number of parameters: $\Delta_i(p) = \Delta_i(p_i)$, where $p_i \in \mathbb{P}_i$, and $\mathbb{P} = \mathbb{P}_1 \oplus \cdots \oplus \mathbb{P}_d$.

Assuming that each p_i has k (potential) components, the total dimension of the entire search space $\mathbb{P}$ is dk. Rather than searching through a dk-dimensional space, access to the enhanced sample data allows us to perform d independent searches in k-dimensions. The gain is immediate: introducing M grid-points in each parameter domain produces M^{dk} points in the first case, but only dM^k points in the latter. This gives a speed-up factor of M^d/d.

We point out that, at present, our proposed computation of the slopes is not very noise-tolerant. There are, however, several possible remedies that we aim to explore in the future. One possibility is to use piece-wise splines with set-valued coefficients. This approach fits well into the framework that we present below. Another option is to simply smooth the data (via e.g. least-squares) before fitting the splines.

2.2 Interval-Valued Slopes

Our approach is a modification of the enhanced data method, and therefore shares the same attractive decomposition property of the global parameter space $\mathbb{P}$. The major improvement is that we now compute ranges of slopes for entire domains of parameters. In essence, we extend the vector field f to a set-valued function F, accepting solid blocks in parameter space as input. The theoretical justification for this type of extension is given shortly. Let $[p_i]$ denote a box in $\mathbb{P}_i$, i.e., each component of $[p_i]$ is an interval. Then, for any point $p_i \in [p_i]$, we have

$$f_i(x(t_j); p_i) \in F_i(x(t_j); [p_i]),$$

i.e., the set $F_i(x(t_j); [p_i])$ contains *all possible* slopes corresponding to parameters taken from the box $[p_i]$. This fact gives us a simple criterion for discarding portions of the search space $\mathbb{P}_i$: if a box $[p_i]$, at a sample point $x(t_j)$, produces a range of slopes such that $s_{i,j} \notin F_i(x(t_j); [p_i])$, then *no* parameter in $[p_i]$ can have generated the sample data. If this situation occurs, we say that the parameter box $[p_i]$ violates the *cone condition* at time t_j, see Figure 2.

Our strategy in reconstructing the target parameter $p^\star$ is to adaptively partition each space $\mathbb{P}_i$ into successively smaller sub-boxes, retaining only those that satisfy the cone condition at all times. At some pre-selected level of coarseness,

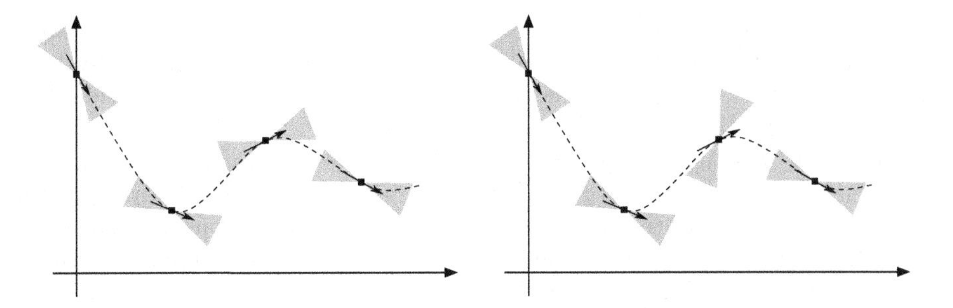

Fig. 2. (a) Cone condition satisfied at t_0, t_1, t_2, and t_3. (b) Violated at time t_2.

we terminate the process, and are left with a collection of boxes $[p_i^{(1)}], \ldots, [p_i^{(n)}]$, each of which satisfies $\mathcal{I}([p_i^{(j)}]) = \texttt{true}$, where

$$\mathcal{I}([p_i]) = \bigwedge_{j=0}^{N} \Big(s_{i,j} \in F_i(x(t_j); [p_i]) \Big) \tag{1}$$

is a boolean function that returns `true` if $[p]$ satisfies the cone condition at all sample times, and `false` otherwise.

2.3 Interval Analysis

Here, we will briefly describe the fundamentals of interval analysis. For a concise reference on this topic, see e.g. [12].

Let $\mathbb{IR}$ denote the set of closed intervals. For any element $[a] \in \mathbb{IR}$, we adapt the notation $[a] = [\underline{a}, \bar{a}]$. Thus "$x \in [x]$" means "the point x belongs to the interval $[x]$". If $\star$ is one of the operators $+, -, \times, \div$, we define the arithmetic on elements of $\mathbb{IR}$ by

$$[a] \star [b] = \{a \star b \colon a \in [a], b \in [b]\},$$

except that $[a] \div [b]$ is undefined if $0 \in [b]$. Working exclusively with closed intervals, we can describe the resulting interval in terms of the endpoints of the operands:

$$\begin{aligned} [a] + [b] &= [\underline{a} + \underline{b}, \bar{a} + \bar{b}] \\ [a] - [b] &= [\underline{a} - \bar{b}, \bar{a} - \underline{b}] \\ [a] \times [b] &= [\min(\underline{ab}, \underline{a}\bar{b}, \bar{a}\underline{b}, \bar{a}\bar{b}), \max(\underline{ab}, \underline{a}\bar{b}, \bar{a}\underline{b}, \bar{a}\bar{b})] \\ [a] \div [b] &= [a] \times [1/\bar{b}, 1/\underline{b}], \quad \text{if } 0 \notin [b]. \end{aligned} \tag{2}$$

When computing with finite precision, directed rounding must also be taken into account (see e.g. [11,13]).

A key feature of interval arithmetic is that it is *inclusion monotonic*, i.e., if $[a] \subseteq [A]$, and $[b] \subseteq [B]$, then

$$[a] \star [b] \subseteq [A] \star [B], \tag{3}$$

where we demand that $0 \notin [B]$ for division.

One of the main reasons for passing to the interval realm is that we want a simple way of enclosing the *range* $R(f; D) = \{f(x) \colon x \in D\}$ of a real-valued function $f \colon D \to \mathbb{R}$. Except for the most trivial cases, mathematics provides few tools to describe this set.

We begin by extending the real functions to *interval functions*. By this, we mean functions that take and return intervals rather than real numbers. Interval arithmetic (2) provides the theory of extending rational functions, i.e., functions on the form $f(x) = p(x)/q(x)$, where p and q are polynomials. Simply substituting all occurrences of the real variable x with the interval variable $[x]$ (and the real arithmetic operators with their interval counterparts) produces a rational interval function $F([x])$, called the *natural* interval extension of f. As long as no singularities are encountered, we have the inclusion $R(f; [x]) \subseteq F([x])$, by property (3). In fact, this type of range enclosure can be achieved for any reasonable function [12].

Higher-dimensional functions $f \colon \mathbb{R}^n \to \mathbb{R}$ can be extended to an interval function $F \colon \mathbb{IR}^n \to \mathbb{IR}$ in a similar manner. The function argument is then an *interval-vector* $[x] = ([x_1], \ldots, [x_n])$, which we also refer to as a *box*.

There exist several open source programming packages for interval analysis [4,8,14].

2.4 S-Systems

An S-system is a system of ordinary differential equations on the form:

$$\dot{x}_i = \alpha_i \prod_{j=1}^{d} x_j^{g_{ij}} - \beta_i \prod_{j=1}^{d} x_j^{h_{ij}} \qquad (i = 1, \ldots, d). \tag{4}$$

Each variable x_i represents the concentration of some reactant, and $\dot{x}_i$ denotes the time derivative of x_i. In a biochemical context, the non-negative parameters α_i and β_i are called *rate constants*. The real-valued parameters g_{ij} and h_{ij} are referred to as the *kinetic orders*. Each component of an S-system is made up of one positive and one negative term, corresponding to the production and consumption of the substance x_i, respectively. In essence, an S-system is a condensed version of a more general GMA – General Mass Action – model, obtained by aggregating individual reactions into the net processes of synthesis and degradation, see [15].

Using the following short-hand notation for the parameters

$$p_i = (\alpha_i, g_{i1}, \ldots, g_{id}, \beta_i, h_{i1}, \ldots, h_{id}) \qquad (i = 1, \ldots, d),$$

we can express (4) more compactly as $\dot{x}_i = f_i(x; p_i)$. The entire S-system then becomes $\dot{x} = f(x; p)$. A d-dimensional S-system has $2d(d+1)$ parameters, so already for small systems the number of parameters becomes unwieldy. We reduce the number of parameters by assuming that no reactant x_j influences both the rate of production *and* the rate of degradation of another reactant x_i (see [1,18]). This assumption can be reformulated more succinctly as:

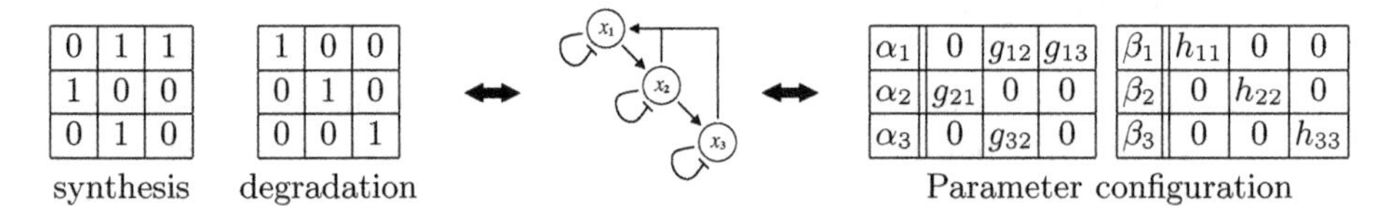

Fig. 3. A boolean topology encodes a metabolic network, and determines the parameter configuration of the S-system in (6)

$$g_{ij} \neq 0 \Rightarrow h_{ij} = 0, \tag{5}$$

and reduces the total number of non-zero parameters to $d(d+2)$, although we now must consider 2^d different parameter configurations for each component of the vector field f_i. Nevertheless, this is a good trade: filling each of the $2(d+1)$ parameter domains of f_i with M grid-points produces $M^{2(d+1)}$ points, compared to $2^d M^{d+2}$ points when using (5). This gives a speed-up factor of $(M/2)^d$.

A simple example of an S-system (appearing in [17] pp. 179-184) is given by:

$$\begin{aligned} \dot{x}_1 &= 7.5 x_2^{-0.1} x_3^{-0.05} - 5 x_1^{0.5} \\ \dot{x}_2 &= 2 x_1^{0.5} - 1.44 x_2^{0.5} \\ \dot{x}_3 &= 3 x_2^{0.5} - 7.2 x_3^{0.5}. \end{aligned} \tag{6}$$

The corresponding metabolic network is illustrated in Figure 3. This is an example of a *cascade* mechanism, which commonly appear in the context of gene regulation and immunology.

2.5 Set-Valued S-Systems

Extending the right-hand side of (4) to accept parameter boxes as input is a simple matter, and produces a vector field $F\colon \mathbb{IR}^d \to \mathbb{IR}^d$ whose components are interval-valued:

$$\dot{x}_i \in F_i(x; [p_i]) = [\alpha_i] \prod_{j=1}^{d} x_j^{[g_{ij}]} - [\beta_i] \prod_{j=1}^{d} x_j^{[h_{ij}]}. \tag{7}$$

It is easy to show ([2], p. 23) that this extension is *sharp*, i.e.,

$$R(f_i(x; \cdot); [p_i]) = F_i(x; [p_i]).$$

This sharpness property is not necessary for our method to work, but it does make it more efficient.

We briefly comment that it is possible to allow for uncertain data in the sense that the exact measurements x_j appearing the right-hand side of (7) be replaced by intervals $[x_j]$. This option will be explored in conjunction with the interval-based slope construction, mentioned in section 2.1.

2.6 The Main Algorithm

Given a collection of enhanced sample data $\{x_{i,j}; s_{i,j}\}_{i,j}$ generated from some target S-system with parameter $p^\star = (p_1^\star, \dots, p_d^\star)$, the search is divided into d independent component-wise searches for $p_1^\star, \dots, p_d^\star$. These $(d+2)$-dimensional searches can be performed as d parallel processes, seeing that they are completely independent. In what follows we fill focus on a single such search. For clarity, we will suppress the component index i.

Each search takes place within a global parameter region $\mathbb{P}$, which is initialized as a box $\mathbb{P} = ([P_1], \dots, [P_{2(d+1)}])$. The bounds for this box are determined by biochemical knowledge (e.g. [17]). Utilizing the constraints (5), we initialize all 2^d possible different parameter configurations $\tilde{\mathbb{P}}_1, \dots, \tilde{\mathbb{P}}_{2^d}$, each having $d+2$ non-zero parameters, and corresponding to different network topologies. Having done this, we examine each $\tilde{\mathbb{P}}_i$ separately (or all $\tilde{\mathbb{P}}_i$ in parallel). As a first step, we initialize a list `parameterList` with the unique element $\tilde{\mathbb{P}}_i$. This list is then passed on to the main loop of our search algorithm.

```
while( isEmpty(parameterList) == false ) {
  parameter = getCurrent(parameterList);
  if ( coneCondition(parameter) == true ) {
    if ( diameter(parameter) > Tol )
      splitAndStore(parameter, parameterList);
    else
      store(parameter, resultList);
  }
}
```

Within this loop, each member of `parameterList` is tested via the cone condition (1). If the condition is satisfied, there are two possibilities: either the diameter of the parameter box is smaller than some pre-assigned tolerance `Tol`, in which case the box is stored in a second list `resultList`; otherwise it is bisected along its widest component, and the two resulting sub-boxes are returned to `parameterList` for further investigation. If, however, the cone condition is not satisfied, the current parameter box is excluded from the remaining search. When the search terminates, `resultList` contains all sub-boxes of size $\approx$ `Tol` satisfying the cone condition. If this list is empty, we have established that this particular network topology does not match our data.

Often, we have have access to sample data from several trajectories, that is, trajectories emanating from different initial points $x^{(1)}(t_0), \dots, x^{(M)}(t_0)$. We can then modify the cone condition (1) to take this additional information into account:

$$\mathcal{I}([p_i]) = \bigwedge_{j=0}^{N} \bigwedge_{k=1}^{M} \left(s_{i,j}^{(k)} \in F_i(x^{(k)}(t_j); [p_i])\right)$$

This additional data improves our method, seeing that it becomes easier to discard parameter regions.

3 Computational Results

Starting with sample data $\{t_j; x_{i,j}\}_{i,j}$ generated from some target S-system with parameter $p^\star$, we first generate the slope data $\{s_{i,j}\}_{i,j}$, as described earlier. These computations are performed by a collection of `Matlab` scripts, using its built-in spline functionality. This allows us to differentiate the reconstructed trajectories, and recover the slopes. It should be pointed out that the data itself is generated within `Matlab`, and that the sample times $t_0, \ldots, t_N$ are non-uniformly distributed. We choose a logarithmic distribution of the sample times, in order to capture the more vivid motion occurring for small times. In the examples presented below we use noise-free sample data $\{x_{i,j}\}_{i,j}$, as discussed earlier.

The actual parameter reconstruction is carried out by a prototype `C++` program, utilizing a modified version of the PROFIL/BIAS interval package [14]. The computations were performed on a single 1200MHz Intel Pentium M processor using 384MB of RAM.

3.1 A Fixed-Topology Cascade

Our first example is the S-system (6) corresponding to the network presented in Figure 3. Note that, since we are given the network topology a priori, the computational complexity of parameter reconstruction is significantly reduced.

For the computations, we used five sets of initial conditions, and each trajectory was sampled at 20 points in time. Following [17], the search region for each of the kinetic orders g_{ij}, and h_{ij} was set to contain $[-1, +1]$, whereas the rate orders α_i and β_i were sought for within $[0, 15]$. The stopping tolerance (i.e., the diameters of the final parameter intervals) was set to 1×10^{-3}. In Table 1, we present the target parameters together with the final result of our reconstruction. We use the notation "—" to indicate the, a priori, non-present parameters. The agreement is seen to be almost perfect. The entire search took 1 minute and 6 seconds.

The reconstructed parameters appearing in Table 1 were obtained as follows: when the global search has terminated, we are left with a collection of parameter

Table 1. The original parameter values (A) and their reconstructions (B) for the S-system (6)

i	α_i	g_{i1}	g_{i2}	g_{i3}	β_i	h_{i1}	h_{i2}	h_{i3}
A								
1	7.5	—	−0.1	−0.05	5.0	0.5	—	—
2	2.0	0.5	—	—	1.44	—	0.5	—
3	3.0	—	0.5	—	7.2	—	—	0.5
B								
1	7.49	—	−0.100	−0.0503	4.99	0.501	—	—
2	2.00	0.501	—	—	1.44	—	0.502	—
3	3.00	—	0.500	—	7.20	—	—	0.500

boxes $[p_1], \ldots, [p_K]$, all satisfying the cone condition. We reduce these boxes to one single box $[\mathbb{P}^\star]$ by forming their *hull* – the smallest box containing all parameter boxes $[p_1], \ldots, [p_K]$. We then have an enclosure of the target parameter $p^\star \in [\mathbb{P}^\star]$. Of course, taking the hull of all parameter boxes is a rather crude measure. We get a better feeling for where the center of mass of the boxes is located by computing the average of the collection of parameter boxes. In order to get a single point in parameter space as our "best guess", we simply take the midpoint of the average:

$$\bar{\mathbb{P}}^\star = \mathtt{Mid}\left(\frac{1}{K}\sum_{i=1}^{K}[p_i]\right).$$

It is the components of the resulting $\bar{\mathbb{P}}^\star$ that are presented in Table 1. Note, however, that any choice of parameters from one of the resulting boxes $[p_1], \ldots, [p_K]$ is consistent with our sample data.

Also note that our computations *prove* that parameters outside the produced ranges do not match the sample data. Considering e.g. the h_{33}-parameter of (6), we found that $h_{33} \in [0.496, 0.503]$. The remaining parameters were enclosed as follows:

$$\begin{aligned}
(\alpha_1, \alpha_2, \alpha_3) &\in ([7.34, 7.62], [1.96, 2.03], [2.98, 3.03]) \\
(g_{12}, g_{13}) &\in ([-0.103, -0.0982], [-0.0527, -0.0486]) \\
(g_{21}, g_{32}) &\in ([0.492, 0.509], [0.493, 0.506]) \\
(\beta_1, \beta_2, \beta_3) &\in ([4.84, 5.13], [1.40, 1.46], [7.18, 7.23]) \\
(h_{11}, h_{22}) &\in ([0.485, 0.519], [0.489, 0.517]).
\end{aligned}$$

Interestingly, some of the parameters reconstructed in [17] did not fall in the parameter intervals that we computed. For example, even when starting the search with initial guesses close to the true values, the "quasi-Newton" algorithm used in [17] produced e.g. $\alpha_1 = 9.237$, $\beta_3 = 3.236$, $h_{22} = 0.0397$, all of which our algorithm has *proved* to be unsuitable. This example was also studied in [16] using four different methods. The outcomes for e.g. α_3 were 1.25, 7.70, 7.3, and 1.45, respectively. Note that none of these values belong to the parameter enclosure $\alpha_3 \in [2.98, 3.03]$ produced by our method.

3.2 A 4-Dimensional S-System

Our second example appears in [18]:

$$\begin{aligned}
\dot{x}_1 &= 12x_3^{-0.8} - 10x_1^{0.5} \\
\dot{x}_2 &= 8x_1^{0.5} - 3x_2^{0.75} \\
\dot{x}_3 &= 3x_2^{0.75} - 5x_3^{0.5}x_4^{0.2} \\
\dot{x}_4 &= 2x_1^{0.5} - 6x_4^{0.8}.
\end{aligned} \tag{8}$$

In this example we are not given the network topology, which makes the parameter reconstruction significantly more demanding. For the computations, we used five sets of initial conditions, and each trajectory was sampled at 20 points in time. Following [18], the search region for each of the kinetic orders g_{ij}, and h_{ij} was set to contain $[-1, +1]$, whereas the rate orders α_i and β_i were sought for within $[0, 20]$. The stopping tolerance was set to 2×10^{-3}. Note that, although each component of the vector field has 10 parameters to be determined, we use the constraints (5) to bring the number of non-zero parameter values down to six, which can be arranged in 16 different network topologies.

Table 2. The parameter values (and their reconstructions) of the S-system (8)

i	α_i	g_{i1}	g_{i2}	g_{i3}	g_{i4}	β_i	h_{i1}	h_{i2}	h_{i3}	h_{i4}
Original										
1	12	0.0	0.0	−0.8	0.0	10	0.5	0.0	0.0	0.0
2	8	0.5	0.0	0.0	0.0	3	0.0	0.75	0.0	0.0
3	3	0.0	0.75	0.0	0.0	5	0.0	0.0	0.5	0.2
4	2	0.5	0.0	0.0	0.0	6	0.0	0.0	0.0	0.8
Reconstructed										
1	12.00	0.0	0.0	−0.802	0.0	9.98	0.501	0.0	0.0	0.0
2	7.96	0.502	0.0	0.0	0.0	2.96	0.0	0.757	0.0	0.0
3	2.95	0.0	0.759	0.0	0.0	4.95	0.0	0.0	0.504	0.202
4	2.00	0.501	0.0	0.0	0.0	6.00	0.0	0.0	0.0	0.800

In Table 2, we present the target parameters together with the final result of our reconstruction. Once again, the agreement is seen to be almost perfect. The entire search took 3 hours, 29 minutes, and 27 seconds. This great increase in time, compared to the three-dimensional example, appears to indicate that the method scales very badly. Note, however, that this increase is mostly due to the fact that we were not given the topology of the four-dimensional system.

The reconstructed parameters appearing in Table 2 were obtained as in the previous example, but with one additional twist: after having computed the midpoint of the average, we set any parameter with value less than 5×10^{-4} to zero:

$$\bar{\mathbb{P}}^\star = \texttt{cutOff}\left(\texttt{Mid}\left(\frac{1}{K}\sum_{i=1}^{K}[p_i]\right); 5 \times 10^{-4}\right).$$

This *skeletalizing* procedure promotes sparse network topologies; in [18], the cut-off level is set to 1×10^{-1}.

4 Discussion

We have presented a novel method for reconstructing parameters using interval analysis. In particular, we have applied it to reconstruct metabolic networks using S-systems, and obtained very encouraging results. We stress that the proposed method is very general, and can be applied to *any* system of finitely parameterized differential equations.

Our method differs in a fundamental way from the main-stream reconstruction methods in that we solve the problem by a pruning scheme based on a boolean function (the cone condition), rather than recasting the parameter reconstruction as a global minimization problem. This has several advantages: First, it is well-known that global minimization is an intractable problem, in the sense that numerical solutions often converge to a local, rather than a global, minimum, and there is no way of telling the two cases apart. Second, the quantity to be minimized is often chosen to be a (weighted) least-square error. This implicitly pre-assumes rather strong statistical properties of the underlying data, assumptions that can not easily be verified. Our method simply retains the parameters that are consistent with the underlying data, avoiding both above-mentioned problems.

The transition to set-valued vector fields also allows us to dismiss, with a mathematical certainty, unrealistic network topologies. In particular, this allows us to detect when the model we are trying to fit to the provided data is not appropriate. At a sufficiently low tolerance, our method would then discard *all* parameter values.

In future work, we will refine the process of parameter exclusion, and exploit the problem's great potential for parallelization. This is an essential step towards exploring the scalability of our proposed method. We will also allow for noisy sample data, using interval-valued cubic splines in the generation of the slopes. We also plan to put our method to test on a larger class of problems (including generalized mass action models).

Acknowledgement

The authors would like to thank Korbinian Strimmer for introducing them to S-systems, and for helpful discussions.

References

1. Akutsu, T., Miyano, S. & Kuhara, S. Inferring qualitative relations in genetic networks and metabolic pathways, *Pacific Symposium on Biocomputing* **5** (2000) 120–301.
2. Alefeld, G. & Herzberger, J. *Introduction to Interval Computations.* Academic Press, New York (1983).
3. Alves, R. & Savageau, M. A. Comparing systemic properties of ensembles of biological networks by graphical and statistical methods, *Bioinformatics* **16:6** (2000) 527–533.
4. CXSC – C++ eXtension for Scientific Computation, version 2.0. Available from `www.math.uni-wuppertal.de/org/WRST/xsc/cxsc.html`
5. de Jong, H. Modeling and Simulation of Genetic Regulatory Systems: A Literature Review, *J. Comp. Biol.* **9:1** (2002) 67–103.
6. Enright, W. H. A New Error-Control for Initial Value Solvers, *Applied Mathematics and Computation* **31** (1998) 288–301.

7. Hlavacek, W. S. & Savageau, M. A. Rules for Coupled Expressions of Regulator and Effector Genes in Inducible Circuits, *J. Mol. Biol.* **255** (1996) 121–139.
8. INTLAB – INTerval LABoratory, version 4.1.2. Available from `www.ti3.tu-harburg.de/~rump/intlab/`
9. Kell, D. *Current Opinion in Microbiology* **7** (2004) 296–307.
10. Kikuchi, S., Tominaga, D., Arita, M., Takahashi, K. & Tomita, M. Dynamic modeling of genetic networks using genetic algorithm and S-system, *Bioinformatics* **19:5** (2003) 643–650.
11. Kulisch, U. W. & Miranker, W. L. *Computer Arithmetic in Theory and Practice.* Academic Press, New York (1981).
12. Moore, R. E. *Interval Analysis.* Prentice-Hall, Englewood Cliffs, New Jersey (1966).
13. Moore, R. E. *Methods and Applications of Interval Analysis.* SIAM Studies in Applied Mathematics, Philadelphia (1979).
14. PROFIL/BIAS – Programmer's Runtime Optimized Fast Interval Library/Basic Interval Arithmetic Subroutines. Available from `www.ti3.tu-harburg.de/Software/PROFILEnglisch.html`
15. Torres, N. V & Voit, E. O. *Pathway Analysis and Optimization in Metabolic Engeneering.* Cambridge University Press, Cambridge (2002).
16. Tsai, K. & Wang, F. Evolutionary optimization with data collocation for reverse engineering of biological networks *Bioinformatics* Advance Access published online on October 28 (2004).
17. Voit, E. O. *Computational Analysis of Biochemical Systems.* Cambridge University Press, Cambridge (2000).
18. Voit, E. O. & Almeida, J. Decoupling dynamical systems for pathway identification from metabolic profiles, *Bioinformatics* **20:11** (2004) 1670–168

A 1.375-Approximation Algorithm for Sorting by Transpositions

Isaac Elias[1] and Tzvika Hartman[2]

[1] Dept. of Numerical Analysis and Computer Science,
Royal Institute of Technology, Stockholm, Sweden
isaac@nada.kth.se
[2] Dept. of Molecular Genetics*,
Weizmann Institute of Science, Rehovot 76100, Israel
tzvi.hartman@weizmann.ac.il

Abstract. Sorting permutations by transpositions is an important problem in genome rearrangements. A transposition is a rearrangement operation in which a segment is cut out of the permutation and pasted in a different location. The complexity of this problem is still open and it has been a ten-year-old open problem to improve the best known 1.5-approximation algorithm. In this paper we provide a 1.375-approximation algorithm for sorting by transpositions. The algorithm is based on a new upper bound on the diameter of 3-permutations. In addition, we present some new results regarding the transposition diameter: We improve the lower bound for the transposition diameter of the symmetric group, and determine the exact transposition diameter of 2-permutations and simple permutations.

1 Introduction

When estimating the evolutionary distance between two organisms using genomic data one wishes to reconstruct the sequence of evolutionary events that transformed one genome into the other. In the 1980's, evidence was found that some species have essentially the same set of genes, but that their gene order differs [17,13]. This suggests that global rearrangement events, such as reversals and transpositions of genome segments, can be used to trace the evolutionary path between genomes. As opposed to local point mutations (i.e., insertions, deletions, and substitutions of nucleotides) global rearrangements are rare and may therefore provide more accurate and robust clues to the evolution.

In the last decade, a large body of work was devoted to genome rearrangement problems. Genomes are represented by permutations, with the genes appearing as elements. Circular genomes (such as bacterial and mitochondrial genomes) are represented by circular permutations. The basic task is, given two permutations, to find a shortest sequence of rearrangement operations that transforms one permutation into the other. Assuming that one of the permutations is the identity

* Work done while at the Dept. of Computer Science and Applied Mathematics, Weizmann Institute of Science.

R. Casadio and G. Myers (Eds.): WABI 2005, LNBI 3692, pp. 204–215, 2005.

permutation, the problem is to find the shortest way of sorting a permutation using a given rearrangement operation, or set of operations. For more background on genome rearrangements the reader is referred to [18,19,20].

The problem of sorting permutations by reversals has been studied extensively. It was shown to be NP-hard [8], and several approximation algorithms have been suggested [4,7,9]. On the other hand, for signed permutations (every element of the permutation has a sign, + or -, which represents the direction of the gene) a polynomial algorithm for sorting by reversals was first given by Hannenhalli and Pevzner [11]. Subsequent work improved the running time of the algorithm, and simplified the underlying theory [14,6,3,21].

There has been significantly less progress on the problem of sorting by transpositions. A transposition is a rearrangement operation, in which a segment is cut out of the permutation, and pasted in a different location. The complexity of sorting by transpositions is still open. It was first studied by Bafna and Pevzner [5], who devised a 1.5-approximation algorithm, which runs in quadratic time. The algorithm was simplified by Christie [9] and further by Hartman [12], which also proved that the analogous problem for circular permutations is equivalent. Eriksson et al. [10] provided an algorithm that sorts any given permutation on n elements by at most $2n/3$ transpositions, but has no approximation guarantee.

The transposition diameter of the symmetric group S_n is unknown. Bafna and Pevzner [5] proved an upper bound of $\frac{3}{4}n$, which was improved to $\frac{2}{3}n$ by the algorithm of Eriksson et al. [10]. A lower bound of $\lfloor\frac{n-1}{2}\rfloor + 1$ (for circular permutations) is given in [9,10,16], where it was conjectured to be the transposition diameter, except for $n = 14$ and $n = 16$.

In this paper we study the problem of sorting permutations by transpositions. We begin with some results regarding the transposition diameter. We prove a lower bound of $\lfloor\frac{n}{2}\rfloor + 1$ on the transposition diameter of the symmetric group of circular permutations, which shows that the conjecture of [9,10,16] is not accurate. Next, we deal with three subsets of the symmetric group (that have been considered in the genome rearrangement literature): simple permutations, 2-permutations, and 3-permutations. We show that the diameter for 2-permutations is $\frac{n}{2}$ (for circular permutations of size n), and for simple permutations is $\lfloor\frac{n}{2}\rfloor$. We prove an upper bound of $11\left\lfloor\frac{n}{24}\right\rfloor + \left\lfloor 3\frac{(n/3 \mod 8)}{2}\right\rfloor + 1$ on the diameter of 3-permutations. Then we derive our main result: A 1.375-approximation algorithm for sorting by transpositions, improving on the ten-year-old 1.5 ratio.

Our main result, like many other results in genome rearrangements, is based on a rigorous case analysis. However, since the number of cases is huge, we developed a computer program that systematically generates the proof. Each case in the proof is discrete and consists of a few elementary steps that can be verified by hand and thus it is a proof in the conventional mathematical sense. Since it is not practical to manually verify the proof as a whole, we have written a verification program, which takes the proof as an input and verifies that each elementary step in the proof is correct. The proof, along with the program, is presented in a user-friendly web interface [1]. A well-known example of a computer assisted proof is that of the *Four Color Theorem* [2] (see [22] for a list of other proofs).

2 Preliminaries

Let $\pi = (\pi_1 \ \dots \ \pi_n)$ be a permutation on n elements. Define a *segment* A in π as a sequence of consecutive elements $\pi_i, \dots, \pi_k$ ($k \geq i$). Two segments $A = \pi_i, \dots, \pi_k$ and $B = \pi_j, \dots, \pi_l$ are *contiguous* if $j = k + 1$ or $i = l + 1$. A *transposition* τ on π is an exchange of two disjoint contiguous segments. If the segments are $A = \pi_i, \dots, \pi_{j-1}$ and $B = \pi_j, \dots, \pi_{k-1}$, then the result of applying τ on π, denoted $\tau \cdot \pi$, is $(\pi_1 \ \dots \ \pi_{i-1} \ \pi_j \ \dots \ \pi_{k-1} \ \pi_i \ \dots \ \pi_{j-1} \ \pi_k \ \dots \ \pi_n)$ (note that the end segments can be empty if $i = 1$ or $k - 1 = n$).

The problem of finding a shortest sequence of transpositions, which transforms a permutation into the identity permutation, is called *sorting by transpositions.* The *transposition distance* of a permutation π, denoted by $d(\pi)$, is the length of the shortest sorting sequence.

The problem of sorting linear permutations of size n is equivalent to sorting circular permutations of size $n + 1$ [12]. Many of the following definitions, as well as the presentation of the algorithm, are more clear for circular permutations. Therefore we present our results for circular permutations and, due to the equivalence, they are true also for linear ones. In a circular permutation there is an element 0, and the equivalent linear permutation can be obtained by simply removing this element.

Breakpoint Graph. The breakpoint graph [5] is a graph representation of a permutation, which is classical in the genome rearrangements literature. In this graph every element of the permutation is represented by a left and a right vertex. As defined below, every vertex is connected to one black and one gray edge. The intuitive idea is that the black edges describe the order in the permutation and the gray edges describe the order in the identity permutation. Throughout the paper all permutations are circular and therefore, for both indices and elements, we identify n and 0.

Definition 1. *Let* $\pi = (\pi_0 \ \dots \ \pi_{n-1})$ *be a permutation. The* breakpoint graph $G(\pi)$ *is a edge-colored graph on* $2n$ *vertices* $\{l_0, r_0, l_1, r_1, \dots, l_{n-1}, r_{n-1}\}$. *For every* $0 \leq i \leq n - 1$, *connect* r_i *and* l_{i+1} *by a gray edge, and for every* π_i, *connect* l_{π_i} *and* $r_{\pi_{i-1}}$ *by a black edge, denoted by* b_i.

It is convenient to draw the breakpoint graph on a circle, such that black edges are on the circumference and gray edges are chords (see Figure 1(a)).

Cycles. Since the degree of each vertex is exactly 2, the graph uniquely decomposes into cycles. Denote the number of cycles in $G(\pi)$ by $c(\pi)$. The *length* of a cycle is the number of black edges it contains. A *k-cycle* is a cycle of length k, and it is *odd* if k is odd. The number of odd cycles is denoted by $c_{odd}(\pi)$, and let $\Delta c_{odd}(\pi, \tau) = c_{odd}(\tau \cdot \pi) - c_{odd}(\pi)$. Bafna and Pevzner proved the following useful lemma:

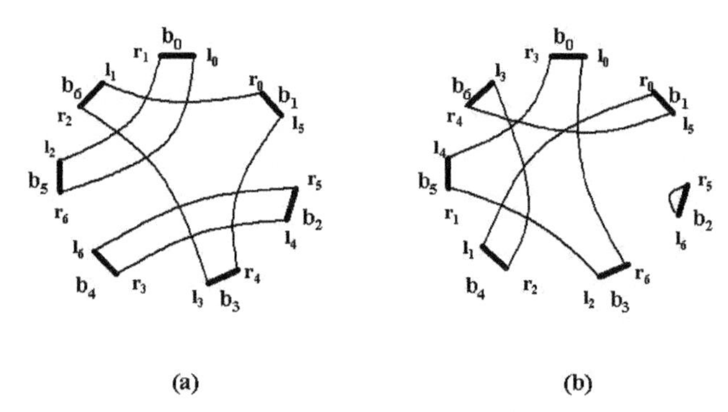

Fig. 1. (a) The circular breakpoint graph of the permutation $\pi = (0\ 5\ 4\ 3\ 6\ 2\ 1)$. Black edges are represented as thick lines on the circumference, and gray edges are chords. The three cycles are $(b_1\ b_3\ b_6)$, $(b_2\ b_4)$, and $(b_5\ b_0)$. (b) The circular breakpoint graph of π after applying a transposition on black edges b_0, b_2 and b_5.

Lemma 1. (Bafna and Pevzner [5]) *For all permutations π and transpositions τ, $\Delta c_{odd}(\pi, \tau) \in \{-2, 0, 2\}$.*

Let $n(\pi)$ denote the number of black edges in $G(\pi)$. The maximum number of cycles is obtained iff π is the identity permutation. In that case, there are $n(\pi)$ cycles, and all of them are odd (in particular, they are all of length 1). Starting with π with c_{odd} odd cycles, Lemma 1 implies the following lower bound on $d(\pi)$:

Theorem 2. (Bafna and Pevzner [5]) *For all permutations π, $d(\pi) \geq (n(\pi) - c_{odd}(\pi))/2$.*

By definition, every transposition must cut three black edges. The transposition that cuts black edges b_i, b_j and b_k is said to *apply on* these edges (see Figure 1). If these black edges are in cycle C, then the transposition is said to apply on C. A transposition τ is a *k-move* if $\Delta c_{odd}(\pi, \tau) = k$. A cycle is called *oriented* if there is a 2-move that is applied on three of its black edges; otherwise, it is *unoriented.*

Throughout the paper, we use the term permutation also when referring to the breakpoint graph of the permutation (as will be clear from the context). For example, when we say that π contains an oriented cycle, we mean that $G(\pi)$ contains an oriented cycle.

Simple Permutations. A k-cycle in the breakpoint graph is called *short* if $k \leq 3$; otherwise, it is called *long.* A breakpoint graph is *simple* if it contains only short cycles. A permutation π is *simple* if $G(\pi)$ is simple, and is a *2-permutation* (resp. *3-permutation*) if $G(\pi)$ contains only 2-cycles (3-cycles).

A common technique in genome rearrangement literature is to transform permutations with long cycles into simple permutations. This transformation consists of inserting new elements into the permutations and thereby splitting the long cycles. The reader is referred to [12] for a thorough description. If $\hat{\pi}$ is the permutation attained by inserting elements into π then $d(\pi) \leq d(\hat{\pi})$, since inserting new elements only can result in a permutation that requires more moves to be sorted. Such a transformation is called *safe* if it maintains the lower bound of Theorem 2, i.e., if $n(\pi) - c_{odd}(\pi) = n(\hat{\pi}) - c_{odd}(\hat{\pi})$.

Lemma 3. (Lin and Xue [15]) *Every permutation can be transformed safely into a simple one.*

Note that the transformation only maintains the lower bound, not the exact distance[1]. We say that permutation π is *equivalent* to permutation $\hat{\pi}$ if $n(\pi) - c_{odd}(\pi) = n(\hat{\pi}) - c_{odd}(\hat{\pi})$.

Lemma 4. (Hannenhalli and Pevzner [11]) *Let $\hat{\pi}$ be a simple permutation that is equivalent to π, then every sorting of $\hat{\pi}$ mimics a sorting of π with the same number of operations.*

The 1.375-approximation given in this paper first transforms the given permutation π into an equivalent simple permutation $\hat{\pi}$, then it finds a sorting sequence for $\hat{\pi}$, and, finally, the sorting of $\hat{\pi}$ is mimicked on π. Therefore, throughout most of the paper we will be concerned with simple permutations and short cycles.

Configurations and Components. Given a cyclic sequence of elements $i_1, \ldots, i_k$, an *arc* is an interval in the cyclic order, i.e., a set of contiguous elements in the sequence. The pair (i_j, i_l) $(j \neq l)$ defines two disjoint arcs: $i_j, \ldots, i_{l-1}$ and $i_l, \ldots, i_{j-1}$. Similarly, a triplet defines a partition of the cyclic sequence into three disjoint arcs. We say that *two pairs* of black edges (a, b) and (c, d) are *intersecting* if a and b belong to different arcs defined by the pair (c, d). A pair of black edges intersects with cycle C, if it intersects with a pair of black edges that belong to C. Cycles C and D intersect if there is a pair of black edges in C that intersect with D (see Figure 2c). *Two triplets* of black edges are *interleaving* if each of the edges of one triple belongs to a different arc of the second triple. Two 3-cycles are interleaving if their edges interleave (see Figure 2e).

A *configuration* of cycles is a subgraph of the breakpoint graph that is induced by one or more cycles. There are only two possible configurations of a 3-cycle in a breakpoint graph, which are shown in Figure 2 (a and b). It is easy to verify that the 3-cycle in (a) is oriented, and (b) is unoriented. A configuration A is a *sub-configuration* of a configuration B if the cycles in A form a subset of the cycles in B. A configuration A is *connected* if for any two cycles c_1 and c_k of A there are cycles $c_2, \ldots, c_{k-1}$ such that, for each $i \in [1, k-1]$, c_i intersects with c_{i+1}. A *component* is a maximal connected configuration in a breakpoint graph. The *size* of configurations and components is the number of cycles they contain, and are said to be *unoriented* if all their cycles are unoriented. They are called *small* if their size is at most 8; otherwise they are *big*.

In a configuration, an *open gate* is a pair of black edges of a 2-cycle or an unoriented 3-cycle that does not intersect with another cycle. The following is an important lemma by Bafna and Pevzner.

[1] Unlike in the problem of sorting by reversals [11], in which the analogous transformation maintains the exact distance.

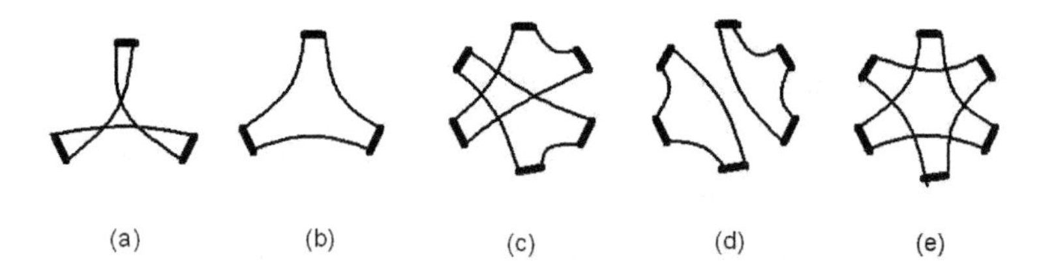

Fig. 2. Configurations of 3-cycles. (a) An oriented 3-cycle. (b) An unoriented 3-cycle. (c) Intersecting 3-cycles. (d) Non-intersecting 3-cycles. (e) Interleaving 3-cycles.

Lemma 5. (Bafna and Pevzner [5]) *Every open gate intersects with some other cycle in the breakpoint graph.*

A configuration not containing open gates is referred to as a *full configuration.* For example, the configuration in 2(e) is full, whereas 2(c) has two open gates.

Sequence of Transpositions. An (x, y)*-sequence* of transpositions on a simple permutation (for $x \geq y$) is a sequence of x transpositions, such that at least y of them are 2-moves and that leaves a simple permutation at the end. For example, a 0-move followed by two consecutive 2-moves (which is called a $(0, 2, 2)$-sequence in previous papers [9,12]) is a $(3, 2)$-sequence. A configuration (or component or permutation) *has* a (x, y)-sequence, if it is possible to apply such a sequence on its cycles.

The following result is the basis of the previous 1.5-approximation algorithms and will be used throughout the paper.

Lemma 6. (Christie [9] and Hartman [12]) *For every permutation (except for the identity permutation) there exists either a 2-move or a* $(3, 2)$*-sequence.*

Corollary 7. *For every permutation that has an oriented cycle and contains at least three 3-cycles there exists a* $(4, 3)$*-sequence.*

Transposition Diameter. The *transposition diameter*, $\mathrm{TD}(n)$, of the symmetric group is the maximum value of $d(\pi)$ taken over all permutations of n elements, i.e., $\mathrm{TD}(n) \triangleq \max_{\pi} d(\pi)$. Similarly, the transposition diameter of simple permutations TDS, 2-permutations TD2, and 3-permutations TD3, is the longest distance for any such permutation to the identity. [2]

3 Transposition Diameter Results

In this section, we first provide a lower bound on the transposition diameter. Then, we determine the exact transposition diameter of 2-permutations and simple permutations, and find an upper bound for the diameter of 3-permutations.

[2] The term *diameter* is somehow misleading for subsets of the symmetric group which are not a sub-group. However, we will stick to this term for the sake of consistency.

Recall that throughout the paper by permutations we mean circular permutations. However, sorting linear permutations of size n is equivalent to sorting circular permutations of size $n+1$ [12]. Therefore all bounds can be applied directly to linear permutations by replacing n with $n+1$.

Previous works [10,9,16] on the transposition diameter have conjectured[3] that the most distant permutation is the reversed permutation $(0\ n-1\ \dots\ 1)$. That is, it was believed that the transposition diameter is $\lfloor \frac{n-1}{2} \rfloor + 1$. However, the theorem below, which is proved in the full version of the paper, disproves this conjecture. Although the improvement of the bound is minor, we believe that this result is important since lower bounds on transposition problems are quite rare and hard to obtain.

The proofs of the following theorems are given in the full version of the paper.

Theorem 8. *$TD(n) \geq \lfloor \frac{n}{2} \rfloor + 1$.*

For linear permutations of size n the lower bound is given by $\lfloor \frac{n+1}{2} \rfloor + 1$.

Theorem 9. *$TD2(n) = \frac{n}{2}$.*

Theorem 10. *$TDS(n) = \lfloor \frac{n}{2} \rfloor$.*

3.1 Diameter for 3-Permutations

The main result given in this section is an upper bound for the diameter of 3-permutations, which is the basis of the 1.375-approximation algorithm for sorting by transpositions (Section 4). This result, like many other results in genome rearrangements, is based on a rigorous case analysis. However, since the number of cases is huge, we developed a computer program that systematically analyzes all the cases. Below we describe the case analysis.

Our goal is to show that every 3-permutation with at least 8 cycles has an (x, y)-sequence such that $x \leq 11$ and $\frac{x}{y} \leq \frac{11}{8}$. Such a sequence is referred to as an $\frac{11}{8}$*-sequence*. By Corollary 7, we need only consider unoriented configurations, since a $(4, 3)$-sequence is an $\frac{11}{8}$-sequences. Thus, in the sequel, when we say configurations we refer to unoriented configurations. The case analysis is done in two steps. In the first step, below, all big components are shown to have an $\frac{11}{8}$-sequence. In the second step, which is described ion the full version of the paper, we consider permutations with at least 8 cycles such that all components are small and prove that also these permutations have an $\frac{11}{8}$-sequence.

Analysis of Unoriented Configurations. The enumeration over all components starts from the basic building blocks: connected configurations consisting of two unoriented cycles. There are only two such configurations, the *unoriented interleaving pair* (Figure 2e) and the *unoriented intersecting pair* (Figure 2c). From these two configurations it is possible to build any other unoriented connected configuration by successively *adding* new unoriented cycles to the configuration. Adding a cycle to a configuration is done by inserting its black edges

[3] [10] conjectured that this was the case with exceptions for $n = 14$ and $n = 16$.

somewhere in the configuration. If it is possible to create a configuration B by adding a cycle to a configuration A, then B is said to be an *extension* of A. For example, both the unoriented interleaving and intersecting pair are extensions of the configuration of only one unoriented 3-cycle.

Consider a full configuration C and let A be a sub-configuration of C. Then C can be constructed by a series of extensions of A. In particular, this means that A can be extended into configuration B, that also is a sub-configuration of C. If A has an open gate then there is such extension B that is closing the open gate (since by definition, C has no open gates), i.e., the pair of black edges constituting the open gate in A intersects with a pair of black edges in B. Otherwise, there is an extension B with at most one open gate that is also a sub-configuration of C.

From the discussion above it follows that there are two types of extensions that are sufficient for building any component. These *sufficient extensions* are (1) extensions closing open gates and (2) extensions of full configurations, such that the extended configuration has at most one open gate. We refer to configurations that are realizable through a series of sufficient extensions from either the unoriented interleaving pair or the unoriented intersecting pair as *sufficient configurations*. Note that in particular, this means that every sufficient configuration has at most two open gates.

The following lemma is proved by our computerized case analysis:

Lemma 11. *Every unoriented sufficient configuration of 9 cycles has an* $\frac{11}{8}$*-sequence.*

By definition every big component has a sufficient configuration of size 9. Therefore the above lemma states that if a permutation contains a big component then there is an $\frac{11}{8}$-sequence.

One way of proving Lemma 11 would be to give a sorting for each of the sufficient configurations of 9 cycles. Such a case analysis would be too time consuming even for a computer. Instead, we utilize the notion of sufficiency and the fact that a sorting sequence for a configuration is also a sub-sorting for every extension of it. In Figure 3 we describe the case analysis which intuitively can be thought of as a breadth first search. When performing the analysis it turns out that no configuration of 10 cycles is added to the queue. This means that all sufficient configurations of 9 cycles have an $\frac{11}{8}$-sequence.

1. Initiate a queue of configurations to contain the unoriented interleaving pair and the unoriented intersecting pair.
2. While the queue is non-empty do:
   (a) Remove the first configuration, A, from the queue.
   (b) For each sufficient extension B of A do:
      i. If B does not have an $\frac{11}{8}$-sequence add it to the queue.
      ii. Otherwise give the sorting sequence for B.

Fig. 3. A brief description of the case analysis

It should be stressed that the program itself is not a proof of the lemma. The proof is the case analysis which is the output of the program. Although each separate case can be verified by hand it is not an appealing thought to verify 80.000 such cases. To remedy this, the case analysis is presented in a user-friendly web interface [1] facilitating a general understanding of its correctness. Moreover, to affirm the correctness we have written a small verification program. This program verifies the proof by verifying (1) that every given sorting is a correct sorting and (2) that all sufficient extensions are considered. Thus the proof as a whole can be checked by verifying the correctness of this small program.

To complete the analysis we now consider small components. Small components that do not have an $\frac{11}{8}$-sequence are called *bad small components.* Our computerized enumeration found that there are only five such components. The second step of the case analysis, which is described in the full version of the paper, is to show that permutations with at least 8 cycles that contain only bad small components have an $\frac{11}{8}$-sequence.

Lemma 12. *Let π be a permutation with at least 8 cycles that contains only bad small components. Then, π has an $\frac{11}{8}$-sequence.*

The conclusion of the case analysis in this section is the corollary below. It follows from Lemmas 11 and 12 and is the basis of the $\frac{11}{8} = 1.375$ approximation algorithm.

Corollary 13. *Every 3-permutation with at least 8 cycles has an $\frac{11}{8}$-sequence.*

The Diameter for 3-Permutations. Here we present an upper bound on the diameter for 3-permutations (the proof is given in the full version). In 3-permutations of size n the number of cycles is $c = n/3$. Let $g(c) \triangleq 11\lfloor c/8 \rfloor + \lfloor 3(c \mod 8)/2 \rfloor$ and define f as follows:

$$f(c) \triangleq \begin{cases} g(c) + 1 & \text{if } c \mod 8 = 1 \\ g(c) & \text{otherwise} \end{cases} \tag{1}$$

Note that $f(8l + r) = 11l + f(r)$. This function gives the upper bound on the diameter for 3-permutations.

Theorem 14. *$TD3(n) \leq f(\frac{n}{3}) \leq 11\left\lfloor \frac{n}{24} \right\rfloor + \left\lfloor 3\frac{(n/3 \mod 8)}{2} \right\rfloor + 1$.*

4 The Approximation Algorithm

Now we are ready to present our main result: Algorithm *Sort*, which is a 1.375-approximation algorithm for sorting by transpositions (Figure 4). Intuitively, the algorithm sorts the permutation by repeatedly applying (11, 8)-sequences and since $\frac{11}{8} = 1.375$ we get the desired approximation ratio (based on the lower bound of Theorem 2). The following lemma, whose proof is deferred to the full version of the paper, analyzes the time complexity of the algorithm.

Algorithm ***Sort*** **(π)**

1. Transform permutation π into a simple permutation $\hat{\pi}$ (Lemma 3).
2. Check if there is a $(2,2)$-sequence. If so, apply it.
3. While $G(\hat{\pi})$ contains a 2-cycle, apply a 2-move (Christie [9]).
4. While $G(\hat{\pi})$ contains at least 8 cycles apply an $\frac{11}{8}$-sequence (Corollary 13).
5. While $G(\hat{\pi})$ contains a 3-cycle, apply a $(3,2)$-sequence (Lemma 6).
6. Mimic the sorting of π using the sorting of $\hat{\pi}$ (Lemma 4).

Fig. 4. A high-level description of the approximation algorithm *Sort*.

Lemma 15. *The time complexity of Algorithm Sort is $O(n^2)$.*

Theorem 16. *Algorithm Sort is a* 1.375*-approximation algorithm for sorting permutations by transpositions, and it runs in quadratic time.*

Proof. The running time is shown in Lemma 15. We now prove the approximation ratio. Depending on Step 2, there are two cases: either there is a $(2,2)$-sequence or not. Let c_3 (resp. c_2) represent the number of 3-cycles (2-cycles) in $G(\hat{\pi})$ after Step 2.

Case 1 In Step 2 if a $(2,2)$-sequence exists. According to the lower bound in Theorem 2 the best possible sorting is that using only 2-moves. Specifically this means that $\hat{\pi}$ can not be sorted better than first applying two 2-moves and then another $c_3 + c_2$ 2-moves to sort the remaining cycles. Therefore a lower bound for any sorting of $\hat{\pi}$ is $c_3 + c_2 + 2$. The algorithm gives a sorting using $2 + \frac{c_2}{2} + f(c_3 + \frac{c_2}{2})$ moves; 2 moves in Step 2, $c_2/2$ moves in Step 3 creating $c_2/2$ 3-cycles, and by the proof of Theorem 14 at most $f(c_3 + \frac{c_2}{2})$ moves in Steps 4 and 5. Thus the approximation ratio of the algorithm is

$$\frac{2 + \frac{c_2}{2} + f(c_3 + \frac{c_2}{2})}{c_3 + c_2 + 2} = \frac{2 + y + f(x)}{x + y + 2} \leq \frac{f(x) + 2}{x + 2}, \tag{2}$$

where $x = c_3 + \frac{c_2}{2}$ and $y = \frac{c_2}{2}$. In Table 1 the last expression of Equation 2 is shown to be bounded from above by $\frac{11}{8}$.

Case 2 In Step 2 if a $(2,2)$-sequence does not exist then there are $c_2 + c_3$ cycles in $G(\hat{\pi})$ and at least $c_3 + c_2 + 1$ moves are required to sort $\hat{\pi}$; at least one 0-move and by Theorem 2 at least $c_3 + c_2$ 2-moves are required. The algorithm gives a sorting using $\frac{c_2}{2} + f(c_3 + \frac{c_2}{2})$ moves. Thus the approximation ratio of the algorithm is

$$\frac{\frac{c_2}{2} + f(c_3 + \frac{c_2}{2})}{c_3 + c_2 + 1} = \frac{f(x) + y}{x + y + 1} \leq \frac{f(x)}{x + 1}, \tag{3}$$

where again $x = c_3 + \frac{c_2}{2}$ and $y = \frac{c_2}{2}$. In Table 1 the last expression of Equation 3 is shown to be bounded from above by $\frac{11}{8}$. □

Table 1. Function table showing that the approximation ratio of the algorithm is $\frac{11}{8} = 1.375$. In the table $x = 8l + r$ and thus $f(x) = 11l + f(r)$.

r	0	1	2	3	4	5	6	7
f(r)	0	2	3	4	6	7	9	10
$\frac{f(x)+2}{x+2}$	$\frac{11l+2}{8l+2}$	$\frac{11l+4}{8l+3}$	$\frac{11l+5}{8l+4}$	$\frac{11l+6}{8l+5}$	$\frac{11l+8}{8l+6}$	$\frac{11l+9}{8l+7}$	$\frac{11l+11}{8l+8}$	$\frac{11l+12}{8l+9}$
$\frac{f(x)}{x+1}$	$\frac{11l}{8l+1}$	$\frac{11l+2}{8l+2}$	$\frac{11l+3}{8l+3}$	$\frac{11l+4}{8l+4}$	$\frac{11l+6}{8l+5}$	$\frac{11l+7}{8l+6}$	$\frac{11l+9}{8l+7}$	$\frac{11l+10}{8l+8}$

5 Discussion and Open Problems

The main result of this paper is a 1.375-approximation algorithm for sorting by transpositions. In addition, there are some new advances regarding the transposition diameter. The main open problems are to determine the complexity of sorting by transpositions, and to find the transposition diameter. We believe that our results give new insights for further investigation of these problems. In particular, our characterization of components which are "hard-to-sort" may be a key to better lower bounds and approximation algorithms.

Empirical evidence indicate that the upper bound given for the diameter of 3-permutations is very close to the true diameter. If this is correct, then there are permutations at distance 1.375 times the lower bound of Theorem 2. That is, finding a better lower bound is essential for improving the approximation ratio.

Acknowledgments. We are grateful to Prof. Ron Shamir, Prof. Jens Lagergren, and Elad Verbin for many invaluable discussions. We also wish to thank Prof. Jens Lagergren for reading early drafts. The first author is grateful to Prof. Benny Chor for hosting him at Tel Aviv University.

References

1. `http://www.sbc.su.se/~cherub/SBT1375_proof.`
2. K. Appel and W. Haken. Every planar map is four colorable part I: Discharging. *Illinois Journal of Mathematics*, 21:429–490, 1977.
3. D.A. Bader, B. M.E. Moret, and M. Yan. A linear-time algorithm for computing inversion distance between signed permutations with an experimental study. *Journal of Computational Biology*, 8(5):483–491, 2001.
4. V. Bafna and P. A. Pevzner. Genome rearragements and sorting by reversals. *SIAM Journal on Computing*, 25(2):272–289, 1996.
5. V. Bafna and P. A. Pevzner. Sorting by transpositions. *SIAM Journal on Discrete Mathematics*, 11(2):224–240, May 1998. Preliminary version in the *Proceedings of SODA*, 1995.
6. A. Bergeron. A very elementary presentation of the Hannenhalli-Pevzner theory. In *Combinatorial Pattern Matching (CPM '01)*, pages 106–117, 2001.
7. P. Berman, S. Hannanhalli, and M. Karpinski. 1.375-approximation algorithm for sorting by reversals. In *Proc. of 10th Eurpean Symposium on Algorithms (ESA '02)*, pages 200–210. Springer, 2002. LNCS 2461.

8. A. Caprara. Sorting permutations by reversals and Eulerian cycle decompositions. *SIAM Journal on Discrete Mathematics*, 12(1):91–110, February 1999.
9. D. A. Christie. *Genome Rearrangement Problems*. PhD thesis, University of Glasgow, 1999.
10. H. Eriksson, K. Eriksson, J. Karlander, L. Svensson, and J. Wastlund. Sorting a bridge hand. *Discrete Mathematics*, 241(1-3):289–300, 2001.
11. S. Hannenhalli and P. Pevzner. Transforming cabbage into turnip: Polynomial algorithm for sorting signed permutations by reversals. *Journal of the ACM*, 46:1–27, 1999.
12. T. Hartman. A simpler 1.5-approximation algorithm for sorting by transpositions. In *Combinatorial Pattern Matching (CPM '03)*, volume 2676, pages 156–169, 2003.
13. S. B. Hoot and J. D. Palmer. Structural rearrangements, including parallel inversions, within the chloroplast genome of Anemone and related genera. *J. Molecular Evooution*, 38:274–281, 1994.
14. H. Kaplan, R. Shamir, and R. E. Tarjan. Faster and simpler algorithm for sorting signed permutations by reversals. *SIAM Journal of Computing*, 29(3):880–892, 2000.
15. G. H. Lin and G. Xue. Signed genome rearrangements by reversals and transpositions: Models and approximations. *Theoretical Computer Science*, 259:513–531, 2001.
16. J. Meidanis, M. E. Walter, and Z. Dias. Transposition distance between a permutation and its reverse. In *Proceedings of the 4th South American Workshop on String Processing*, pages 70–79, 1997.
17. J. D. Palmer and L. A. Herbon. Tricircular mitochondrial genomes of Brassica and Raphanus: reversal of repeat configurations by inversion. *Nucleic Acids Research*, 14:9755–9764, 1986.
18. P. A. Pevzner. *Computational Molecular Biology: An Algorithmic Approach*. MIT Press, 2000.
19. D. Sankoff and N. El-Mabrouk. Genome rearrangement. In *Current Topics in Computational Molecular Biology*. MIT Press, 2002.
20. R. Shamir. Algorithms in molecular biology: Lecture notes, 2002. Available at http://www.math.tau.ac.il/~rshamir/algmb/01/algmb01.html.
21. E. Tannier and M.F. Sagot. Sorting by reversals in subquadratic time. In *Combinatorial Pattern Matching (CPM '04)*, volume 3109, pages 1–13, 2004.
22. U. Zwick. Computer assisted proof of optimal approximability results. In *Symposium On Discrete Mathematics (SODA-02)*, pages 496–505, 2002.

A New Tight Upper Bound on the Transposition Distance

Anthony Labarre*

Université Libre de Bruxelles,
Département de Mathématique, CP 216,
Service de Géométrie, Combinatoire et Théorie des Groupes,
Boulevard du Triomphe, B-1050 Bruxelles, Belgium
alabarre@ulb.ac.be

Abstract. We study the problem of computing the minimal number of adjacent, non-intersecting block interchanges required to transform a permutation into the identity permutation. In particular, we use the graph of a permutation to compute that number for a particular class of permutations in linear time and space, and derive a new tight upper bound on the so-called transposition distance.

1 Introduction

The problem we study is a particular case of a problem called *genome rearrangement* [1,2], which is motivated by applications in biology. The genome rearrangement problem can be formulated as follows: given two genomes, find the minimum number of evolutionary events transforming one into the other. This number is defined as the *distance* between the two genomes.

The model we are interested in applies to the case where the order of genes is known and where all genomes share the same set and number of genes (without duplications), which allows us to represent them by *permutations*. It is easy to show that what we have defined as a distance is indeed a distance on the set of all permutations (i.e. it satisfies the three usual axioms).

We will consider only one operation on permutations: biological *transpositions*, which consist in moving a block of contiguous elements from one place to another one. This problem was first introduced in 1995 by Bafna and Pevzner [3], and is generally considered harder than similar problems. In particular, neither its complexity, nor even the *diameter* of the transposition distance (i.e. the maximal value it can reach), is known, which has led several authors to design polynomial-time approximation algorithms (whose best known ratio[1] is $\frac{3}{2}$ [3,5,6]) as well as using and comparing heuristics [5,7,8,9]. An interesting property of this distance is that the transposition distance between any two permutations π, σ is the same as the distance between $\sigma^{-1} \circ \pi$ and the *identity permutation*

* Funded by the "Fonds pour la Formation à la Recherche dans l'Industrie et dans l'Agriculture" (F.R.I.A.).

[1] A 1.375−approximation has recently been proposed in [4].

R. Casadio and G. Myers (Eds.): WABI 2005, LNBI 3692, pp. 216–227, 2005.

$\iota = (1\ 2\ \cdots\ n)$. Therefore, the problem of transforming a permutation into another one using as few transpositions as possible is the same as that of *sorting* a permutation using the minimum number of transpositions. In what follows, we refer to the latter number as the *distance of* π, noted $d(\pi)$. While other authors have tried to find the shortest possible sequences of transpositions that sort a permutation, we have chosen to focus on computing their length.

In this paper, we make use of the common graph of a permutation rather than of the "cycle graph" introduced in [3]. A step in this direction was mentioned in [10] and successfully used to compute another rearrangement distance in [11]. As we suspected, it proved fruitful for our problem too: we were able to show that the distance of some nicely characterized permutations, namely those who fix even or odd elements and another class derived from those two, can be computed in linear time, using a formula that completely bypasses any graph structure used so far. Furthermore, we use those permutations to derive a tight upper bound on the transposition distance of every permutation.

This paper is organized as follows. In Section 2, we review previous results and typical notations. In Section 3, we introduce a graph that we use in Section 4 to provide a formula for computing the distance of some special permutations in linear time. In Section 5, we use those to derive an upper bound on the transposition distance of every permutation. Finally, we discuss our results in Section 6 and suggest some open questions of interest.

2 Notations and Preliminaries

Permutations are denoted by lower case Greek letters, typically π, and S_n is the set of all permutations of $\{1, 2, ..., n\}$. For any permutation π in S_n, the *transposition* $\tau(i, j, k)$ with $1 \leq i < j < k \leq n + 1$ applied to π exchanges the closed intervals determined respectively by i and $j - 1$ and by j and $k - 1$, transforming π into $\pi \circ \tau(i, j, k)$. So $\tau(i, j, k)$ is the following permutation:

$$\begin{pmatrix} 1 \cdots i-1 & \boxed{i\ i+1 \cdots j-2\ j-1} & \boxed{j\ j+1 \cdots k-1} & k \cdots n \\ 1 \cdots i-1 & \boxed{j\ j+1 \cdots k-1} & \boxed{i\ i+1 \cdots j-2\ j-1} & k \cdots n \end{pmatrix}.$$

The usual notation is shorter than the one we have just used to describe transpositions, i.e. we write a permutation π in S_n as $(\pi_1\ \pi_2\ \cdots\ \pi_n)$. Bafna and Pevzner [3] define the *cycle graph* of π as the bicoloured directed graph $G(\pi)$, whose vertex set consists of the elements of π plus two new elements $\pi_0 = 0$ and $\pi_{n+1} = n + 1$, and whose edge set consists of:

- *black* edges (π_i, π_{i-1}) for $1 \leq i \leq n + 1$;
- *gray* edges $(i, i + 1)$ for $0 \leq i \leq n$.

The set of black and gray edges decomposes in a single way into *alternate cycles*, i.e. cycles which alternate black and gray edges, and we note the number of such cycles $c(G(\pi))$. An alternate cycle of $G(\pi)$ is *odd* (resp. even) if it contains an *odd* (resp. even) number of black edges, and we note $c_{odd}(G(\pi))$ (resp.

$c_{even}(G(\pi)))$ the number of odd (resp. even) alternate cycles of $G(\pi)$. Bafna and Pevzner proved the following lower bound on the transposition distance.

Theorem 1. *[3]* $\forall\ \pi \in S_n : d(\pi) \geq \frac{n+1-c_{odd}(G(\pi))}{2}$.

For a permutation π, define an ordered pair (π_i, π_{i+1}) as a *breakpoint* if $\pi_{i+1} \neq \pi_i + 1$. The number of breakpoints of π is denoted by $b(\pi)$. Christie [5] decomposes permutations into *strips*, which he defines as maximal intervals containing no breakpoint. He denotes $gl(\pi)$ the *reduced* version of π, obtained as follows: assuming π has r strips, remove strip 1 if it begins with 1, strip r if it ends with n, replace every other strip with its minimal element and finally, renumber the resulting sequence so as to obtain a permutation of S_r $(r \leq n)$.

Theorem 2. *[5]* $\forall\ \pi \in S_n : d(\pi) = d(gl(\pi))$.

We say that π and σ are *equivalent by reduction* if $gl(\pi) = gl(\sigma)$, which we also write as $\pi \equiv_r \sigma$. Since we are presenting a new upper bound on the transposition distance of every permutation, it is only fair that we conclude this section with the ones that were previously shown.

Theorem 3. *[3]* $\forall\ \pi \in S_n$:

$$d(\pi) \leq \frac{3(n+1-c_{odd}(G(\pi)))}{4} . \tag{1}$$

Theorem 4. *[12]* $\forall\ \pi \in S_n$:

$$d(\pi) \leq \frac{3}{4}\ b(\pi) . \tag{2}$$

Theorem 5. *[13]* $\forall\ \pi \in S_n$:

$$d(\pi) \leq \begin{cases} \left\lceil \frac{2n}{3} \right\rceil & \textit{if } n < 9 ; \\ \left\lfloor \frac{2n-2}{3} \right\rfloor & \textit{if } n \geq 9 . \end{cases} \tag{3}$$

3 Another Useful Graph

We will make use of a variant of the well-known *graph of a permutation*. The *$\Gamma-$graph* of a permutation π in S_n is the directed graph $\Gamma(\pi)$ with vertex set $\{(1, \pi_1), (2, \pi_2), ..., (n, \pi_n)\}$ and edge set $\{((i, \pi_i), (j, \pi_j)) \mid j = \pi_i\}$.

If $C = (i_1, i_2, ..., i_k)$ is a cycle of π (i.e. π maps i_l onto i_{l+1} for $1 \leq l \leq k-1$ and i_k onto i_1), we obtain a cycle (i_1, π_{i_1}), (i_2, π_{i_2}), ..., (i_k, π_{i_k}), which we also denote C, in $\Gamma(\pi)$, and call it a *$k-$cycle*. We say that such a cycle is *positively oriented* if $k \geq 3$ and its elements can be written as a strictly increasing sequence, *negatively oriented* if $k \geq 3$ and its elements can be written as a strictly decreasing sequence, and *unoriented* otherwise.

For instance, in Fig. 1, cycle (4, 2, 1) is negatively oriented, cycle (5) is unoriented, and cycle (3, 6, 7) is positively oriented. Note that every 1$-$cycle

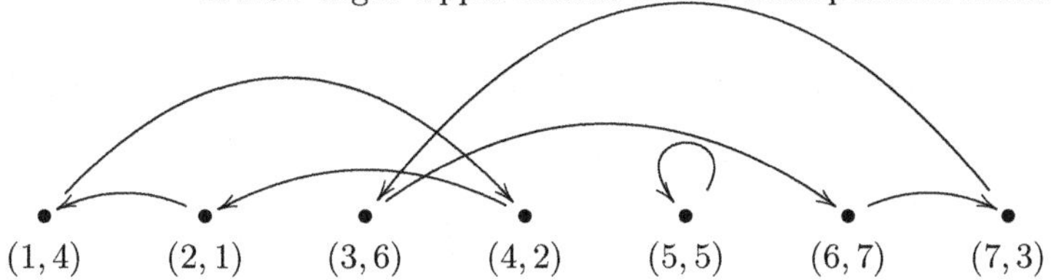

Fig. 1. The Γ−graph of the permutation (4 1 6 2 5 7 3)

and every 2−cycle is unoriented. In a quite similar fashion to the parity of cycles defined in the context of $G(\pi)$, we say that a k−cycle of $\Gamma(\pi)$ is *odd* (resp. *even*) if k is odd (resp. even). Likewise, we note $c(\Gamma(\pi))$ the number of cycles of $\Gamma(\pi)$, and $c_{odd}(\Gamma(\pi))$ (resp. $c_{even}(\Gamma(\pi))$) the number of odd (resp. even) cycles of $\Gamma(\pi)$.

4 An Explicit Formula for Some Permutations

We define a γ−*permutation* as a reduced permutation that fixes even elements (thus n must be odd), and show (Theorem 6) that the distance of such a permutation, and several others, can be computed quickly, without the need of the cycle graph.

Proposition 1. *For every γ−permutation π in S_n:*

$$\begin{cases} c_{even}(G(\pi)) = 2\ c_{even}(\Gamma(\pi))\ ; \\ c_{odd}(G(\pi)) \ \ = 2\left(c_{odd}(\Gamma(\pi)) - \frac{n-1}{2}\right)\ . \end{cases}$$

Proof. Each vertex (i, π_i) of $\Gamma(\pi)$ corresponding to an odd element π_i is both the starting point of an edge $((i,\pi_i),(\pi_i,\pi_{j_1}))$ and the ending point of an edge $((j_2,i),(i,\pi_i))$. Since π_i is odd and π is a γ−permutation, π_i+1 is mapped onto itself, and π_{j_1} precedes π_i+1 in π. In $G(\pi)$, those edges are each transformed into one sequence of two edges (gray-black for the first one, black-gray for the second one):

$$\begin{cases} ((i,\pi_i),(\pi_i,\pi_{j_1}))\ \text{ becomes }\ (\pi_i,\pi_i+1),(\pi_i+1,\pi_{j_1})\ ; \\ ((j_2,i),(i,\pi_i))\qquad \text{ becomes }\ (\pi_i,\pi_{i-1}),(\pi_{i-1},\pi_{j_2})\ . \end{cases}$$

I.e. the outgoing edge of (i,π_i) in $\Gamma(\pi)$ is transformed in one of the following ways (according to the relative positions of π_i and π_{j_1}):

a) (i,π_i) $\cdots$ (π_i,π_{j_1}) (π_i+1,π_i+1) $\longrightarrow$ π_i $\cdots$ π_{j_1} π_i+1

b) (π_i,π_{j_1}) (π_i+1,π_i+1) $\cdots$ (i,π_i) $\longrightarrow$ π_{j_1} π_i+1 $\cdots$ π_i

Similarly, the incoming edge of (i, π_i) in $\Gamma(\pi)$ is transformed in one of the following ways (according to the relative positions of π_i and π_{j_2}):

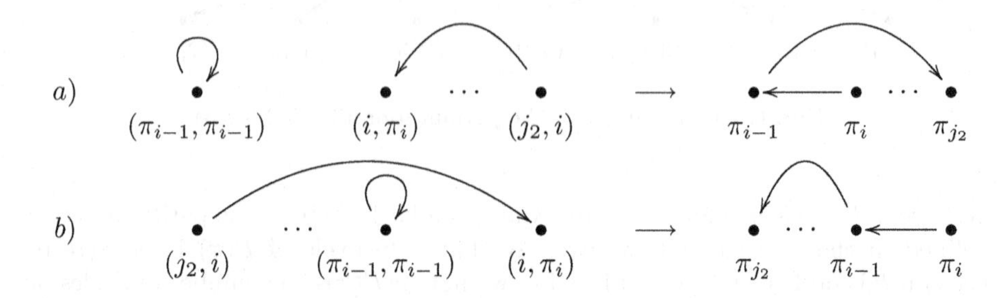

Each $k-$cycle ($k \geq 2$) of $\Gamma(\pi)$ provides an alternate cycle with k black edges in $G(\pi)$; moreover, for every such cycle of $\Gamma(\pi)$, a new cycle is created in $G(\pi)$, which actually corresponds to the cycle of $\Gamma(\pi)$ followed in the opposite direction. Parity of cycles of $\Gamma(\pi)$ is obviously preserved in $G(\pi)$, since to each vertex of a $k-$cycle ($k \geq 2$) of $\Gamma(\pi)$ corresponds a black edge in $G(\pi)$. Finally, $1-$cycles of $\Gamma(\pi)$ are not preserved in $G(\pi)$, and there are $\frac{n-1}{2}$ of them. □

We derive the following lower bound from Proposition 1 and Theorem 1.

Lemma 1. *For every $\gamma-$permutation π in S_n, we have $d(\pi) \geq n - c_{odd}(\Gamma(\pi))$.*

Proof. Straightforward. □

We will first study permutations such that odd elements form only one cycle in $\Gamma(\pi)$, distinguishing the case of oriented cycles and that of unoriented ones.

4.1 Oriented Cycles

We define an $\alpha-$*permutation* as a reduced permutation that fixes even elements and whose odd elements form one oriented cycle in the graph Γ, and refer to the long cycle formed by the $\frac{n+1}{2}$ odd elements as its *main cycle.*

Proposition 2. *For every $\alpha-$permutation π in S_n, we have $d(\pi) = \frac{n+1}{2} - \left(\frac{n+1}{2} \bmod 2\right)$.*

Proof. Since every $\alpha-$permutation is a $\gamma-$permutation, Lemma 1 yields $d(\pi) \geq \frac{n+1}{2} - \left(\frac{n+1}{2} \bmod 2\right)$. We assume that the main cycle of $\Gamma(\pi)$ is positively oriented (a similar proof is easily obtained in the negative case).

1. if $\frac{n+1}{2}$ is odd, consider transpositions $\tau_1(2, 4, n+1)$, $\tau_2(1, 3, n)$; then an optimal sequence of $\frac{n+1}{2} - 1$ transpositions that sorts π is

$$(\tau_1 \circ \tau_2)^{\frac{\frac{n+1}{2}-1}{2}} .$$

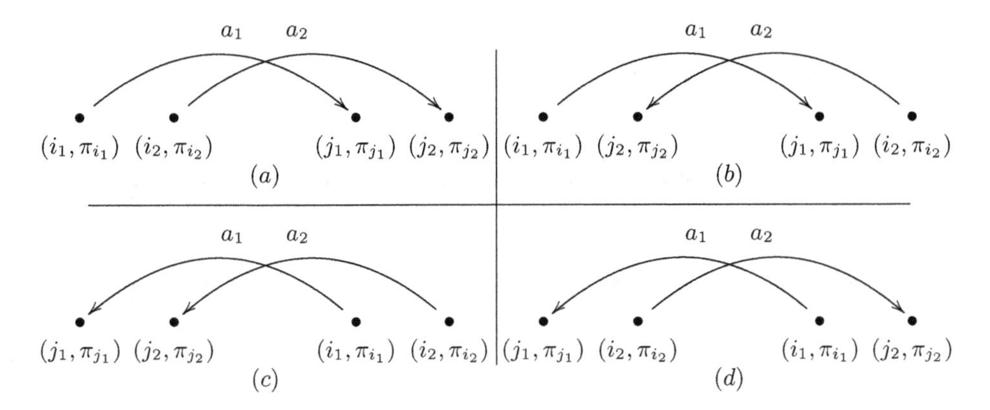

Fig. 2. The four possible configurations for crossing edges in $\Gamma(\pi)$

2. if $\frac{n+1}{2}$ is even, consider again transpositions τ_1, τ_2 defined above, and also transpositions $\tau_3(2,3,n+1)$, $\tau_4(1,2,n+1)$; then an optimal sequence of $\frac{n+1}{2}$ transpositions that sorts π is

$$(\tau_1 \circ \tau_2)^{\frac{\frac{n+1}{2}-2}{2}} \circ \tau_3 \circ \tau_4 \ .$$

Short of space, we omit the proof that those sequences indeed sort our permutations, but this can be easily shown by induction. □

4.2 Unoriented Cycles

We now show that the orientation of a cycle does not matter, i.e. Proposition 2 still holds if the main cycle of $\Gamma(\pi)$ is unoriented. We will make use of so-called *exchanges* to simplify the proofs, namely bypassing the construction of optimal sequences. An *exchange* $exc(i,j)$ is the permutation that exchanges elements in positions i and j, thus transforming every permutation π into the permutation $\pi \circ exc(i,j)$.

$$exc(i,j) = \begin{pmatrix} 1 \cdots i-1 \boxed{i} \; i+1 \cdots j-1 \boxed{j} \; j+1 \cdots n \\ 1 \cdots i-1 \boxed{j} \; i+1 \cdots j-1 \boxed{i} \; j+1 \cdots n \end{pmatrix} .$$

We will only use exchanges of the form $exc(i,i+2k)$ with $k \geq 1$; such an exchange can be simulated by two transpositions, but the correspondence between those two types of operations is not that straightforward when exchanges are composed.

We say that two edges $a_1 = ((i_1, \pi_{i_1}), (j_1, \pi_{j_1}))$ and $a_2 = ((i_2, \pi_{i_2}), (j_2, \pi_{j_2}))$ of $\Gamma(\pi)$ *cross* if intervals $[i_1, j_1]$ and $[i_2, j_2]$ do not contain each other and have a non-empty intersection. The four possible configurations for crossing edges are shown in Fig. 2.

We define a β*−permutation* as a reduced permutation that fixes even elements and whose odd elements form one unoriented cycle in the graph Γ.

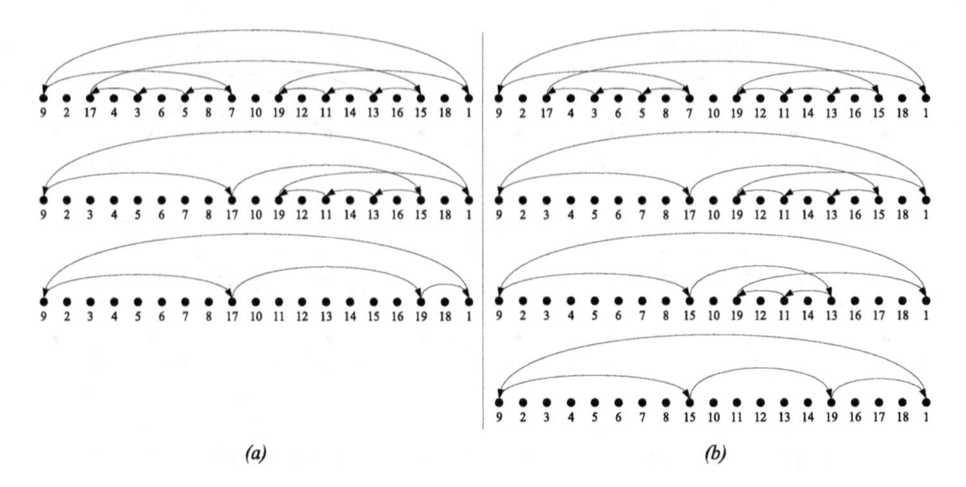

Fig. 3. Two possible ways of contracting paths in a β−permutation (1−cycles and indices omitted for clarity)

Clearly, the main cycle of the Γ−graph of every β−permutation π (except (3 2 1)) contains crossing edges. We are going to transform π into a permutation σ that reduces to an α−permutation, and this will be achieved through the removal of crossing edges using a certain sequence $\mathscr{E}$ of exchanges. We thus get the following upper bound on the distance of a β−permutation π:

$$d(\pi) \leq f(\mathscr{E}) + d(\sigma)$$

where $f(\mathscr{E})$ gives the minimum number of transpositions having the same effect on π as $\mathscr{E}$ does. Finding σ is not difficult, but we have to find a σ such that our upper bound is minimized.

To eliminate a crossing, we just have to make the ending point of one edge become the starting point of the one it crosses, and this will be achieved using a sequence of exchanges of the form described in the following proposition. By a *path*, we mean the sequence of edges joining the extremities of the crossing edges as mentioned above, and we will refer to the elimination of this path as the *contraction* of it. The following proposition will be useful.

Proposition 3. *For both sequences* $\mathscr{E} = exc(i, i+2) \circ exc(i, i+4) \circ \cdots \circ exc(i, i+2t)$ *and* $\mathscr{F} = exc(i, i+2t) \circ \cdots \circ exc(i, i+4) \circ exc(i, i+2)$ *of* t *exchanges:*

$$f(\mathscr{E}) = f(\mathscr{F}) = t + (t \bmod 2) .$$

Proof. For any valid i, t, $\mathscr{E}$ and $\mathscr{F}$ reduce to an α−permutation π whose main cycle is a $(t+1)$−cycle, and by Proposition 2:

$$d(\pi) = t + 1 - ((t+1) \bmod 2) = t + (t \bmod 2) . \qquad \square$$

Proposition 4. *For every β−permutation π in S_n, we have* $d(\pi) = \frac{n+1}{2} - \left(\frac{n+1}{2} \bmod 2\right)$.

Proof. Since every β−permutation is a γ−permutation, Lemma 1 yields $d(\pi) \geq \frac{n+1}{2} - \left(\frac{n+1}{2} \bmod 2\right)$. If $\pi = (3\ 2\ 1)$, the thesis is easily verified; else the main cycle of $\Gamma(\pi)$ contains at least one crossing.

If the main cycle of $\Gamma(\pi)$ contains only one crossing, then there is a path of t edges joining the two crossing edges; this path can be contracted by a sequence of t exchanges, sorting the elements belonging to that part of the cycle. For instance, in case (a) of Fig. 2, it suffices to apply the sequence $exc(i_2, j_1) \circ \cdots \circ exc(i_2, i_2+4) \circ exc(i_2, i_2+2)$, and those t exchanges correspond to exactly $t + (t \bmod 2)$ transpositions (Proposition 3).

Once this path has been contracted, t vertices have been removed from the main cycle of $\Gamma(\pi)$ and this results in a permutation σ reducible to an α−permutation. We have:

$$d(\pi) \leq d(\pi, \sigma) + d(\sigma) = t + (t \bmod 2) + \frac{n+1}{2} - t - \left(\left(\frac{n+1}{2} - t\right) \bmod 2\right)$$
$$= \frac{n+1}{2} - \left(\frac{n+1}{2} \bmod 2\right)$$

which verifies our thesis.

In the case where several crossings exist, one must be careful not to contract paths "individually". Indeed, if we were to contract p such paths of t_g edges $(1 \leq g \leq p)$ in that way, we would have to use $\sum_{g=1}^{p} t_g$ exchanges to contract them all, which would correspond to $\sum_{g=1}^{p}(t_g + (t_g \bmod 2))$ transpositions. This can actually be improved by exchanging the last exchanged element in the first contracted path with the first element of the next path to contract, then continue the contraction of the latter with dependent exchanges as before, and repeating the same process whenever need be. For instance, Fig. 3 shows the transformation of a β−permutation into a permutation reducible to an α−permutation in two different ways. Scenario (a) uses 3+3 exchanges = 8 transpositions (Proposition 3), whereas scenario (b) uses the same number of exchanges, but requiring this time only 6 transpositions.

Every β−permutation π containing p paths of t_g edges to contract $(1 \leq g \leq p)$ can thus be transformed into a permutation σ reducible to an α−permutation such that $d(\pi, \sigma) = T + T \bmod 2$, where $T = \sum_{g=1}^{p} t_g$. The transpositions representing those exchanges will eliminate T vertices from the main cycle of $\Gamma(\pi)$, which yields the following upper bound:

$$d(\pi) \leq d(\pi, \sigma) + d(\sigma)$$
$$= T + (T \bmod 2) + \left(\frac{n+1}{2} - T - \left(\left(\frac{n+1}{2} - T\right) \bmod 2\right)\right)$$
$$= \frac{n+1}{2} - \left(\frac{n+1}{2} \bmod 2\right)$$

which equals the lower bound given above. □

4.3 Distance of $\gamma-$Permutations

Every permutation π can be sorted by eliminating each cycle of $\Gamma(\pi)$ individually using exchanges (and therefore also using only transpositions), so that each elimination does not modify other cycles. This strategy yields the following upper bound on $d(\pi)$.

Lemma 2. *For every permutation π, consider its disjoint cycle decomposition $\Gamma(\pi) = C_1 \cup C_2 \cup \cdots \cup C_{c(\Gamma(\pi))}$. Denote $d(C)$ the minimum number of transpositions required to transform $C = (i_1, i_2, ..., i_k)$ into $(i_1), (i_2), ..., (i_k)$; then:*

$$d(\pi) \leq \sum_{i=1}^{c(\Gamma(\pi))} d(C_i) .$$

We now show that the lower bound of Lemma 1 is reached.

Proposition 5. *For every $\gamma-$permutation π in S_n :*

$$d(\pi) = n - c_{odd}(\Gamma(\pi)) . \tag{4}$$

Proof. Each cycle of $\Gamma(\pi)$ is either oriented or unoriented, and the distance of both kinds of cycles is known (Propositions 2 and 4). Denote $odd(\Gamma(\pi))$ (resp. $even(\Gamma(\pi)))$ the set of odd (resp. even) cycles of $\Gamma(\pi)$; by Lemma 2, we have:

$$\begin{aligned} d(\pi) &\leq \sum_{i=1}^{c(\Gamma(\pi))} |C_i| - (|C_i| \bmod 2) \\ &= \sum_{C_{i_1} \in \; odd(\Gamma(\pi))} (|C_{i_1}| - 1) + \sum_{C_{i_2} \in \; even(\Gamma(\pi))} |C_{i_2}| \\ &= \sum_{i=1}^{c(\Gamma(\pi))} |C_i| - c_{odd}(\Gamma(\pi)) . \end{aligned}$$

And since every element belongs to exactly one cycle, the last sum equals n and Lemma 1 verifies the thesis. □

Note that Proposition 5 can be expressed in a more general way: by Theorem 2, we know that π needs not be reduced, and adding k 1−cycles to $\Gamma(\pi)$ at any position increases both n and $c_{odd}(\Gamma(\pi))$ by k, so Equation 4 still holds. The same Theorem allows us to ask for *odd* elements to be fixed, instead of *even* ones; we then have $\pi_1 = 1$, and we can reduce π to a $\gamma-$permutation (e.g. $(1\ 4\ 3\ 6\ 5\ 8\ 7\ 2) \equiv_r (3\ 2\ 5\ 4\ 7\ 6\ 1)$). This result can also be extended using *toric permutations*, introduced in [13] and further studied in [14]. We borrow the latter author's notations; let us note, for $x \in \{0, 1, 2, ..., n\}$:

1. $\overline{x}^m = (x + m) \pmod{n + 1}$;
2. $\underline{x}_m = (x - m) \pmod{n + 1}$.

Define a *circular permutation* obtained from a permutation π in S_n as $\pi^\circ = 0\ \pi_1\ \pi_2\ \cdots\ \pi_n$, where $0 = \pi_0^\circ = \pi_{n+1}^\circ$. This circular permutation can be read starting from any position, and the original linear permutation is reconstructed by taking the element following 0 as π_1 and removing 0. Define the following operation on circular permutations:

$$m + \pi^\circ = \overline{0}^m\ \overline{\pi_1}^m\ \overline{\pi_2}^m\ \cdots\ \overline{\pi_n}^m\ .$$

Then for π in S_n, the corresponding *toric permutation* is $\pi_\circ^\circ$, which is the set of permutations obtained from $m + \pi^\circ$ with $0 \leq m \leq n$.

Lemma 3. *[13]* $\forall\ \pi, \sigma \in S_n : \sigma \in \pi_\circ^\circ \Rightarrow d(\sigma) = d(\pi)$.

Therefore, if n is odd and all odd elements of π in S_n occupy odd positions and form an increasing subsequence modulo $n+1$, then $\pi \in \sigma_\circ^\circ$ where σ satisfies the conditions described right after Proposition 5. Indeed, if n is odd, adding 1 (mod $n+1$) to each element of π° transforms n into 0, i.e. the new starting point of the resulting permutation, and all odd elements into even ones; therefore π is transformed into a permutation whose even elements are all fixed, and we have the following.

Theorem 6. *For every π in S_n that fixes even or odd elements:*

$$d(\pi) = n - c_{odd}(\Gamma(\pi))\ .$$

Moreover, every permutation σ with n odd and whose odd elements occupy odd positions and form an increasing subsequence modulo $n+1$ can be transformed in linear time into a permutation π such that $d(\sigma) = d(\pi) = n - c_{odd}(\Gamma(\pi))$.

5 A New Upper Bound

We now show that the right-hand side of Equation 5 is an upper bound on the transposition distance. First we show why $\gamma-$permutations are so important.

Theorem 7. *Every permutation π in S_n, except ι, can be obtained from a $\gamma-$permutation σ in S_{n+k} by removing k even elements in σ.*

Proof. If $\pi \neq \iota$ is no $\gamma-$permutation, just add a $1-$cycle to $\Gamma(\pi)$ between every ordered pair (π_i, π_{i+1}) $(1 \leq i \leq n-1)$ of non-fixed elements and reduce the resulting permutation π' in order to obtain a $\gamma-$permutation $\sigma \in S_{n+k}$. This operation can clearly be reverted, and this completes the proof. □

We can now prove our main result.

Theorem 8. $\forall\ \pi \in S_n$:

$$d(\pi) \leq n - c_{odd}(\Gamma(\pi))\ . \tag{5}$$

Proof. Let σ be the γ−permutation from which $\pi \neq \iota$ is obtained by deletion of k even elements. Cycles in $\Gamma(\sigma)$ can all be sorted individually by sequences of exchanges for which we know the corresponding minimal number of transpositions (Proposition 5), and all these exchanges still work on $\Gamma(\pi)$ (after accordingly adapting some of them). Therefore, we can claim that $d(\pi) \leq d(\sigma)$, and:

$$d(\pi) \leq d(\sigma) = n + k - c_{odd}(\Gamma(\sigma)) = n + k - c_{odd}(\Gamma(\pi)) - k = n - c_{odd}(\Gamma(\pi)).$$

And even though ι cannot be obtained from a γ−permutation, it is clear that our thesis holds for it too, since $d(\iota) = 0 \leq n - n$. □

6 Conclusions and Future Plans

Using a well-known graph, we were able to show that the transposition distance of some nicely characterized permutations can be computed in linear time, bypassing the classical structure introduced in [3]. In fact, no graph at all is needed, since decomposing a permutation into "classical cycles" is quite a trivial algorithm, running in linear time. Such an approach has proved most successful for computing another rearrangement distance in [11], and we are confident it is of great interest, certainly not just for the transposition distance. Furthermore, we also proved that the formula used to compute this distance is actually an upper bound on the transposition distance of every permutation.

Several questions arise.

Firstly, this new upper bound in Equation (5) is sometimes better, sometimes worse than the bounds in Equations (1), (2) and (3), and we want to tighten it. Table 1 compares our result with previous ones, giving the number of cases where it is at least as good as that of Theorems 3, 4, and 5. Apart from a fast approximation of the transposition distance, this could also help determine the maximal value of $d(\pi)$, which is still an open problem, as is the complexity of sorting by transpositions.

Secondly, the permutations characterized in Theorem 6 are not the only ones to reach our upper bound. Can the set of all such permutations be characterized?

Finally, there might be other useful permutation-related notions in combinatorics that are as well-known and eluded in the theory of genome rearrangement as is the graph we used. Although we do not think that those classical notions

Table 1. Comparison of our new upper bound with previous results

n	Number of permutations	(5)≤(1)	(5)≤(2)	(5)≤(3)
3	6	2	1	6
4	24	8	8	15
5	120	45	24	31
6	720	304	49	495
7	5040	2055	722	1611
8	40320	17879	3094	4355
9	362880	104392	60871	10243

can model each and every notion of this problem (in particular, it has been shown [14] that the structure of $G(\pi)$ is much more stable than that of $\Gamma(\pi)$ under the toric equivalence class), we feel that some of them could be of interest.

References

1. Meidanis, J., Setubal, J.: Introduction to Computational Molecular Biology. Brooks-Cole (1997)
2. Pevzner, P.A.: Computational molecular biology. MIT Press, Cambridge, MA (2000)
3. Bafna, V., Pevzner, P.A.: Sorting by transpositions. SIAM J. Discrete Math. **11** (1998) 224–240 (electronic)
4. Elias, I., Hartman, T.: A 1.375−Approximation Algorithm for Sorting by Transpositions (2005) Submitted.
5. Christie, D.A.: Genome Rearrangement Problems. PhD thesis, University of Glasgow, Scotland (1998)
6. Hartman, T.: A simpler 1.5-approximation algorithm for sorting by transpositions. In: Combinatorial pattern matching. Volume 2676 of Lecture Notes in Computer Science. Springer, Berlin (2003) 156–169
7. Guyer, S.A., Heath, L.S., Vergara, J.P.: Subsequence and run heuristics for sorting by transpositions. In: Fourth DIMACS Algorithm Implementation Challenge, Rutgers University (1995)
8. Vergara, J.P.C.: Sorting by Bounded Permutations. PhD thesis, Virginia Polytechnic Institute, Blacksburg, Virginia, USA (1997)
9. Walter, M.E.M.T., Curado, L.R.A.F., Oliveira, A.G.: Working on the problem of sorting by transpositions on genome rearrangements. In: Combinatorial pattern matching. Volume 2676 of Lecture Notes in Computer Science. Springer, Berlin (2003) 372–383
10. Dias, Z., Meidanis, J.: An Alternative Algebraic Formalism for Genome Rearrangements. Comparative Genomics (2000) 213–223
11. Dias, Z., Meidanis, J.: Genome Rearrangements Distance by Fusion, Fission, and Transposition is Easy. In: Proceedings of SPIRE'2001 - String Processing and Information Retrieval, Laguna de San Rafael, Chile (2001) 250–253
12. Dias, Z., Meidanis, J., Walter, M.E.M.T.: A New Approach for Approximating The Transposition Distance. In: Proceedings of SPIRE'2000 - String Processing and Information Retrieval, La Coruna, Espagne (2000)
13. Eriksson, H., Eriksson, K., Karlander, J., Svensson, L., Wästlund, J.: Sorting a bridge hand. Discrete Mathematics **241** (2001) 289–300 Selected papers in honor of Helge Tverberg.
14. Hultman, A.: Toric Permutations. Master's thesis, Dept. of Mathematics, KTH, Stockholm, Sweden (1999)

Perfect Sorting by Reversals Is Not Always Difficult
(Extended Abstract*)

Sèverine Bérard[1], Anne Bergeron[2], Cedric Chauve[2], and Christophe Paul[3]

[1] INRA Toulouse, Dépt. de Mathématique et Informatique Appliquée, France
Severine.Berard@toulouse.inra.fr
[2] LaCIM et Dépt. d'Informatique, Université du Québec à Montréal, Canada
{anne, chauve}@lacim.uqam.ca
[3] CNRS, LIRMM, Montpellier, France
paul@lirmm.fr

Abstract. This paper investigates the problem of conservation of combinatorial structures in genome rearrangement scenarios. We characterize a class of signed permutations for which one can compute in polynomial time a reversal scenario that conserves all common intervals, and that is parsimonious among such scenarios. Figeac and Varré (WABI 2004) announced that the general problem is NP-hard. We show that there exists a class of permutations for which this computation can be done in linear time with a very simple algorithm, and, for a larger class of signed permutations, the computation can be achieved in subquadratic time. We apply these methods to permutations obtained from the X chromosomes of the human, mouse and rat.

1 Introduction

The reconstruction of evolution scenarios based on genome rearrangements has proven to be a powerful tool in understanding the evolution of close species [9,14]. The computation of such evolution scenarios relies on the problem of *sorting signed permutations by reversals*: given two chromosomes, represented as sequences of genomic segments, find a parsimonious sequence of reversals that transforms a chromosome into the other one. However, the number of parsimonious sequences of reversals can be exponential [5]. It is then natural to ask for some additional criteria that can help to select putative scenarios. We are interested in scenarios that do not break combinatorial structures that are present in both chromosomes. In this work, the combinatorial structures that we consider are *common intervals* [22,15]. A rearrangement scenario is said to be *perfect* if it does not break any common interval. It was claimed in [13] that computing a parsimonious perfect scenario is difficult, but recent works [2,18] showed that in some non-trivial cases, such scenarios can be computed efficiently. In this paper we describe a class of instances that allow efficient computation of a parsimonious perfect scenario.

* A complete Version, Including Proofs, is Available in [3].

R. Casadio and G. Myers (Eds.): WABI 2005, LNBI 3692, pp. 228–238, 2005.

In Section 2, we define the links between sorting by reversals and structure conservation, and we state precisely the problem we address in this paper. In Section 3, we relate common intervals and perfect scenarios to a basis of common intervals, the *strong intervals*, that can be represented with a classical data structure in graph theory, the PQ-trees. These observations are consequences of deep relationships between common intervals and the modular decomposition of permutation graphs [11,17]. This point is central in our approach since we rely, in Section 4, on the combinatorial structure of strong intervals trees to design algorithms that are both efficient – subquadratic time, even linear time in some cases – and simple. We apply these results to the comparison of the human, mouse and rat X chromosomes, based on data of [14].

2 Sorting by Reversals and Common Intervals

A *signed permutation* on n elements is a permutation on the set of integers $\{1, 2, \ldots, n\}$ in which each element has a sign, positive or negative. Negative integers are represented by placing a bar over them. An *interval* of a signed permutation is a segment of consecutive elements of the permutation. One can define an interval by giving the set of its unsigned elements, called the *content* of the interval.

The *reversal* of an interval of a signed permutation reverses the order of the elements of the interval, while changing their signs. Note that every reversal is an interval of the permutation on which it is performed, which lead us to often treat reversals as intervals, and to represent a reversal by the corresponding interval. If P is a permutation, we denote by $\overline{P}$ the permutation obtained by reversing the complete permutation P.

Definition 1. Let P and Q be two signed permutations on n elements. A *scenario* between P and Q is a sequence of distinct reversals that transforms P into Q, or P into $\overline{Q}$. The *length* of such a scenario is the number of reversals it contains. When Q is the identity permutation, a scenario between P and Q is called a *scenario* for P.

Given a signed permutation P on n elements, the problem of *sorting by reversals*, defined in [19], asks for a scenario for P of minimal length among all possible scenarios, also called a *parsimonious* scenario. Currently, the best known algorithm for this problem runs in $O(n^{3/2}\log(n))$ worst-case time [21].

Definition 2. Two distinct intervals I and J *commute* if either $I \subset J$, or $J \subset I$, or $I \cap J = \emptyset$. If intervals I and J do not commute, they *overlap*.

Definition 3. Let P be a signed permutation on n elements. A *common interval* of P is a set of one or more integers that is an interval in both P and the identity permutation Id_n. Note that any such set is also an interval of $\overline{P}$ and of $\overline{Id_n}$. The singletons and the set $\{1, 2, \ldots, n\}$ are called *trivial* common intervals.

The notion of *common interval* was introduced in [22]. It was studied, among others, in [15], to model the fact that a group of genes can be rearranged in a genome but still remain connected.

Definition 4. Let P be a signed permutation. A scenario S for P is called a *perfect scenario* if every reversal of S commutes with every common interval of P. A perfect scenario of minimal length is called a *parsimonious perfect scenario.*

Example 1. Let $P = (\overline{3}\ \overline{2}\ \overline{5}\ \overline{4}\ 1)$ be a signed permutation.

1. Reversing the interval $(\overline{2}\ \overline{5}\ \overline{4}\ 1)$, or equivalently the set $\{1, 2, 4, 5\}$, yields the signed permutation $(\overline{3}\ \overline{1}\ 4\ 5\ 2)$.
2. The common intervals of P are $\{2, 3\}, \{4, 5\}, \{2, 3, 4, 5\}, \{1, 2, 3, 4, 5\}$ and the singletons $\{1\}$, $\{2\}$, $\{3\}$, $\{4\}$ and $\{5\}$.
3. $(\{1, 2, 3, 4, 5\}, \{2, 3, 4, 5\}, \{2, 3\}, \{4, 5\}, \{1\})$ is a perfect scenario that transforms P into Id_5.
4. $(\{2, 3, 4, 5\}, \{2, 3\}, \{4, 5\}, \{1\})$ is a parsimonious perfect scenario for P, transforming P into the reverse $\overline{Id_5}$ of the identity.
5. $(\{1, 4, 5\}, \{1, 2, 3\})$ is a parsimonious scenario, but it is not perfect since the reversal $\{1, 4, 5\}$ overlaps the common interval $\{2, 3, 4, 5\}$.

As shown in [13], given a signed permutation P, there exists at least one perfect scenario for P. However, the authors of [13] claim that the construction of parsimonious perfect scenarios is computationally difficult: they state that computing a parsimonious perfect scenario between two signed permutations is NP-hard in general. Hence the difficulty of the problem relies in the parsimonious aspect.

The main goal of this paper is to propose algorithms for computing parsimonious perfect scenarios that are efficient for large classes of signed permutations (Section 4). Our results rely on the *strong intervals tree* of a signed permutation described in the next section.

3 Strong Intervals Trees

As the number of common intervals of a permutation P on n elements can be quadratic in n, an efficient algorithm (i.e. subquadratic time) for computing perfect scenarios should rely on a space efficient encoding of the set of common intervals. In [17], the author pointed out a correspondence between common intervals of permutations and the concept, well studied in graph theory, of *modules* of graphs. Inspired from the *modular decomposition* theory[1], this section describes structural properties of the set of common intervals of a permutation P that are central in the design of the algorithms in Section 4.

Let us first remark that being a common interval for a set I has nothing to do with the sign of the elements of I. Therefore all the structural results presented

[1] All the results presented in this section can be seen as direct consequences or corollaries of well known graph theoretical results (see [11] for example).

in this section are valid for both signed and unsigned permutations. For the sake of simplicity, we omit the signs which will be reintroduced in the next section.

Definition 5. A common interval I of a permutation P is a *strong* interval of P if it commutes with every common interval of P.

For example, the strong intervals of $P = (1\ 4\ 2\ 5\ 3\ 7\ 8\ 6\ 9)$ are $\{2,3,4,5\}$, $\{7,8\}$, $\{6,7,8\}$, $\{1,2,3,4,5,6,7,8,9\}$ and $\{1\}$,..., $\{9\}$. The singletons and $\{1,2, \ldots, 9\}$ are the *trivial* strong intervals of P.

It follows from Definition 5 that the inclusion order of the set of strong intervals defines an n-leaves tree, denoted by $T_s(P)$, whose leaves are the singletons, and whose root is the interval containing all elements of the permutation. We call the tree $T_s(P)$ the *strong intervals tree* of P (Fig. 1), and we identify a node of $T_s(P)$ with the strong interval it represents.

Since each strong interval with more than one element, or equivalently each internal node of $T_s(P)$, has at least two children in $T_s(P)$, we have immediately:

Proposition 1. *A permutation on n elements has $O(n)$ strong intervals.*

Let I be a common interval of a permutation P on n elements and $x \in \{1, 2, \ldots, n\}$ such that $x \notin I$. It follows from the definition of common interval that either x is larger than all elements of I or x is smaller than all elements of I. Hence, for two *disjoint* common intervals I and J, we can define the relation $I < J$ by extending the order relation on integers that belong to I and J.

Definition 6. Let P be a permutation and $\mathcal{I} = \{I_1, \ldots, I_k\}$ be a partition of the elements of P into strong intervals. The *quotient permutation* of P with respect to $\mathcal{I}$, denoted $P_{|\mathcal{I}}$, is the permutation on k elements such that i precedes j if and only if $I_i < I_j$.

For example, for the permutation $P = (1\ 4\ 2\ 5\ 3\ 7\ 8\ 6\ 9)$ of Fig. 1, $\mathcal{I} = \{\{1\}, \{2,3,4,5\}, \{7,8\}, \{6\}, \{9\}\}$ is a partition of P into strong intervals, and $P_{|\mathcal{I}} = (1\ 2\ 4\ 3\ 5)$.

Theorem 1. *Let P be a permutation on n elements and $\mathcal{I} = \{I_1, \ldots, I_k\}$ be the partition of P into strong intervals given by the children of the root of $T_S(P)$. Then exactly one of the following is true:*

1. *$P_{|\mathcal{I}}$ is Id_k (the identity permutation on k elements).*
2. *$P_{|\mathcal{I}}$ is $\overline{Id_k}$ (the reverse of the identity permutation on k elements).*
3. *The only common intervals of $P_{|\mathcal{I}}$ are trivial.*

Theorem 1 induces a classification of the nodes of the strong intervals tree $T_s(P)$ that is central in the design of our algorithms: let P_I be the quotient permutation defined by the children of an internal node I of $T_s(P)$. The node I, or equivalently the strong interval I of P, is either:

1. *Increasing linear*, if P_I is the identity permutation, or
2. *Decreasing linear*, if P_I is the reverse of the identity permutation, or
3. *Prime*, otherwise.

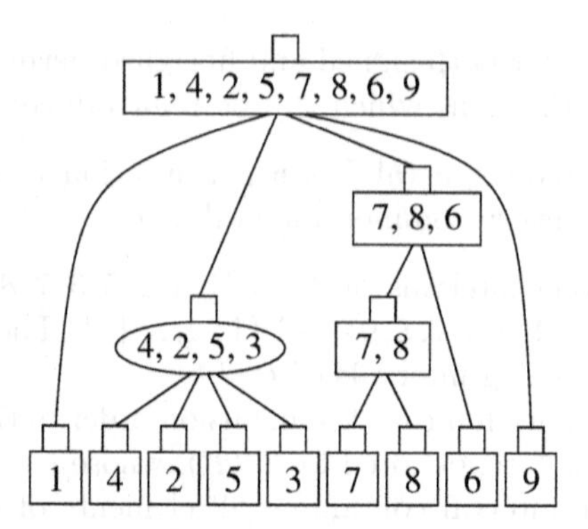

Fig. 1. The strong intervals tree $T_s(P)$ of the permutation $P = (1\ 4\ 2\ 5\ 3\ 7\ 8\ 6\ 9)$. Prime and linear nodes are distinguished by their shape. There are two non-trivial linear nodes, the rectangular nodes: $(7, 8)$ is increasing and $(7, 8, 6)$ is decreasing. There is only one prime node, the round node $(4, 2, 3, 5)$.

For example, in Fig. 1, the rectangular nodes are the linear nodes, and the round node $(4, 2, 5, 3)$ is the unique prime node. The only decreasing linear node in this tree is $(7, 8, 6)$.

Property 1. In a strong intervals tree, a child of an increasing (resp. a decreasing) linear node is either a prime node or a decreasing (resp. an increasing) linear node.

Finally, we show that the strong intervals tree is a compact representation – it only requires $O(n)$ space – of the set of all common intervals, which is possibly a set of quadratic size.

Proposition 2. *Let P be a signed permutation. An interval I of P is a common interval of P if and only if it is either a node of $T_S(P)$, or the union of consecutive children of a linear node of $T_S(P)$.*

This representation for strong intervals was first given implicitly in [15], and explicitly in [16], where it was shown that $T_s(P)$ can be related to a data structure widely used in graph theory, called PQ-tree. It can be computed in $O(n)$ worst-case time using algorithms similar to the ones given in [15,16]. A formal link between *PQ*-trees and conserved structures in signed permutations was first proposed in [4], in the context of conserved intervals, a subset of common intervals.

4 Computing Perfect Scenarios

We now turn to the description of efficient algorithms for computing parsimonious perfect scenarios for large classes of signed permutations. The central point is a characterization of perfect scenarios in terms of $T_S(P)$. We assume that $T_S(P)$ is given (see [6] for a simple algoritm to build it).

Proposition 3. *A scenario S for a permutation P is perfect if and only if each of the reversals of S is either a node of $T_S(P)$, or the union of children of a prime node of $T_S(P)$.*

Computing a perfect scenario S thus amounts to identify leaves, linear nodes and union of children of prime nodes of $T_S(P)$ that are the reversals of S. Even if the general problem of computing parsimonious perfect scenarios was claimed to be difficult [13], it can be done efficiently for a large class of signed permutations, defined in terms of the structure of their strong intervals tree.

A strong intervals tree $T_S(P)$ is *unambiguous* if every prime node has a linear parent, and *definite* if it has no prime nodes. For definite trees, there is essentially a unique perfect scenario for P (Theorem 2), and for unambiguous trees, we can compute a parsimonious perfect scenario in subquadratic time (Theorem 3).

Remark 1. Note that definite strong intervals trees are also known as *co-trees* in the theory of modular decomposition of graphs [11].

A *signed* tree is a tree in which each node has a sign, + or −. We associate to an unambiguous tree $T_S(P)$ the following signed tree $T'_S(P)$:

1. The sign of a leaf x is the sign of the corresponding element in P.
2. The sign of a linear node is +, if the node is increasing, and − if the node is decreasing.
3. The sign of a prime node is the sign of its parent.

Fig. 2, Fig. 3 and Fig. 4 show signed strong intervals trees associated to the permutations obtained by comparing 16 synteny blocks of the human, mouse and rat X chromosomes [14]. In Fig. 2 the labels of the nodes are given with respect to the order of the blocks of the mouse chromosome.

$$\begin{array}{rcl}
\text{Human} &=& 1\ 2\ 3\ 4\ 5\ 6\ 7\ 8\ 9\ 10\ 11\ 12\ 13\ 14\ 15\ 16 \\
\text{Mouse} &=& \bar{6}\ \bar{5}\ 4\ 13\ 14\ \overline{15}\ 16\ 1\ \bar{3}\ 9\ \overline{10}\ 11\ 12\ \bar{7}\ 8\ \bar{2} \\
\text{Rat} &=& \overline{13}\ \bar{4}\ 5\ \bar{6}\ \overline{12}\ \bar{8}\ \bar{7}\ 2\ 1\ \bar{3}\ 9\ 10\ 11\ 14\ \overline{15}\ 16
\end{array}$$

Theorem 2. *If $T_S(P)$ is definite, the set of nodes having a sign different from the sign of their parent is a parsimonious perfect scenario for P.*

Given the tree $T_S(P)$, Theorem 2 implies that, when $T_S(P)$ is definite, computing a parsimonious perfect scenario for P is almost immediate. The comparison of the rat and mouse X chromosomes yields a definite tree, Fig. 2, and the corresponding scenario can be obtained by comparing the signs of the $O(n)$ nodes. When such a scenario exists, it is unique up the order of the reversals, since each of them commutes with all the others.

We next turn to the more general case of unambiguous trees. Recall that a prime node inherits its sign from its parent, and that any reversal that is a union of children of a prime node commutes with all common intervals, thus may belong to a perfect scenario.

$$\begin{array}{rcccccccccccccccc} \text{Mouse} = & 1 & 2 & 3 & 4 & 5 & 6 & 7 & 8 & 9 & 10 & 11 & 12 & 13 & 14 & 15 & 16 \\ \text{Rat} = & \overline{4} & \overline{3} & \overline{2} & 1 & \overline{13} & \overline{15} & 14 & \overline{16} & 8 & 9 & 10 & \overline{11} & 12 & 5 & 6 & 7 \end{array}$$

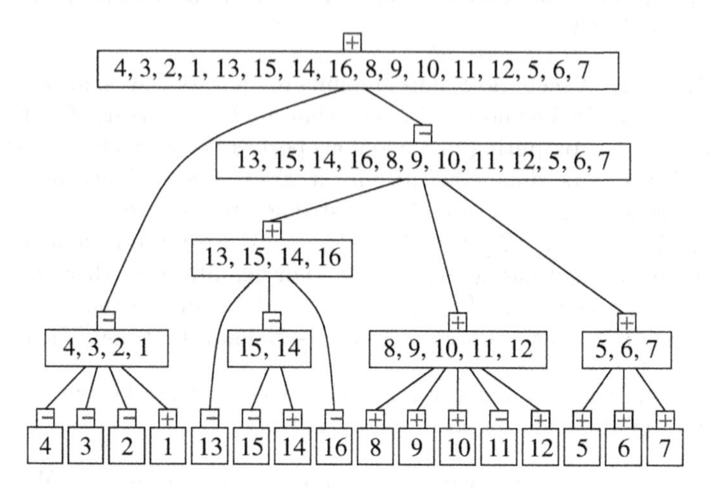

Fig. 2. Comparing the rat and mouse X chromosomes. The set of nodes having a sign different from the sign of their parent form a parsimonious perfect scenario that transforms the rat X chromosome into the mouse X chromosome: $(4, 3, 2, 1)$, (1), $(13, 15, 14, 16, 8, 9, 10, 11, 12, 5, 6, 7)$, $(13, 15, 14, 16)$, (13), $(15, 14)$, (14), (16), $(8, 9, 10, 11, 12)$, (11), $(5, 6, 7)$.

Algorithm 1 describes how to obtain a parsimonious perfect scenario in the case of unambiguous trees. The basic idea is to compute, for each prime node I of the tree, any parsimonious scenario that sorts the children of node I in increasing or decreasing order, depending on the sign of I. Then, it suffices to deal with linear nodes whose parent is linear in the same way than for a definite tree.

Fig. 3 shows the signed tree associated to the permutations of the human and rat X chromosomes. This tree is unambiguous: it has one prime node $(4, 5, 6, 12, 8, 7, 2, 1, 3, 9, 10, 11)$ whose parent is a decreasing linear node. The quotient permutation of this node over its five children is $P_I = (2\ \overline{5}\ \overline{3}\ 1\ 4)$, and a parsimonious scenario that sorts P_I to $\overline{Id}$ is given by: $\{1, 3, 4\}$, $\{1, 3\}$, $\{1\}$, $\{2, 3, 4, 5\}$, $\{3, 4, 5\}$. Note that if the corresponding five reversals are applied to the rat chromosome, the resulting permutation has a definite tree.

The time complexity of Algorithm 1 depends on the time complexity of the sorting by reversals algorithm used to compute a reversal scenario that sorts the children of a prime node. Using the $O(n^{3/2} \log n)$ algorithm described in [21], we have:

Theorem 3. *If $T_S(P)$ is unambiguous, Algorithm 1 computes a parsimonious perfect scenario for P in subquadratic time.*

When $T_S(P)$ is ambiguous, the sign of some prime nodes is undefined. A general algorithm to compute parsimonious perfect scenario would repeatedly apply Al-

Algorithm 1: Computing a parsimonious perfect scenario for unambiguous $T_S(P)$

S is an empty scenario.
For each prime node I of $T_s(P)$
    P_I is the quotient permutation of I over its children
e   *If the sign of I is positive*
        Then compute any parsimonious scenario T from P_I to Id
        Else compute any parsimonious scenario T from P_I to $\overline{Id}$
    Deduce the corresponding scenario T' on the children of P_I
    Add the reversals of T' to scenario S
Add to S the linear nodes and leaves having a linear parent and a sign different from the sign of their parent.

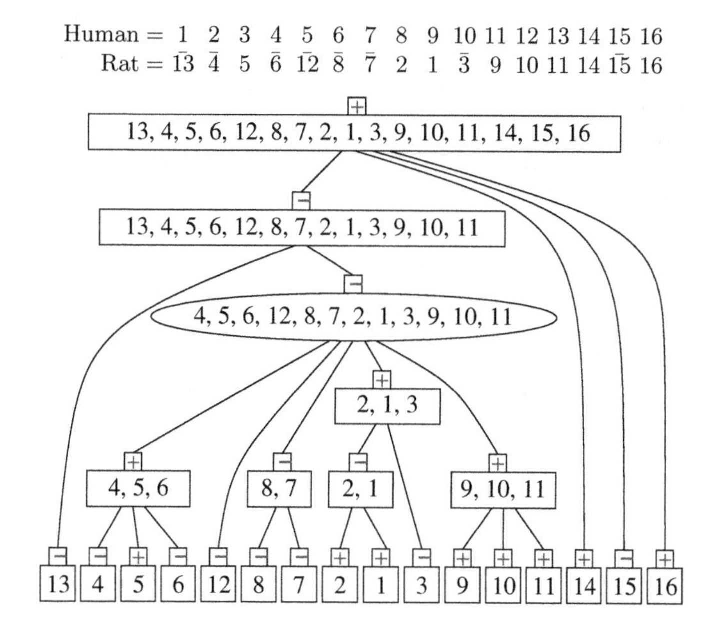

Fig. 3. Comparing the human and rat X chromosomes: a parsimonious perfect scenario is obtained by sorting the five children $(4,5,6)$, (12), $(8,7)$, $(2,1,3)$ and $(9,10,11)$ in decreasing order using any parsimonious scenario that sorts the quotient permutation $P_I = (2\ \bar{5}\ \bar{3}\ 1\ 4)$, and then reversing the linear nodes and leaves whose linear parent have a different sign: $(13,4,5,6,12,8,7,2,1,3,9,10,11)$, (4), (6), $(2,1)$, (2), (1), (3), (15). The length of the scenario is 13.

gorithm 1 to all possible sign assignments to prime nodes that do not have linear parents. As an example, consider Fig. 4 that shows the signed tree associated to the permutations of the human and mouse X chromosomes.

This tree is ambiguous since its root is a prime node, and we must try to sort this node both to Id and to $\overline{Id}$. In this case, both parsimonious scenarios have the same length. Such an algorithm is a generalization of the algorithm we described in this section for unambiguous trees, and was described using another

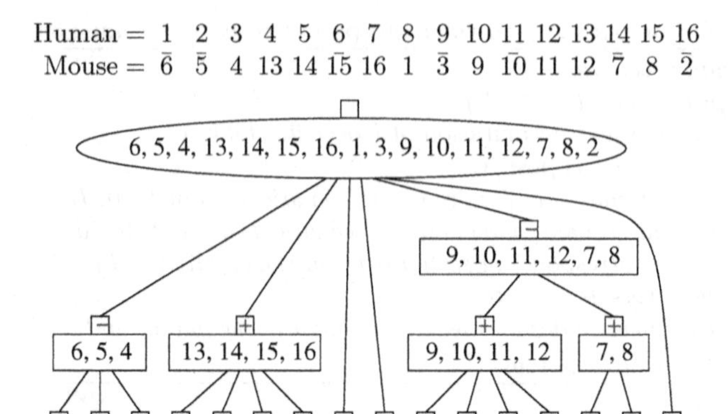

Fig. 4. Comparing the human and mouse X chromosomes: the root has no sign but its children can be sorted to Id or $\overline{Id}$ in 6 reversals using a parsimonious scenario that sorts the quotient permutation $P_I = (\bar{4}\ 6\ 1\ \bar{3}\ \bar{5}\ 2)$, A parsimonious perfect scenario would also contain the reversals: (4), (15), (9, 10, 11, 12), (10), (7, 8), (7). The total length of the scenario is 12.

formalism in [13]. It has a worst-case time complexity that is exponential in the number of prime nodes whose parent is prime, and thus is efficient if the number of such nodes is small.

Permutations that arise from the comparison of genomic sequences are not "random", and this could partly explain why perfect sorting is not difficult for the permutations we considered. Constructing permutations that are hard to sort perfectly requires to break almost any structure in a given permutation. The smallest example of a hard to sort permutation is given by the permutation $P = (2\ \bar{5}\ 7\ 4\ \bar{6}\ 1\ 3)$.

5 Conclusion

From the algorithmic point of view, the central aspect of our work is the detailed description of the link between computing perfect scenarios and the strong intervals tree of a signed permutation. We gave a new description of the exponential time algorithm of [13] that highlights many of its properties. In particular, in Section 4, we characterized classes of signed permutations for which the computation of a parsimonious perfect scenario can be done efficiently.

In [18], it was shown that when there exists a parsimonious scenario that is also a perfect scenario for a given signed permutation, computing such a scenario can be done in subquadratic time, extending a previous result of [2]. In the present work, one can, once a parsimonious perfect scenario has been computed, check whether this scenario is also parsimonious, using for example one of the linear time algorithm for computing the reversal distance proposed in [1,7]. However, the computation of a parsimonious perfect scenario can require an exponential time depending on the strong intervals tree. In order to close the gap between these two approaches in computing perfect scenarios, it would be interesting to characterize, in terms of strong intervals tree, the class of signed permutations for which a parsimonious perfect scenario is also parsimonious among all possible scenarios.

From a practical point of view, it is worth to recall that the interest in computing scenarios that do not break any common intervals relies on the assumption that genes, or other genomic markers, cluster in such groups for functional reasons, like co-transcription for example. Of course, it is possible that clusters of genes exist by "chance", or are not supported by any functional evidence, and it would not be relevant to impose that such intervals should not be broken. Note, however, that the algorithms developed in this work can be used without any modification: given a set of common intervals that are believed to be pertinent from the evolutionary point of view, they define a set of strong intervals, and then a PQ-tree, and one can apply our method on this PQ-tree.

Among other future directions, it would be interesting to consider the median problem, that is one of the main reasons for computing reversals scenarios [8]. This problem has been shown to be NP-hard [10] in the general case, but the question has not been settled if one restricts every scenario to be perfect. Finally, the most natural extension would be to consider signed sequences instead of signed permutations. Some work has been done on common intervals of signed sequences [12,20], but structures that could play the role of strong intervals are yet to be discovered in this case.

References

1. D. A. Bader, B. M. E. Moret and M. Yan. A linear-time algorithm for computing inversion distance between signed permutations with an experimental study. *J. Comp. Biol.*, 8(5):483–491, 2001.
2. S. Bérard, A. Bergeron and C. Chauve. Conserved structures in evolution scenarios. In *RCG 2004*, volume 3388 of *Lecture Notes in Bioinformatics*, p. 1–15, 2005.
3. S. Bérard, A. Bergeron, C. Chauve and C .Paul. Perfect sorting by reversals is not always difficult. Technical Report LIRMM RR-05042 (Montpellier, France), 2005.
4. A. Bergeron, M. Blanchette, A. Chateau and C. Chauve Reconstructing ancestral gene orders using conserved intervals. In *WABI 2004*, volume 3240 of *Lecture Notes in Bioinformatics*, p. 14–25, 2004.
5. A. Bergeron, C. Chauve, T. Hartman and K. St-Onge. On the properties of sequences of reversals that sort a signed permutation. In *JOBIM 2002*, p. 99–108, 2002.
6. A. Bergeron, C. Chauve, F. de Montgolfier and M. Raffinot. Computing common intervals of K permutations, with applications to modular decomposition of graphs. To appear in *ESA 2005*, (to be published in *Lecture Notes in Comput. Sci.*), 2005.
7. A. Bergeron, J. Mixtacki and J. Stoye. Reversal distance without hurdles and fortresses. In *CPM 2004*, volume 3109 of *Lecture Notes in Comput. Sci.*, p. 388–399, 2004.
8. G. Bourque and P. A. Pevzner. Genome-scale evolution: Reconstructing gene orders in the ancestral species. *Genome Res.*, 12(1):26–36, 2002.

9. G. Bourque, P. A. Pevzner and G. Tesler. Reconstructing the genomic architecture of ancestral mammals: Lessons from human, mouse, and rat genomes. *Genome Res.*, 14(4):507–516, 2004.
10. A. Caprara. Formulations and hardness of multiple sorting by reversals. In *RECOMB'99*, p. 84–94, ACM Press, 1999.
11. M. Chein, M. Habib and M. C. Maurer. Partitive hypergraphs. *Discrete Math.*, 37(1):35–50, 1981.
12. G. Didier. Common intervals of two sequences. In *WABI 2003*, volume 2812 of *Lecture Notes in Bioinformatics*, p. 17–24, 2003.
13. M. Figeac and J.-S. Varré. Sorting by reversals with common intervals. In *WABI 2004*, volume 3240 of *Lecture Notes in Bioinformatics*, p. 26–37, 2004.
14. R. A. Gibbs et al. Genome sequence of the brown norway rat yields insights into mammalian evolution. *Nature*, 428(6982):493–521, 2004.
15. S. Heber and J. Stoye. Finding all common intervals of k permutations. In *CPM 2001*, volume 2089 of *Lecture Notes in Comput. Sci.*, p. 207–218, 2001.
16. G.M. Landau, L. Parida and O. Weimann. Using PQ trees for comparative genomics. In *CPM 2005*, volume 3537 of *Lecture Notes in Comput. Sci.*, p. 128–143, 2005.
17. F. de Montgolfier *Décomposition modulaire des graphes. Théorie, extensions et algorithmes.* Ph.D. thesis, Université Montpellier II (France), 2003.
18. M.-F. Sagot and E. Tannier. Perfect sorting by reversals. In *COCOON 2005*, volume 3595 of *Lecture Notes in Comput. Sci.*, pages 42–52, 2005.
19. D. Sankoff. Edit distance for genome comparison based on non-local operations. In *CPM 1992*, volume 644 of *Lecture Notes in Comput. Sci.*, p. 121–135, 1992.
20. T. Schmidt and J. Stoye. Quadratic time algorithms for finding common intervals in two and more sequences. In *CPM 2004*, volume 3109 of *Lecture Notes in Comput. Sci.*, p. 347–358, 2004.
21. E. Tannier and M.-F. Sagot. Sorting by reversals in subquadratic time. In *CPM 2004*, volume 3109 of *Lecture Notes in Comput. Sci.*, p. 1–13, 2004.
22. T. Uno and M. Yagiura. Fast algorithms to enumerate all common intervals of two permutations. *Algorithmica*, 26(2):290–309, 2000.

Minimum Recombination Histories by Branch and Bound

Rune B. Lyngsø[1], Yun S. Song[2], and Jotun Hein[1]

[1] Dept. of Statistics, Oxford University,
Oxford, OX1 3TG, United Kingdom
{lyngsoe, hein}@stats.ox.ac.uk
[2] Dept. of Computer Science, University of California, Davis, CA 95616, U.S.A.
yssong@cs.ucdavis.edu

Abstract. Recombination plays an important role in creating genetic diversity within species, and inferring past recombination events is central to many problems in genetics. Given a set M of sampled sequences, finding an evolutionary history for M with the minimum number of recombination events is a computationally very challenging problem. In this paper, we present a novel branch and bound algorithm for tackling that problem. Our method is shown to be far more efficient than the only preexisting exact method, described in [1]. Our software implementing the algorithm discussed in this paper is publicly available.

1 Introduction

Recombination is a fundamental biological process that plays a central role in generating genetic diversity within species. A question that has been receiving considerable interest, from biologists and mathematical scientists alike, is determining the minimum number $R_{\min}(M)$ of recombination events needed in the evolution of a given set M of sequences. A closely-related problem of tantamount interest is reconstructing a possible evolutionary history for M with exactly that many recombination events. Finding $R_{\min}(M)$ for an arbitrary data set M is a very difficult problem, however, and therefore several methods of computing lower bounds on $R_{\min}(M)$ have been proposed [2,3,4,5,6,7]. Such deterministic methods have interesting practical applications. For example, it has been shown [8] that efficient lower bound methods can be quite useful for detecting potential recombination hotspots. A challenging algorithmic problem is to develop new efficient methods that can produce good estimates of $R_{\min}(M)$, while being able to handle large data sets at genomic scale. For making significant progress in that direction, it would be of great help if we could study systematically when currently existing methods produce poor bounds on $R_{\min}(M)$.

A point pertinent to all lower bound methods is that, in most cases, it is difficult to know whether the number one obtains is actually equal to $R_{\min}(M)$, or whether one should try harder to obtain a sharper lower bound. To address this problem, a method of computing upper bounds on $R_{\min}(M)$ has recently been proposed [7], the main motivation being that if the upper bound is equal to the lower bound, then one actually knows the minimum number of recombination events for a given data set. Lower and upper bounds constructed in [7] are surprisingly often very close or equal when

R. Casadio and G. Myers (Eds.): WABI 2005, LNBI 3692, pp. 239–250, 2005.

recombination and mutation rates are low, but they begin to diverge as those parameters are increased, and it becomes difficult to know where in-between the two bounds $R_{\min}(M)$ lies.

Unfortunately, no efficient algorithm for computing $R_{\min}(M)$ is currently known. The only work so far that has tried to compute $R_{\min}(M)$ exactly is [1]. Being computationally intensive, both in terms of time and space, that method can analyse at most 9 sequences after some data reduction, and therefore does not have a wide range of application in practise. In this paper, we propose a branch-and-bound-based method of computing $R_{\min}(M)$ exactly that is far more efficient than the method described in [1]; our method can handle tens of sequences, requires far less memory, and runs thousands of times faster. The root sequence can be chosen to be either known or unknown in our method. In addition to finding the minimum number $R_{\min}(M)$, our method explicitly constructs a minimal ancestral recombination graph [9] that represents a possible evolutionary history with exactly $R_{\min}(M)$ recombination events; that most parsimonious (or minimal) history can be viewed using open source graphics interfaces. Our method has been fully implemented in the programming language C. The software is called `beagle` and is publicly available.

Our work should have a number of important applications. For example, using simulated data generated under various mutation and recombination rates, one can use our method to find out, when lower and upper bounds on $R_{\min}(M)$ do not match, which bound is closer to the minimum number. Such study should prove useful for devising new efficient recombination detection methods that can produce qualitatively better estimates of $R_{\min}(M)$. The minimal ARG constructed by our method should also be of interest to many researchers who wish to see an explicit graphical representation of the most parsimonious history. For instance, the upper bound method proposed in [7] explicitly constructs a class of ARGs, which may not contain minimal ARGs. By studying the minimal ARG constructed by our method when the upper bound is not equal to $R_{\min}(M)$, one may be able to learn how to capture the structure of minimal ARGs more accurately.

The organisation of this paper is as follows. In Sect. 2 we introduce the model assumed for the inference of $R_{\min}(M)$. In Sect. 3 we present our method for computing $R_{\min}(M)$. In Sect. 4 we apply our method to the data set analysed in [1] and compare performances. Finally, in Sect. 5 we discuss some future directions and open problems.

2 The Infinite Sites Model

We assume the standard infinite sites model of mutation [10]. This model is applicable to phased single nucleotide polymorphism (SNP) data sampled from a population with a relatively low mutation rate. The infinite sites model restricts the occurrence of mutation event to at most one per site. Hence, each polymorphism, or *segregating site*, is caused by exactly one mutation event, and sampled sequences can be represented as binary sequences, as each segregating site contains only two of the four possible nucleotides.

In our work, we allow sequences to contain unresolved positions. Some SNP data sets do contain sequences with unresolved positions (i.e. missing data). More importantly, as described in Sect. 3.2, recombination introduces uncertainty regarding the state of ancestral sequences. Hence, we assume that data sets consist of m sequences of length n over $\{0, 1, *\}$ (or equivalently an $m \times n$ binary matrix with entries possibly left unspecified). A "$*$" at a site corresponds to an unresolved character, while 0 and 1 correspond to the two observed nucleotides. If the grand most recent common ancestor is known, we adopt the convention that, at each site, 0 corresponds to the ancestral type while 1 corresponds to the mutant type, i.e., the grand most recent common ancestor is the all-0 sequence.

In what follows, we use M to denote the data set being analysed, with m denoting the number of sequences and n the number of segregating sites. We assume that the observed sequences have evolved from a single common ancestor through a succession of three types of events.

- a mutation event at a site that has not been subjected to mutation before, changing the state of the site in a sequence from 0 to 1 (or possibly from 1 to 0, if the most recent common ancestor is not known).
- a coalescent event, creating an extra copy of a sequence
- a recombination event, replacing two sequences with a recombinant consisting of a prefix from one sequence concatenated with the corresponding suffix of the other sequence.

As exactly one mutation event occurs in each segregating site, any valid history for M contains exactly n mutation events. Each coalescent event increases the number of sequences by one, while each recombination event decreases the number of sequences by one. Obtaining m sequences from a single common ancestor therefore require $m-1+r$ coalescent events if r recombination events occur. So the histories with the minimum number of events are the ones with the minimum number of recombination events. We denote the minimum number of recombinations required to explain a data set M by $R_{\min}(M)$.

3 Finding a Minimum Recombination History

Given a data set M, it is relatively straightforward to determine whether it can be explained by an evolutionary history with no recombinations, and to reconstruct such a history [11]. Two sites are said to be *conflicting* if they contain all four possible gametic types 00, 01, 10, and 11 (or the three types 01, 10, and 11, if the ancestral sequence is known). At least one recombination event must have occurred between two conflicting sites. Testing for the presence of the above gametic types is known as the four (or three, if the ancestral sequence is known) gamete test. Based on this test, [2] proposes a method for finding a lower bound on the number of recombinations required: a set of pairs of conflicting sites spanning non-overlapping regions implies a lower bound on $R_{\min}(M)$ equal to the set size, as at least one recombination is required between each pair of conflicting sites. In [2] it is shown how to find such a set of maximum size. However, determining the exact value of $R_{\min}(M)$ for an arbitrary data set M is **NP** hard when the most recent common ancestor is known [12].

Recently, improved methods for computing lower bounds on $R_{\min}(M)$ have been proposed [3, 4, 5, 6, 7], and in certain restricted cases, efficient methods exist for finding an evolutionary history with the minimum number of recombinations [13, 7]. We are only aware of one implemented method for computing $R_{\min}(M)$ exactly [1]. This method is based on scanning M from left to right, detecting recombinations that are needed to change local tree topologies. The main result of our present paper is a branch and bound algorithm that searches for an evolutionary history with the minimum number of recombinations. Our method constructs histories backward in time starting from the input data M, until a single common ancestor is reached. For ease of exposition, we assume that the ancestral sequence is known, but everything immediately extends to the case in which the ancestral sequence is unknown.

3.1 Reimplementation of the Haplotype Bound

Crucial elements of a branch and bound algorithm are the quality of the bounds computed and the speed with which they are obtained. We have chosen to use the haplotype bound introduced in [3]. It can provide quite powerful bounds as documented in [7], tremendously improving on the lower bound method of [2]. Though **NP** hard to compute [6], heuristics can significantly reduce computation time without seriously sacrificing power. It should be mentioned that the branch and bound algorithm implementation makes no assumptions about the lower bound computation, hence it is straightforward to substitute in any other lower bound method or implementation.

The haplotype bound can be seen as a generalisation of the method in [2]. Whereas the method of [2] looks only at pairs of segregating sites to establish a lower bound on the number of recombinations required in a region, the haplotype bound looks at all subsets of sites in the region. The key observation is that each mutation event and recombination event can introduce at most one extra sequence type. The number of mutation events equals the number of sites, due to the infinite sites assumption. Hence, if a subset of a sites gives rise to b distinct sequences, at least $b - a - 1$ recombination events must have occurred in the interval spanned by the sites. Local lower bounds obtained this way are then combined to produce a global lower bound for $R_{\min}(M)$ by determining the maximum sum of lower bounds on a set of disjoint intervals (in [3] this second step is called the composite bound, while the term haplotype bound is reserved for the first step). The authors of [3] originally implemented the haplotype bound in a program called `RecMin`. In the following we discuss our reimplementation of the haplotype bound.

If M contains identical sites, some subsets of sites can be discarded a priori: in general, we want subsets to contain as few sites and span as short intervals as possible. Let c_i denote site i of M, and let $C \subseteq \{1, \ldots, n\}$ be a set of sites. We can ignore C if there exists $i, j \in C$ with $c_i = c_j$. The set $C \setminus \{i\}$ (or equivalently $C \setminus \{j\}$) will give rise to as many sequence types as C using one site less. This provides a better lower bound in any region containing C. Let span(C) denote the interval spanned by a set of sites C, i.e. span$(C) = [\min_{i \in C} .. \max_{i \in C}]$. Assume there exists another set of sites C' containing the same types of sites, i.e. $\{c_i \mid i \in C\} = \{c_i \mid i \in C'\}$. If span$(C') \subset$ span(C) we can ignore C as C' provides an identical lower bound in any region containing C.

It follows that it suffices to determine the minimal intervals spanned by sets of sites corresponding to a set of site types, for all sets of type sites. We do this by incrementally

Algorithm 1. Minimal intervals combining a position from s and an interval from t

while s and t are non-empty **do**
  while $|s| > 1$ **and** $s[2] < t[1]['right']$ **do**
    $s = s[2:]$
  while $|t| > 1$ **and** $t[2]['right'] < s[1]$ **do**
    $t = t[2:]$
  if $s[1] < t[1]['left']$ **then**
    Report minimal interval $\{'left' : s[1], 'right' : t[1]['right']\}$
    $s = s[2:]$
  else
    if $t[1]['right'] < s[1]$ **then**
      Report minimal interval $\{'left' : t[1]['left'], 'right' : s[1]\}$
    else
      Report minimal interval $t[1]$
    $t = t[2:]$

0000
0010
0001
1000
0100
1010
0101
1234

Fig. 1. An example where our heuristic fails. The exact haplotype lower bound on the set is $7 - 4 - 1 = 2$ recombinations. The only conflicting pairs of sites in the data are $\{1, 3\}$ and $\{2, 4\}$. Therefore, our heuristic fails to build the full set of all four sites, and only establish lower bounds of one recombination on the three non-disjoint regions from site 1 to site 3, from site 2 to site 4, and from site 1 to site 4.

constructing sets of type sites, adding one type site at a time. Algorithm 1 shows how we maintain the ordered set of minimal intervals spanned under the addition of a new site type, in time linear in the total number of previous minimal intervals and occurrences of the new site type.

To limit computation time, the default setting of `RecMin` is to only consider subsets of up to five sites spanning regions of width at most 12. Depending on the data, this heuristic may significantly decrease the computed bound. The user is advised to start with reasonable values of these parameters, and then increase them until the computed bound appears stable or the run time becomes too large. This approach is not well suited as a subprocess of a branch and bound algorithm. Instead, we use a more adaptive heuristic. Recall that a subset of a sites giving rise to b distinct sequences results in a bound of $b - a - 1$. If adding a new site type only increases the number of distinct sequences by at most one, the bound does not improve. We use this as a stopping criteria, ceasing further expansion of a set of site types the first time an increased bound is not obtained. It significantly reduces running time but in most cases results in the same lower bound on the minimum number of recombinations as the exact haplotype bound. As shown by the example in Fig. 1, however, data sets causing this heuristic to perform arbitrarily poorly can be constructed.

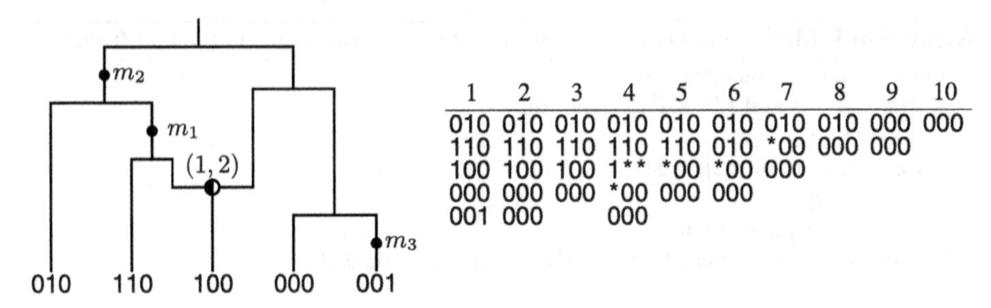

1	2	3	4	5	6	7	8	9	10
010	010	010	010	010	010	010	010	000	000
110	110	110	110	110	010	*00	000	000	
100	100	100	1**	*00	*00	000			
000	000	000	*00	000	000				
001	000		000						

Fig. 2. One possible evolutionary history, or ARG, explaining the small data set $\{010, 110, 100, 000, 001\}$. Mutations are represented by •s, where m_i denotes a mutation at the i'th site; coalescent events are represented by two lineages merging; recombination events are represented by one lineage forking out into two lineages at a ◐, where $(i, i+1)$ denotes that the crossover point is between site i and site $i+1$ and the orientation of the black half of the ◐ indicates the lineage supplying the prefix to the recombinant sequence. Listed in the table are the 10 ancestral states encountered in this ARG, with the most recent first.

3.2 Basic Branch and Bound Algorithm

We trace the evolution of M backward in time. The configurations encountered are called *ancestral states* and the aim is to reach the ancestral state consisting of only the all-0 common ancestor sequence. For a configuration ψ, the possible predecessors ψ' can be deduced from the event types listed in Sect. 2:

Mutation can occur in site i if exactly one sequence $s \in \psi$ carries the mutant type in site i; i.e. if $s[i] = 1$ and $\forall t \in \psi \setminus \{s\} : t[i] \neq 1$, then $\psi' = \psi \setminus \{s\} \cup \{s[1..i-1]0s[i+1..n]$ is a possible predecessor of ψ.

Coalescence can occur if $s, t \in \psi$ are *compatible*; i.e. if $\forall 1 \leq i \leq n : \{s[i], t[i]\} \neq \{0, 1\}$, then $\psi' = \psi \setminus \{s, t\} \cup \{u\}$ is a possible predecessor of ψ, where $u[i] = s[i]$ if $s[i] \neq *$ and $u[i] = t[i]$ otherwise.

Recombination can always occur in any sequence; if $s \in \psi$ and $0 \leq i \leq n$, then $\psi' = \psi \setminus \{s\} \cup \{s[1..i]*^{n-i}, *^i s[i+1..n]\}$ is a possible predecessor of ψ. Note that we only trace the ancestral material observed in M, so sites not passing ancestral material on to ψ are left unresolved indicated by the $*$ character. Moreover, it is easy to see that a recombination introducing the all-$*$ sequence cannot be part of an evolutionary history with a minimum number of recombinations. Hence, we implicitly assume that only recombinations splitting the recombinant into two sequences both carrying ancestral material are considered.

A sequence of events, or evolutionary history, can graphically be represented by an *ancestral recombination graph* (ARG) as illustrated in Fig. 2.

Based on this description of the consequences of each type of evolutionary event, it is easy to search for an evolutionary history with at most r recombinations. To avoid repeating previous work, we maintain a hash table containing the best lower bound established for all ancestral states already encountered. So when we arrive at an ancestral state ψ with r' recombinations remaining to meet the overall target of r recombinations we

- first check whether ψ is already present in the hash table with a lower bound exceeding r'; if this is the case we backtrack the search
- if ψ is not present in the hash table, we compute a lower bound on the minimum number of recombinations required for ψ as described in Sect. 3.1; if this lower bound exceeds r' we insert ψ in the hash table with this lower bound and backtrack the search
- otherwise, we make a depth-first search starting at the possible predecessors of ψ, with r' recombinations remaining for predecessors obtained through coalescence and mutation events and $r' - 1$ recombinations remaining for predecessors obtained through recombination events; if this search fails, we update the hash table to contain ψ with lower bound $r' + 1$.

To determine the minimum number of recombinations required for a data set M, we start by computing a lower bound r on this number as described in Sect. 3.1. We then try to find an evolutionary history with at most r recombinations. If this fails, we increase r by one. We repeat this until r is increased to $R_{\min}(M)$, at which stage we will find a minimum recombination history.

This approach essentially develops the true minimum number of recombinations required by improving a lower bound. An alternative would be to start with an upper bound r on the number of recombinations required – obtained by heuristic means, see e.g. [7], or possibly even ∞ – and updating it whenever a new evolutionary history with a lower number of recombinations is discovered. This continues until it is established that no evolutionary history with fewer recombinations than the current value of r is possible. This alternative approach would essentially develop the true minimum number of recombinations required by successive improvements of an upper bound. We would expect these two alternatives to have comparable run times, although our approach would probably exhibit a smaller variance; the total time will usually be dominated by the time used to establish that no evolutionary history exists with $R_{\min}(M)-1$ recombinations.

3.3 Refinements of the Branch and Bound Algorithm

In the previous section we described a vanilla flavoured branch & bound algorithm for determining $R_{\min}(M)$, utilising the haplotype bound. This forms the basis of our full branch & bound algorithm, but would by itself not be sufficiently efficient to yield a major improvement in efficiency compared to [1]. Hence, we apply three tricks to speed up the computation and decrease the amount of memory used: one based on data set reduction, one based on only considering certain orderings of events, and one based on requiring splits caused by recombinations to be maximal.

Reduced Data Sets. As observed in [3, 1], an ancestral state ψ can be *reduced* without changing the value of $R_{\min}(\psi)$. We use l to refer to the sequence length in ψ, which may differ from n as the reduction process may remove sites. We reduce ψ by repeatedly applying the following operations until no operations are possible:

1. collapse identical sequences into one; i.e. if $s, t \in \psi$ and $s = t$, then ψ can be replaced by $\psi' = \psi \setminus \{s\}$

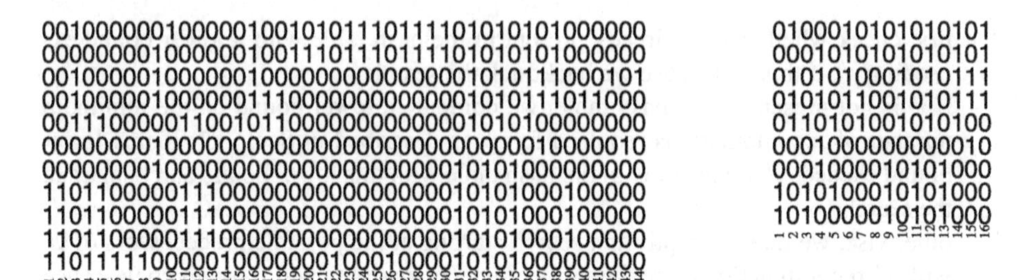

Fig. 3. The *Drosophila melanogaster* alcohol dehydrogenase data from [14], and the same data after having been reduced

2. remove sites where at most one sequence carries the mutant type, also known as *uninformative* sites; i.e. if $\exists 1 \leq i \leq l : |\{s \in \psi \mid s[i] = 1\}| = 1$, then ψ can be replaced by $\psi' = \{s[1..i-1]s[i+1..l] \mid s \in \psi\}$
3. collapse neighbouring identical sites; i.e. if $\exists 1 \leq i < l, \forall s \in \psi : s[i] = s[i+1]$, then ψ can be replaced by $\psi' = \{s[1..i-1]s[i+1..l] \mid s \in \psi\}$

In the presence of unresolved sites we can slightly strengthen operation 1. We will say that a sequence s is *subsumed* in another sequence t if $s[i] \in \{*, t[i]\}$ for all $1 \leq i \leq l$. That is, if s and t are compatible and s does not carry ancestral material in any site where t does not carry ancestral material. Operation 1 can be applied to any pair of sequences where one is subsumed in the other. We can define a site being subsumed in another site in a similar manner, and strengthen operation 3 to apply to neighbouring sites where one is subsumed in the other. That the effect of this reduction can be significant is illustrated in Fig. 3.

Restricted Event Orderings. Reduction can be seen as imposing restrictions on the order in which events occur. Operations 1 and 2 can be seen as equivalent to only considering evolutionary histories where coalescent events, that as a net result removes all traces of one of the sequences, and mutations are carried out at the first given opportunity. Operation 3 can be seen as requiring the mutations in the collapsed sites to happen consecutively. We can impose further restrictions on the sequences of events considered, while guaranteeing that we will still consider at least one sequence of events containing a minimum number of recombinations.

As operation 2 of ancestral state reduction forces all mutations possible, we never perform a recombination event on an ancestral state where a mutation is possible. It is easy to see that a recombination event cannot make a new mutation event possible. So after a sequence of recombination events starting from ancestral state ψ, a coalescence event must necessarily follow. This event will coalesce sequences $t = *^i s[i+1..j]*^{l-j}$ and $t' = *^{i'} s'[i'+1..j']*^{l-j'}$ (with l denoting sequence length in ψ), i.e. segments of ancestral material carried by sequences $s, t \in \psi$. Note that i, i' may equal 0 and j, j' may equal l. Evidently, any recombination not involved in creating the segments of t and t' can be postponed until after coalescing t and t'.

Even a recombination involved in creating t or t' can be postponed. Only the at most two recombinations defining the boundaries of overlap of ancestral material between t

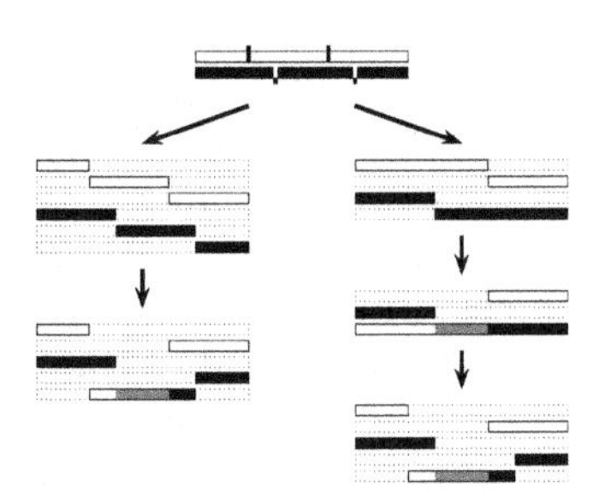

```
0100010101010101
0001010101010101
0101010001010111
0101011001010111
0111010100101010 0
0001000000000010
0001000010101000
1010100010101000
1010000010101000
1 2 3 4 5 6 7 8 9 10 11 12 13 14 15 16
```

Fig. 4. Illustration of the event ordering and maximality principles used in the refined branch & bound algorithm. The right-hand illustration shows that of the four splits defining two segments being coalesced, two can be postponed till after the coalescence as illustrated in the right-hand sequence of events. In the right-hand illustration, the maximum subsumed prefix and suffix is underlined for each sequence.

and t' are required to make the coalescence possible, as illustrated in Fig. 4. Hence, we will only consider sequences of events with at most two consecutive recombination events. Moreover, the recombination event or events have to be involved in creating the sequences of the ensuing coalescence event.

Maximality. Finally, we require recombinations to split off segments that are in a certain sense maximal. We discuss this for the case where a single recombination event is followed by a coalescence involving the prefix split off by the recombination event, but the principle is applied to all recombination events. Assume the recombination splits sequence $s \in \psi$ after site i, and that the prefix $s[1..i]*^{l-i}$ (where l is the sequence length in ψ) coalesces with sequence $t \in \psi$ to create sequence t'. The ensuing ancestral state is $\psi' = \psi \cup \{*^i s[i+1..l], t'\} \setminus \{s, t\}$. If $s[i+1] = t[i+1]$ then $s[1..i+1]*^{l-i-1}$ will also coalesce with t to create t'. So by splitting s after site $i+1$ instead, we obtain the alternative ancestral state $\psi'' = \psi \cup \{*^i s[i+2..l], t'\} \setminus \{s, t\}$. Evidently, $R_{\min}(\psi'') \leq R_{\min}(\psi')$ as $*^i s[i+2..l]$ does not contain any ancestral material not present in $*^i s[i+1..l]$. Hence, we only consider recombinations that split off segments that cannot be extended without changing the outcome of the ensuing coalescence event.

More generally, for two ancestral states ϕ and ϕ' we will say that $\phi \prec \phi'$ if we can obtain ϕ from ϕ' by changing one or more characters in ϕ' to $*$. We do not pursue events leading to an ancestral state ϕ whenever it can be established that we can reach another ancestral state $\phi' \prec \phi$ using the same number of recombinations. In particular, let the *maximum subsumed prefix* of a sequence s in ancestral state ψ be the longest prefix of s that is subsumed in some other sequence of ψ. Similarly define the *maximum subsumed suffix*. Then no recombination events occurring within the maximum subsumed prefix or suffix of a sequence are considered (with the exception of one recombination event if they overlap). This is illustrated in Fig. 4.

4 Application

We have implemented the branch & bound algorithm outlined in the previous section in a C program called `beagle` (as branch & bound shares its acronym with the infamous Beagle Boys). The source code is available under the GNU public license from `http://www.stats.ox.ac.uk/~lyngsoe/beagle/`. The program can produce an evolutionary history that established the value of $R_{\min}(M)$, and accompanying scripts can convert such a history to three different formats used by different programs for drawing networks. The program can also compute the minimum number $R_{\min}(I)$ of recombinations required in a sub-interval I of the input data M. Note that these local values $R_{\min}(I)$ may not be attainable by any minimal ARG for the entire data M.

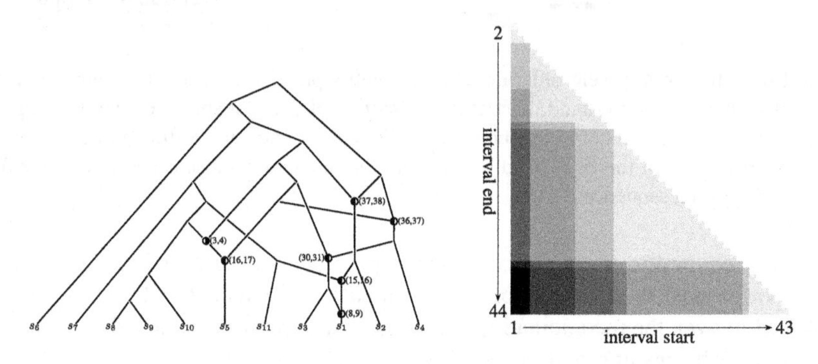

Fig. 5. Results obtained from applying `beagle` to the data of [14]. On the left is the ARG representing the evolutionary history with $R_{\min}(M) = 7$ found by `beagle`, with mutations left out for clarity. On the right is the matrix of local values of $R_{\min}(I)$ for all sub-intervals I of the data, colour coded from light grey for no recombinations to black for 7 recombinations.

We have applied `beagle` to Kreitman's data of the alcohol dehydrogenase locus from 11 chromosomes of *Drosophila melanogaster*. This data set was also analysed in [1], and therefore it provides a useful benchmark for comparing our method with the only previously existing method for computing $R_{\min}(M)$ exactly. The data is shown in Fig. 3. An ARG recreated from the history and the matrix of local values of $R_{\min}(I)$ are shown in Fig. 5. The implementation of the method described in [1] required about 30 minutes on a 1.26GHz Pentium III processor and 1.5GB of memory to determine that $R_{\min}(M) = 7$ for this data set. On a similar processor, a 1.4GHz Athlon processor, `beagle` obtained the same result in less than a second using less than 100kB of memory. On this particular data set, we have thus managed an improvement of three to four orders of magnitude in both time and space requirements. This significantly expands the range of data for which an exact computation of $R_{\min}(M)$ is feasible.

To illustrate how the performance of `beagle` depends on the size and complexity of the input data, we have also applied `beagle` to the human LPL data from [15], with missing or non-SNP sites removed. These sequences are sampled from three geographically distinct regions, Jackson, North Karelia, and Rochester. Following [3] we further partition the sites into three regions. Analysing the full data set is beyond the capabilities of `beagle`, but based on the above partitions we obtain nine smaller data sets of varying size and complexity. For eight of the nine data sets `beagle` was able to determine $R_{\min}(M)$ and the results are summarised in Table 1. Not surprisingly, a key aspect influencing the time requirement seems to be how far the initial lower bound

Table 1. Results of applying `beagle` to the human LPL data. In each entry, we first list the size of the data set in terms of number of sequences/number of sites; we then list $R_{\min}(M)$ with the initial lower bound obtained from our haplotype heuristic in parentheses; finally, we list the time required to determine $R_{\min}(M)$ on a 1.4GHz Athlon processor.

Population	region 1	region 2	region 3
Jackson	40/14; 13(10); 18m52s	40/14; 10(8); 26s	40/22; 15(12); > 15h
North Karelia	31/15; 2(2); < 1s		31/24; 8(7); < 1s
Rochester	27/17; 1(1); < 1s	27/13; 14(12); 8m00s	27/28; 8(7); < 1s

is from $R_{\min}(M)$. Only two of these data sets are sufficiently small that it would be feasible to use the method of [1] to determine $R_{\min}(M)$.

5 Discussion

The framework of our branch & bound algorithm works independently of the lower bound method utilised. It should be quite easy to replace our implementation of the haplotype bound with other lower bound implementations available, to explore whether any of these can further improve the efficiency of `beagle`. Two crucial aspects are the quality of the bound produced and the time taken to produce it. A qualitatively improved lower bound will in general result in `beagle` having to explore fewer ancestral states, while a faster lower bound method will result in `beagle` having to spend less time on each ancestral state encountered. Profiling data obtained from running `beagle` indicates that most of CPU time is spent on computing lower bounds. A lower bound of similar quality that can be computed faster can lead to a significant reduction in computation time, but not a reduction by orders of magnitude. To obtain such a reduction we would need a method producing better lower bounds. Therefore, we are particularly interested in testing `beagle` with the improved haplotype bound discussed in [7].

It would be interesting and worthwhile to use `beagle` to produce a lower bound on $R_{\min}(M)$. We can use `beagle` to compute the minimum number of recombinations for all small regions of the data, while using an efficient lower bound method for large regions. These local bounds can then be combined using the composite bound method of [3] to produce a lower bound on $R_{\min}(M)$ for the entire data. We expect `beagle` to work quite well on data sets of short sequences, for which the search space reduction based on ignoring recombinations in maximum subsumed prefixes and suffixes should dramatically limit the number of recombination events pursued. In this context, it should be noted that the method of [1] may be more efficient for data sets consisting of very few long sequences, as the running time of that method depends only linearly on sequence length, albeit super-exponentially on the number of sequences.

The computation of a local value $R_{\min}(I)$ by `beagle` is oblivious to the region surrounding I, i.e. the local value may not be attainable in any history with a globally minimum number $R_{\min}(M)$ of recombinations. It would be of interest to compute local values in the context of the surrounding data, not least to study the effects of ignoring context. It is not clear to us, however, whether `beagle` can be modified to compute local $R_{\min}(I)$ values without sacrificing the refinements discussed in Sect. 3.3

Acknowledgements

YS is supported by grant EIA-0220154 from the National Science Foundation. JH and RL are supported by EPSRC grant HAMJW, and MRC grant HAMKA.

References

1. Song, Y.S., Hein, J.: Parsimonious reconstruction of sequence evolution and haplotype blocks. In: Proceedings of the 3rd Workshop on Algorithms in Bioinformatics (WABI). (2003) 287–302
2. Hudson, R.R., Kaplan, N.L.: Statistical properties of the number of recombination events in the history of a sample of DNA sequences. Genetics **111** (1985) 147–164
3. Myers, S.R., Griffiths, R.C.: Bounds on the minimum number of recombination events in a sample history. Genetics **163** (2003) 375–394
4. Gusfield, D., Hickerson, D.: A new lower bound on the number of needed recombination nodes in both unrooted and rooted phylogenetic networks. Technical Report UCD-ECS-06, University of California, Davis (2004)
5. Song, Y.S., Hein, J.: On the minimum number of recombination events in the evolutionary history of DNA sequences. Journal of Mathematical Biology **48** (2004) 160–186
6. Bafna, V., Bansal, V.: Improved recombination lower bounds for haplotype data. In: Proceedings of the 9th Annual International Conference on Computational Molecular Biology (RECOMB). (2005)
7. Song, Y.S., Wu, Y., Gusfield, D.: Efficient computation of close lower and upper bounds on the minimum number of recombinations in biological sequence evolution. In: Proceedings of the 13th International Conference on Intelligent Systems for Molecular Biology (ISMB). (2005) in press.
8. Fearnhead, P., Harding, R.M., Schneider, J.A., Myers, S., Donnelly, P.: Application of coalescent methods to reveal fine-scale rate variation and recombination hotspots. Genetics **167** (2004) 2067–2081
9. Griffiths, R.C., Marjoram, P.: An ancestral recombination graph. In: Progress in Population Genetics and Human Evolution. Volume 87 of IMA Volumes in Mathematics and its Applications. Springer Verlag (1997) 257–270
10. Kimura, M.: The number of heterozygous nucleotide sites maintained in a finite population due to steady flux of mutations. Genetics **61** (1969) 893–903
11. Gusfield, D.: Efficient algorithms for inferring evolutionary trees. Networks **21** (1991) 19–28
12. Wang, L., Zhang, K., Zhang, L.: Perfect phylogenetic networks with recombination. Journal of Computational Biology **8** (2001) 69–78
13. Gusfield, D., Eddhu, S., Langley, C.: Optimal, efficient reconstruction of phylogenetic networks with constrained recombination. Journal of Bioinformatics and Computational Biology **2** (2004) 173–213
14. Kreitman, M.: Nucleotide polymorphism at the alcohol dehydrogenase locus of Drosophila melanogaster. Nature **304** (1983) 412–417
15. Nickerson, D.A., Taylor, S.L., Weiss, K.M., Clark, A.G., Hutchinson, R.G., Stengard, J., Salomaa, V., Vartiainen, E., Boerwinkle, E., Sing, C.F.: DNA sequence diversity in a 9.7-kb region of the human lipoprotein lipase gene. Nature Genetics **19** (1998) 216–7

A Unifying Framework for Seed Sensitivity and Its Application to Subset Seeds
(Extended Abstract)

Gregory Kucherov[1], Laurent Noé[1], and Mikhail Roytberg[2]

[1] INRIA/LORIA, 615, rue du Jardin Botanique,
B.P. 101, 54602, Villers-lès-Nancy, France
{Gregory.Kucherov, Laurent.Noe}@loria.fr
[2] Institute of Mathematical Problems in Biology,
Pushchino, Moscow Region, 142290, Russia
roytberg@impb.psn.ru

Abstract. We propose a general approach to compute the seed sensitivity, that can be applied to different definitions of seeds. It treats separately three components of the seed sensitivity problem – a set of target alignments, an associated probability distribution, and a seed model – that are specified by distinct finite automata. The approach is then applied to a new concept of *subset seeds* for which we propose an efficient automaton construction. Experimental results confirm that sensitive subset seeds can be efficiently designed using our approach, and can then be used in similarity search producing better results than ordinary spaced seeds.

1 Introduction

In the framework of pattern matching and similarity search in biological sequences, seeds specify a class of short sequence motif which, if shared by two sequences, are assumed to witness a potential similarity. Spaced seeds have been introduced several years ago [1,2] and have been shown to improve significantly the efficiency of the search. One of the key problems associated with spaced seeds is a precise estimation of the sensitivity of the associated search method. This is important for comparing seeds and for choosing most appropriate seeds for a sequence comparison problem to solve.

The problem of seed sensitivity depends on several components. First, it depends on the *seed model* specifying the class of allowed seeds and the way that seeds match (*hit*) potential alignments. In the basic case, seeds are specified by binary words of certain length (*span*), possibly with a constraint on the number of 1's (*weight*). However, different extensions of this basic seed model have been proposed in the literature, such as multi-seed (or multi-hit) strategies [3,4,2], seed families [5,6,7,8,9,10], seeds over non-binary alphabets [11,12], vector seeds [13,10].

The second parameter is the class of *target alignments* that are alignment fragments that one aims to detect. Usually, these are *gapless* alignments of a

R. Casadio and G. Myers (Eds.): WABI 2005, LNBI 3692, pp. 251–263, 2005.

given length. Gapless alignments are easy to model, in the simplest case they are represented by binary sequences in the match/mismatch alphabet. This representation has been adopted by many authors [2,14,15,16,17,18]. The binary representation, however, cannot distinguish between different types of matches and mismatches, and is clearly insufficient in the case of protein sequences. In [13,10], an alignment is represented by a sequence of real numbers that are *scores* of matches or mismatches at corresponding positions. A related, but yet different approach is suggested in [12], where DNA alignments are represented by sequences on the ternary alphabet of match/transition/transversion. Finally, another generalization of simple binary sequences was considered in [19], where alignments are required to be *homogeneous*, i.e. to contain no sub-alignment with a score larger than the entire alignment.

The third necessary ingredient for seed sensitivity estimation is the probability distribution on the set of target alignments. Again, in the simplest case, alignment sequences are assumed to obey a Bernoulli model [2,16]. In more general settings, Markov or Hidden Markov models are considered [17,15]. A different way of defining probabilities on binary alignments has been taken in [19]: all homogeneous alignments of a given length are considered equiprobable.

Several algorithms for computing the seed sensitivity for different frameworks have been proposed in the above-mentioned papers. All of them, however, use a common dynamic programming (DP) approach, first brought up in [14].

In the present paper, we propose a general approach to computing the seed sensitivity. This approach subsumes the cases considered in the above-mentioned papers, and allows to deal with new combinations of the three seed sensitivity parameters. The underlying idea of our approach is to specify each of the three components – the seed, the set of target alignments, and the probability distribution – by a separate finite automaton.

A deterministic finite automaton (DFA) that recognizes all alignments matched by given seeds was already used in [17] for the case of ordinary spaced seeds. In this paper, we assume that the set of target alignments is also specified by a DFA and, more importantly, that the probabilistic model is specified by a *probability transducer* – a probability-generating finite automaton equivalent to HMM with respect to the class of generated probability distributions.

We show that once these three automata are set, the seed sensitivity can be computed by a unique general algorithm. This algorithm reduces the problem to a computation of the total weight over all paths in an acyclic graph corresponding to the automaton resulting from the product of the three automata. This computation can be done by a well-known dynamic programming algorithm [20,21] with the time complexity proportional to the number of transitions of the resulting automaton. Interestingly, all above-mentioned seed sensitivity algorithms considered by different authors can be reformulated as instances of this general algorithm.

In the second part of this work, we study a new concept of *subset seeds* – an extension of spaced seeds that allows to deal with a non-binary alignment alphabet and, on the other hand, still allows an efficient hashing method to locate seeds. For this definition of seeds, we define a DFA with a number of states independent of the size of the alignment alphabet. Reduced to the case of ordinary spaced seeds, this DFA construction gives the same worst-case number of states as the Aho-Corasick DFA used in [17]. Moreover, our DFA has always no more states than the DFA of [17], and has substantially less states on average.

Together with the general approach proposed in the first part, our DFA gives an efficient algorithm for computing the sensitivity of subset seeds, for different classes of target alignments and different probability transducers. In the experimental part of this work, we confirm this by running an implementation of our algorithm in order to design efficient subset seeds for different probabilistic models, trained on real genomic data. We also show experimentally that designed subset seeds allow to find more significant alignments than ordinary spaced seeds of equivalent selectivity.

2 General Framework

Estimating the seed sensitivity amounts to compute the probability for a random word (target alignment), drawn according to a given probabilistic model, to belong to a given language, namely the language of all alignments matched by a given seed (or a set of seeds).

2.1 Target Alignments

Target alignments are represented by words over an alignment alphabet $\mathcal{A}$. In the simplest case, considered most often, the alphabet is binary and expresses a match or a mismatch occurring at each alignment column. However, it could be useful to consider larger alphabets, such as the ternary alphabet of match/transition/transversion for the case of DNA (see [12]). The importance of this extension is even more evident for the protein case ([10]), where different types of amino acid pairs are generally distinguished.

Usually, the set of target alignments is a finite set. In the case considered most often [2,14,15,16,17,18], target alignments are all words of a given length n. This set is trivially a regular language that can be specified by a deterministic automaton with $(n+1)$ states. However, more complex definitions of target alignments have been considered (see e.g. [19]) that aim to capture more adequately properties of biologically relevant alignments. In general, we assume that the set of target alignments is a finite regular language $L_T \in \mathcal{A}^*$ and thus can be represented by an acyclic DFA $T =< Q_T, q_T^0, q_T^F, \mathcal{A}, \psi_T >$.

2.2 Probability Assignment

Once an alignment language L_T has been set, we have to define a probability distribution on the words of L_T. We do this using probability transducers.

A probability transducer is a finite automaton without final states in which each transition outputs a *probability*.

Definition 1. *A* probability transducer G *over an alphabet* $\mathcal{A}$ *is a 4-tuple* $< Q_G, q_G^0, \mathcal{A}, \rho_G >$, *where* Q_G *is a finite set of states,* $q_G^0 \in Q_G$ *is an initial state, and* $\rho_G : Q_G \times \mathcal{A} \times Q_G \to [0,1]$ *is a real-valued probability function such that* $\forall q \in Q_G, \sum_{q' \in Q_G, a \in \mathcal{A}} \rho_G(q, a, q') = 1$.

A *transition* of G is a triplet $e =< q, a, q' >$ such that $\rho(q, a, q') > 0$. Letter a is called the *label* of e and denoted $label(e)$. A probability transducer G is *deterministic* if for each $q \in Q_G$ and each $a \in \mathcal{A}$, there is at most one transition $< q, a, q' >$. For each path $P = (e_1, ..., e_n)$ in G, we define its *label* to be the word $label(P) = label(e_1)...label(e_n)$, and the associated probability to be the product $\rho(P) = \prod_{i=1}^{n} \rho_G(e_i)$. A path is *initial*, if its start state is the initial state q_G^0 of the transducer G.

Definition 2. *The* probability *of a word* $w \in \mathcal{A}^*$ *according to a probability transducer* $G =< Q_G, q_G^0, \mathcal{A}, \rho_G >$, *denoted* $\mathcal{P}_G(w)$, *is the sum of probabilities of all initial paths in* G *with the label* w. $\mathcal{P}_G(w) = 0$ *if no such path exists. The probability* $\mathcal{P}_G(L)$ *of a finite language* $L \subseteq \mathcal{A}^*$ *according a probability transducer* G *is defined by* $\mathcal{P}_G(L) = \sum_{w \in L} \mathcal{P}_G(w)$.

Note that for any n and for $L = A^n$ (all words of length n), $\mathcal{P}_G(L) = 1$.

Probability transducers can express common probability distributions on words (alignments). Bernoulli sequences with independent probabilities of each symbol [2,16,18] can be specified with deterministic one-state probability transducers. In Markov sequences of order k [17,6], the probability of each symbol depends on k previous symbols. They can therefore be specified by a deterministic probability transducer with at most $|\mathcal{A}|^k$ states.

A Hidden Markov model (HMM) [15] corresponds, in general, to a non-deterministic probability transducer. The states of this transducer correspond to the (hidden) states of the HMM, plus possibly an additional initial state. Inversely, for each probability transducer, one can construct an HMM generating the same probability distribution on words. Therefore, non-deterministic probability transducers and HMMs are equivalent with respect to the class of generated probability distributions. The proofs are straightforward and are omitted due to space limitations.

2.3 Seed Automata and Seed Sensitivity

Since the advent of spaced seeds [1,2], different extensions of this idea have been proposed in the literature (see Introduction). For all of them, the set of possible alignment fragments matched by a seed (or by a set of seeds) is a finite set, and therefore the set of matched alignments is a regular language. For the original spaced seed model, this observation was used by Buhler et al. [17] who proposed an algorithm for computing the seed sensitivity based on a DFA defining the language of alignments matched by the seed. In this paper, we extend this approach to a general one that allows a uniform computation of seed sensitivity for a wide class of settings including different probability distributions on target alignments, as well as different seed definitions.

Consider a seed (or a set of seeds) π under a given seed model. We assume that the set of alignments L_π matched by π is a regular language recognized by a DFA $S_\pi =< Q_S, q_S^0, Q_S^F, \mathcal{A}, \psi_S >$. Consider a finite set L_T of target alignments and a probability transducer G. Under this assumptions, the sensitivity of π is defined as the conditional probability

$$\frac{\mathcal{P}_G(L_T \cap L_\pi)}{\mathcal{P}_G(L_T)}. \tag{1}$$

An automaton recognizing $L = L_T \cap L_\pi$ can be obtained as the product of automata T and S_π recognizing L_T and L_π respectively. Let $K =< Q_K, q_K^0, Q_K^F, \mathcal{A}, \psi_K >$ be this automaton. We now consider the product W of K and G, denoted $K \times G$, defined as follows.

Definition 3. *Given a DFA $K =< Q_K, q_K^0, Q_K^F, \mathcal{A}, \psi_K >$ and a probability transducer $G =< Q_G, q_G^0, \mathcal{A}, \rho_G >$, the product of K and G is the* probability-weighted automaton *$W =< Q_W, q_W^0, Q_W^F, \mathcal{A}, \rho_W >$ (for short,* PW-automaton*) such that $Q_W = Q_K \times Q_G$, $q_W^0 = (q_K^0, q_G^0)$, $q_W^F = \{(q_K, q_G) | q_K \in Q_K^F\}$, $\rho_W((q_K, q_G), a, (q'_K, q'_G)) = \rho_G(q_G, a, q'_G)$ if $\psi_K(q_K, a) = q'_K$, and 0 otherwise.*

W can be viewed as a non-deterministic probability transducer with final states. $\rho_W((q_K, q_G), a, (q'_K, q'_G))$ is the *probability* of the $< (q_K, q_G), a, (q'_K, q'_G) >$ transition. A path in W is called *full* if it goes from the initial to a final state.

Lemma 1. *Let G be a probability transducer. Let L be a finite language and K be a deterministic automaton recognizing L. Let $W = G \times K$. The probability $\mathcal{P}_G(L)$ is equal to sum of probabilities of all full paths in W.*

Proof. Since K is a deterministic automaton, each word $w \in L$ corresponds to a single accepting path in K and the paths in G labeled w (see Definition 1) are in one-to-one correspondence with the full path in W accepting w. By definition, $\mathcal{P}_G(w)$ is equal to the sum of probabilities of all paths in G labeled w. Each such path corresponds to a unique path in W, with the same probability. Therefore, the probability of w is the sum of probabilities of corresponding paths in W. Each such path is a full path, and paths for distinct words w are disjoint. The lemma follows.

2.4 Computing Seed Sensitivity

Lemma 1 reduces the computation of seed sensitivity to a computation of the sum of probabilities of paths in a PW-automaton.

Lemma 2. *Consider an alignment alphabet $\mathcal{A}$, a finite set $L_T \subseteq \mathcal{A}^*$ of target alignments, and a set $L_\pi \subseteq \mathcal{A}^*$ of all alignments matched by a given seed π. Let $K =< Q_K, q_t^0, Q_K^F, \mathcal{A}, \psi_Q >$ be an acyclic DFA recognizing the language $L = L_T \cap L_\pi$. Let further $G =< Q_G, q_G^0, \mathcal{A}, \rho >$ be a probability transducer defining a probability distribution on the set L_T. Then $\mathcal{P}_G(L)$ can be computed in time $\mathcal{O}(|Q_G|^2 \cdot |Q_K| \cdot |\mathcal{A}|)$ and space $\mathcal{O}(|Q_G| \cdot |Q_K|)$.*

Proof. By Lemma 1, the probability of L with respect to G can be computed as the sum of probabilities of all full paths in W. Since K is an acyclic automaton, so is W. Therefore, the sum of probabilities of all full paths in W leading to final states q_W^F can be computed by a classical DP algorithm [20] applied to acyclic directed graphs ([21] presents a survey of application of this technique to different bioinformatic problems). The time complexity of the algorithm is proportional to the number of transitions in W. W has $|Q_G| \cdot |Q_K|$ states, and for each letter of $\mathcal{A}$, each state has at most $|Q_G|$ outgoing transitions. The bounds follow.

Lemma 2 provides a general approach to compute the seed sensitivity. To apply the approach, one has to define three automata:

- a deterministic acyclic DFA T specifying a set of target alignments over an alphabet $\mathcal{A}$ (e.g. all words of a given length, possibly verifying some additional properties),
- a (generally non-deterministic) probability transducer G specifying a probability distribution on target alignments (e.g. Bernoulli model, Markov sequence of order k, HMM),
- a deterministic DFA S_π specifying the seed model via a set of matched alignments.

As soon as these three automata are defined, Lemma 2 can be used to compute probabilities $\mathcal{P}_G(L_T \cap L_\pi)$ and $\mathcal{P}_G(L_T)$ in order to estimate the seed sensitivity according to equation (1).

Note that if the probability transducer G is deterministic (as it is the case for Bernoulli models or Markov sequences), then the time complexity is $\mathcal{O}(|Q_G| \cdot |Q_K| \cdot |\mathcal{A}|)$. In general, the complexity of the algorithm can be improved by reducing the involved automata. Buhler et al. [17] introduced the idea of using the Aho-Corasick automaton [22] as the seed automaton S_π for a spaced seed. The authors of [17] considered all binary alignments of a fixed length n distributed according to a Markov model of order k. In this setting, the obtained complexity was $\mathcal{O}(w2^{s-w}2^k n)$, where s and w are seed's span and weight respectively. Given that the size of the Aho-Corasick automaton is $\mathcal{O}(w2^{s-w})$, this complexity is automatically implied by Lemma 2, as the size of the probability transducer is $\mathcal{O}(2^k)$, and that of the target alignment automaton is $\mathcal{O}(n)$. Compared to [17], our approach explicitly distinguishes the descriptions of matched alignments and their probabilities, which allows us to automatically extend the algorithm to more general cases.

Note that the idea of using the Aho-Corasick automaton can be applied to more general seed models than individual spaced seeds (e.g. to multiple spaced seeds, as pointed out in [17]). In fact, all currently proposed seed models can be described by a finite set of matched alignment fragments, for which the Aho-Corasick automaton can be constructed. We will use this remark in later sections.

The sensitivity of a spaced seed with respect to an HMM-specified probability distribution over binary target alignments of a given length n was studied by Brejova et al. [15]. The DP algorithm of [15] has a lot in common with the algorithm implied by Lemma 2. In particular, the states of the algorithm of [15] are triples $< w, q, m >$, where w is a prefix of the seed π, q is a state of the HMM, and $m \in [0..n]$. The states therefore correspond to the construction implied by Lemma 2. However, the authors of [15] do not consider any automata, which does not allow to optimize the preprocessing step (counterpart of the automaton construction) and, on the other hand, does not allow to extend the algorithm to more general seed models and/or different sets of target alignments.

A key to an efficient solution of the sensitivity problem remains the definition of the seed. It should be expressive enough to be able to take into account properties of biological sequences. On the other hand, it should be simple enough to be able to locate seeds fast and to get an efficient algorithm for computing seed sensitivity. According to the approach presented in this section, the latter is directly related to the size of a DFA specifying the seed.

3 Subset Seeds

3.1 Definition

Ordinary spaced seeds use the simplest possible binary "match-mismatch" alignment model that allows an efficient implementation by hashing all occurring combinations of matching positions. A powerful generalization of spaced seeds, called *vector seeds*, has been introduced in [13]. Vector seeds allow one to use an arbitrary alignment alphabet and, on the other hand, provide a flexible definition of a hit based on a cooperative contribution of seed positions. A much higher expressiveness of vector seeds lead to more complicated algorithms and, in particular, prevents the application of direct hashing methods at the seed location stage.

In this section, we consider *subset seeds* that have an intermediate expressiveness between spaced and vector seeds. It allows an arbitrary alignment alphabet and, on the other hand, still allows using a direct hashing for locating seed, which maps each string to a unique entry of the hash table. We also propose a construction of a seed automaton for subset seeds, different from the Aho-Corasick automaton. The automaton has $\mathcal{O}(w2^{s-w})$ states *regardless of the size of the alignment alphabet*, where s and w are respectively the span of the seed and the number of "must-match" positions. From the general algorithmic framework presented in the previous section (Lemma 2), this implies that the seed sensitivity can be computed for subset seeds with same complexity as for ordinary spaced seeds. Note also that for the binary alignment alphabet, this bound is the same as the one implied by the Aho-Corasick automaton. However, for larger alphabets, the Aho-Corasick construction leads to $\mathcal{O}(w|\mathcal{A}|^{s-w})$ states. In the experimental part of this paper (section 4.1) we will show that even for the binary alphabet, our automaton construction yields a smaller number of states in practice.

Consider an alignment alphabet $\mathcal{A}$. We always assume that $\mathcal{A}$ contains a symbol 1, interpreted as "match". A *subset seed* is defined as a word over a *seed alphabet* $\mathcal{B}$, such that

- letters of $\mathcal{B}$ denote subsets of the alignment alphabet $\mathcal{A}$ containing 1 ($\mathcal{B} \subseteq \{1\} \cup 2^{\mathcal{A}}$),
- $\mathcal{B}$ contains a letter # that denotes subset $\{1\}$,
- a subset seed $b_1 b_2 \dots b_m \in \mathcal{B}^m$ matches an alignment fragment $a_1 a_2 \dots a_m \in \mathcal{A}^m$ if $\forall i \in [1..m]$, $a_i \in b_i$.

The *#-weight* of a subset seed π is the number of # in π and the *span* of π is its length.

Example 1. [12] considered the alignment alphabet $\mathcal{A} = \{1, \mathtt{h}, 0\}$ representing respectively a match, a transition mismatch, or a transversion mismatch in a DNA sequence alignment. The seed alphabet is $\mathcal{B} = \{\#, @, \_\}$ denoting respectively subsets $\{1\}$, $\{1, \mathtt{h}\}$, and $\{1, \mathtt{h}, 0\}$. Thus, seed π = #@_# matches alignment s = 10h1h1101 at positions 4 and 6. The span of π is 4, and the #-weight of π is 2.

Note that unlike the weight of ordinary spaced seeds, the #-weight cannot serve as a measure of seed selectivity. In the above example, symbol @ should be assigned weight 0.5, so that the weight of π is equal to 2.5 (see [12]).

3.2 Subset Seed Automaton

Let us fix an alignment alphabet $\mathcal{A}$, a seed alphabet $\mathcal{B}$, and a seed $\pi = \pi_1 \pi_2 \dots \pi_m \in \mathcal{B}^*$ of span m and #-weight w. Let R_π be the set of all non-# positions in π, $|R_\pi| = r = m - w$. We now define an automaton $S_\pi = < Q, q_0, Q_f, \mathcal{A}, \psi : Q \times \mathcal{A} \to Q >$ that recognizes the set of all alignments matched by π.

The states Q of S_π are pairs $< X, t >$ such that $X \subseteq R_\pi, t \in [0, \dots, m]$, with the following invariant condition. Suppose that S_π has read a prefix $s_1 \dots s_p$ of an alignment s and has come to a state $< X, t >$. Then t is the length of the longest suffix of $s_1 \dots s_p$ of the form 1^i, $i \leq m$, and X contains all positions $x_i \in R_\pi$ such that prefix $\pi_1 \cdots \pi_{x_i}$ of π matches a suffix of $s_1 \cdots s_{p-t}$.

(a) π = #@#_##_###

(b) s = 111h1011h11...

(c)
```
                 s9 t
         111h1011h11...
  π1..7 =#@#_##_
      π1..4 =#@#_
        π1..2 =#@
```

Fig. 1. Illustration to Example 2

Example 2. In the framework of Example 1, consider a seed π and an alignment prefix s of length $p = 11$ given on Figure 1(a) and 1(b) respectively. The length t of the last run of 1's of s is 2. The last mismatch position of s is $s_9 = \mathtt{h}$. The set R_π of non-# positions of π is $\{2, 4, 7\}$ and π has 3 prefixes ending at positions of R_π (Figure 1(c)). Prefixes $\pi_{1..2}$ and $\pi_{1..7}$ do match suffixes of $s_1 s_2 \dots s_9$, and prefix $\pi_{1..4}$ does not. Thus, the state of the automaton after reading $s_1 s_2 \dots s_{11}$ is $< \{2, 7\}, 2 >$.

The initial state q_0 of S_π is the state $< \emptyset, 0 >$. The final states Q_f of S_π are all states $q =< X, t >$, where $max\{X\} + t = m$. All final states are merged into one state.

The transition function $\psi(q, a)$ is defined as follows: If q is a final state, then $\forall a \in \mathcal{A}$, $\psi(q, a) = q$. If $q =< X, t >$ is a non-final state, then

- if $a = 1$ then $\psi(q, a) =< X, t + 1 >$,
- otherwise $\psi(q, a) =< X_U \cup X_V, 0 >$ with
  - $X_U = \{x | x \leq t + 1 \text{ and } a \in \pi_x\}$
  - $X_V = \{x + t + 1 | x \in X \text{ and } a \in \pi_{x+t+1}\}$

Lemma 3. *The automaton S_π accepts the set of all alignments matched by π.*

Proof. It can be verified by induction that the invariant condition on the states $< X, t > \in Q$ is preserved by the transition function ψ. The final states verify $max\{X\} + t = m$, which implies that π matches a suffix of $s_1 \ldots s_p$.

Lemma 4. *The number of states of the automaton S_π is no more than $(w+1)2^r$.*

Proof. Assume that $R_\pi = \{x_1, x_2, \ldots, x_r\}$ and $x_1 < x_2 \cdots < x_r$. Let Q_i be the set of non-final states $< X, t >$ with $max\{X\} = x_i$, $i \in [1..r]$. For states $q =< X, t > \in Q_i$ there are 2^{i-1} possible values of X and $m - x_i$ possible values of t, as $max\{X\} + t \leq m - 1$. Thus, $|Q_i| \leq 2^{i-1}(m - x_i) \leq 2^{i-1}(m - i)$ and $\sum_{i=1}^{r} |Q_i| \leq \sum_{i=1}^{r} 2^{i-1}(m - i) = (m - r + 1)2^r - m - 1$. Besides states Q_i, Q contains m states $< \emptyset, t >$ $(t \in [0..m - 1])$ and one final state. Thus, $|Q| \leq (m - r + 1)2^r = (w + 1)2^r$.

Note that if π starts with #, which is always the case for ordinary spaced seeds, then $X_i \geq i+1$, $i \in [1..r]$, and previous bound rewrites to $2^{i-1}(m-i-1)$. This results in the same number of states $w2^r$ as for the Aho-Corasick automaton [17]. The construction of automaton S_π is optimal, in the sense that no two states can be merged in general. A straightforward generation of the transition table of the automaton S_π can be performed in time $\mathcal{O}(r \cdot w \cdot 2^r \cdot |\mathcal{A}|)$. A more complicated algorithm allows one to reduce the bound to $\mathcal{O}(w \cdot 2^r \cdot |\mathcal{A}|)$. In the next section, we demonstrate experimentally that on average, our construction yields a very compact automaton, close to the minimal one. Together with the general approach of section 2, this provides a fast algorithm for computing the sensitivity of subset seeds and, in turn, allows to perform an efficient design of spaced seeds well-adapted to the similarity search problem under interest.

4 Experiments

Several types of experiments have been performed to test the practical applicability of the results of sections 2,3. We focused on DNA similarity search, and set the alignment alphabet $\mathcal{A}$ to $\{1, h, 0\}$ (match, transition, transversion). For subset seeds, the seed alphabet $\mathcal{B}$ was set to $\{\#, @, \_\}$, where $\# = \{1\}, @ = \{1, h\}, \_ = \{1, h, 0\}$ (see Example 1). The weight of a subset seed is computed by assigning weights 1, 0.5 and 0 to symbols #, @ and _ respectively.

4.1 Size of the Automaton

We compared the size of the automaton S_π defined in section 3 and the Aho-Corasick automaton [22], both for ordinary spaced seeds (binary seed alphabet) and for subset seeds (ternary seed alphabet). The Aho-Corasick automaton for spaced seeds was constructed as defined in [17]. For subset seeds, a straightforward generalization was considered: the Aho-Corasick construction was applied to the set of alignment fragments matched by the seed.

Tables 1(a) and 1(b) present the results for spaced seeds and subset seeds respectively. For each seed weight w, we computed the average number of states (*avg. s.*) of the Aho-Corasick automaton and our automaton S_π, and reported the corresponding ratio (δ) with respect to the average number of states of the minimized automaton. The average was computed over all seeds of span up to $w+8$ for spaced seeds and all seeds of span up to $w+5$ with two @'s for subset seeds. Interestingly, our automaton turns out to be more compact than the Aho-Corasick automaton not only on non-binary alphabets (which was expected), but also on the binary alphabet (cf Table 1(a)). Note that for a given seed, one can define a surjective mapping from the states of the Aho-Corasick automaton onto the states of our automaton. This implies that our automaton has *always* no more states than the Aho-Corasick automaton.

Table 1. Comparison of the average number of states of Aho-Corasick automaton, automaton S_π of section 3 and minimized automaton

Spaced	Aho-Corasick		S_π		Minimized
w	*avg. s.*	δ	*avg. s.*	δ	*avg. s.*
9	345.94	3.06	146.28	1.29	113.21
10	380.90	3.16	155.11	1.29	120.61
11	415.37	3.25	163.81	1.28	127.62
12	449.47	3.33	172.38	1.28	134.91
13	483.27	3.41	180.89	1.28	141.84

(a)

Subset	Aho-Corasick		S_π		Minimized
w	*avg. s.*	δ	*avg. s.*	δ	*avg. s.*
9	1900.65	15.97	167.63	1.41	119.00
10	2103.99	16.50	177.92	1.40	127.49
11	2306.32	16.96	188.05	1.38	135.95
12	2507.85	17.42	198.12	1.38	144.00
13	2709.01	17.78	208.10	1.37	152.29

(b)

4.2 Seed Design

In this part, we considered several probability transducers to design spaced or subset seeds. The target alignments included all alignments of length 64 on alphabet $\{1, h, 0\}$. Four probability transducers have been studied (analogous to those introduced in [23]):

- *B*: Bernoulli model
- *DT*1: deterministic probability transducer specifying probabilities of $\{1, h, 0\}$ at each codon position (extension of the $M^{(3)}$ model of [23] to the three-letter alphabet),
- *DT*2: deterministic probability transducer specifying probabilities of each of the 27 codon instances $\{1, h, 0\}^3$ (extension of the $M^{(8)}$ model of [23]),
- *NT*: non-deterministic probability transducer combining four copies of *DT*2 specifying four distinct codon conservation levels (called HMM model in [23]).

Table 2. Best seeds and their sensitivity for probability transducer B

w	spaced seeds	Sens.	subset seeds, two @	Sens.
9	###___#_#_##_##	0.4183	###_#__#@#_@##	0.4443
10	##_##___##_#_###	0.2876	###_@#_@#_#_###	0.3077
11	###_###_#__#_###	0.1906	##@#__##_#_#_@###	0.2056
12	###_#_##_#__##_###	0.1375	##@#_#_##__#@_####	0.1481

Table 3. Best seeds and their sensitivity for probability transducer $DT1$

w	spaced seeds	Sens.	subset seeds, two @	Sens.
9	###___##_##_##	0.4350	##@___##_##_##@	0.4456
10	##_##____##_##_##	0.3106	##_##___@##_##@#	0.3173
11	##_##____##_##_###	0.2126	##@#@_##_##__###	0.2173
12	##_##____##_##_####	0.1418	##_@###__##_##@##	0.1477

Table 4. Best seeds and their sensitivity for probability transducer $DT2$

w	spaced seeds	Sens.	subset seeds, two @	Sens.
9	#_##____##_##_##	0.5121	#_#@_##_@__##_##	0.5323
10	##_##_##____##_##	0.3847	##_@#_##__@_##_##	0.4011
11	##_##__#_#___#_##_##	0.2813	##_##_@#_#___#_#@_##	0.2931
12	##_##_##_#___#_##_##	0.1972	##_##_#@_##_@__##_##	0.2047

Table 5. Best seeds and their sensitivity for probability transducer NT

w	spaced seeds	Sens.	subset seeds, two @	Sens.
9	##_##_##____##_#	0.5253	##_@@_##____##_##	0.5420
10	##_##____##_##_##	0.4123	##_##____##_@@_##_#	0.4190
11	##_##____##_##_##_#	0.3112	##_##____##_@@_##_##	0.3219
12	##_##____##_##_##_##	0.2349	##_##____##_@@_##_##_#	0.2412

Models $DT1$, $DT2$ and NT have been trained on alignments resulting from a pairwise comparison of 40 bacteria genomes. For each of the four probability transducers, we computed the best seed of weight w ($w = 9, 10, 11, 12$) among two categories: ordinary spaced seeds of weight w and subset seeds of weight w with two @. Ordinary spaced seeds were enumerated exhaustively up to a given span, and for each seed, the sensitivity was computed using the algorithmic approach of section 2 and the seed automaton construction of section 3. Each such computation took between 10 and 500ms on a Pentium IV 2.4GHz computer depending on the seed weight/span and the model used. In each experiment, the most sensitive seed found has been kept. The results are presented in Tables 2-5.

In all cases, subset seeds yield a better sensitivity than ordinary spaced seeds. The sensitivity increment varies up to 0.04 which is a notable increase. As shown in [12], the gain in using subset seeds increases substantially when the transition probability is greater than the transversion probability, which is very often the case in related genomes.

5 Discussion

We introduced a general framework for computing the seed sensitivity for various similarity search settings. The approach can be seen as a generalization of methods of [17,15] in that it allows to obtain algorithms with the same worst-case complexity bounds as those proposed in these papers, but also allows to obtain efficient algorithms for new formulations of the seed sensitivity problem. This versatility is achieved by distinguishing and treating separately the three ingredients of the seed sensitivity problem: a set of target alignments, an associated probability distributions, and a seed model.

We then studied a new concept of *subset seeds* which represents an interesting compromise between the efficiency of spaced seeds and the flexibility of vector seeds. For this type of seeds, we defined an automaton with $\mathcal{O}(w2^r)$ states regardless of the size of the alignment alphabet, and showed that its transition table can be constructed in time $\mathcal{O}(w2^r|\mathcal{A}|)$. Projected to the case of spaced seeds, this construction gives the same worst-case bound as the Aho-Corasick automaton of [17], but results in a smaller number of states in practice. Different experiments we have done confirm the practical efficiency of the whole method, both at the level of computing sensitivity for designing good seeds, as well as using those seeds for DNA similarity search.

As far as the future work is concerned, it would be interesting to study the design of efficient spaced seeds for protein sequence search (see [10]), as well as to combine spaced seeds with other techniques such as seed families [5,6,8] or the group hit criterion [12].

Acknowledgments. G. Kucherov and L. Noé have been supported by the *ACI IMPBio* of the French Ministry of Research. M. Roytberg has been also supported by the Russian Foundation for Basic Research (projects 03-04-49469, 02-07-90412) and by grants from the RF Ministry of Industry, Science and Technology (20/2002, 5/2003) and NWO (Netherlands Science Foundation).

References

1. Burkhardt, S., Kärkkäinen, J.: Better filtering with gapped q-grams. Fundamenta Informaticae **56** (2003) 51–70 Preliminary version in Combinatorial Pattern Matching 2001.
2. Ma, B., Tromp, J., Li, M.: PatternHunter: Faster and more sensitive homology search. Bioinformatics **18** (2002) 440–445
3. Altschul, S., Madden, T., Schäffer, A., Zhang, J., Zhang, Z., Miller, W., Lipman, D.: Gapped BLAST and PSI-BLAST: a new generation of protein database search programs. Nucleic Acids Research **25** (1997) 3389–3402
4. Kent, W.J.: BLAT–the BLAST-like alignment tool. Genome Research **12** (2002) 656–664

5. Li, M., Ma, B., Kisman, D., Tromp, J.: PatternHunter II: Highly sensitive and fast homology search. Journal of Bioinformatics and Computational Biology (2004) Earlier version in GIW 2003 (International Conference on Genome Informatics).
6. Sun, Y., Buhler, J.: Designing multiple simultaneous seeds for DNA similarity search. In: Proceedings of the 8th Annual International Conference on Computational Molecular Biology, ACM Press (2004)
7. Yang, I.H., Wang, S.H., Chen, Y.H., Huang, P.H., Ye, L., Huang, X., Chao, K.M.: Efficient methods for generating optimal single and multiple spaced seeds. In: Proceedings of the IEEE 4th Symposium on Bioinformatics and Bioengineering, IEEE Computer Society Press (2004) 411–416
8. Kucherov, G., Noé, L., Roytberg, M.: Multiseed lossless filtration. IEEE Transactions on Computational Biology and Bioinformatics **2** (2005) 51 – 61
9. Xu, J., Brown, D., Li, M., Ma, B.: Optimizing multiple spaced seeds for homology search. In: Proceedings of the 15th Symposium on Combinatorial Pattern Matching. Volume 3109 of Lecture Notes in Computer Science., Springer (2004)
10. Brown, D.: Optimizing multiple seeds for protein homology search. IEEE Transactions on Computational Biology and Bioinformatics **2** (2005) 29 – 38
11. Chen, W., Sung, W.K.: On half gapped seed. Genome Informatics **14** (2003) 176–185 preliminary version in the 14th International Conference on Genome Informatics (GIW).
12. Noé, L., Kucherov, G.: Improved hit criteria for DNA local alignment. BMC Bioinformatics **5** (2004)
13. Brejova, B., Brown, D., Vinar, T.: Vector seeds: an extension to spaced seeds allows substantial improvements in sensitivity and specificity. In Benson, G., Page, R., eds.: Proceedings of the 3rd International Workshop in Algorithms in Bioinformatics. Volume 2812 of Lecture Notes in Computer Science., Springer (2003)
14. Keich, U., Li, M., Ma, B., Tromp, J.: On spaced seeds for similarity search. to appear in Discrete Applied Mathematics (2002)
15. Brejova, B., Brown, D., Vinar, T.: Optimal spaced seeds for homologous coding regions. Journal of Bioinformatics and Computational Biology **1** (2004) 595–610
16. Choi, K., Zhang, L.: Sensitivity analysis and efficient method for identifying optimal spaced seeds. Journal of Computer and System Sciences **68** (2004) 22–40
17. Buhler, J., Keich, U., Sun, Y.: Designing seeds for similarity search in genomic DNA. In: Proceedings of the 7th Annual International Conference on Computational Molecular Biology, ACM Press (2003) 67–75
18. Choi, K.P., Zeng, F., Zhang, L.: Good Spaced Seeds For Homology Search. Bioinformatics **20** (2004) 1053–1059
19. Kucherov, G., Noé, L., Ponty, Y.: Estimating seed sensitivity on homogeneous alignments. In: Proceedings of the IEEE 4th Symposium on Bioinformatics and Bioengineering, IEEE Computer Society Press (2004) 387–394
20. Ullman, J.D., Aho, A.V., Hopcroft, J.E.: The Design and Analysis of Computer Algorithms. Addison-Wesley, Reading (1974)
21. Finkelstein, A., Roytberg, M.: Computation of biopolymers: A general approach to different problems. BioSystems **30** (1993) 1–19
22. Aho, A.V., Corasick, M.J.: Efficient string matching: An aid to bibliographic search. Communications of the ACM **18** (1975) 333–340
23. Brejova, B., Brown, D., Vinar, T.: Optimal spaced seeds for Hidden Markov Models, with application to homologous coding regions. In R. Baeza-Yates, E. Chavez, M.C., ed.: Proceedings of the 14th Symposium on Combinatorial Pattern Matching. Volume 2676 of Lecture Notes in Computer Science., Springer (2003) 42–54

Generalized Planted (l,d)-Motif Problem with Negative Set*

Henry C.M. Leung and Francis Y.L. Chin

Department of Computer Science,
The University of Hong Kong, Pofulam, Hong Kong
{cmleung2, chin}@cs.hku.hk

Abstract. Finding similar patterns (motifs) in a set of sequences is an important problem in Computational Molecular Biology. Pevzner and Sze [18] defined the planted (l,d)-motif problem as trying to find a length-l pattern that occurs in each input sequence with at most d substitutions. When d is large, this problem is difficult to solve because the input sequences do not contain enough information on the motif. In this paper, we propose a generalized planted (l,d)-motif problem which considers as input an additional set of sequences without any substring similar to the motif (negative set) as extra information. We analyze the effects of this negative set on the finding of motifs, and define a set of unsolvable problems and another set of most difficult problems, known as "challenging generalized problems". We develop an algorithm called VANS based on voting and other novel techniques, which can solve the (9,3), (11,4),(15,6) and (20,8)-motif problems which were unsolvable before as well as challenging problems of the planted (l,d)-motif problem such as (9,2), (11,3), (15,5) and (20,7)-motif problems.

1 Introduction

A genome is a sequence consisting of four symbols 'A', 'C', 'G' and 'T'. Along the genome are substrings, called *genes*, which are blueprints of proteins. In order to decode a gene (*gene expression*) to produce the corresponding protein, a molecule called a *transcription factor* binds to a short region (6 - 20 base pairs), called the *binding site*, in the promoter region of the gene. One kind of transcription factor can bind to the binding sites of several genes to cause these genes to coexpress. These binding sites, which should have similar lengths and patterns, can be represented by a pattern called *motif*. The motif discovering problem [14,18] is predicting the motif given a set of coexpressed genes, i.e., when given a set of sequences T, each of which contains at least one binding site. Pevzner and Sze [18] gave a precise definition of this problem.

Planted (l,d)-Motif Problem: Suppose there is a fixed but unknown string M (the motif) of length l. Given t length-n sequences, each of which contains

* This research is supported in part by an RGC grant HKU 7135/04E.

R. Casadio and G. Myers (Eds.): WABI 2005, LNBI 3692, pp. 264–275, 2005.

a planted *d-variant* of M, we want to determine M without a priori knowledge of the positions of the planted d-variants. A *d-variant* (or simply *variant*) is a string derivable from M with at most d symbol substitutions.

Many algorithms [1,3,4,5,8,9,10,11,12,13,15,16,17,18,19] have been developed to solve this problem and have predicted some motifs successfully. However, this problem model will fail to find a solution when d is large, because there will be many length-l strings having at least one variant in each input sequence and no algorithm is likely to distinguish the motif from these strings. Buhler and Tompa [4] found the maximum d such that a planted (l,d)-motif problem is still solvable by calculating the expected number $E_t(l,d)$ of length-l strings with at least one variant in each input sequence. When $E_t(l,d)$ is small, say $E_t(l,d) \leq 1$, the problem is theoretically solvable. When $E_t(l,d)$ is large, no algorithm is likely to discover M without extra information. For example, when $t = 20$ and $n = 600$, the planted (9,3), (11,4), (15,6) and (20,8)-motif problems are unsolvable as the values of $E_t(l,d)$ for these problems are huge (2.5×10^5, 3.3×10^6, 1.8×10^8 and 3.1×10^4 respectively).

In biological experiments, besides getting a set of sequences bound by the transcription factor, we may have as a by-product another set of sequences F which are not bound by the transcription factor [2,6,7,20]. We may assume sequences in F contain no d-variant of the motif M. Based on this extra information, we can modify the planted (l,d)-motif problem as follows.

Generalized Planted (l,d)-Motif Problem: Suppose there is a fixed but unknown string M (the motif) of length l. Given t length-n sequences, each of which contains a planted d-variant of M, and f length-n sequences which contains no d-variant of M, we want to determine M without a priori knowledge of the positions of the planted d-variants.

Note that when $f = 0$, the generalized planted (l,d)-motif problem (or simply generalized (l,d)-problem) is reduced to the planted (l,d)-motif problem (or simply (l,d)-problem). The extra information provided by F might make some of the previously unsolvable problems based only on information in T, e.g. (9,3), (11,4), (15,6) and (20,8)-motif problems, solvable.

In this paper, we analyze the information provided by set T and set F (Section 2) and how they are related (Section 3). We define a new set of unsolvable and also another set of "challenging" generalized (l,d)-problems (most difficult solvable problems). In Section 4, we develop an algorithm called VANS (Voting Algorithm with Negative Set) to solve this generalized (l,d)-problem under different situations by employing, in additional to voting, other simple but novel techniques, such as filtering, projection with merging and local search. In particular, VANS can solve those challenging (l,d)-problem, such as (9,2), (11,3), (15,5) and (20,7)-problems, when F is empty. Experimental results (Section 5) show that VANS can solve all theoretically solvable generalized (l,d)-problems when $d \leq 20$ and works well on some real data.

2 Calculation the Expected Value $E_b(l,d)$

Let T be the set of t length-n input sequences, each of which contains a variant of M and let F be the set of f length-n input sequences with no variant of M. Assume the occurrence probabilities of 'A', 'C', 'G' and 'T' are equal. Buhler and Tompa [4] studied the limitation of the (l,d)-problem by calculating the expected number $E_t(l,d)$ of length-l strings with at least one d-variant in each sequence in T. Their calculation is described as follows.

Given a length-l string P and a length-l substring σ in the input sequence, the probability that P and σ are at most d symbol substitution apart is

$$p(l,d) = \sum_{i=0}^{d} C_i^l \left(\frac{3}{4}\right)^i \left(\frac{1}{4}\right)^{l-i}$$

The probability that a length-l string P has at least one d-variant in each sequence in T is

$$\left(1 - (1-p(l,d))^{n-l+1}\right)^t$$

Consider the 4^l possible length-l strings, the expected number of length-l strings with at least one d-variant in each sequence in T is approximately

$$E_t(l,d) = 4^l \left(1 - (1-p(l,d))^{n-l+1}\right)^t \quad (1)$$

When $E_t(l,d)$ is much larger than 1, that means there are many random length-l strings which have the same characteristics as motif M on T, i.e. have at least one d-variant in each sequence in T. Under this situation, no algorithm can distinguish the motif M from this set of random length-l strings. On the other hand if $E_t(l,d)$ is smaller than 1, the smaller the value of $E_t(l,d)$, the more plausible that the found pattern is M and not an artifact. Thus $E_t(l,d)$ can be used to estimate the amount of information contained in the set of sequences T; the larger is $E_t(l,d)$, the less is the information and vice versa. Given the parameters t, n and l, we can find the range of d such that the (l,d)-problem is unsolvable, i.e. with $E_t(l,d)$ much larger than 1, e.g. $(9,\geq 3)$, $(11,\geq 4)$, $(15,\geq 6)$ and $(20,\geq 8)$-problems.

Similarly, we can estimate the amount of information contained in F by the expected number $E_f(l,d)$ of length-l strings with no variant in any sequences in F, and also the amount of information of both T and F by the expected number $E_b(l,d)$ of length-l strings with at least one variant in each sequence in T and no variant in the sequences of F. If $E_b(l,d)$ is smaller than 1, the generalized (l,d)-problem is theoretically solvable, otherwise, it is unsolvable.

The probability that a length-l string P has no variant in F is

$$\left((1-p(l,d))^{n-l+1}\right)^f$$

Consider the 4^l possible length-l strings, we have

$$E_f(l,d) = 4^l \left((1-p(l,d))^{n-l+1}\right)^f \quad (2)$$

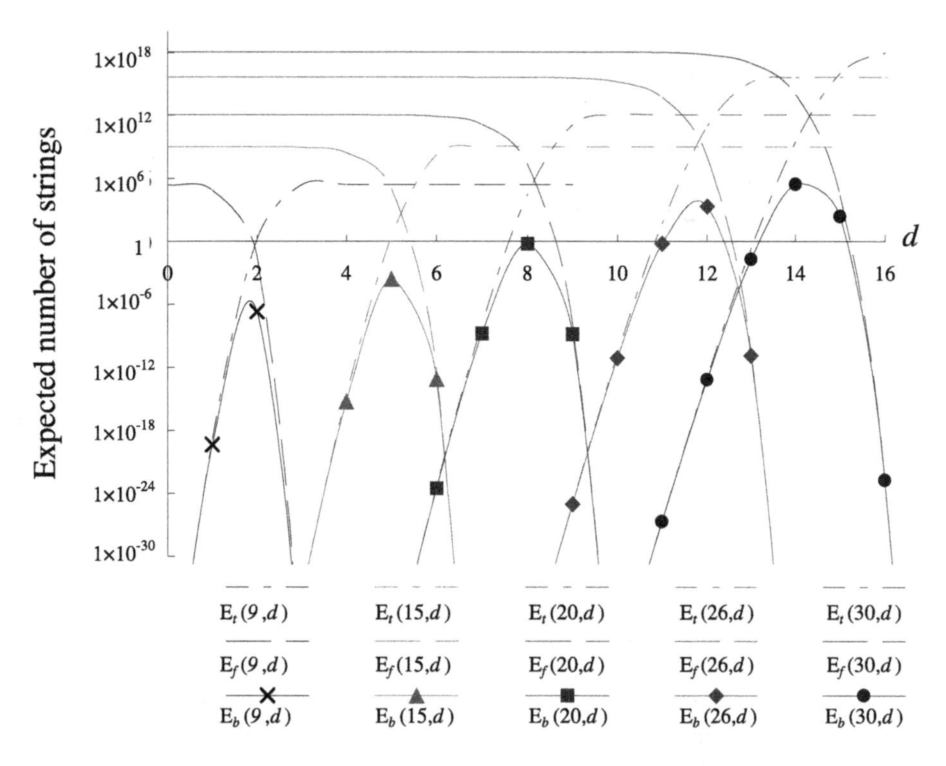

Fig. 1. Expected number of strings against d for different motif length l

$$E_b(l,d) = 4^l \left(1 - (1 - p(l,d))^{n-l+1}\right)^t \left((1 - p(l,d))^{n-l+1}\right)^f \tag{3}$$

Figure 1 shows the values of $E_t(l,d)$, $E_f(l,d)$ and $E_b(l,d)$ for different values of l and d when $t = f = 20$ and $n = 600$. We have the following observations which match with our intuition.

1. The (l,d)-problem is easier to solve for a smaller d because T has more information for smaller d. Thus $E_t(l,d)$ increases with d. By the same argument, F has more information for larger d, thus $E_f(l,d)$ decreases with d.
2. The value of $E_b(l,d)$ is always less than $E_t(l,d)$ and $E_f(l,d)$. $E_b(l,d) \approx E_t(l,d)$ when d is small and $E_b(l,d) \approx E_f(l,d)$ when d is large. $E_b(l,d)$ is peaked or the amount of information is the least for some d, $0 < d < l$. It can be shown that $E_b(l,d)$ is maximum when $p(l,d) = p_{thres}$, where

$$p_{thres} = 1 - e^{\frac{\ln f - \ln(t+f)}{n-l+1}}$$

Figure 2 shows the value of $p(l,d)$ against d and $p_{thres} = 0.0012$ when $t = f = 20$ and $n = 600$. The intersections between p_{thres} and each curve represent those problems with the least amount of information, e.g. the generalized (9,2), (15,5), (20,8), (26,12) and (30,14)-problems which match the results given in Figure 1.

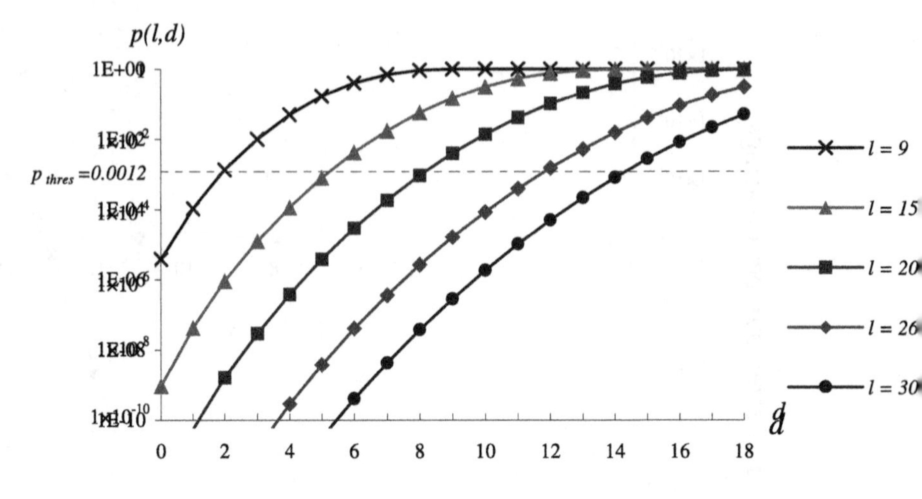

Fig. 2. The value of $p(l, d)$ against d for different motif length l

3. As a result, some previously unsolvable (l,d)-problems, e.g. the (9,3), (11,4) and (15,6)-problem, become solvable after adding the set F. In fact, all (l,d)-problems with $0 \leq d \leq l$ and $l \leq 20$ are solvable. However, when l increases, there are still some generalized (l,d)-problems having a large value of $E_b(l, d)$ (e.g. the generalized (26,12), (30,14) and (30,15)-problems) which means they are theoretically unsolvable.

Buhler and Tompa [4] defined those solvable (l,d)-problems with the largest d as "challenging problems" (i.e. if the (l,d)-problem is a challenging problem, the (l,$d+1$)-problem should be unsolvable). Similarly, we can define a set of "challenging problems" for the generalized planted (l,d)-motif problems. A generalized planted (l,d)-motif problems is *challenging* if it is solvable and either (l,$d-1$) or (l,$d+1$) is unsolvable. For example, the generalized (26,11),(26,13),(30,13) and (30,16)-problems are "challenging problems".

3 Trade Off Between t and f

Although sequences in both T and F contain information of the motif, the amounts of information in these two sets vary with the values of n and $p(l, d)$. One question we want to know is "If we reduce the number of sequences in T by Δt, how many sequences should we add to F so that the input data retains the same amount of information?" This question can be answered by comparing the value of $E_b(l, d)$ before and after changing the number of input sequences. The amount of information is retained if and only if the new value of $E_b(l, d)$ is no larger than the original value.

Table 1. Trade off between t and f when $n = 600$

	R	l: 9	11	15	20
	1	44.16	1118	4.22×10^5	5.35×10^8
	2	0.7547	35.41	1.39×10^4	1.48×10^7
	3	0.0004	0.9763	680.6	6.37×10^5
	4	4.27×10^{-15}	0.0026	40.36	3.74×10^4
d	5	0	5.44×10^{-11}	2.119	2760
	6	0	0	0.0362	237.6
	7	0	0	3.54×10^{-6}	21.46
	8	0	0	4.23×10^{-17}	1.598
	9	0	0	0	0.0463

$$\frac{4^l \left(1-(1-p(l,d))^{n-l+1}\right)^t \left((1-p(l,d))^{n-l+1}\right)^f}{4^l \left(1-(1-p(l,d))^{n-l+1}\right)^{t-\Delta t} \left((1-p(l,d))^{n-l+1}\right)^{f+\Delta f}} \geq 1$$

$$\Delta t \log\left(1-(1-p(l,d))^{n-l+1}\right) - \Delta f \log\left((1-p(l,d))^{n-l+1}\right) \geq 1$$

$$\Delta f \geq R\Delta t$$

where $R = \frac{\log\left(1-(1-p(l,d))^{n-l+1}\right)}{\log\left((1-p(l,d))^{n-l+1}\right)}$

Since a random sequence containing d-variants of the motif M is independent of another random sequence containing a d-variants of M, the information of M contained in each input sequence is independent of the input size t and f. Therefore, the value of R is independent of the number of sequences t and f in the data set as shown in the above equation.

Table 1 shows the values of R for different values of l and d when $n = 600$. For example, when $l = 20$ and $d = 8$, if we remove one sequence from T, we should add at least $\lceil 1.598 \rceil = 2$ sequences in F to retain the same amount of information in the input data. When d is large, the probability that a random sequence contains a d-variant of the motif M is large while the probability that a random sequence contains no d-variant of the motif M is small, therefore the amount of information in each sequence in T is much less than that in each sequence in F. If we remove a sequence in T, even no sequence is added to F, the total amount of information in the data set remains almost the same(e.g. $l = 9$, $d = 9$). On the other hand, when d is small, the amount of information in each sequence in T is much more than that in each sequence in F. If we remove a sequence in T, many sequences should be added to F in order to retain the same amount of information (e.g. $l = 20$, $d = 1$).

4 Voting Algorithm with Negative Set (VANS)

Our Voting Algorithm with Negative Set (VANS) is based on a simple idea that if a substring σ is a planted d-variant of the motif M, M is also a d-variant of σ. In order to find the motif, each length-l substring in T and F gives one vote to its d-variant. The motif M would receive at least one vote from each sequence in T and no vote from any length-l substring in F. Although the idea used in the Voting algorithms [5] is simple and enumerative, they are so far the fastest algorithms than the other methods based on brute-force [3,10,17], finding the maximum clique [15,18] and heuristic search [1,4,8,9,11,12,13] for solving the (l,d)-problem without F. The running times of the brute-force and the clique search algorithms ($O(ntl4^l)$ and $O((nt)^{t+2.376})$ respectively) are much longer than that of the Voting algorithms ($O(ntC^l_d4^d)$). The brute-force algorithms can only solve the (l,d)-problems with $l \leq 11$ and the clique search algorithms can tackle those problem with small d. Thus, they have difficulties to deal with those challenging (11,3), (15, 5) and (20,7)-problems. Heuristic algorithms can solve the (l,d)-problems for larger l, say $l \leq 20$, but they do not guarantee finding the motifs all the time. The Voting algorithms, [5] on the other hand, can solve not only the challenging (9,2), (11,3), (15,5) and (20,7)-motif problems, but also (30,11) and (40,15)-problems.

As indicated in Figure 1, the generalized (l,d)-problem can be solved for small d and large d when $E_b(l,d)$ is much less than 1. Moreover, since we have shown in Sections 2 and 3, the amount of information in the generalized (l,d)-problem is mainly derived from T when d is small and from F when d is large, our algorithm VANS will first identify a set of candidate motifs by voting from T when d is small and from F when d is large. VANS will then filter out the false candidate motifs by F or T accordingly. Based on the value of d, VANS applies different strategies to solve the generalized (l,d)-problem.

4.1 Voting by Sequences in T

Since each length-l substring σ has $C^i_l3^i$ variants with exactly i substitutions, σ has $\sum_{i=0}^{d} C^i_i3^i$ variants. By considering each length-l substring in each sequence in T, the voting process by T takes

$$O\left(nt\sum_{i=0}^{d}(C^l_i3^i)\right) = O\left(ntC^l_d4^d\right) \text{ time}$$

The candidate motifs are those variants (length-l strings) which receive at least one vote from each sequence in T. It is shown in [5] that this simple algorithm can solve the challenging (9,2), (11,3), (15,5) and (20,7)-motif problems in time less than a few minutes. The filtering process removes those candidate motifs having variants in F and this filtering step takes $O(nlf)$ time for each candidate motif. Since the expected number of length-l strings with at least one variant in each sequence in T is $E_t(l,d)$, the expected running time is

$$O\left(ntC^l_d4^d + nlfE_t(l,d)\right)$$

This approach works well on T when d or $E_t(l,d)$ is small. Table 2 shows the values of d for which this approach works well.

Table 2. Values of d at which voting by T works when $t = f = 20$ and $n = 600$

l	9	11	15	20	26	30
d	≤ 2	≤ 3	≤ 5	≤ 7	≤ 10	≤ 13
$E_t(l,d)$	≤ 1.6	≤ 4.7	≤ 2.8	$\leq 1.4 \times 10^{-8}$	$\leq 2.1 \times 10^{-11}$	≤ 0.22

4.2 Voting by Sequences in F

When d increases, both 4^d and $E_t(l,d)$ increase exponentially such that the running time of the voting process becomes unacceptable. Since the amount of information in F is much more than the amount of information in T when d is large, we should focus on F instead of T.

As motif M has no variant in F, M should not be a variant of any length-l substring in F. If each length-l substring in F gives one vote to its variants, we can find a set of candidate motifs which get no vote from any substring in F. The expected number of candidate motifs getting zero vote is $E_f(l,d)$. The filtering process removes those candidate motifs which have a variant in each sequence in T. When d is large, $E_f(l,d) \approx E_b(l,d)$ is small, therefore the expected running time of the filtering process ($O(nltE_f(l,d))$) should be small too. However, the running time of the voting process by sequences in F, i.e.

$$O\left(nf\sum_{i=0}^{d}(C_i^l 3^i)\right) = O\left(nfC_d^l 4^d\right)$$

might be prohibitively long for large d.

Our approach is to reduce this generalized (l,d)-problem to a smaller generalized (l',d')-problem with $l' < l$, $d' < d$ and d' small enough to be solvable. Let us consider a generalized (l,d)-problem. Since M has no d-variant in F, the length-l' prefix of M and the length-l' suffix of M should not have any d_s-variant in F either, where $d_s = d - (l - l')$. Let

$$d' = \min_{l-d \leq l' \leq l} \{d_s | E_f(l', d_s) \leq 1\}$$

For example, the generalized (20,11)-problem can be reduced to the generalized (14,5)-problem where $E_f(14,5) = 0.0027$. Since $E_f(l',d')$ is small and solvable, the reduced generalized (l', d')-problem is much easier to solve because d' is much smaller. The set of length-l' candidate motifs should contain any length-l' substrings of any length-l candidate motifs for the generalized (l,d)-problem, in particular, the length-l' prefix and length-l' suffix of motif M. If the length-$(2l' - l)$ suffix of a length-l' candidate motif is the same as the length-$(2l' - l)$ prefix of another length-l' candidate motif, we can combine them to form a length-l candidate motif. It can be shown that any length-l candidate motif for the generalized (l,d)-problem can be formed by combining two candidate length-l' motifs of the generalized (l',d')-problem. Thus the expected number of length-l candidate motifs by merging two candidate (l',d')-motifs is at most $[E_f(l',d') + (l - l')]^2$ and the expected running time is

Table 3. Values of d for which voting by F works when $t = f = 20$ and $n = 600$

l	9	11	15	20	26	30
d	≥ 3	≥ 4	≥ 6	≥ 11	≥ 16	≥ 20
d'	1	2	4	5	6	6

$$O\left(nfC_{d'}^{l'}4^{d'} + nlt[E_f(l',d') + (l-l')]^2\right)$$

Again this Voting algorithm based on a reduced size generalized (l',d')-problem works very well for large d as long as d' is reasonably small. Table 3 shows the values of d for which this approach works.

4.3 Local Search

The voting technique discussed in Section 4.1 and 4.2 works fine for all d when $l \leq 15$. However, when $l > 15$, there are cases that this voting technique will fail. In particular when d is not too small or too large, we cannot solve the generalized (l,d)-problems by voting from T or from F. For example, the generalized (20,9)-problem cannot be solved by voting directly from sequences in T or directly from sequences in F because of the long running time ($c_9^{20}4^9 \approx 4.4 \times 10^{10}$ is a big number). On the other hand, we cannot reduce the generalized (20,9)-problem to another smaller generalized (l',d')-problem with a small value of $E_f(l', d')$.

In order to solve these problems, a local searching method is proposed. The motif M has no d-variant in F and this information in F should be more useful than the information in T for finding M by local search. Assume we have a length-l seed string S. For each length-l neighboring string N, i.e. 1-variant of S, we find the number of d-variants of N occurring in set F. We replace string S by the neighboring string N if N has the least number of d-variants in F and has less d-variants than S in F. We repeat this process several times. If seed S and motif M are within a few symbol substitutions, we may hopefully refine S to M.

The seeds can be generated randomly or selected by voting from T and F. The probabilities that S can be refined to M after k iterations when S and M differ by k symbols for some generalized (l,d)-motif problems were shown in Table 4. It is evident from Table 4 that there is a high probability that we can find motif M from seed S when S and M differ by no more than 5 symbols.

Table 4. Probabilities for refining the seeds successfully. k is the number of symbol substitutions between seed S and motif M.

k	(20,8)	(20,9)	(20,10)
1	0.9895	1	1
2	0.9784	0.9707	0.9394
3	0.9563	0.9282	0.8652
4	0.9074	0.8603	0.7588
5	0.8483	0.4554	0
6	0	0	0

5 Experimental Results

We have implemented VANS in C++ and tested it on a computer with P4 2.4 GHz CPU and 4GB memory on the simulated and real data. For the simulated data, we picked a length-l motif M randomly and also generated 20 length-600 sequences in F randomly with the 0.25 occurrence probability of 'A', 'C', 'G' and 'T' at each position. Each of these sequences in F would be regenerated if it had a variant of M. Similarly we generated 20 length-600 sequences in T and planted a variant of M at a random position in each of these sequences. For each pair of l and d values, we ran 50 test cases and checked whether our program could discover the motif. Our program discovered the motif in all cases and the average running time is shown in Table 5. Some simulated data is missed (e.g. (9,6), (9,7) and (11,7)) because d is so large that any randomly generated length-600 sequence always contain a variant of any motif.

We have also tested VANS on real biological sequences stored in the public database SCPD. For each set of genes, we chose the 600 bp upstream of the genes as the input sequences in T. We also randomly picked the same number of genes and chose the 600 bp upstream of these genes as the input sequences in F. The lengths of the motifs l were the same as those of the published motifs

Table 5. VANS' Experimental results on simulated data

running time		l			
		9	11	15	20
d	2	0.4s	2s	201s	9.4s
	3	tends to 0s	9s	240s	10.6s
	4	tends to 0s	1s	382s	11.2s
	5	0.2s	3s	113.6s	27.1s
	6	-	4s	17m	107.1s
	7	-	-	9s	111.4s
	8	-	-	47s	3.4hr
	9	-	-	-	80m
	10	-	-	-	8.6m
	11	-	-	-	7.7m
	12	-	-	-	-

Table 6. VANS' Experimental results on real data

Transcription Factor	Published Motif pattern	Motif Pattern Found
GCR1	CWTCC	CTTCC
GATA	CTTATC	CTTAT
CCBF,SCB,SWI6	CNCGAAA	CGCGAAA
CuRE,MAC1	TTTGCTC	TTTGCTC
GCFAR	CCCGGG	CCCGGG
GCN1	TAATCTAATC	TAATCTAATC

and d was 1. Experimental results are shown in Table 6. VANS could find the motifs for these data sets within one second for each data set.

6 Conclusion

Since the (l,d)-problem has a limitation that no algorithm can discover the motif when d is large, we define the generalized (l,d)-problem which treats those sequences without variants of motif M as additional input. With this extra information, the motif discovering problem with large d becomes theoretically solvable. We also developed the VANS algorithm to solve the generalized (l,d)-problem. Experimental results showed that VANS performed well on most problem instances including the challenging (9,2), (11,3), (15,5), (20,7)-motif problem when F is empty.

The challenging generalized (l,d)-problems for $l > 20$, e.g. (26,11) and (26,13), remain unsolvable because of its long running time. Local search might work if we can reduce the number of seeds by generating "good" seeds efficiently.

References

1. Bailey, T., Charles Elkan, C.: Unsupervised learning of multiple motifs in biopolymers using expectation maximization. Machine Learning. **21** (1995) 51–80
2. Barash, Y., Bejerano, G., Friedman, N.: A Simple Hyper-Geometric Approach for Discovering Putative Transcription Factor Binding Sites. Workshop on Algorithms in Bioinformatics WABI **1** (2001) 278–293
3. Brazma, A., Jonassen, I., Eidhammer, I., Gilbert, D.: Approaches to the automatic discovery of patterns in biosequences. Jour. Comp. Biol. **5** (1998) 279–305
4. Buhler, J., Tompa, M.: Finding motifs using random projections. Research in Computational Molecular Biology RECOMB **1** (2001) 69–76
5. Chin, F., Leung, H.: Voting Algorithms for Discovering Long Motifs. Asia-Pacific Bioinformatics Conference APBC **3** (2005) 261–271
6. Chin, F., Leung, H., Yiu, S.M., Lam, T.W., Rosenfeld, R., Tsang, W.W., Smith, D., Jiang, Y.: Finding Motifs for Insufficient Number of Sequences with Strong Binding to Transcription Factor. Research in Computational Molecular Biology RECOMB **4** (2004) 125–132
7. Chin, F., Leung, H., Yiu, S.M., Rosenfeld, R., Tsang, W.W.: Finding Motifs with Insufficient Number of Strong Binding Sites. Jour. Comp. Biol. (to appear)
8. Fraenkel, Y., Mandel, Y., Friedberg, D., Margalit. H.: Identification of common motifs in unaligned dna sequences: application to Escherichia coli Lrp regulon. Bioinformatics **11** (1995) 379–387
9. Gelfand, M., Koonin, E., Mironov, A.: Prediction of transcription regulatory sites in archaea by a comparative genomic approach. Nucl. Acids Res. **28** (2000) 695–705
10. J. van Helden, B. Andre, Vides, J.C.: Extracting regulatory sites from the upstream region of yeast genes by computational analysis of oligonucleotide frequencies. Journal of Molecular Biology **281(5)** (1998) 827–842
11. Hertz, G.Z., Stormo, G.D.: Identification of consensus patterns in unaligned dna and protein sequences: a large-deviation statistical basis for penalizing gaps. International Conference on Bioinformatics and Genome Research **3** (1995) 201–216

12. Lawrence, C., Altschul, S., Boguski, M., Liu, J., Neuwald, A., Wootton, J.: Detecting subtule sequence signals: a Gibbs sampling strategy for multiple alignment. Science **262** (1993) 208–214
13. Lawrence, C., Reilly, A.: An expectation maximization (em) algorithm for the identification and characterization of common sites in unaligned biopolymer sequences. Proteins: Structure, Function and Genetics **7** (1990) 41–51
14. Leung, H., Chin, F.: Finding Exact Optimal Motif in Matrix Representation by Partitioning. European Conference on Computational Biology ECCB (2005) (to appear)
15. Liang, S.: cWINNOWER Algorithm for Finding Fuzzy DNA Motifs. Computer Society Bioinformatics Conference **2** (2003) 260–265
16. Marsan L., Sagot, M.F.: Algorithms for extracting structured motifs using a suffix tree with an application to promoter and regulatory site consensus identification. Jour. Comp. Biol. **7(3-4)** (2000) 345–362
17. Pesole, G., Prunella, N., Liuni, S., Attimonelli, M., Saccone, C.: Wordup: an efficient algorithm for discovering statistically significant patterns in dna sequences. Nucl. Acids. Res. **20(11)** (1992) 2871–2875
18. Pevzner, P., Sze, S.H.: Combinatorial approaches to finding subtle signals in dna sequences. International Conference on Intelligent Systems for Molecular Biology **8** (2000) 269–278
19. Sagot, M.F.: Spelling approximate repeated or common motifs using a suffix tree. Latin'98: Theoretical informatics, Lecture Notes in Computer Science **1380** (1998) 111–127
20. Sinha, S.: Discriminative motifs. Jour. Comp. Biol. **10** (2003) 599–616
21. Zhu, J., Zhang, M.: SCPD: a promoter database of the yeast Saccha-romyces cerevisiae. http://cgsigma.cshl.org/jian/ Bioinformatics **15** (1999) 563–577

Alignment of Tandem Repeats with Excision, Duplication, Substitution and Indels (EDSI)

Michael Sammeth[1,*], Thomas Weniger[2], Dag Harmsen[2], and Jens Stoye[1]

[1] Technische Fakultät, Universität Bielefeld, Germany
[2] Department of Periodontology, University Hospital Münster, Germany

Abstract. Traditional sequence comparison by alignment applies a mutation model comprising two events, substitutions and indels (insertions or deletions) of single positions (SI). However, modern genetic analysis knows a variety of more complex mutation events (e.g., duplications, excisions and rearrangements), especially regarding DNA. With the ever more DNA sequence data becoming available, the need to accurately compare sequences which have clearly undergone more complicated types of mutational processes is becoming critical.

Herein we introduce a new model, where in total four mutational events are considered: excision and duplication of tandem repeats, as well as substitutions and indels of single positions (**EDSI**). Assuming the EDSI model, we develop a new algorithm for pairwisely aligning and comparing DNA sequences containing tandem repeats. To evaluate our method, we apply it to the *spa* VNTR (variable number of tandem repeats) of *Staphylococcus aureus*, a bacterium of great medical importance.

1 Introduction

Sequence alignment is a rather well established tool to compare biological sequences. To align sequences, so-called *edit operations* have been defined which represent the atomic steps of the biological phenomenon called evolution. By successively applying such edit operations, the compared sequences can be converted into each other and - assuming parsimony as a major characteristic of evolution - good sequence alignments minimize the number of operations for these conversions or, more precisely, the assigned *costs*. In the classical model of mutation, two different edit operations are considered: the substitution and the insertion or deletion (together indel) of single characters in a sequence. In accordance with other literature [2], we will refer to this as the **SI model** (for substitution and indels) further on.

In general sequence alignments, the SI model has proven to work well. However, modern genetics knows more complex sources of mutation, especially when regarding the evolution of DNA. These mechanisms affect no longer only single

* The work of Michael Sammeth was supported by a doctoral scholarship of the Ernst-Schering Research Foundation.

R. Casadio and G. Myers (Eds.): WABI 2005, LNBI 3692, pp. 276–290, 2005.

positions but complete subareas of a sequence. Common other edit operations are *duplications* (insertions of copied subsequences, in case of *tandem* duplications, immediately adjacent to the original), *excisions* (deletions of subareas of a sequence), and *rearrangements* (relocations or reorientations of substrings within the sequence, e.g. *transpositions* or *inversions*).

In recent years quite some work has been invested in the algorithmic investigation of tandem repeats. Tandem duplications and excisions follow different rules than regular, character-based indels. On one hand the inserted or deleted substrings are usually much bigger in duplications and excisions, and on the other hand they contain the pattern of the tandem repeats in the corresponding sequence. Preliminary work in this field roughly is categorized into (1) tandem repeat *detection*, (2) *alignment* of sequences containing tandem repeats (with or without knowledge of their positions), and (3) reconstruction of a tandem *repeat history* where the phylogenetic history of the tandem repeats of one sequence is tracked down to a single ancestor repeat. (1) concerns the detection of tandem repeat copies with an unknown pattern [9]. In the context of (2) various works extended the SI mutation model to additionally respect tandem duplication events (**DSI model**), e.g. in [2,5,1]. The research of (3) investigates possible duplication histories of the tandem repeats in a sequence. These are represented by duplication phylogenies which under certain conditions can be turned into rooted duplication trees, see [3,10,7,4].

Staphylococcus aureus (*S. aureus*), a bacterium responsible for a wide range of human diseases (e.g., endocarditis, toxic shock syndrome, skin, soft tissue and bone infections etc. [16]), contains polymorphic 24-bp variable-number tandem repeats (VNTRs) in the 3' coding region of the staphylococcal protein A (the *spa* protein) [6]. The tandem repeats in this region undergo a mutational process including duplication and excision events in addition to nucleotide-based substitutions and indels [11], probably caused by slipped strand mispairing [17]. Further on, the microvariation of the *spa* VNTR cluster [12] seems to support the phylogenetic signal reported by other methods (e.g., by [14]). Therefore, an automated method to compare strains of *S. aureus* and classify them according to the microvariation of the *spa* tandem repeats is critical in order to determine the types of newly acquired sequences rapidly and accurately.

In this paper, we introduce a novel model of evolution, the **EDSI model** (excisions, duplicatons, substitutions and indels), which in addition to the DSI model includes repeat excision operations. Moreover, the restrictions on the order of mutation events are relaxed: all four edit operations may occur arbitrarily cascaded with each other. In Section 2 we formalize the EDSI model and give an overview of the problem addressed. Next, in Section 3, we propose an exact algorithm to align and compare a pair of sequences under the EDSI model of mutation. Finally, in Section 4, we give some practical examples for comparing *spa* sequences of *S. aureus* with the novel method and Section 5 summarizes the benefit of the EDSI model and outlines its potential for accurate phylogenetic investigations. Additional material including the proofs of all theorems is provided in the online supplement `http://www.sammeth.net/pub/05wabi_suppl.pdf`.

2 Description of the EDSI Model

Let s be a sequence of characters over the DNA alphabet $\Sigma = \{A,C,G,T\}$, and let s contain tandem repeats. If the boundaries of the repeats are known, s can be written directly as a sequence s' over the macro alphabet of the different repeat types $\Sigma' = \{-, A, B, C, D, \ldots\}$. The additional gap character ($- \in \Sigma'$) is used later on when aligning repeat sequences (Section 3). $(\Sigma')^+$ denotes the set of all nonempty strings over the repeat alphabet Σ'. On s' we define the EDSI evolution, allowing duplication and excision of repeats (characters in s'), as well as substitutions and indels of nucleotides within the repeats. Note that the commonly used substitution and indel operations work on the DNA bases of s, and therefore are comprised in the term *mutation* of a tandem repeat. In contrast, the duplication and excision events affect complete repeats of s' (Fig. 1).

Precisely, the duplication events occurring in the evolution of the *spa* repeat cluster are *multi-copy* duplications (1-duplication, 2-duplication, etc.) copying one or more adjacent repeat copies at a time. The *boundaries* of the duplicated repeats are restricted to the boundaries of tandem repeats on the nucleotide sequence s. However, on s' the duplication boundaries are *free* in the characters of the macro alphabet Σ', i.e., duplicated substrings may start and end anywhere in s'. Finally, the duplication operation in the EDSI model is *single-step*, denoting that no more than one copy of a duplicated substring is produced in one evolutionary step. In the same manner, the excision operation of the model is characterized as multi-copy (1-excision, 2-excision, etc.) with free boundaries on s'. The order of events in EDSI is unrestricted. To be specific, all four edit operations described by the model may be applied arbitrarily cascaded with each other (Fig. 1).

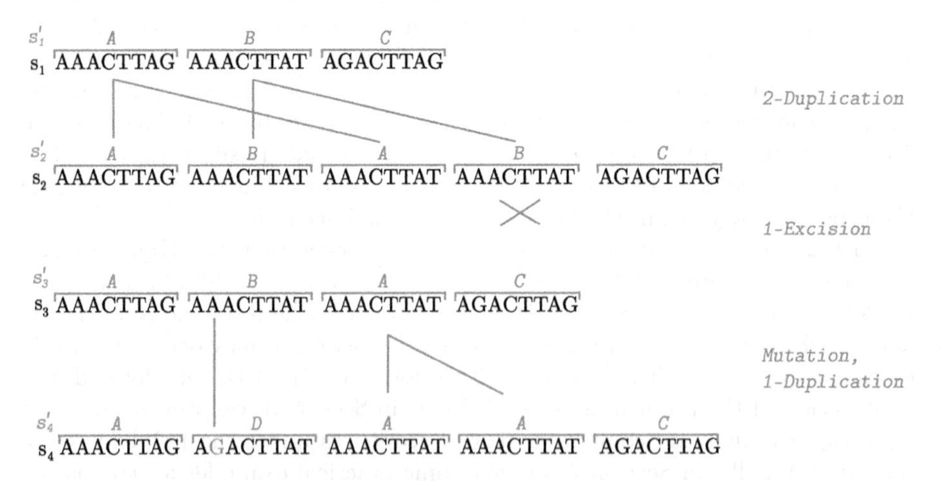

Fig. 1. An example for cascaded duplication, excision, and mutation events. Shown are DNA sequences s_i and the corresponding sequences s_i' on the macro alphabet Σ' of repeat types (superimposed on s_i in grey). Some edit operations (as given to the right) successively are performed on the sequence. It can be easily seen that after a couple of cascaded operations the sequence of characters is rather scrambled.

In order to assess the evolutionary distance between two given sequences, we assign *costs* to all operations comprised in the EDSI model: $cost_e(w)$ for the excision of the tandem repeat copies in string w, $cost_d(w)$ for a duplication of the tandem repeats in string w, and $cost_m(w_1, w_2)$ for a mutation of a repeat type w_1 into the repeat type w_2. The **cost model** of EDSI evolution then can be freely adjusted[1] with respect to the following criteria:

- Excision costs should be positive, $cost_e(w) > 0$ for all $w \in (\Sigma')^+$, since excision events can replace all other operations. To be specific, any non-identical pair of sequences (s', t') can be derived from a concatenated ancestor string $s't'$ by two excisions (once excising s' to reconstruct t' and once exising t' to deduce s' from the ancestor). Hence, finding the minimum distance for sequences in a cost model with $cost_e(w) = 0$ is trivial.
- Duplication costs should be non-negative, $cost_d(w) \geq 0$ for all $w \in (\Sigma')^+$.
- Mutation costs should comply with the properties of a metric: symmetry ($cost_m(w_1, w_2) = cost_m(w_2, w_1)$ for all $(w_1, w_2) \in \Sigma'$), zero property ($cost_m(w, w) = 0$ for all $w \in \Sigma'$) and the triangle inequality ($cost_m(w_1, w_2) + cost_m(w_2, w_3) \geq cost_m(w_1, w_3)$ for all $w_1, w_2, w_3 \in (\Sigma')^+$).

The problem of sequence evolution comprising EDSI operations can now be formulated as an optimization problem with the goal of minimizing the EDSI distance defined in the following: Given two sequences (s', t') and cost measures $cost_e$, $cost_d$, and $cost_m$, find the distance $d(s', t')$ under the EDSI evolution that is the minimum sum of costs of all series of operations possible to reproduce one sequence from the other. This can be interpreted such that both sequences s' and t' are subjected to evolutionary operations in order to transform them into a common string – a possible common ancestor according to the biological model. The operations that produce a common ancestor of s' and t' with the least costs define the EDSI distance $d(s', t')$.

Theorem 1 (finiteness). *The edit operations under EDSI evolution and their unrestricted order basically force us to explore an infinite search space of possible ancestor sequences. However, the space of operation sets to be explored in order to find the minimal distance between two sequences $d(s', t')$ is finite.*

Proof. When reconstructing possible evolutionary histories from a given pair of sequences (s', t'), theoretically there could have been present an arbitrarily large number of repeats between two adjacent positions x and $x + 1$ of s' (or, symmetrically, t') which later were deleted with cost $cost_e > 0$. There are two possible sources for these deleted repeats in the tandem-repeat history: (i) they may have emerged from duplication events, or (ii) they may have been repeats from non-duplicated sequence areas. Cnsider case (i). If a deleted repeat has originated from a duplication event, the corresponding excision can be detected by investigating all possible duplication events to the left (in $s'[1, x]$) and to the

[1] For a definition of the costs used for the *Staphylococcus aureus* evolution see Section 4.

right (in $s'[x+1,|s'|]$). The number of single duplication events on a finite string is limited, and so is the number of possibly excised repeat units between x and $x+1$. Moreover, all character insertions between x and $x+1$ induced by possible duplication events with consecutive excisions are collected (Section 3.1 and 3.2) and taken into account in the final comparison between s' and t' (Section 3.3). Consider now case (ii). If deleted substrings have originated from non-duplicated sequence areas (or whenever the second repeat copy of a duplication has been excised as well), they are not relevant in the search for a minimal distance: in the comparison of s' with t' an appropriate excision event will be detected (Section 3.3), whenever t' has a substring that aligns between $s'[x]$ and $s'[x+1]$. However, if the alignment with t' does not indicate any presumptive excision between x and $x+1$ in s', all such theoretically possible excisions are not contained in the operations determining the minimum distance since additional excision costs $cost_e > 0$ produced an ancestor sequence that is not closer to t' than the original sequence s'. □

3 Pairwise Alignment Under the EDSI Model

After the definition of the EDSI model, we can describe an exact algorithm to compare and align sequences with respect to the four edit operations. The main idea of our method is to find possible repeat histories, assign costs to them according to the edit operations, and consider them as alternatives during an alignment procedure. Thereby the alignment possibility between both sequences with the least cost is selected, regarding the original sequences with all contracted substrings generated by the repeat histories. Assuming the parsimony principle for nature we take these costs as a distance measure for the compared sequences. So basically our algorithm works in two steps: first it finds possible duplications on each sequence under the rules given by the EDSI model. Afterwards, we determine the distance between a sequence pair in a high-dimensional multiple sequence alignment (MSA) using the duplication events found before as alternative alignment possibilities between the compared sequences.

Although not observed in biology, we also use the term *contraction* for the mathematically inverse process of a duplication. Our technique is based on *contramers* $C = (s', b, m, e, A)$, representing contraction units. These are substrings of s' (the macro alphabet representation of the input string s) on which a contraction is performed. The substring to be contracted, $s'[b, e]$, is located within s' by its beginning b and its end position e. The *meridian* $m, (b < m \leq e)$ splits the contramer into two *segments*, also called the *prefix* (the first segment $s'[b, m-1]$) and the *suffix* (the second segment $s'[m, e]$). Finally, the alignment A of the prefix and the suffix describes how the characters of both segments are evolutionarily related according to the contramer. To be specific, aligned repeats correspond to each other (with respect to possible mutation events) and gaps indicate the excision of repeats. An example of a contramer representing a duplication event (including mutation and excision) is given in Fig. 2.

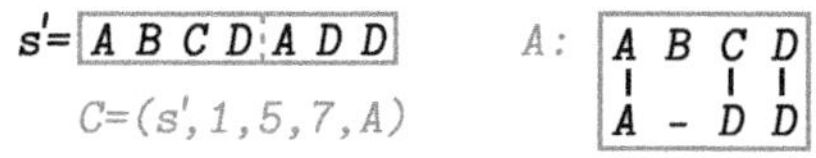

s'= A B C D A D D   A: A B C D / A - D D

C=(s',1,5,7,A)

Fig. 2. A contramer $C = (s', 1, 5, 7, A)$ that implies the duplication of substring $s'[1, 4] = ABCD$ and its post-duplicational modification into ADD. The alignment (grey box) shows that repeat B was excised while repeat C mutated to repeat D. All vertically adjacent repeat pairs (i.e., non-gap characters) in an alignment layout correspond to each other w.r.t. possible mutations. These links (black lines) are not explicitly visualized in further representations of an alignment.

3.1 Primary Library of Contramers

The initial set of contramers is extracted directly from the repeat sequence s'. For each meridian position in s', $1 < m \leq |s'|$, all alignment possibilities of available non-empty prefixes $s'[b, m-1]$, $1 \leq b < m$, and non-empty suffixes $s'[m, e]$, $m \leq e \leq |s'|$, are generated. The contramers inferred thereby form the *primary library*. Note that at this stage the similarity of the aligned segments is not optimized by any objective function since links between amalgamated contramers later on can involve new repeat copies (i.e., characters of s'). Contramers in the primary library represent possible duplication events, i.e. links of positions in neighboring segments.

Algorithm 1. (Generate the contramers for the primary library L)

```
 1: L ← ∅
 2: for m ← 2 to |s'| do
 3:    for b ← 1 to (m − 1) do
 4:       for e ← m to |s'| do
 5:          AP[] ← GeneratePossibleAlignments(b, m, e)
 6:          for all A in AP[] do
 7:             Store(L, C = (s', b, m, e, A))
 8:          end for
 9:       end for
10:    end for
11: end for
```

Algorithm 1 outlines the technique used to assemble the primary library of contramers. As input serves a sequence s' on the alphabet of tandem repeats Σ'. The resulting list L contains each possible contramer $C = (s', b, m, e, A)$ of s'. The *cost* of a contramer may be derived directly from the associated alignment A by adding, for each column of the alignment, the costs of mutations or excisions. In addition, to reflect the costs for the duplication event, $cost_d(s'[b, (m-1)])$ is added, yielding the final cost of contramer $C = (s', b, m, e, A)$:

$$cost(C) = cost_d(s'[b,(m-1)]) + \sum_{i=1}^{|A|} \begin{cases} cost_m(A_{1i}, A_{2i}) & \text{for each mutation } (A_{1i} \neq - \neq A_{2i}) \\ cost_e(A_{1i}) & \text{for each excision in the prefix } (A_{1i} \neq -,\ A_{2i} = -) \\ cost_e(A_{2i}) & \text{for each excision in the suffix } (A_{1i} = -,\ A_{2i} \neq -). \end{cases}$$

Theorem 2 (completeness of the primary library). *Contramers contained in the primary library exhaustively generate all ancestor strings that can be derived from a sequence by reversing exactly one duplication event (Supplement).*

3.2 The Secondary Library

In order to infer cascaded duplication histories, overlapping primary contramers C_1 and C_2 are to be merged to form cascaded duplication events. Abusing notation, we define a contramer intersection (union) as the intersection (union) of the corresponding segments of s', i.e. $C_1 \cap C_2 = \{b_1, \ldots, e_1\} \cap \{b_2, \ldots, e_2\}$ ($C_1 \cup C_2 = \{b_1, \ldots, e_1\} \cup \{b_2, \ldots, e_2\}$).

If $C_1 \cap C_2$ comprises positions of both segments of C_1, we call C_1 a *contained* contramer (and C_2 a *containing* contramer). Otherwise, C_1 and C_2 are *connected* contramers. However, not all overlapping duplication events are necessarily *compatible* with each other. The precondition for a pair of compatible contramers (C_1, C_2) is that they can be realized in a common *evolutionary order*, i.e., there exists at least one repeat history tree comprising both described duplication events.

Evolutionary order of compatible contramers. The common evolutionary realizability can be deduced from analyzing the intersection of the two contramers $C_1 \cap C_2$. In evolution, the duplication events described by contained contramers must have happened before the duplication events of the contramer they are contained in (Fig. 3a), whereas for a pair of connected contramers the evolutionary order does not matter (Fig. 3b). Two contramers mutually contained in each other are not realizable in a common repeat history (Fig. 3c), even if they share the same meridian position m (Fig. 3d).

Lemma 1 (merging conditions). *Two contramers* $C_1 = (s', b_1, m_1, e_1, A_1)$ *and* $C_2 = (s', b_2, m_2, e_2, A_2)$ *are compatible and can be merged if:*

1. *they overlap,* $C_1 \cap C_2 \neq \emptyset$ (connectibility) *and*
2. *one of the reflected duplication events has happened after the other one. Therefore at least the contramer* C_1 *needs to have a segment outside of the intersection area,* $(m_1 - 1) \notin C_1 \cap C_2$ *or* $m_1 \notin C_1 \cap C_2$ (compatibility)*.*

Lemma 1 describes the preconditions that are to be met to merge a pair of contramers. Afterwards, C_1 and C_2 are merged into $C_1 \cup C_2$ by combining their respective alignments: any repeat $s'[x]$ in the overlapping area $C_1 \cap C_2$ may be linked to two other repeat copies $s'[y] \in (C_1 \setminus C_2)$ and $s'[z] \in (C_2 \setminus C_1)$ by the alignments A_1 and A_2. Thereby a *transitive link* between both of the not directly associated repeats $s'[y]$ and $s'[z]$ is created. All three repeats $(s'[x], s'[y], s'[z])$

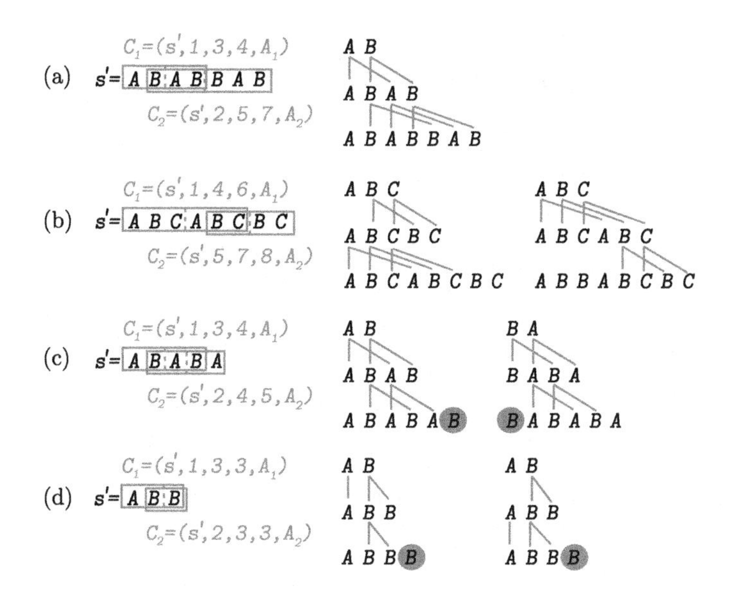

Fig. 3. Restrictions on compatible contramer pairs $C_1 = (s', b_1, m_1, e_1, A_1)$ and $C_2 = (s', b_2, m_2, e_2, A_2)$ (grey boxes, meridian position indicated by a dashed line). Possible repeat histories expressed by the amalgamation $C_1 \cup C_2$ are depicted on the right. (a) C_1 is *contained* in C_2, therefore the evolutionary order is fixed and the duplication captured in C_1 must have happened before the one described by C_2 (only one possible repeat history). (b) Merging two *connected* contramers imposes no order on the evolution (i.e., the duplication of C_1 or C_2 may have happened first). (c) and (d) If none of the contramers has a non-intersecting segment, $\{m_1 - 1, m_1, m_2 - 1, m_2\} \notin C_1 \cap C_2$, *no repeat history* can be found incorporating both duplication events captured by the contramers. This holds even if the meridians coincide, $m_1 = m_2$, see (d).

are then written in one column of the merged alignment (Fig. 4a). Problems arise when both contramers comprise excisions in between corresponding positions of the overlapping area (Fig. 4b). In this case the contramers do not provide a unique information about the transitive relation between the excised characters. One possibility would be to exhaustively generate all the alignment possibilities between the respective characters. However, since we are only interested in finding a "good" combination of characters minimizing the distance to another sequence, we let these ambiguous repeats unaligned for the moment and search for a combination similar to the compared sequence later on in the comparison step (Section 3.3).

The merging strategy is straightforwardly extendable to deal with more than two contramers. A set of combinable contramers $\{C_1, C_2, \ldots, C_k\}$ obviously requires that each of the contramers C_i, $i \in \{1, 2, \ldots, k\}$ has to fulfill the precondition of connectibility with at least one other contramer $j \in \{1, 2, \ldots, k\}, j \neq i$. Otherwise C_i is isolated and cannot be joined. Furthermore, it is required that each pair of overlapping contramers (C_i, C_j), $i, j \in \{1, 2, \ldots, k\}$, is compatible. The order in which the contramers are joined is not important:

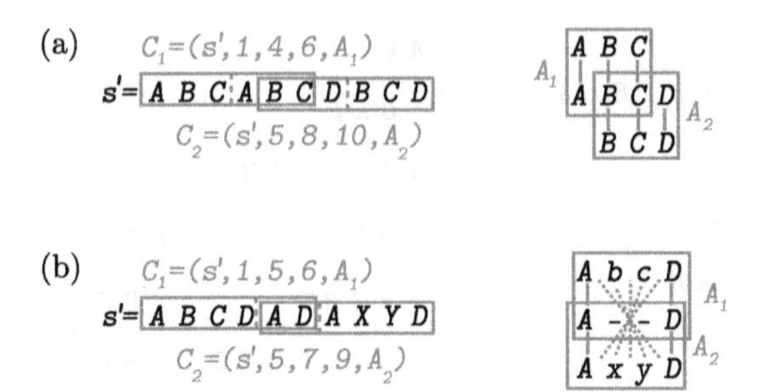

Fig. 4. Transitive links when merging contramers. (a) A pair of partially overlapping contramers where e.g. C_1 connects the positions $(2, 5)$ and C_2 links position 5 with 8. The transitive link created when merging A_1 with A_2 links all three B-characters together (2nd column of the merged alignment to the right). (b) The amalgamation of a pair of contramers (C_1, C_2) which are both inducing characters in the same area. Consequently, the phylogenetic relation of the characters (lowercase) cannot be exactly determined (possible relations are indicated by the dotted grey lines).

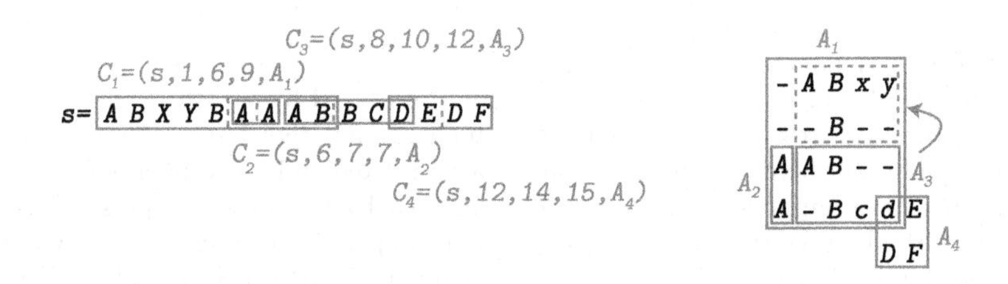

Fig. 5. A set of multiply merged contramers $(C_1 \cup C_2 \cup C_3 \cup C_4)$ and the respective concatenated alignment. Note that lowercase characters are not uniquely aligned by the transitive links of the contramers, and their position is determined later during the comparison process (Section 3.3).

Theorem 3 (commutativity). *The pairwise merging steps of multiply joined contramers are* commutative, $(C_1 \cup C_2) = (C_2 \cup C_1)$ *(Supplement).*

Figure 5 demonstrates that when performing a multiple amalgamation with all preconditions met by the contramers $\{C_1, C_2, \ldots, C_k\}$, we perform successively the merging step for each pair of overlapping contramers (C_i, C_j) such that $i \neq j$, $C_i \cap C_j \neq \emptyset$.

Algorithm 2 describes the construction of contramers in the secondary library. Initially, L comprises the contramers already included in the primary library. The set of contramers with beginning b, meridian m, and end e can be accessed via the function GETC(L, b, m, e). The functions FINDCONNECTEDC() and FINDCONTAINEDC() extract contramers in a given subarea (specified by the start and end point). For each pair of overlapping contramers the preconditions are checked before returned (set $DP[]$). In the end, compatible contramers F are merged with C, and the result is added to L.

Algorithm 2. (Amalgamate contramers to build the secondary library L)

```
1: L ← PrimaryLibrary()
2: for m ← 2 to |s'| do
3:    for b ← 1 to (m − 1) do
4:       for e ← m to |s'| do
5:          CP[] ← GETC(L, b, m, e)
6:          for all C in CP[] do
7:             DP[] ← FINDCONNECTEDC(L, m, e)
8:             for all D in DP[] do
9:                F ← MERGE(C,D)
10:               STORE(L,F)
11:            end for
12:            DP[] ← FINDCONTAINEDC(L, b, m)
13:            for all D in DP[] do
14:               F ← MERGE(C,D)
15:               STORE(L,F)
16:            end for
17:         end for
18:      end for
19:   end for
20: end for
```

Theorem 4 (completeness of the secondary library). *Contramers contained in the secondary library generate all ancestor strings that can be derived from a sequence under the EDSI model containing one or more duplication events (Supplement).*

3.3 Contramer Alignment

In the final alignment phase, the possible tandem repeat histories of two sequences (s', t') are used as alternative character combinations when comparing s' to t'. To this end, we extend the well established technique of dynamic programming (DP) for sequence alignment to additionally take into account the (cascaded) duplications. The contramers found along both sequences to be compared serve as additional alignment possibilities, i.e., cells extending the regular DP matrix. For each cell (i, j) to be computed in the DP recursion of the main matrix M of size $|s'| \times |t'|$, (merged) contramers ending at position i in s' (or at position j in t, respectively) are considered. The alignment profile of each comprimer $C = (s, b, m, e, A)$ can substitute the characters of the original sequence in the area $s'[b, e]$. Note that each cell (i, j) of the matrix M is connected by multiple contramers with any of the cells computed earlier during the DP process. Therefore, the resulting alignment is high-dimensional with multiple alternative submatrices for each contramer in both sequences (Fig. 6).

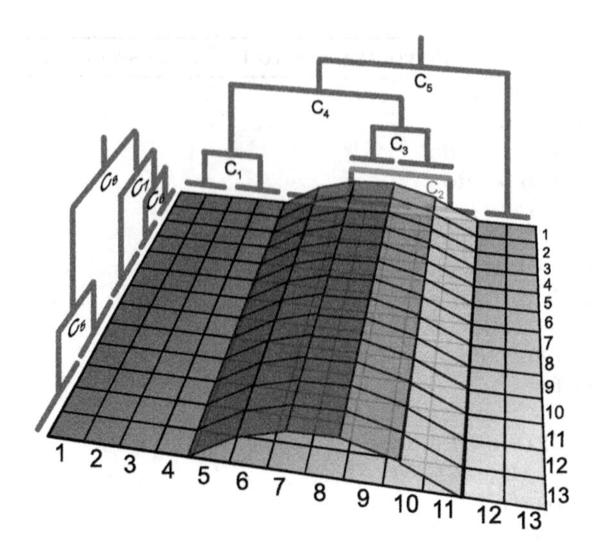

Fig. 6. An example for an alternative submatrix within a DP matrix $M = |s'| \times |t'|$. $C_2 = (t', 5, 10, 11, A_2)$ substitutes the substring $t'[5, 11]$ with the alignment A_2. Projected into M, C_2 spans the submatrix shown. During the DP process paths within the original and within the submatrix are taken into account when determining the optimum of the cells in column 12. Note that only contramers of one possible repeat history are depicted here, but all cascaded duplication events of the secondary library are investigated.

The matrix M can be computed by the following recursion formula:

$$M(i,j) = \min \begin{cases} M(i-1, j-1) + cost_m(s'[i], t'[j]) & //mutation \\ M(i-1, j) + cost_e(s'[i]) & //excision\ in\ s' \\ M(i, j-1) + cost_e(t'[i]) & //excision\ in\ t' \\ M(i - b_x, 0) + cost(C_x) + align(t'[0, j], A_x) & \\ \quad \text{for all } C_x = (s', b_x, m_x, i, A_x) & //duplication\ in\ s' \\ M(0, j - b_y) + cost(C_y) + align(s'[0, i], A_y) & \\ \quad \text{for all } C_y = (s', b_y, m_y, j, A_y) & //duplication\ in\ t' \\ M(i - b_x, j - b_y) + cost(C_x) + cost(C_y) + align(A_x, A_y) & \\ \quad \text{for all } C_x = (s', b_x, m_x, i, A_x) & \\ \quad \text{and } C_y = (s', b_y, m_y, j, A_y) & //duplication\ in\ s'\ and\ t' \end{cases}$$

At each stage (i, j) of the alignment, the minimum cost is calculated for all edit operations of the EDSI model: mutation (line 1) of repeats (comprising substitutions and indels on the DNA alphabet Σ), excisions (lines 2 and 3) of repeat copies (on the macro alphabet Σ') or duplication events (lines 4, 5 and 6). The preference of the algorithm is in the same order as given, and to optimize the performance a bounding step was added to only assess the alignment of contramers C which are not already exceeding the costs found earlier for a cell (i, j) by their cost $cost(C)$.

As found earlier (Section 3.2, Fig. 4), in merged contramers not necessarily all of the transitive relations are clear. These positions are to be aligned within the amalgamated contramer taking also into account the sequence the contramer is compared to. To this end we use a stable re-implementation of the hyperspace multiple sequence alignment procedure [15], which was modified to use the scoring function for repeat evolution, when aligning the amalgamated contramers with the corresponding compared sequence: all positions already aligned between the duplication events of the contramers are provided as constraints, whereas the ambiguous positions finally are aligned optimally according to the sequence information (Supplement).

Theorem 5 (correctness of the method). *The distance $d(s', t')$ found by the DP recursion is the minimum distance possible in the comparison of s' and t', assuming the model of EDSI evolution.*

Proof. In the primary library all possible links between repeats of s' and t' that can originate from single duplication events, are generated (Theorem 2). Thus, by merging overlapping duplication events in the secondary library, we explore all possible cascades of duplications collected in the primary library (with restrictions to the biological model as given in Lemma 1, Theorem 4). On each of these cascades, excision events are tried before and after the respective duplication in order to yield the minimum costs according to the EDSI evolution. Finally, in a high dimensional alignment all contramers extracted from s' and t' are used as alternative substrings imposing replacement costs as calculated. The minimum distance is then finally found by a DP recursion in a high dimensional alignment (Section 3.3). □

Obviously the time and space complexity of the method are exponential w.r.t. the sequence length. Note that the input of the algorithm are sequences of already annotated repeats and the input size therefore is much smaller than the original sequences.

4 Results

To test our method, we applied it to the DNA sequences of *Staphylococcus aureus*. To be specific, the 5'-VNTR clusters in the gene encoding the *spa* protein were used as input for pairwise alignment under the EDSI model. Since the repeat patterns for all hitherto isolated strains (so-called *spa* types) are known, the sequences are provided in characters of the macro alphabet Σ'. To this point, we use the *Kreiswirth* notation defined by identify the repeats $\Sigma' = \{A, A_2, B, B_2, C, C_2, \ldots, V, V_2, W, X, Y, Y_2, Z, Z_2\}$. In addition to the simplified alphabet used to introduce the model, in the Kreiswirth notation each letter may be used more than once in conjunction with a unique index [12].

Comparison of ST-254 spa *types.* We set up a simple cost scheme for the comparison of *spa* types: since we are interested in a distance to measure evolutionary

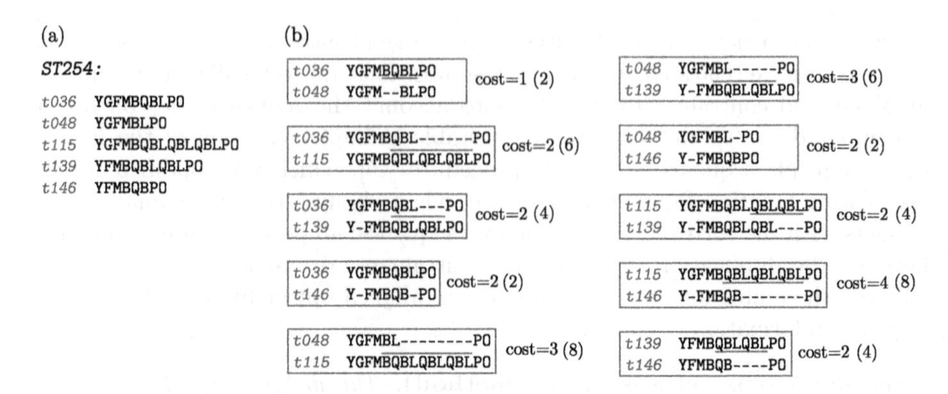

Fig. 7. Sequence comparison of the MLST sequence type ST-254. (a) a list of *spa* types found to have the ST-254 pattern. The data was acquired by sequencing from the same laboratory strains the VNTR cluster of the *spa* protein and the MLST loci [13]. (b) One alignment that scores minimal costs for each pair of *spa* types from the ST-254 group. Substrings involved in duplication events leading to the minimum distance are underlined. To the right of the alignments the costs are given w.r.t. the EDSI model and in parentheses the costs under the SI model (without taking into account duplications). Under the EDSI model, the costs in the comparison of $t036$ and $t048$ are composed of a duplication event of the substring $t048[5,6] = BL$ and the mutation of repeat L into $t036[6] = Q$ with distance $d(t036, t048) = cost_d(t048[5,6]) + cost_m(L, Q) = 1 + 0 = 1$.

steps, we assign a unit cost u corresponding to the number of operations needed to perform the change. A duplication costs one operation ($cost_d(s'[x,y]) = u$), regardless of its length. The objective function to score mutation events (substitutions and indels of nucleotides) is based on the alignment of the repeat types (Supplement). In order to contribute to the fact that the repeat cluster is coding, nucleotide substitutions changing effectively the corresponding codon are weighted with a cost of u, while silent mutations are omitted. In the same manner indels are penalized according to the number of codons x missing (xu). If not further specified, we set $u = 1$ in the tests. The mutation costs $cost_m(x, y)$ for $x, y \in \Sigma'$ are summed up along the pairwise DNA alignment of x and y which is projected from the global alignment of all repeats. Excisions are treated differently, we penalize them according to their length, such that $cost_c(s'[x, y]) = (y - x)$. The linear cost function prevents the algorithm from replacing all evolutionary events by excisions when repeat copies are no longer exact. From another point of view, the scoring biases the algorithm to favor duplications and mutations and prefer them – up to a certain threshold – over possible excisions.

Since, to our knowledge, this is the first time the VNTR data of *spa* types is used to infer distance measures, we focus on one sequence type (ST-254), which by definition pools strains with the same types of the seven housekeeping genes used for MLST [8]. However, the resolution of STs found by MLST is lower than the microvariation within the *spa* repeat cluster. Thus, a ST group with an identical MLST pattern can pool several strains with diverging *spa* types (named by "t" and a 3-digit code), while the other way around a *spa* type may have evolved in different ST groups. *Spa* types used in here to investigate the micro-variation of the repeats (i.e., t036, t048, t115, t139, and t146) were isolated in the laboratory from identical strain stocks [13]. Therefore, the microvariation of these *spa* types can be assumed to bear a phylogenetic marker (Fig. 7a [11]).

Figure 7b summarizes the differences of applying the novel method based on the EDSI evolution in contrast to standard scoring functions for SI model. We adapted the scoring function of the SI alignment to the same values given for the EDSI evolution (xu for the insertion of x gaps and substitution costs according to non-synonymous mutations, Supplement). We want to stress on the fact, that the alignments shown are only one example from a set of alignments that can reproduce the minimal costs shown. Minimal costs of the other alignments in Fig. 7b can be calculated as follows (mutations of cost 0 are omitted):

$$
\begin{aligned}
d(t036, t115) &= 2cost_d(t036[6,8]) = 2 \\
d(t036, t139) &= cost_d(t036[6,8]) + cost_e(t036[2,2]) = 2 \\
d(t036, t146) &= cost_e(t036[2,2]) + cost_e(t036[8,8]) = 2 \\
d(t048, t115) &= cost_d(t048[5,6]) + 2cost_d(QBL) = 3 \\
d(t048, t139) &= cost_e(t048[2,2]) + cost_d(t048[4,5]) + cost_d(QBL) = 3 \\
d(t048, t146) &= cost_e(t048[2,2]) + cost_e(t146[6,6]) = 2 \\
d(t115, t139) &= cost_e(t115[2,2]) + cost_e(t115[6,8]) = 2 \\
d(t115, t146) &= cost_d(t146[2,2]) + cost_e(t115[8,8]) + 2cost_d(QBL) = 4 \\
d(t139, t146) &= cost_d(t139[7,7]) + cost_d(QBL) = 2
\end{aligned}
$$

5 Conclusion

The EDSI model of evolution joins the events of tandem duplication, tandem copy excision, point mutation and deletion that may happen in arbitrary order throughout evolution. To our knowledge, this is the first time an evolutionary model of that complexity has been described for sequence comparison. Taking into account operations as captured in the EDSI model, we described an exact method to compare a pair of repeat sequences and to assign them a distance. In first tests we could show that the pairwise comparison under the EDSI model efficiently captures cascades of duplication events and expresses them in the distance measure. Regular scoring functions (based on the SI or DSI model) cannot resolve these distances, which already have been demonstrated *in vivo* studies to be essential mechanisms in the evolution of *S. aureus* [11].

References

1. B. Behzadi and J.-M. Steyaert. The minisatellite transformational problem revisted. In *Proc. of WABI 2004*, volume 3240 of *LNBI*, pages 310–320, 2004.
2. G. Benson. Sequence alignment with tandem duplication. *J. Comput. Biol.*, 4:351–367, 1997.
3. G. Benson and L. Dong. Reconstructing the duplication history of a tandem repeat. In *Proc. of ISMB 1999*, pages 44–53, 1999.

4. D. Bertrand and O. Gascuel. Topological rearrangements and local search method for tandem duplication trees. *IEEE/ACM Trans. Comput. Biol. Bioinformatics*, 2:1–13, 2005.
5. S. Bérard and E. Rivals. Comparison of minisatellites. In *Proc. of RECOMB 2002*, pages 67–76, 2002.
6. M. de Macedo Brígido, C. R. M. Barardi, C. A. Bonjardin, C. L. S. Santos, M. de Lourdes Junqueira, and R. R. Brentani. Nucleotide sequence of a variant protein a of *Staphylococcus aureus* suggests molecular heterogeneity among strains. *J. Basic Microbiol.*, 31:337–345, 1991.
7. O. Elemento, O. Gascuel, and M.-P. Lefranc. Reconstructing the duplication history of tandemly repeated genes. *Mol. Biol. Evol.*, 19:278–288, 2002.
8. M. Enright, N. Day, C. Davies, S. Peacock, and B. Spratt. Multilocus sequence typing for characterization of methicillin-resistant and methicillin-susceptible clones of *Staphylococcus aureus. J. Clin. Microbiol.*, 38:1008–1015, 2000.
9. R. Groult, M. Léonard, and L. Mouchard. A linear algorithm for the detection of evolutive tandem repeats. In *The Prague Stringology Conference 2003*, 2003.
10. D. Jaitly, P. Kearney, G.-H. Lin, and B. Ma. Reconstructing the duplication history of tandemly repeated genes. *J. Comput. Sys. Sci.*, 65:494–507, 2002.
11. B. Kahl, A. Mellmann, S. Deiwick, G. Peters, and D. Harmsen. Variation of the polymorphic region X of the protein A gene during persistent airway infection of cystic fibrosis patients reflects two independent mechanisms of genetic change in *Staphylococcus aureus. J. Clin. Microbiol.*, 43:502–505, 2005.
12. L. Koreen, S. V. Ramaswamy, E. A. Graviss, S. Naidich, J. M. Musser, and B. N. Kreiswirth. spa typing method for discriminating among staphylococcus aureus isolates: implications for use of a single marker to detect genetic micro- and macrovariation. *J. Clin. Microbiol.*, 47:792–799, 2004.
13. Ridom nomenclature server. `http://www.ridom.de/spaserver/nomenclature.shtml`.
14. D. A. Robinson and M. C. Enright. Evolutionary models of the emerge of methicillin-resistant staphylococcus aureus. *Antimicrob. Agents Chemother.*, 47:3926–3934, 2003.
15. M. Sammeth, J. Rothgänger, W. Esser, J. Albert, J. Stoye, and D. Harmsen. Qalign: quality based multiple alignments with dynamic phylogenetic analysis. *Bioinformatics*, 19:1592–1593, 2003.
16. N. N. I. S. System. National nocosomial infections surveillance (nnis) system report, data summary from january 1990-may 1999. *Am. J. Infect. Control*, 27:520–532, 1999.
17. A. van Belkum, S. Scherer, L. van Alphen, and H. Verbrugh. Short-sequence dna repeats in prokaryotic genomes. *Microbiol. Mol. Biol. Rev.*, 62:275–293, 1998.

The Peres-Shields Order Estimator for Fixed and Variable Length Markov Models with Applications to DNA Sequence Similarity

Daniel Dalevi and Devdatt Dubhashi

Department of Computing Science, Chalmers University,
SE 412 96 Göteborg, Sweden
{dalevi, dubhashi}@cs.chalmers.se

Abstract. Recently Peres and Shields discovered a new method for estimating the order of a stationary fixed order Markov chain [15]. They showed that the estimator is consistent by proving a threshold result. While this threshold is valid asymptotically in the limit, it is not very useful for DNA sequence analysis where data sizes are moderate. In this paper we give a novel interpretation of the Peres-Shields estimator as a *sharp transition* phenomenon. This yields a precise and powerful estimator that quickly identifies the core dependencies in data. We show that it compares favorably to other estimators, especially in the presence of noise and/or variable dependencies. Motivated by this last point, we extend the Peres-Shields estimator to Variable Length Markov Chains. We give an application to the problem of detecting DNA sequence similarity using genomic signatures.

Abbreviations: Mk = Fixed order Markov model of order k, PST = Prediction suffix tree, MC = Markov chain, VLMC = Variable length Markov chain.

1 Introduction

Markov chains (MCs) are often used for analysis of biological sequence data, such as DNA [8]. The predominant model is a fixed order MC of discrete state space. The order, k, represents the number of nucleotides (symbols) taken into account when predicting the next state. These models have been applied successfully when identifying transcription factor binding sites [9] and detecting new microbial genes [3].

A fundamental problem in using fixed order MCs is the so called *order estimation* problem, where the unknown order is inferred from observed data. There are many order estimators described in the literature. Some of the better known include the AIC [1] and BIC [19] methods based on maximum likelihood. The consistency of the BIC method was only recently established by a complicated analysis [6]. More recently, Peres and Shields [15] discovered a new method for the order estimation problem and established its consistency by a significantly simpler proof. In this article we present *the first experimental analysis and comparative evaluation of the new method of order estimation due to Peres and Shields.*

R. Casadio and G. Myers (Eds.): WABI 2005, LNBI 3692, pp. 291–302, 2005.

One drawback with the fixed order MCs is that the number of free parameters, the transition probabilities, grows very rapidly as the dependencies get longer. The number of free parameters in a fixed order MC of order k with alphabet Γ is $O(k) := |\Gamma|^k(|\Gamma| - 1)$ More specifically, if the alphabet is DNA, the resulting number of parameters for order $k = 0, 1, 2, 3, 4, 5$ is $O(k) = 3, 12, 48, 192, 768, 3072$ respectively. The amount of data required to estimate these parameters increases rapidly resulting in the "curse of dimensionality": This problem is accentuated in the presence of noise. If the number of data available was unlimited this would not be a problem; however, this is not the case in biology.

The class of fixed order MCs is also structurally poor: to quote Bühlmann *et al.* [4], "There are no models "in between" ... it is impossible to fit a model with, say, 72 parameters. Such a very discontinuous increase in dimensionality of the model does not allow a good trade-off between bias (being low with many parameters) and variance (being low with few parameters)".

Both the "curse of dimensionality" and the structural poverty of fixed order MCs are addressed by allowing the context to vary as in VLMC [17]. These models constitute a powerful tool when transition-matrices are sparse and the memory long in certain directions. VLMCs have been applied successfully in biology, e.g. identification of transcription factor binding sites [20] and classification of protein sequences [2]. The fundamental problem in VLMC, analogous to the order estimation problem, is to estimate (identify) the context describing the transitions. This is captured by the minimal state space description as a *Prediction Suffix Tree* (PST) [17].

We show how to extend the Peres-Shields idea to the problem of estimating the underlying PST of a VLMC. We prove that the resulting estimator eventually *almost surely* identifies the underlying PST. We also demonstrate how to engineer algorithms for implementing the estimator using sharp transition phenomena and give a comparison to the VLMC module in the statistical software package R (http://www.R-project.org).

This paper is organized as follows: In Sect. 2 we describe the fixed order MC and discuss practical issues in terms of using the Peres-Shields estimator for identifying the order. We give an extensive experimental analysis comparing it to other estimators and identify the strength of the new estimator. In particular, we show that it is significantly more robust to noise and that it is more sensitive to mixture of models. In Sect. 4 we describe VLMCs and show how to extend the Peres-Shields idea to identify the underlying PST. In Sect. 5 we give a biological application for detecting sequence similarity using genomic signatures.

2 Fixed Order Markov Chains

2.1 Higher Order Markov Chains

Let Γ be a finite alphabet (for biological applications, $\{A, C, G, T\}$ is a common choice). Let $X := \{X_n\}$ denote a stationary Γ-valued process with distribution $P = P_X$ defined by: $P(a_m^n) := \mathbf{P}(X_m^n = a_m^n)$, for all $m \leq n$ and all sequences

$a_m^n := a_m, a_{m+1}, \cdots, a_n \in \Gamma$. We will also use the standard conditional probability notation, e.g. $P(a_{m+1} \mid a_1^m) := P(a_1^{m+1})/P(a_1^m)$.

The process $X = \{X_n\}$ is said to be *Markov of order* 0 if it is i.i.d. and *Markov of order* m if m is the least positive integer k such that

$$P(a_1 \mid a_{-\infty}^0) = P(a_1 \mid a_{-k+1}^0), \tag{1}$$

for all sequences $a_{-\infty}^0 := a_0, a_{-1}, \cdots \in \Gamma$.

2.2 Peres-Shields Fluctuation Formula and Estimator

Let $N_x(v) := |\{i \in [n] \mid x_i^{i+\ell} = v\}|$, denote the frequency of occurrence of the word v of length $\ell \leq n$ in x_1^n.

Using these frequencies, we can form the *empirical transition probabilities*: for $v \in \Gamma^\ell$ and $a \in \Gamma$,

$$\tilde{p}_{v,a} := \begin{cases} N_x(va)/N_x(v) & \text{if } N_x(v) \neq 0, \\ 0 & \text{otherwise.} \end{cases} \tag{2}$$

The basic observation of Peres and Shields is that if the source is of true order k, then for any $v \in \Gamma^\ell$, the empirical probabilities $\tilde{p}_{v,a}$ should be very close to the empirical probabilities $\tilde{p}_{\tau_k(v),a}$, where $\tau_k(v)$ denotes the k-suffix of v i.e. $\tau_k(v) := v_{\ell-k+1}^\ell$.

To quantify this, introduce the *Peres–Shields Fluctuation* function

$$\Delta_x^k(v) := \max_{a \in \Gamma} \left| N_x(va) - \frac{N_x(\tau_k(v)a)}{N_x(\tau_k(v))} N_x(v) \right|. \tag{3}$$

If the true order is k or less, one expects this fluctuation to be "small", otherwise "large".

Theorem 1 (Peres-Shields 2004). *The estimator*

$$k_{\mathtt{PS}}(x_1^n) := \min\{k \geq 0 \mid \max_{k<|v|<\log\log n} \Delta_x^k(v) < n^{3/4}\} \tag{4}$$

is a consistent Markov order estimator i.e. it is almost surely correct in the limit $n \to \infty$.

3 Experimental Analysis: Fixed Order

3.1 Sharp Transitions

The threshold in (4) is valid in the asymptotic limit. As we discovered, just taken as such, it is not very useful in a typical practical setting. The actual thresholds at finite values of n are different for different order models. So one is forced to the cumbersome resort of finding a different threshold for every k to use in practice.

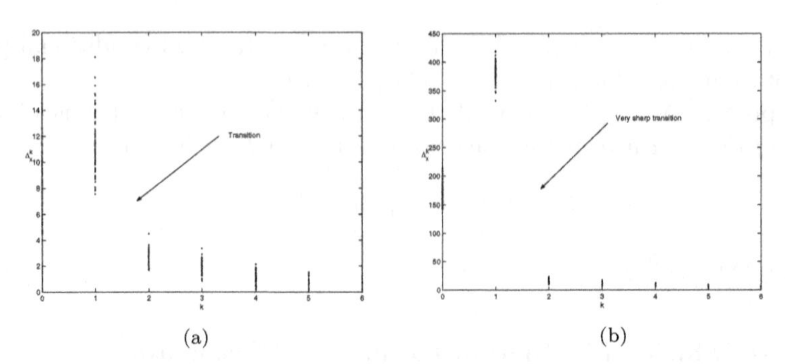

(a) (b)

Fig. 1. The sharp transitions. In, (a) and (b) the Peres-Shields fluctuation. In (a) the simulation is for $n = 300$ nucleotides. The cut-off is already clear, but in (b) the simulation is for $n = 10000$ and the transition is much sharper. 100 simulated DNA sequences were used for each plot using an M1 model.

Even then, it is quite difficult (if not impossible) to come up with a consistent set of thresholds that work together for all k simultaneously.

We avoid the obstacle by making the observation that even though the absolute thresholds are different for different order models (at finite values of n), nevertheless, *all order models display the same distinct trend: there is a sharp transition from a high fluctuation to a low fluctuation value* at the correct order. Figure 1 shows the sharp transition phenomenon used by the Peres-Shields method to pinpoint the true order for two different sizes of data. The rule,

$$\operatorname*{argmax}_{k \geq 0} \frac{\Delta_{\mathrm{x}}^{\mathrm{k}}}{\Delta_{\mathrm{x}}^{\mathrm{k+1}}}. \tag{5}$$

was tested as a simple criterion determining for which k the sharp transition occurs. Using this criterion may be viewed actually as using a new order estimator, albeit one very closely related to the Peres-Shields estimator. The criterion successfully determines the correct order when many nucleotides are present.

3.2 Comparison to Other Estimators

In order to compare the performance of the Peres-Shields estimator (PS) to some of the more well-known estimators we simulated DNA sequences under different MCs according to the following settings: for an order k model, the transition probabilities $p_{u,a}$ are generated independently at random to a specified target $G + C$ content, i.e. a stationary distribution which the MC will converge to as time goes to infinity. The $G + C$ content is known to be an important factor in the genome evolution of bacteria (e.g. [11]). The target value will be denoted X and is a normally distributed stochastic variable of mean $\mu := 0.5$ and standard deviation $\sigma := 0.15$, i.e. $X \sim N(0.5, 0.15)$. The values were found by studying a set of bacteria.

Simulating 1000 artificial sequences from the above schema for $k = 1, 2, 3, 4, 5$ we compared the PS estimator, implemented as in Sect. 3.1, to the following

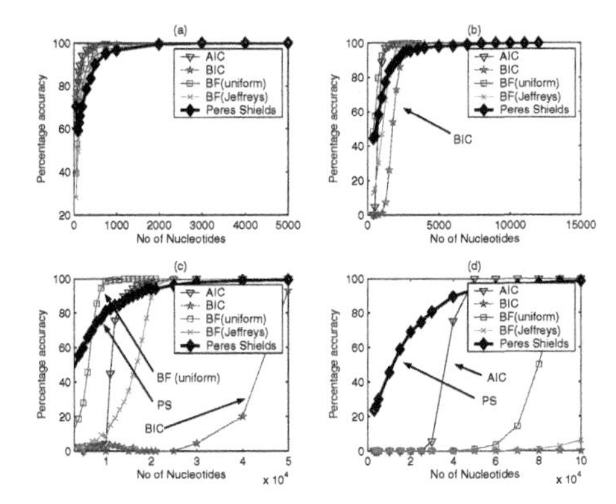

Fig. 2. Efficiency of different estimators. 1000 simulated sequences per sequence length were used. In (a) data was generated using a $k = 1$ model and the percentage of $\hat{k} = 1$ of each estimator is presented on the y-axis. Similar in (b) $k = 3$, (c) $k = 5$. In (d) an M5 model ($k = 5$) was first used and the output passed through a *noisy channel* where each site is randomly mutated into another with $p_{\mathtt{noise}} = 0.4$.

estimators [1,19,6,10]: Akaike Information Criterion (AIC), Bayesian Information Criterion (BIC), Bayes Factor with uniform prior (BF1) and Bayes Factor with Jeffreys prior (BF2). The results indicate that the PS estimator is comparable in terms of efficiency with the others. It performs slightly worse for data under an M1 model, Fig. 2 (a), but performs comparable for larger k. Fig. 2 (b-c). The BIC estimator tend to under-estimate the order as k gets larger for moderate data-sizes. This trend has been observed earlier (eg. [10]).

Knowing that all models are rough approximations to the biological reality, we decided to add a fraction of noise to the sequences produced according to the above models. The generated sequences are passed through a *noisy channel* where each site randomly mutates into another with probability $p_{\mathtt{noise}} > 0$. It appears that the PS estimator is more robust to these changes and when the level of noise reaches higher values than 50% it is *much more robust* than the others. Fig. 2 (d) shows the case of 40% noise of the M5 data. The AIC estimator performs quite well but no other estimator is near the PS.

The critical reader might STOP here and draw the conclusion that this is just yet another estimator that is comparable to the others but not significantly better. *It works better in some conditions but worse in others!* This is a mistake and we shall try to convince him/her otherwise. One of the reasons the estimator falls a bit behind in some cases is that it seeks the longest dependencies in data and not the average. It identifies the maximal dependencies and this is also its strength when exposed to noise. This strength will also be seen in another, perhaps even more important category of data — mixtures of models.

3.3 Mixture of Models

An Mk model can be sparse meaning that many of the transition probabilities are of lower order than k. One example is if $P(A|TT) = P(A|T)$. This is often

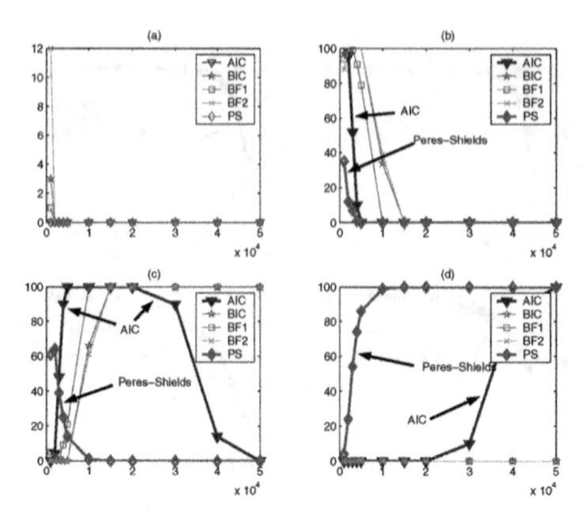

Fig. 3. Percentage of times of estimating models of different order k when using a variable Markov model as underlying model. The model used has the longest dependence of order 4 and hence can only fully be represented with an M4 model in a fixed context. In (a) the number of times each estimator assigns the order to one, $\hat{k} = 1$. In (b) similarly but $\hat{k} = 2$. In (c) $\hat{k} = 3$. In (d) $\hat{k} = 4$, which is the true order.

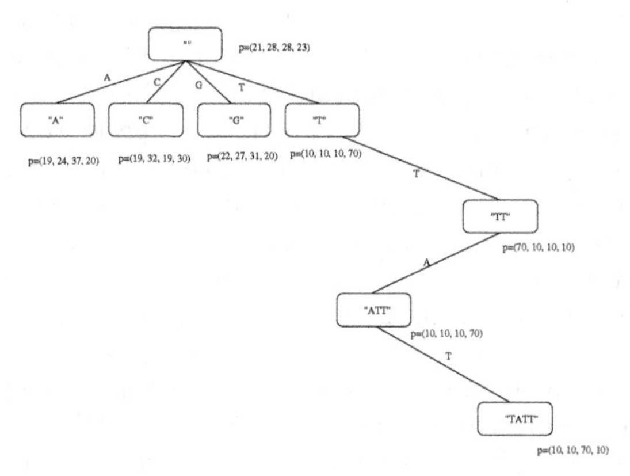

Fig. 4. An example of a PST which can be used for representing a variable length Markov model. The corresponding fixed order Markov model would need to be M4 to model this structure.

the case in biology where genes, repetitive regions and other patterns shape the structure of the chromosomes. *Memory is long in certain directions.* The estimated k of any consistent order estimator should output k equal to the longest dependence when data is unlimited. To investigate this setting, we construct two types of models.

1. Two *nested* models are used simultaneously to generate data. One of lower order and one of higher order. When a new site is to be generated the lower or higher order model is used with probabilities according to $p_{\texttt{lower}} > 0$ and

$1 - p_{\texttt{lower}} > 0$. We use the word *nested* meaning that the lower order model can be derived from the higher order without any additional information.
2. A variable length Markov model (see Sect. 4) corresponding to a non-sparse fixed lower order model (in other words complete in terms of having a distinct set of all transition-probabilities) with an additional single long branch of length k, where k will be the order of any fixed Markov chain to model this dependence (An example of such is seen in Fig. 4).

Using both the above categories of simulated data, we find that the PS estimator is much more efficient at finding longer dependencies in data than all other estimators. A typical trend observed from the first category, when $1 - p_{lower}$ is low, is that most estimators initially select the lower order model. The PS estimator tends to select the higher model at a much lower number of data. In the second category of simulated data, VLMCs, we observe a similar trend. Fig. 3 shows four plots illustrating the percentage of estimating the order 1, 2, 3 and 4 where the true order, or the longest branch in the PST, equals 4. Note how *AIC* picks up the trace gradually. In (b) it believes the order is 2 but abandons this hypothesis as data grows. In (c) it believes the order is 3 and also here abandons the result. Finally in (d) AIC finds the correct order (not shown in plot, all estimators find the correct $k = 4$ order as the size of data grows).

The simulations in this section show that the Peres-Shield estimator is superior for identifying the order of Markov models when the context is variable or comprised of several Markov models: *It is more efficient and identifies the true model at a significantly lower number of data when the source is* **sparse** — as often is the case in biology.

4 Variable Length Markov Chains

The basic idea behind variable length Markov chains (VLMC, eg. [17,4]) is that the memory of the process at any point in time is allowed to depend on the current context i.e. the preceding history. A convenient way to formalize this is by a representation as a finite rooted tree, a so called Prediction Suffix Tree (PST) [17]. The edges of the tree are labeled by letters from the alphabet Γ and each vertex by a string over Γ. The root is labeled by the empty string. Each vertex may have up to $|\Gamma|$ children, each connected by an edge labeled with a distinct letter from Γ. If a vertex v is labeled $\ell(v)$, then a child w connected by an edge with label a is labeled $\ell(w) := \ell(v)a$. A vertex v is called *maximal* if there is a $a \in \Gamma$ and no child of v with label $\ell(v)a$. Finally, with each maximal vertex v, is associated a set of probabilities $P(a \mid \ell(v))$ for all $a \in \Gamma$, i.e. the next-symbol probabilities given the string $\ell(v)$.

Now, the analogue of (1) is as follows: given the current context $a_{-\infty}^{0}$, follow the path from the root taking the edges labeled $a_0, a_{-1}, \cdots$ successively until we arrive at a maximal node. (Note that reading down from the root gives the context when reading *backwards in time*). Then $P(a_1 \mid a_{-\infty}^{0}) = P(a_1 \mid \ell(v))$.

Example 1. Consider the PST in Fig. 4. This is the underlying PST for a VLMC whose transitions are performed as follows. If the current context is "A", "C" or "G", then there are corresponding leaves with the transition probabilities $P(\cdot \mid A)$, $P(\cdot \mid C)$ and $P(\cdot \mid G)$ respectively. If the current context is "AT", then the transition probabilities correspond to the maximal node labeled by "T" and given by by $P(\cdot \mid T)$. If the context is "CATT", the transitions correspond to the maximal node labeled "ATT" and are given by $P(\cdot \mid ATT)$ (Remember that in the current context, we read backwards in time.)

Note that a Markov chain of a fixed order m is a special case where the tree is the complete $|\Gamma|$-ary tree of depth m.

4.1 Extending Peres-Shields to VLMC

The Peres-Shields estimator can be readily extended to estimating the underlying tree of a VLMC from the observed output. In analogy to the fluctuation $\Delta_x^k(v)$ in (3) introduce, for every suffix w of v, the fluctuation

$$\Delta_x(v, w) := \max_{a \in \Gamma} \left| N_x(va) - \frac{N_x(wa)}{N_x(w)} N_x(v) \right| . \tag{6}$$

To infer the tree, we infer equivalently the set of labels of the maximal vertices. A word w will be included as the label of a maximal vertex in the tree if for all extensions v of w of length at most $\log \log n$, $\Delta_x(v, w) < n^{3/4}$, and this is not true for any prefix of w. This is the analogue of (4).

The following theorem extends one of the main results of Peres and Shields [15]. The proof we give is more direct (avoiding the law of the iterated logarithm).

Theorem 2. *The extended Peres-Shields procedure infers the underlying tree of a VLMC with high probability, and eventually almost surely.*

Proof. The proof that almost surely "underestimation" does not occur is the same ergodic argument as in [15]. To show that eventually almost surely "overestimation" does not occur, we must show that for every maximal word in the true PST, $\Delta_x(v, w) < n^{3/4}$ for all extensions v of w of length at most $\log \log n$. Let v be any extension of w including w itself. Consider the function

$$f(x) := N_x(va) - p(a \mid w) N_x(v).$$

This is a Lipschitz function, and moreover,

$$\begin{aligned} \mathbf{E}[f] &= \mathbf{E}[N_x(va) - p(a \mid w) N_x(v)] \\ &= \mathbf{E}[\mathbf{E}[N_x(va) - p(a \mid w) N_x(v) \mid N_x(v)]] \\ &= 0, \end{aligned}$$

since $\mathbf{E}[N_x(va) \mid N_x(v)] = p(a \mid w) N_x(v)$. Thus, by the Method of Bounded Differences [14],

$$\mathbf{P}[|f| > n^{3/4}] < e^{-c\sqrt{n}},$$

for some constant $c > 0$. Applying this successively to w and an arbitrary extension v of w, we conclude that with probability at least $1 - 2e^{-c\sqrt{n}}$,

$$|N_x(wa) - p(a \mid w)N_x(w)| \leq n^{3/4}, \tag{7}$$

and

$$|N_x(va) - p(a \mid w)N_x(v)| \leq n^{3/4}. \tag{8}$$

Combining (7) and (8), with probability at least $1 - 2e^{-c\sqrt{n}}$,

$$N_x(va) - \frac{N_x(wa)}{N_x(w)}N_x(v) \leq n^{3/4}\left(1 - \frac{N_x(v)}{N_x(w)}\right) \leq n^{3/4},$$

and similarly, $N_x(va) - \frac{N_x(wa)}{N_x(w)}N_x(v) \geq -n^{3/4}$. Thus,

$$\mathbf{P}\left[\left|N_x(va) - \frac{N_x(wa)}{N_x(w)}N_x(v)\right| > n^{3/4}\right] < 2e^{-c\sqrt{n}},$$

for some constant c. Since there are at most $o(\log n)$ extensions v of length at most $\log\log n$, we conclude that with high probability, all the fluctuations $\Delta(v, w)$ are small and so "overestimation" does almost surely not occur as $n \to \infty$.

The Peres-Shields estimator for VLMC may be implemented either *top-down* or *bottom-up*:

Top-down. We grow the tree from the root. We start at the root labeled with the empty string. To decide whether to extend a vertex w with label $\ell(w)$, we check for all extensions v of w of length at most $\log\log n$ if the fluctuation $\Delta_x(v, w) > n^{3/4}$. If there is such an extension v, we extend the tree by adding a child of w which is a suffix of v. The advantage of the top-down method is it does not need a prior bound on the depth of the tree.

Bottom-up. If we know a prior bound L on the depth of the tree, we can implement the method bottom up. First we generate the full $|\Gamma|$-ary tree of depth L. Then we *prune* the tree bottom-up. To decide whether to prune an edge connecting a vertex w to a child v, we check if the fluctuation $\Delta_x(v, w) < n^{3/4}$. If so, we delete the edge. This is repeated until no more edges can be deleted.

In practice, the method may be implemented without absolute thresholds, but by identifying a sharp transition. Thus, we study each of the child-branches of a node individually and observe if the next branch is the major transition compared to all possible transitions in the subtree. If it is, this branch is removed, and if not it is considered next — a recursive algorithm. For the purpose of illustration we show the fluctuation Δ calculated in each of the sub-branches of the root in the tree and its children from the example in Fig. 4. Figure 5 shows these transitions and note how the three short branches have a sharp transition at $k = 1$ while the fourth continues all the way to $k = 4$, the correct order.

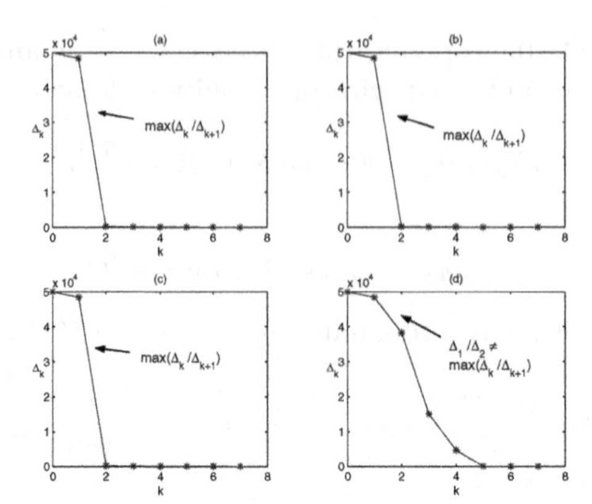

Fig. 5. Shows the first 4 branches from the root in the PST obtained from simulating data under the model in Fig. 4. In (a), the branch towards the letter "A", the proceeding branches should be truncated since we observe the "sharpest" transition in along towards the terminal leaves. In (b) the branch towards "C", also branches proceeding this should be truncated. In (c) similarly towards "G". In (d), branch towards "T" where children must be kept until next step in recursion. (d) also shows that the order of this branch corresponds to $k = 4$.

Table 1. Showing the number of correct and additional nodes identified for different cut-offs using the method implemented in the vlmc package in R

Cut-off	#correct	#additional
2.0	8	26
3.0	8	6
3.1	7	1
3.5	7	1
3.9	6	0

4.2 Comparison to Other Methods for Estimating the Minimal State Space

The algorithm in Sect. 4.1 was implemented and tested on a variety of topologies. The result looks promising and the estimator finds the topologies of a mixture of simulated data. It was also compared to the one implemented in R [4,13,5] that relies on a threshold that is either set manually or by the software. The threshold suggested by R often seems more conservative than necessary and the resulting topology of the PST has less parameters than the true one. To illustrate the problem, data was generated from the model shown in Fig. 4. The Peres-Shields estimator correctly identifies the full model at 400 nucleotides. At the same number of data, threshold=3.9, the method in R does not identify two nodes in the topology, among these, the longest dependency. Table 1 shows the number of correct nodes identified together with the number of additional nodes.

The threshold, in order to get all correct nodes, ought to be somewhere between 3.0 and 3.1.

A more systematic testing of the Peres-Shields method is needed to compare methods of identifying the minimal state-space. However, it is worth noting that a major advantage of our method is that the user *does not* need to specify any cut-off and that it runs in a fully automated fashion.

5 Application: Detecting Sequence Similarity Using Genomic Signatures

The usage of short oligomers, over-lapping DNA words, constitute a genomic signature that has proven to be species-specific (e.g. [12,16,18]), even in sequences as short as 1000 nt. Sandberg *et al.* [18] built a naïve classifier for modeling an arbitrary size of oligomers. A natural extension to this approach is by using Markov models instead. A fixed order Markov chain can model oligomers exactly as their length l naturally translates into using an $l-1$ order model. We have investigated the use of both fixed and variable Markov chains applied to this problem [7]. The result is an improvement in classification. However, the gain in using variable length Markov models compared to fixed order Markov models is heavily dependent on finding a good representation of the state-space, the PST. Ideally, a VLMC of the same number of parameters as a fixed order Markov chain should never be a worse representation. The Peres-Shield estimator will help move a step closer in identifying a more parsimonious set of probabilities.

6 Conclusions

We have given an extensive experimental analysis of the new method for estimating the order of a Markov chain due to Peres and Shields. While comparable in general to other estimators with respect to accuracy, we show that the Peres-Shields estimator is superior to the previous estimators in several respects. First, it is simpler and faster. Second, it is much more robust against noise in the data. Third, it is able to identify variable dependencies very quickly — for example, in mixture models. We showed that the same basic idea can be extended to inferring the underlying PST of a VLMC. The resulting estimator seems very promising.

Acknowledgments

We wish to thank Niklas Eriksen for critically reading the manuscript and Paul Shields for communicating his results directly by email. Thanks to WABI referees for useful feedback.

References

1. Akaike, H.: A new look at the statistical model identification. IEEE Trans. Auto. Cont. **19** (1974) 716–723
2. Bejerano, G., Yona, G.: Variations on probabilistic suffix trees: statistical modeling and prediction of protein families. Bioinformatics, **17(1)** (2001) 23–43
3. Borodovsky, M., McIninch, J.: Recognition of genes in DNA sequence with ambiguities. Biosystems, **30** (1993) 161–171
4. Bühlmann, P., Wyner A.: Variable length Markov chains, Ann. Statist. **27(2)** (1999) 480–513
5. Bühlmann, P., Wyner A.: Model selection for variable length Markov chains and tuning the context algorithm. Annals of the Inst. of Stat. Math. **52(2)** (2000) 287–315
6. Csiszàr, I., Shields, P.: The Consistency of the BIC Markov Order Estimator. The Annals of Statistics. **28(6)** (2000) 1601–1619
7. D. Dalevi and D. Dubhashi.: Bayesian Classifiers for Detecting HGT Using Fixed and Variable Length Markov Chains. submitted.
8. R. Durbin, S. Eddy, A. Krogh, G. Mitchison.: Biological Sequence Analysis. Cambridge University Press (2004)
9. Ellrott, K., Yang, C., Saldek, M. , Jiang, T.: Identifying transcription binding sites through Markov chain optimization. Bioinformatics **18(2)** (2002) 100–109
10. Fan, T-H., Tsai, C.: A Bayesian Method in Determining the Order of a Finite State Markov Chain. Comm. Statist. Theory and Methods **28(7)** (1999) 1711–1730.
11. Forsdyke, D.: Different Biological Species "Broadcast" Their DNAs at Different (G+C)% "Wavelengths". J. Theor. Biol. **178** (1996) 405–417.
12. Karlin, S., Burge, C.: Dinucleotide relative abundance extremes: a genomic signature. Trends Genet **11(7)** (1995) 283–90
13. Mächler, M., Bühlmann, P.: Variable Length Markov Chains: Methodology, Computing, and Software. J Comp Graph Stat, **13(2)** (2004) 435–455
14. McDiarmid, C.: Concentration. Probabilistic Methods for Algorithmic Discrete Mathematics Series: Algorithms and Combinatorics, Vol. 16, Habib, M.; McDiarmid, C.; Ramirez-Alfonsin, J.; Reed, B. (Eds.) (1998), 195–248, Springer, Berlin.
15. Peres, Y., Shields, P.: Two New Markov Order Estimators, to appear, see `http://www.math.utoledo.edu/~pshields/latex.html`.
16. Pride, D., Meinersmann, R., Wassenaar, T., Blaser, M.: Evolutionary implications of microbial genome tetranucleotide frequency biases Genome Res. **13** (2003) 145–158
17. Ron, D., Singer, Y., Tishby, N.: The Power of Amnesia: Learning Probabilistic Automata with Variable Memory Length. Machine Learning. **25(2-3)** (1996) 117–149
18. Sandberg, R., Winberg, G., Branden, CI., Kaske, A., Ernberg, I., Coster, J.: Capturing whole-genome characteristics in short sequences using a naïve Bayesian classifier. Genome Res. **11(8)** (2001) 1404–1409
19. Schwartz, G. Estimating the dimension of a model, Annals of Statistics **6** (1978) 461–464
20. Zhao, X., Huang, H., Speed, T.: Finding Short DNA motifs using Permuted Markov models. RECOMB (2004): 68–75

Multiple Structural RNA Alignment with Lagrangian Relaxation
Extended Abstract

Markus Bauer[1,2], Gunnar W. Klau[3], and Knut Reinert[1]

[1] Institute of Computer Science, Free University of Berlin, Germany
[2] International Max Planck Research School for Computational Biology and Scientific Computing
[3] Institute of Mathematics, Free University of Berlin, Germany

Abstract. Many classes of functionally related RNA molecules show a rather weak sequence conservation but instead a fairly well conserved secondary structure. Hence, it is clear that any method that relates RNA sequences in form of (multiple) alignments should take structural features into account. Since multiple alignments are of great importance for subsequent data analysis, research in improving the speed and accuracy of such alignments benefits many other analysis problems.

We present a formulation for computing provably *optimal*, structure-based, multiple RNA alignments and give an algorithm that finds such an optimal (or near-optimal) solution. To solve the resulting computational problem we propose an algorithm based on Lagrangian relaxation which already proved successful in the two-sequence case. We compare our implementation, `mLARA`, to three programs (`clustalW`, `MARNA`, and `pmmulti`) and demonstrate that we can often compute multiple alignments with consensus structures that have a significant lower minimum free energy term than computed by the other programs. Our prototypical experiments show that our new algorithm is competitive and, in contrast to other methods, is applicable to long sequences where standard dynamic programming approaches must fail. Furthermore, the Lagrangian method is capable of handling arbitrary pseudoknot structures.

1 Introduction

Similarity searches based on primary sequence or the detection of structural features using multiple alignments are usually the first steps in analyzing the sequences of biomolecules. Unfortunately, many functional classes of RNA show little sequence conservation, but rather a conserved secondary structure which is formed by folding in space and forming hydrogen bonds between its bases. Among such RNAs are tRNA, rRNA, snoRNAs, and SRP RNA [10].

Hence, algorithms to compute (multiple) alignments ought to take not only the sequence, but also the secondary structure into account. Washietl and Hofacker [18] back up this consideration by showing that sequence based alignments

R. Casadio and G. Myers (Eds.): WABI 2005, LNBI 3692, pp. 303–314, 2005.

are significantly worse than sequence-structure based alignments if their pairwise sequence identity sinks below $\approx 60\%$. Thus, the problem of producing RNA alignments that find a common structure has become the bottleneck in the computational study of functional RNAs. To date, the available tools for computing structural alignments are either based on heuristic approaches and thus produce suboptimal alignments or cannot attack instances of reasonable input size. In this paper we deal with the computation of a *multiple* RNA sequence-structure alignment, given a number of RNA sequences together with their secondary structure. Our formulation and algorithms are able to deal with pseudoknots, although their presence makes the problem algorithmically harder—the problem becomes NP-hard even when only two sequences have to be aligned; Evans gives an NP-hardness proof for a special case of this problem [6].

Previous Work. The computational problem of considering sequence and structure of an RNA molecule simultaneously was first addressed by Sankoff [16] who proposed a dynamic programming algorithm that simultaneously aligns and folds a set of RNA sequences. Bafna *et al.* [2] improved the dynamic programming algorithm to a running time of $O(n^4)$ which still does not make it applicable to many instances of realistic size. Common motifs among several sequences are searched by Waterman [19]. Eddy and Durbin [5] describe probabilistic models for measuring the secondary structure and primary sequence consensus of RNA sequence families. They present algorithms for analyzing and comparing RNA sequences as well as database search techniques. Since the basic operation in their approach is an expensive dynamic programming algorithm, their algorithms cannot analyze sequences longer than 150-200 nucleotides. Gorodkin *et al.* [8], Mathews and Turner [14], and Hofacker *et al.* [10] published banded versions of Sankoff's original algorithm. We will make use of the proposed similarity function for two RNA sequences with structures proposed in [10] which uses the base pair probability matrices to search for a common structure of maximal weight.

This function can be directly used to weight edges in the structural alignment graph introduced in Lenhof *et al.* [13] where the authors presented a branch-and-cut algorithm for structurally aligning two RNA sequences. The underlying formulation is flexible and allows for pseudoknots. Previous work on contact map alignment in the area of proteomics by Caprara and Lancia [4] and for the two-sequence case of the sequence-structure alignment problem by Bauer and Klau [3] indicates, however, that Lagrangian relaxation is better suited to obtain good solutions to this ILP than a direct branch-and-cut approach in terms of running time.

Contribution. We extend the formulation of Lenhof *et al.* [13] to the case of multiple sequences and show how to solve it efficiently using Lagrangian relaxation. While progressive approaches can only approximate the usual sum-of-pair score for multiple alignments, we can compute for the first time a solution of a true sum-of-pair multiple sequence-structure alignment.

We tested a first version of our algorithm with a dataset used in a recent study [7] on different alignment programs for functional RNAs. We compared our results to those of other sequence-structure programs by comparing the *minimum free energy* values of the consensus secondary structure as given by the multiple alignment.

Our experiments show that computing the best multiple RNA alignment is worthwhile, since the obtained alignments are often better than the ones obtained by heuristic approaches. Furthermore, our approach can deal with arbitrary pseudoknots without additional costs and with long sequences where standard dynamic programming approaches fail due to their space requirements.

2 Approach

We first describe the graph-theoretical model we use, which is based on the description in [3] and [13]. Then, we present an integer linear programming formulation for this model and devise a solution approach based on Lagrangian relaxation.

Graph-Theoretical Model for Structural RNA Alignment. Let S be a sequence $s_1, \ldots, s_n$ of length n over the alphabet $\Sigma = \{A, C, G, U\}$. A paired base (i, j) is called an *interaction* if (i, j) forms a Watson-Crick-pair. The set P of interactions is called the *annotation* of sequence S. Two interactions are said to be in *conflict*, if they share one base; they form a *pseudoknot* if they cross each other. A pair (S, P) is called an *annotated sequence*. Note that a structure where no pair of interactions is in conflict with each other forms a valid secondary structure of an RNA sequence, possibly with pseudoknots.

We are given a set of k annotated sequences $\{(S_1, P_1), \ldots, (S_k, P_k)\}$ and model the input as a mixed graph $G = (V, L \cup I \cup A)$. The set V denotes the vertices of the graph, in this case the bases of the sequences, and we write v_j^i for the jth base of the ith sequence. The set L contains undirected *alignment edges* between vertices of two different input sequences (for sake of better distinction called *lines*) whereas the set I codes the annotation of the sequence by means of *interaction edges* between vertices of the same sequence. In addition to the undirected edges the graph has directed arcs A representing *consecutivity* of characters within the same string that run from each vertex to its "right" neighbor, *i.e.*, $A = \{(v_j^i, v_{j+1}^i) : 1 \leq i \leq k, 1 \leq j < |S_i|\}$. A *path* in a mixed graph is an alternating sequence $v_1, e_1, v_2, e_2, \ldots$ of vertices v_i and lines or edges $e_i \in L \cup A$. It is a *mixed path* if it contains at least one arc in A and one line in L. A mixed path is called a *mixed cycle* if the start and end vertex are the same. A mixed cycle represents an ordering conflict of the letters in the sequences. In the two-sequence case a mixed cycle represents lines crossing each other. A subset $\mathcal{L} \subset L$ corresponds to an *alignment* of the sequences $S_1, \ldots S_k$ if $\mathcal{L} \cup A$ does not contain a mixed cycle. In this case, we use the term alignment also for $\mathcal{L}$.

Two interaction edges $(i_1, i_2) \in P_i$ and $(j_1, j_2) \in P_j$ are said to be *realized* by an alignment $\mathcal{L}$ if and only if $\mathcal{L}$ contains the alignment edges $l = (i_1, j_1)$ and $m = (i_2, j_2)$. The pair (l, m) is called an *interaction match.* Note that we define (l, m) as an ordered tuple, that is, (l, m) is distinct from (m, l). Figure 1 illustrates the above definitions by means of an example.

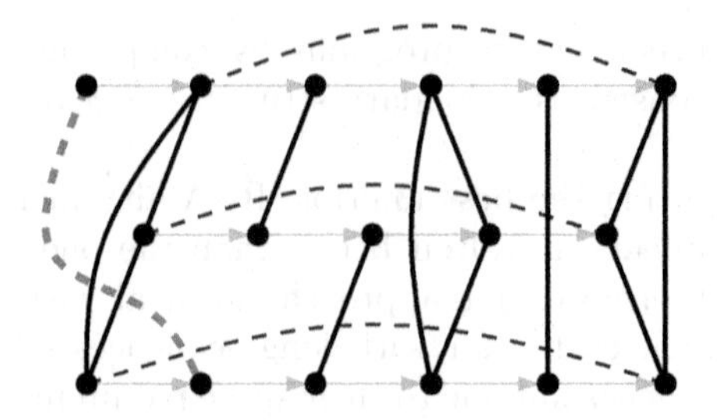

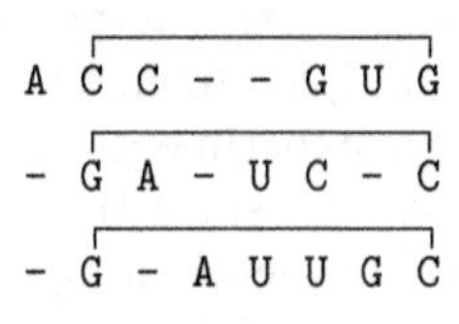

Fig. 1. Graph-theoretic concept of alignment. The right side shows a structural alignment of three annotated sequences, the left side the corresponding graph. Thicker lines represent alignment edges in $\mathcal{L}$, adding the grey, dotted line creates a mixed cycle. Lines $\mathcal{L}$ realize six interaction matches (mind that interaction matches are ordered tuples).

We assign positive weights w_l and w_{ij} to each line l and each interaction match (i, j), respectively, that represent the benefit of realizing the line or the match. The weights are given, for example, by mutation score matrices or— in the case of interaction matches— by the base pair probability computed by McCaskill's algorithm [15]. Note that since each interaction edge occurs in two interaction matches (m, l) and (l, m) we divide the weight of these edges by two.

Approaches for traditional sequence alignment aim at maximizing the score of edges in an alignment $\mathcal{L}$. Structural alignments, however, must also take the structural information encoded in the interaction edges into account.The problem of structurally aligning a set of annotated sequences $\{(S_1, P_1), \ldots, (S_k, P_k)\}$ calls for an alignment such that the weight of the lines plus the weight of the realized interaction matches is maximal. More formally, we seek to maximize $\sum_{l \in \mathcal{L}} w_l + \sum_{(i,j) \in \mathcal{I}} w_{ij}$ where $\mathcal{L}$ is an alignment and $\mathcal{I}$ contains the interaction matches realized by $\mathcal{L}$. In graph-theoretic terms, this corresponds to finding a maximally weighted subset of lines and interaction edges in the input graph such that no mixed cycles occur, each interaction match has to be realized, and no vertex is incident to more than one interaction edge for each pair of sequences.

Integer Linear Programming Formulation. The graph-theoretic model lets us very conveniently state the following integer linear program (ILP):

$$\max \sum_{l \in L} w_l x_l + \sum_{l \in L} \sum_{m \in L} w_{lm} y_{lm} =: z \tag{1}$$

$$\text{s.t.} \sum_{l \in C} x_l \leq |C \cap L| - 1 \qquad \forall \text{ mixed cycles } C \tag{2}$$

$$y_{lm} = y_{ml} \qquad \forall l, m \in L, l < m \tag{3}$$

$$\sum_{m \in L} y_{lm} \leq x_l \qquad \forall l \in L \tag{4}$$

$$0 \leq x \leq 1, \quad 0 \leq y \leq 1 \qquad \text{integer} \tag{5}$$

The variable x_l equals one, if line l is part of the alignment, whereas $y_{lm} = 1$ holds, if lines l and m realize the interaction match (l, m). One can easily verify that all properties for a multiple structural alignment are satisfied: (3) and (4) guarantee that interaction matches are realized by lines and that every vertex is incident to at most one interaction edge, whereas (2) ensures that the selection of lines forms a multiple alignment. The order $l < m$ within the equality constraints (3) denotes an arbitrary order defined on the elements of A.

Lagrangian Relaxation. Following the Lagrangian optimization method (see, *e.g.*, [20]), we drop the constraints that complicate the original problem and incorporate them into the objective function with a penalty term for their violation. We obtain the *relaxed problem* by dropping constraint (3). Although it is still NP-hard, medium instance sizes can be solved to optimality in reasonable computation time with a sophisticated branch-and-cut approach as proposed by Althaus *et al.* [1].

Lemma 1. *The relaxed problem is equivalent to the general multiple sequence alignment problem.*

Proof. We distinguish two cases, depending on whether a line l is part of an alignment or not. First, assume $x_l = 0$. In this case, due to (4), all y_{lm} must be zero as well, and the contribution of line l to the objective function is zero. If, however, a line is part of an alignment, its maximal contribution to the score is given by solving

$$p_l := \max \; w_l + \sum_{m \in L} w_{lm} y_{lm} \tag{6}$$

$$\text{s.t.} \; \sum_{m \in L} y_{lm} \leq 1 \tag{7}$$

$$\sum y_{lm} \leq 0 \qquad \forall m \in L \text{ crossing } l \tag{8}$$

$$0 \leq x \leq 1, \quad 0 \leq y \leq 1 \qquad \text{integer} \tag{9}$$

Inequality (7) states that only one interaction match can be chosen. According to the objective function (6) it is clear that this will be the one with the largest weight w_{lm}. Inequality (8) constrains this choice by excluding interaction matches with lines m that are in conflict to l. This ILP is easily solvable by just selecting the most profitable interaction match $(l, \hat{m})$ such that l and $\hat{m}$ do not cross each other, which can be done in linear time. Thus, the profit p_l a line l can realize is given by its own weight w_l plus the weight $w_{l\hat{m}}$ of such an interaction match. In the second step, we compute the optimal overall profit by solving the multiple sequence alignment problem

$$\max \sum_{l \in L} p_l x_l$$

$$\text{s.t.} \sum_{l \in C} x_l \leq |C \cap L| - 1 \qquad \forall \text{ mixed cycles } C$$

$$0 \leq x \leq 1 \qquad \text{integer}$$

Let x^* be the solution of this problem. We claim that an optimal solution of the relaxed problem is given by (x^*, y^*) with $y^*_{lm} = x^*_m y_{l\hat{m}}$. First, it is easy to see that (x^*, y^*) is indeed a feasible solution of the relaxed problem, since x^* represents an alignment and our choice of y^* does not violate the restrictions given in (4). To see that (x^*, y^*) is optimal, observe that its value is

$$\sum_{l \in \mathcal{L}} p_l x^*_l = \sum_{l \in \mathcal{L}} (w_l + w_{l\hat{m}}) x^*_l = \sum_{l \in L} x^*_l w_l + \sum_{l \in L} \sum_{m \in L} w_{lm} y^*_{lm} \ .$$

Thus, the optimal solution x^* of the sequence-based multiple alignment yields an optimal solution (x^*, y^*) of the relaxed problem. □

Having demonstrated how to formulate the relaxed problem as a pure sequence-based multiple alignment problem we now describe the Lagrangian method. Formally, we introduce appropriate Lagrangian multipliers λ^i with $\lambda^i_{ml} = -\lambda^i_{lm}$ for $l < m$ and with $\lambda^i_{ll} = 0$ and define the Lagrangian problem as

$$\begin{aligned} \max \ & \sum_{l \in L} w_l x_l + \sum_{l \in L} \sum_{m \in L} (\lambda^i_{lm} + w_{lm}) y_{lm} := \bar{z}^i \\ \text{s.t.} \ & \sum_{l \in C} x_l \le |C \cap L| - 1 && \forall \text{ mixed cycles } C \\ & \sum_{m \in L} y_{lm} \le x_l && \forall l \in L \\ & 0 \le x \le 1, \quad 0 \le y \le 1 && \text{integer} \end{aligned}$$

Note that, according to Lemma 1, we can solve instances of the Lagrangian problem by solving a multiple sequence alignment problem where the profits of the interaction matches are coded in the weights of the lines.

The task is now to find Lagrangian multipliers that provide the best bound to the original problem. We do this by employing iterative subgradient optimization as proposed by Held and Karp [9]. This method determines the multipliers of the current iteration by adapting the values from the previous iteration. More formally, we set $\lambda^0_{lm} = 0, \forall m, l \in L$ and

$$\lambda^{i+1}_{lm} = \begin{cases} \lambda^i_{lm} & \text{if } s^i_{lm} = 0 \\ \max(\lambda^i_{lm} - \gamma_i, -w_{lm}) & \text{if } s^i_{lm} = 1 \\ \min(\lambda^i_{lm} + \gamma_i, w_{lm}) & \text{if } s^i_{lm} = -1 \end{cases}$$

$$\text{where} \qquad s^i_{lm} = y^*_{lm} - y^*_{ml} \quad \text{and} \quad \gamma_i = \mu \frac{z_U - z_L}{\sum_{l,m \in L} (s^i_{lm})^2} \ .$$

Here, μ is a common adaption parameter and z_U and z_L denote the best upper and lower bounds, respectively. The closer these bounds are to z (the optimal value of a multiple structural alignment), the faster the method will converge.

Clearly, we can set z_U to $\min\{\bar{z}^j \mid 0 \le j \le i\}$, the lowest objective function value of the Lagrangian problems solved so far. To obtain a high lower bound is

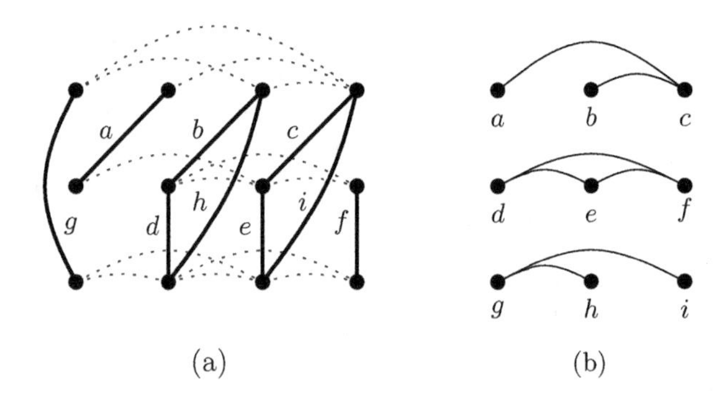

Fig. 2. (a) Multiple alignment (solution of the Lagrangian problem) and (b) interaction matching graph

more involved and we show in the following how to use the information computed in the Lagrangian problem in order to deduce a good feasible solution.

A solution of the Lagrangian problem is a multiple alignment $\mathcal{L}$ plus some information about interaction matches coded by the y-values; see Figure 2 (a). If for all lines l and m the equation $y_{lm} = y_{ml}$ holds, then the solution is a feasible multiple structural alignment, and we have found the optimal solution to the original problem. Otherwise, some pairs y_{lm} and y_{ml} contradict each other. The idea is to select a subset of interaction edges of maximum weight such that the structural information for each sequence is valid, that is, each base is paired with at most one other base (the *structural completion* of $\mathcal{L}$).

We can formulate this problem as a *general weighted matching problem* in an auxiliary graph, the *interaction matching graph*. Consider the edges of $\mathcal{L}$ as vertices and every pair of interaction edges (i_1, i_2) whose endpoints are adjacent to a pair $(l, m) \in \mathcal{L} \times \mathcal{L}$ as the edges of the graph (see Figure 2 (b)).

Lemma 2. *A matching of maximum weight in the interaction matching graph corresponds to the best structural completion of the given alignment $\mathcal{L}$.*

Proof. The equivalence follows directly from the construction of the interaction matching graph and the definition of a matching. □

3 Computational Results

Assessing the Quality. Our tool `mLARA` is a prototypical C++ implementation of the Lagrangian approach. The evaluation of our results is as follows:

- We compare the best lower and upper bound for the Lagrangian relaxed problem. If these values coincide, we have found a provably optimal solution, otherwise, the ratio between lower and upper bound gives a measure for the quality of the best solution found.
- In order to compare our results to other structural alignment programs, we evaluate the computed multiple sequence alignments using `RNAalifold` [11].

In short, given a multiple sequence alignment, this tool computes a consensus folding that does not only take the well-studied *minimum free energy* model (MFE) into account, but also the sequence covariation as given by the multiple alignment. Thus, it incorporates the phylogenetic information contained in the alignment. Lower values of this consensus folding correspond, generally spoken, to more stable structures.

We ran `mLARA` on an AMD server with a 2Ghz Opteron processor, allowing a maximal running time of three hours per computation: If the computation was not finished within this limit, we stopped, taking the best alignment found so far as the final one.

Generating the Input Graph. For generating the set of alignment edges we apply a more sophisticated approach than the one reported in [13]: instead of computing a conventional sequence alignment with affine gap costs and subsequently inserting all alignment edges realized by any suboptimal alignment scoring better than a fixed threshold s below the optimal score, we use a *sliding window technique*—as described in [12]—that adjusts the suboptimality threshold s according to the local quality of the alignment. More precisely, for every nucleotide we compute a *confidence value* evaluating the quality of the local alignment within a certain window. In regions of the sequence where the quality of the conventional sequence alignment appears to be very good, none or only a small number of suboptimal alignment edges are considered. In alignment regions that show little sequence conservation, more alignment edges are generated.

In our experiments we start from a conventional sequence alignment with affine gap costs (gap open and gap extension penalty are set to 6 and 2, the score for a (mis-)match is set to 2 and 1, respectively) and insert alignment edges according to the local quality of the alignment. To count the sequence specific parts less, we use an approach proposed by Hofacker *et al.* [10], namely to multiply the scores for matches and mismatches with a constant factor τ to gain the actual weights of the alignment edges. Convenient values for τ are, for instance, 0.05 (as proposed in [10]) or 0.5 in case of sequences with a high sequence identity.

For generating a set of reasonable interaction edges we resort to the same technique that was already successfully employed in [10]: we compute the base pair probability matrices for each RNA sequence and assign the value $\varphi_{ij} = \log_{10}(\phi_{ij}/\phi_{\min})$ as the weight to the interaction edge between nucleotide i and j (with ϕ_{ij} being the actual base pair probability for (i, j) and $\phi_{\min}$ being the smallest probability found).

It should be noted, however, that the underlying graph-theoretical model gives absolutely no restriction on scoring the interaction and alignment edges: even position-dependent scoring matrices are possible and do not enforce any modification to the approach itself.

Computing an Optimal Multiple Sequence Alignment. In each iteration of our Lagrangian approach we have to compute an optimal multiple sequence alignment. We use the implementation by Althaus *et al.* [1] for this task. Their branch-and-

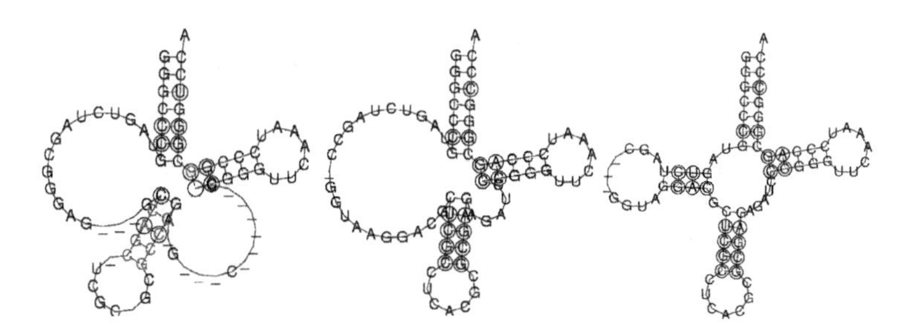

Fig. 3. Consensus secondary structure of three tRNA sequences based on alignments computed by MARNA, pmmulti, and mLARA. Graphics are generated by RNAalifold.

cut approach is based on a similar graph-theoretical model as our algorithm and computes an optimal multiple alignment with arbitrary gap costs and arbitrary scores assigned to the alignment edges.

Experimental Evaluation. The following evaluation should not be understood as a comprehensive study of the performance of different RNA structure alignment programs, but rather serves as an illustration that convincingly demonstrates the advantages of truly optimal multiple structural alignments. We compared mLARA to the latest versions of clustalW, MARNA [17], and pmmulti [10].

As a first example of small and medium–sized instances, we took instances of three randomly chosen tRNA and 5S RNA sequences from BRAliBase [7] and compared the MFE-values of the consensus structure as computed by RNAalifold. Table 1 shows the sequence identity of the instances where mLARA performed well compared to pmmulti (or clustalW for longer sequences) and we give the MFE values of clustalW, MARNA, and pmmulti for comparing the results. As one can see, the results back up the observation given in [18] that for sequences with a sequence identity greater than $\approx 65\%$ pure sequence-based approaches yield similar results to those of sequence-structure based ones.

Figure 3 illustrates the corresponding consensus secondary structures of three tRNA sequences (GenBank: X06054, AE009773, and AE006699), based on the alignments computed by MARNA, pmmulti, and mLARA. For this example, the alignment computed by mLARA is the only one that yields the typical tRNA clover-leaf structure. The structure is supported by many compensatory mutations (indicated through circles around the nucleotides) in each of the four stems.

Two things are quite remarkable: First, the mLARA consensus secondary structure is based on a set of ≈ 1400 candidate lines. This small number of edges is nevertheless sufficient to reconstruct the correct structure. Secondly, there is a substantial gap between the best lower and upper bound, that is, it is not guaranteed that the solution found by the Lagrangian method is actually the optimal one. Nevertheless we can guarantee that it is at most 12% worse than the optimal solution. Again, although the optimal solution was not found, the typical tRNA clover-leaf structure was reconstructed.

Finally, we took longer SRP sequences from [7] and different ncRNA families from the *Rfam* database. Due to the average sequence length of $\approx$ 300-320

Table 1. Results of clustalW, MARNA, pmmulti, and mLARA on tRNA (upper part), 5S (middle part), and SRP (lower part) sequences. Level of optimality given in brackets, lowest RNAalifold value given in bold face.

SeqID	Instance	clustalW	MARNA	pmmulti	mLARA
0.30	aln22	−20.69	−29.47	**−34.64**	−33.97 (0.98)
0.30	aln30	−10.62	−12.79	**−21.48**	−20.18 (0.66)
0.65	aln42	−38.51	−38.51	−38.51	**−38.81** (0.97)
0.66	aln49	−31.18	−25.16	**−39.27**	**−39.27** (0.99)
0.67	aln38	−29.29	−26.72	**−34.60**	**−34.60** (0.99)
0.70	aln39	−33.46	−12.57	−33.46	**−33.66** (0.97)
0.71	aln40	**−35.84**	−24.85	**−35.84**	**−35.84** (0.96)
0.71	aln37	−24.00	−24.34	**−38.78**	−38.42 (0.94)
0.73	aln33	**−35.83**	**−35.83**	**−35.83**	−35.40 (0.98)
0.79	aln34	**−35.09**	−27.38	**−35.09**	−34.86 (0.98)
0.87	aln36	**−32.39**	−25.36	**−32.39**	**−32.39** (0.99)
0.38	aln84	−17.45	−29.63	−38.02	**−38.75** (0.99)
0.38	aln87	−32.86	−41.64	−42.54	**−43.11** (0.98)
0.44	aln62	−36.63	−20.38	**−42.62**	**−42.62** (0.94)
0.46	aln76	−42.64	−26.10	−45.99	**−48.23** (0.90)
0.52	aln30	−47.54	−50.76	−51.80	**−52.16** (1.00)
0.56	aln72	**−28.73**	−20.22	**−28.73**	−28.61 (0.95)
0.71	aln36	**−36.70**	−28.19	−35.01	−35.54 (0.99)
0.76	aln44	−47.06	−45.21	−47.06	**−47.58** (0.99)
0.79	aln58	−43.72	−46.48	**−46.72**	**−46.72** (1.00)
0.87	aln45	−36.55	**−40.21**	**−40.21**	**−40.21** (0.99)
0.43	RF00012	−28.88	−20.00	—	**−39.80** (0.97)
0.53	aln7	−87.15	−75.57	—	**−89.10** (0.87)
0.54	aln20	−86.65	**−97.60**	—	−94.46 (0.93)
0.54	aln31	−83.45	−47.69	—	**−87.10** (0.99)
0.59	aln37	−78.86	−72.05	—	**−82.92** (0.97)
0.59	aln10	−90.02	−77.21	—	**−100.63** (0.99)
0.60	aln23	−93.98	−94.07	—	**−99.57** (0.99)
0.61	RF00017	−75.25	−61.80	—	**−83.22** (0.90)
0.62	RF00229	−51.73	−41.55	—	**−53.26** (0.93)
0.75	aln12	−93.44	−72.38	—	**−106.74** (0.99)
0.78	RF00012	−42.42	−35.63	—	**−42.54** (0.98)

nucleotides, the exact progressive approach pmmulti fails due to its high computational demands (see Table 1).

4 Discussion

It has to be remarked that in a considerable number of instances mLARA did not compute an alignment that yields an MFE value competitive to the one computed by pmmulti. There are three main reasons for that:

1. The number of alignment edges is not sufficient to compute an alignment that yields the smallest MFE value possible, *e.g.*, computing an alignment

of three 5S RNA sequences (GenBank: AJ131595, X06578, X02627) with 1025 alignment edges yields a consensus MFE value of −24.75, computing an alignment with 1602 edges already yields a value of −31.77 (the MFE value of the progressive approach `pmmulti` is actually −37.45). We expect a substantial improvement in terms of speed and quality by combining sets of variables that correspond to local structural motifs in combination with a preprocessing and pricing strategy.

2. A higher number of alignment edges leads to much higher computational costs for computing the exact multiple sequence alignments, which makes each iteration very expensive. In order to decrease the number of iterations, we plan to switch from *subgradient optimization* to the *bundle method*.
3. Finally, the underlying scoring scheme has to be revisited: during our experiments we observed that a higher score of an alignment does not necessarily yield a lower MFE value, *e.g.*, aligning three 5S RNA sequences (GenBank: X01501, X02260, M58416) yields a score of 1447.13 with a corresponding MFE value of −28.61 after three hours. Allowing 5.5 hours of computation time, the score increases to 1450.52, whereas the MFE value drops to −27.63. Therefore, incorporating statistically significant values for scoring RNA nucleotides would be preferable to the scores that are used in the current version of the prototype.

The evaluation shows that `mLARA` is often competitive to or better than `pmmulti` and almost always better than `MARNA`. In most cases, however, `pmmulti` achieves the lowest MFE values and, even when it is worse compared to `mLARA`, it is only in a small amount. Therefore, `pmmulti` can currently be seen as the tool of choice for sequences up to 150 nucleotides. `mLARA`, however, can compute alignments of good quality for sequences longer than 150 nucleotides, where the time and space demands of `pmmulti` are prohibitive. The results obtained by `mLARA` are still better than those of `MARNA` for longer sequences.

As pointed out in the introduction, the Lagrangian method does not always yield an optimal solution. We therefore plan to embed our approach in a branch-and-bound algorithm, such that we can always compute an optimal solution to our ILP formulation. This will allow us to assess whether it is worth to solve the problem to complete optimality. Simultaneously, we are incorporating the two-sequence version of the Lagrangian approach into a progressive alignment framework: we hope to combine the advantages of both approaches—computing multiple alignments based on pairwise structural alignments—to obtain a tool that yields excellent results for sequences up to 1000 nucleotides.

Acknowledgments. This work was supported by the DFG Research Center MATHEON "Mathematics for Key Technologies" in Berlin, the German Federal Ministry of Education and Research, (grant no. 0312705A 'Berlin Center for Genome Based Bioinformatics'), and the IMPRS for Computational Biology and Scientific Computing. The authors thank Ernst Althaus for help with incorporating the code for exact multiple alignments into the prototype of `mLARA` and Ivo Hofacker for helpful discussions.

References

1. E. Althaus, A. Caprara, H.-P. Lenhof, and R. K. Multiple sequence alignment with arbitrary gap costs: Computing an optimal solution using polyhedral combinatorics. *Bioinformatics*, 18(90002):S4–S16, 2002.
2. V. Bafna, S. Muthukrishnan, and R. Ravi. Computing similarity between RNA strings. In *Proc. of CPM'95*, number 937 in LNCS, pages 1–16. Springer, 1995.
3. M. Bauer and G. W. Klau. Structural Alignment of Two RNA Sequences with Lagrangian Relaxation. In *Proc. of ISAAC'04*, number 3341 in LNCS, pages 113–123. Springer, 2004.
4. A. Caprara and G. Lancia. Structural Alignment of Large-Size Proteins via Lagrangian Relaxation. In *Proc. of RECOMB'02*, pages 100–108. ACM Press, 2002.
5. S. P. Eddy and R. Durbin. RNA sequence analysis using covariance models. *Nucl. Acids Research*, 22(11):2079–2088, 1994.
6. P. Evans. Finding common subsequences with arcs and pseudoknots. In *Proc. of CPM'99*, number 1645 in LNCS, pages 270–280. Springer, 1999.
7. P. Gardner, A. Wilm, and S. Washietl. A benchmark of multiple sequence alignment programs upon structural RNAs. *Nucl. Acids Res.*, 33(8):2433–2439, 2005.
8. J. Gorodkin, L. J. Heyer, and G. D. Stormo. Finding the most significant common sequence and structure motifs in a set of RNA sequences. *Nucl. Acids Res.*, 25:3724–3732, 1997.
9. M. Held and R. Karp. The traveling-salesman problem and minimum spanning trees: Part II. *Mathematical Programming*, 1:6–25, 1971.
10. I. L. Hofacker, S. H. F. Bernhart, and P. F. Stadler. Alignment of RNA base pairing probability matrices. *Bioinformatics*, 20:2222–2227, 2004.
11. I. L. Hofacker, M. Fekete, and P. F. Stadler. Secondary structure prediction for aligned RNA sequences. *J. Mol. Biol.*, 319:1059–1066, 2002.
12. J. Kececioglu, H.-P. Lenhof, K. Mehlhorn, P. Mutzel, K. Reinert, and M. Vingron. A polyhedral approach to sequence alignment problems. *Discrete Applied Mathematics*, 104:143–186, 2000.
13. H.-P. Lenhof, K. Reinert, and M. Vingron. A polyhedral approach to RNA sequence structure alignment. *Journal of Comp. Biology*, 5(3):517–530, 1998.
14. D. H. Mathews and D. H. Turner. Dynalign: An algorithm for finding secondary structures common to two RNA sequences. *J. Mol. Biol.*, 317:191–203, 2002.
15. J. S. McCaskill. The Equilibrium Partition Function and Base Pair Binding Probabilities for RNA Secondary Structure. *Biopolymers*, 29:1105–1119, 1990.
16. D. Sankoff. Simultaneous solution of the RNA folding, alignment, and protosequence problems. *SIAM J. Appl. Math.*, 45:810–825, 1985.
17. S. Siebert and R. Backofen. MARNA: Multiple alignment and consensus structure prediction of RNAs based on sequence structure comparisons. *Bioinformatics*, 2005. In press.
18. S. Washietl and I. L. Hofacker. Consensus folding of aligned sequences as a new measure for the detection of functional RNAs by comparative genomics. *J. Mol. Biol.*, 342(1):19–30, 2004.
19. M. S. Waterman. Consensus methods for folding single-stranded nucleic adds. *Mathematical Methods for DNA Sequences*, pages 185–224, 1989.
20. L. A. Wolsey. *Integer Programming.* Wiley-Interscience series in discrete mathematics and optimization. Wiley, 1998.

Faster Algorithms for Optimal Multiple Sequence Alignment Based on Pairwise Comparisons *

Pankaj K. Agarwal[1], Yonatan Bilu[2], and Rachel Kolodny[3]

[1] Department of Computer Science, Box 90129,
Duke University, Durham NC 27708-0129, USA
pankaj@cs.duke.edu

[2] Department of Molecular Genetics,
Weizmann Institute, 76100 Rehovot,Israel
johnblue@cs.huji.ac.il

[3] Department of Biochemistry and Molecular Biophysics, Columbia University
rachel.kolodny@columbia.edu

Abstract. Multiple Sequence Alignment (MSA) is one of the most fundamental problems in computational molecular biology. The running time of the best known scheme for finding an optimal alignment, based on dynamic programming, increases exponentially with the number of input sequences. Hence, many heuristics were suggested for the problem. We consider the following version of the MSA problem: In a preprocessing stage pairwise alignments are found for every pair of sequences. The goal is to find an optimal alignment in which matches are restricted to positions that were matched at the preprocessing stage. We present several techniques for making the dynamic programming algorithm more efficient, while still finding an *optimal* solution under these restrictions. Namely, in our formulation the MSA must conform with pairwise (local) alignments, and in return can be solved more efficiently. We prove that it suffices to find an optimal alignment of sequence segments, rather than single letters, thereby reducing the input size and thus improving the running time.

1 Introduction

Multiple Sequence Alignment (MSA) is one of the central problems in computational molecular biology — it identifies and quantifies similarities among several protein or DNA sequences. Typically, MSA helps in detecting highly conserved motifs and remote homologues. Among its many uses, MSA offers evolutionary insight, allows transfer of annotations, and assists in representing protein families [20].

Dynamic programming (DP) algorithms compute an optimal Multiple Sequence Alignment for a wide range of scoring functions. In 1970, Needleman and Wunsch [19] proposed a DP algorithm for pairwise alignment, which was later improved by Masek and Paterson [13]. Murata *et al.* [17] extended this algorithm to aligning k sequences (each of length n). Their solution constructs a k-dimensional grid of size $O(n^k)$, with

* Part of this work was done while the second author was at the School of Engineering and Computer Science, The Hebrew University of Jerusalem; the third author was at the Department of Computer Science, Stanford University and visiting Duke University.

R. Casadio and G. Myers (Eds.): WABI 2005, LNBI 3692, pp. 315–327, 2005.

each of the sequences enumerating one of its dimensions. The optimal MSA is an optimal path from the furthermost corner (the end of all sequences) to the origin (their beginning). Unfortunately, the $O(n^k)$ running time of this approach makes it prohibitive even for modest values of n and k. There is little hope for improving the worst-case efficiency of algorithms that solve this problem, since the MSA problem is known to be NP-Hard for certain natural scoring functions [10,3]. This is shown by reduction from Max-Cut and Vertex Cover [8], that is, instances of these problems are encoded as a set of sequences. However, the encoding sequences are not representative of protein and DNA sequences abundant in nature, and the alignments are not reminiscent of ones studied in practice. This is the main motivation for our work.

Since MSA is NP-hard, heuristics were devised, including MACAW [22], DIALIGN [14], ClustalW [25], T-Coffee [21], and POA [12]. Many of these methods share the observation that aligned segments of the pairwise alignments are the basis for the multiple alignment process. Lee *et al.* [12] argued that the *only* information in MSA is the aligned sub-sequences and their relative positions. Indeed, many methods (e.g., [14,22,21]) align all pairs of sequences as a preprocessing step and reason about the similar parts; the additional computational cost of $O(n^2k^2)$ is not considered a problem. In progressive methods, this observation percolates to the order of adding the sequences to the alignment [21,25,5]. Other methods assemble an alignment by combining segments in an order dictated by their similarity [22,14]. The Carrillo-Lipman method restricts the full DP according to the pairwise similarities [4]. Unfortunately, none of these methods guarantee an optimal alignment. Another expression of this observation is scoring, and then matching, of full segments rather than single residues [16,14,21,27]. See [20] for recent results on MSA.

Alternatively, researchers designed optimal algorithms for (mostly pairwise) sequence alignment that are faster than building the full DP table. The algorithms of Eppstein *et al.* [6,7] modify the objective function for speedup. Wilbur and Lipman [26,27] designed a pairwise alignment algorithm that offers a tradeoff between accuracy and running time, by considering only matches between identical "fragments". Myer and Miller [18] and Morgenstern [15] designed efficient solutions for special cases of the segment matching problem. In particular, the case considered by Myer and Miller can be solved in polynomial time [18], while the general problem is NP-hard.

In this study, we identify combinatorial properties that are amenable to faster DP algorithms for MSA, and are biologically reasonable. We measure the efficiency of a DP solution by the number of table updates; this number is correlated with both the time and memory required by the algorithm. We suggest a way to exploit the fact that the input sequences are not general, but rather naturally occurring — some of their segments are evolutionary related, while others are not.

We define and study the *Multiple Sequence Alignment from Segments* (MSAS) problem, a generalization of MSA. Intuitively, MSAS accounts for assumptions regarding the pairwise characteristics of the optimal MSA. In MSAS, the input also includes a segmentation of the sequences, and a set of matching segment pairs. As in the original problem, we seek an MSA that optimizes the objective score. However, only corresponding positions in matching segments may be aligned. Trivially, one can segment the sequences into individual letters and specify all possible segment (letter) pairs, each with their substitution matrix score, getting back the original MSA problem. However, for biological sequences we can often postulate that only solutions that conform to some pairwise alignments are valid, e.g. when segments of different sequences clearly match, or clearly do not match. Using these assumptions, we develop a more efficient DP algorithm.

We then prove that the MSAS problem is essentially equivalent to the *segment matching problem*. This equivalence implies that it is enough to match segments, rather than individual positions. In particular, the complexity of DP algorithm for MSA, and, indeed, *any* algorithm for MSA, depends on the number of *segments* in each sequence, rather than the number of letters. We show that in practice this reduces the number of table updates by several orders of magnitude. For example, aligning five human proteins (denoted by their Swiss-Protidentifiers) GBAS, GBI1, GBT1, GB11, and GB12 requires 4.3×10^8 rather than 6.6×10^{12} table updates. Nonetheless, we prove that in general it is NP-hard.

We can make the algorithm even faster, while still guaranteeing an optimal solution, by further decoupling the sub-problems computation. Essentially, this improved DP algorithm avoids some of the nodes in the k-dimensional grid when calculating the optimal path. Indeed, in practice it outperforms naive DP, and the MSA of the example mentioned above requires only 1.5×10^5 table updates.

Lastly, we further study the combinatorial structure of the problem by considering two additional assumptions, and the performance improvement they imply. The following assumptions may hold in some cases of aligning DNA sequences, where a match indicates (near) identity. Here, we assume that the segment matches have a transitive structure, i.e., if segment A matches segment B, and B matches C, then A necessarily matches C. Also, an optimal alignment is one of minimal width, rather than optimal under an arbitrary scoring function. We prove that under these assumptions, an optimal alignment has a specific structure, which leads to a faster algorithm.

The paper is organized as follows: In Section 2 we define the MSA problem and cast it into a graph-theoretic framework; for completeness, we mention the straightforward DP solution. In Section 3 we present the MSAS problem and prove its equivalence to the segment matching problem, leading to a faster algorithm. We improve the running time even more by considering only "relevant directions" in Section 3. We describe our implementation in Section 4, including the conversion of pairwise alignments to the input format of MSAS, and give several examples of the performance when aligning human proteins. Lastly, in Section 5 we show that a transitivity assumption on the matches leads to further improved efficiency.

2 Multiple Sequence Alignment

The input of a *Multiple Sequence Alignment* (MSA) problem is a set $\mathbb{S} = \{\sigma_1, \ldots, \sigma_k\}$ of k sequences of lengths $n_1, \ldots, n_k$ over an alphabet Σ and a scoring function $f : (\Sigma \cup \{-\})^* \rightarrow \mathbb{R}$ (where the gap sign, "$-$", is not in Σ). A multiple alignment of the sequences is a $k \times n$ matrix with entries from $\Sigma \cup \{-\}$. In the ith row the letters of the ith sequence appear in order, possibly with gap signs between them. The score of a column of the matrix is the value of f on the k-tuple that appears in that column. The

FindOptimalPath(x) (Version 0)

1. If $x = \mathbf{0}$ return $\mathbf{0}$
2. For all $\emptyset \neq I \subseteq [k]$
   2.1 If p_{x-e_I} is undefined, compute p_{x-e_I} = FindOptimalPath($x - e_I$)
3. $I = \arg\max_{J \subseteq [k]} s(x, x - e_J) + s(p_{x-e_J})$
4. Return the path x, p_{x-e_I}.

Fig. 1. Basic DP MSA algorithm

score of a multiple alignment of $\mathbb{S}$ is the sum of scores over all columns. The objective in the MSA problem is to find an alignment of $\mathbb{S}$ with optimal score. Without loss of generality, we consider scoring functions whose optimum is a maximum, rather than a minimum. Other formulations of MSA, which have been suggested (e.g. [12,16]), are beyond the scope of this work.

We first define our notation: Let $I \subseteq [k]$, where $[k] := \{1, \ldots, k\}$. We denote by $e_i \in \{0,1\}^k$ the vector that is zero in all coordinates except the ith, where it is 1, and $e_I = \sum_{i \in I} e_i$. For a vector $x = (x_1, \ldots, x_k) \in \mathbb{N}^k$ let $x|_I$ be the projection of x onto the subspace spanned by $\{e_i\}_{i \in I}$, i.e., the ith coordinate of $x|_I$ is x_i if $i \in I$, and 0 otherwise. For two vectors, $x, y \in \mathbb{N}^k$ we say that x *dominates* y, and write $x > y$ if $x_i \geq y_i$ for $i = 1, \ldots, k$. We study the directed graph $\mathbb{G}_0$ — its vertex set is $[n_1] \cup \{0\} \times [n_2] \cup \{0\} \times \ldots \times [n_k] \cup \{0\}$, and there is an edge (x, y) in $\mathbb{G}_0$ if and only if $x > y$ and $x - y = e_I$ for some $\emptyset \neq I \subset [k]$; in this case we call I the *direction* that leads from x to y.

The paths from the vertex $(n_1, \ldots, n_k)$ to $(0, \ldots, 0)$ in $\mathbb{G}_0$ correspond to alignments of the input sequence. Let p be such a path. Consider $(x, x - e_I)$, the jth edge that the path transverses: In the corresponding sequence alignment, the jth column is a k-tuple that aligns positions x_i of sequences $i \in I$, and has a gap in the rest (in this case we say that the path matches position x_i of sequence i and position $x_{i'}$ of sequence i', for all $i, i' \in I$). We define $s : E(\mathbb{G}_0) \to \mathbb{R}$ to be a scoring function over the edges of $\mathbb{G}_0$, based on the scoring function f over the columns of the alignment. The function s assigns to an edge the value that f assigns to the corresponding column. We also extend s to paths, or sets of edges $E' \subseteq E(\mathbb{G}_0)$: $s(E') = \sum_{e \in E'} s(e)$. It is not hard to see that every such path defines a multiple alignment, and that every multiple alignment can be described by such a path.

In MSA we seek a *maximal (scoring) path* from $(n_1, \ldots, n_k)$ to $(0, \ldots, 0)$ in $\mathbb{G}_0$. The well-known DP solution to this problem is straightforward; we sketch it in Figure 1. Most importantly, we store the optimal scores of subproblems that have been solved recursively to avoid recomputing them later. For each vertex $x \in \mathbb{G}_0$, we compute the optimal path from x to the origin, denoted p_x, by considering the optimal scores of all its neighbors that are closer to the origin. Thus, we calculate the optimal MSA by calling FindOptimalPath($n_1, \ldots, n_k$). The time complexity of the algorithm is the number of edges in $\mathbb{G}_0$, i.e., $\Theta(2^k \prod_{j=1}^{k} n_j)$.

3 MSA from Segments

In this section we formulate Multiple Sequence Alignment from Segments (MSAS) — a generalization of MSA. We assume a preprocessing step that partitions the sequences into segments and matches pairs of these segments. These define a subgraph $\mathbb{G}_1 \subseteq \mathbb{G}_0$, and we then consider the restricted problem of finding an optimal path in $\mathbb{G}_1$. Intuitively, $\mathbb{G}_1$ disallows some of the pairwise alignments in $\mathbb{G}_0$ and consequently in the optimal alignment; clearly, we can allow all the diagonals in $\mathbb{G}_0$ (by segmenting the sequences into letters), leaving the MSA problem unchanged. Next, we show that the vertices of $\mathbb{G}_1$ can be condensed, yielding an even smaller graph $\mathbb{G}_2$; the vertices in $\mathbb{G}_2$ correspond to the segments of input sequences computed in the preprocessing step. The problem is now reduced to computing an optimal path in $\mathbb{G}_2$, which we refer to as the *segment matching problem.* Finally, we show that for computing the optimal path at a vertex it suffices to consider a subset of directions – the so-called relevant directions. We discuss the implementation of the algorithm and elaborate on the preprocessing step in Section 4.

Preliminaries.

Definition 1. *For a sequence q of length n, a* segmentation *of q is a sequence of* extremal *points $0 = c_0 \leq c_1 \leq \ldots \leq c_l = n$. The interval $[c_{i-1}+1, c_i]$ is called the ith segment of q. The extremal point c_i is said to be the* entry point *into segment i (for $i = 1, \ldots, l$), and the* exit point *from segment $i+1$ (for $i = 0, \ldots, l-1$). Denote by l_j the number of segments in the jth sequence.*

Definition 2. *A* segment matching graph *(SMG) over k segmented sequences is a k-partite undirected weighted graph with vertex set $\{(j,i) : j \in [k], i \in [l_j]\}$. Each vertex has an edge connecting it to itself. In addition, vertices (j_1, i_1) and (j_2, i_2) may be connected if the i_1th segment of sequence j_1 has the same length as the i_2th segment of sequence j_2, and $j_1 \neq j_2$.*

An edge $e = ((j_1, i_1), (j_2, i_2))$ in the SMG signifies a match between segment i_1 in sequence j_1 and segment i_2 in sequence j_2. Let l be the (same) length of these segments, and x_1 and x_2 their exit points on sequences j_1 and j_2, respectively, then for $t = 1, \ldots, l$, the edge e implies a match between position $x_1 + t$ of sequence j_1 and position $x_2 + t$ of sequence j_2.

The input to the MSAS problem is a set of segmented sequences, and a list of matching segments, described by an SMG M. The objective is still finding the highest scoring sequence alignment, but with the following restrictions. First, two sequence positions may be aligned together only if they appear in matching segments, and in the same relative position therein. Second, the score of a multiple match depends only on the weights of the corresponding edges in the SMG (and not on the letters themselves). In other words, we can think of the domain of the scoring function as being k-tuples of segments, rather than positions.

The intuition behind these restrictions is that the preprocessing stage identifies matching segments, and commits the algorithm to them. Furthermore, it assigns a "confidence level" (or the weight) to each match, and the objective is to find a highest-scoring alignment, with respect to these values. Here, the segments of each sequence

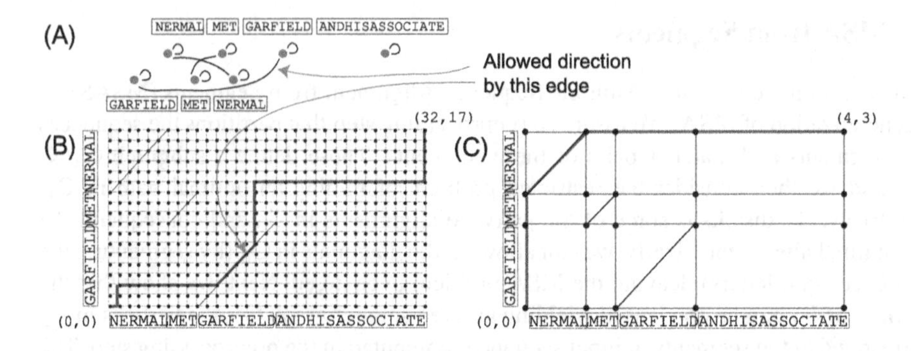

Fig. 2. Example of an SMG for two sequences: Panel (A) shows the sequences, their partitioning and the SMG where each segment corresponds to a (gray) node. Panel (B) shows $\mathbb{G}_1$. Unlike $\mathbb{G}_0$ that has all diagonals, the diagonals in $\mathbb{G}_1$ are defined by the SMG. An allowed path in $\mathbb{G}_1$ is also shown. Panel (C) shows $\mathbb{G}_2$, and an allowed path in it. The directions of the edges are omitted in the illustration for clarity, but are towards the origin.

are non-overlapping. In practice, we derive the segments from aligned portions of two sequences, and these may be overlapping. This is resolved by splitting the overlapping segments to smaller non-overlapping ones, as we discuss in Section 4.

Formally, given a set of segmented sequences, and an SMG M, we define $\mathbb{G}_1(M)$ as follows. It is a subgraph of $\mathbb{G}_0$, containing all vertices. The edge $(x, x - e_I)$ is in $\mathbb{G}_1(M)$ if and only if for all $i, j \in I$ there is an edge $m \in E(M)$, such the position x_i on sequence i is matched to position x_j on sequence j. In this case we say the I is an *allowed direction* at x, and that m is a match *defining* the edge $(x, x - e_I)$. The score of such an edge depends only on the weights of the corresponding edges in M (e.g., the sum-of-pairs scoring function). It is not hard to see that if x and y are vertices such that $x_I = y_I$ and I is allowed at x, then I is also allowed at y. Note also, that because all vertices in M have an edge connecting them to themselves, for $i \in [k]$, $\{i\}$ is an allowed direction at all vertices x such that $x_i > 0$.

As in the MSA problem, the goal in the MSAS problem is to find a highest scoring path from $(n_1, \ldots, n_k)$ to $(0, \ldots, 0)$. Clearly the previously mentioned DP algorithm solves this restricted MSA problem as well. In the following subsections we describe how it can be improved.

MSAS and Segment Matching. The vertices of $\mathbb{G}_1$ correspond to k-tuples of positions along the input sequences, one from each sequence. We now define the graph $\mathbb{G}_2$, a "condensed" version of $\mathbb{G}_1$, whose vertices correspond to k-tuples of *segments*. That is, its vertex set is $[l_1] \cup \{0\} \times [l_2] \cup \{0\} \times \ldots \times [l_k] \cup \{0\}$. There is a directed edge from $z = (z_1, \ldots, z_k)$ to $z - e_I$ in $\mathbb{G}_2$ if for all $i, j \in I$ the z_ith segment of sequence i matches the z_jth segment of sequence j. Define $x \in V(\mathbb{G}_1)$ by taking x_i to be the entry point into the z_ith segment of sequence i. Suppose the length of the segments defining the edge $(z, z - e_I)$ is l (recall that two segment match only if they are of the same length). Observe that $(z, z - e_I) \in E(\mathbb{G}_2)$ implies that $(x, x - e_I), (x - e_I, x - 2e_I), \ldots, (x - (l-1)e_I, x - l \cdot e_I)$ are all edges in $\mathbb{G}_1$. In this sense, $(z, z - e_I)$ is a "condensation" of all these edges. Define the score of the edge $(z, z - e_I)$ as the sum

```
FindOptimalPath(x)    (Version 1)

1. If x = 0 return 0.
2. For all y = x − l · e_I that is extremal with respect to x
   2.1 If p_y is undefined, compute p_y = FindOptimalPath(y)
   2.2 d_y = l · s(x, x − e_I)
3. y* = arg max s(p_y) + d_y
4. Return the path x, p_{y*}.
```

Fig. 3. Segment based DP MSA algorithm

of the scores of all the edges in $\mathbb{G}_1$ that it represents. Since the score depends only on the segments, this is simply $l \cdot s(x, x - e_I)$.

The *segment matching* problem is to find a highest-scoring path from $(l_1, \ldots, l_k)$ to $(0, \ldots, 0)$ in $\mathbb{G}_2$. Clearly the same DP algorithm as above can be used to solve this problem in time $\Theta(2^k \prod_{j=1}^{k} l_j)$. Hence, when the sequences are long, but consist of a small number of segments, DP for solving the segment matching problem may be plausible, while solving the MSA problem might not.

In the sequel of this section we prove that in order to find an optimal solution to the MSAS problem, it is enough to solve the associated segment matching problem. To state this precisely, we need the following definition.

Definition 3. *Let x be a vertex in $\mathbb{G}_1(M)$. We say that x is an* extremal *vertex if for all $i \in [k]$, x_i is an extremal point of sequence i.*
We say that y is extremal with respect to x, *if it is the first extremal vertex reached when starting at x and repeatedly going in direction I, for some allowed direction I. Denote $X(x) = \{y \in V(\mathbb{G}_1(M)) : y$ is extremal w.r.t. $x\}$.*

Theorem 1. *There is an optimal path, $p = p_1, \ldots, p_v$, such that if $x^1, \ldots, x^u$ are the extremal points, in order, through which it passes, then $x^{i+1} \in X(x^i)$.*

Observe that in particular, the theorem says that segments are either matched in their entirety, or not matched at all. Hence, any solution to the segment matching problem defines an optimal solution of the MSAS problem. In other words, it suffices to solve the problem on the "condensed" graph $\mathbb{G}_2$. While Theorem 1 is intuitively clear, the proof is somewhat involved, and omitted from this version. Figure 3 sketches the revision of the DP algorithm based on Theorem 1.

Narrowing the Search Space: Relevant Directions. Consider an input to the MSAS problem that consists of two subsets of k sequences each. Suppose that none of the segments in the first subset match any of those in the second subset. Naively applying the algorithm above will require running time exponential in $2k$. Yet clearly the problem can be solved on each subset independently, in time exponential in k rather than $2k$. Intuitively, this is also the case when there are only few matches between the two subsets. We make this notion explicit in this subsection. Again, we start with some definitions:

FindOptimalPath(z) (Version 2)

1. If $z = \mathbf{0}$ return $\mathbf{0}$.
2. D = minimal set of directions that intersect a subset of independent relevance.
3. For all $\emptyset \neq I \in D$
   3.1 If p_{z-e_I} is undefined, compute p_{z-e_I} = FindOptimalPath($z - e_I$)
4. $I = \arg\max_{J \in D} s(z, z - e_J) + s(p_{z-e_J})$
5. Return the path z, p_{z-e_I}.

Fig. 4. Version 2 of MSA algorithm. Details on how to compute D are given in the full version

Definition 4. *Let x be a vertex in $\mathbb{G}_2(M)$. Let $((i, y_i), (j, y_j))$ be a match in the SMG. We say that such a match is* relevant *for x at coordinate i, if $x_i = y_i$ and $x_j > y_j$. We say that a subset of indices $S \subset [k]$ is of* independent relevance *at x if for all $i \in S$ the match $((i, y_i), (j, y_j))$ is relevant for x at coordinate i implies $j \in S$.*

Theorem 2. *Let p be an optimal path in $\mathbb{G}_2$, and x a vertex on it. Let S be a subset of indices of independent relevance at x. Then there is an optimal path p' that is identical to p up to x, and from x goes to $x - e_I$ for some $I \subset [k]$ such that $I \cap S \neq \emptyset$.*

Proof: Let y be the first vertex on p after x, such that $y_i = x_i - 1$ for some $i \in S$. Define p' to be the same as p up to x, and from y onwards. We will define a different set of allowed directions that lead from x to y. Let $I_1, \ldots, I_t$ be the directions followed from x to y. Let $i \in I_t \cap S$. For all $i \neq j \in I_t$, there is a match between (i, x_i) and $(j, y_j + 1)$. Hence, either $j \in S$, or $y_j + 1 = x_j$. Since y is the first vertex in p that differs from x on a coordinate in S, if $j \in S$, then $j \notin I_1, \ldots I_{t-1}$. Clearly, if $y_j + 1 = x_j$ then again $j \notin I_1, \ldots I_{t-1}$. In other words, for all $h < t$, we have $I_h \cap I_t = \emptyset$. Define p' to follow directions $I_t, I_1, \ldots, I_{t-1}$ from x. As I_t is disjoint from the other directions, this indeed defines an allowed path from x to y, and $i \in I_t \cap S$. ∎

The theorem implies that in the DP there is no need to look in *all* directions. Let S be a subset of independent relevance at a point x, then to compute the optimal path from x to the origin it is enough to consider paths from $x - e_I$ to the origin for $I \subset [k]$ such that $I \cap S \neq \emptyset$. This suggests the DP algorithm sketched in Figure 4 (this time think of z as a vertex in $\mathbb{G}_2$). Note that to implement this algorithm there is no need to keep a table of size $|V(\mathbb{G}_2)|$. The vertices that are actually visited by the algorithm can be kept in a hash table.

4 Implementing the Algorithm

We have implemented Version 2 of our algorithm, described in Figure 4. Using our implementation of the algorithm, we investigate its efficiency (measured in the number of vertices it visits, or table updates) on real biological sequences. We first describe the preprocessing step that constructs the SMG, and then discuss the performance of the algorithm on a few examples. We stress that *efficiency* is indeed the property of interest here, as the multiple alignment found is an *optimal* solution for the MSAS problem.

Generating a Segment Matching Graph (SMG). Existing tools, such as BLAST [2] or DIALIGN [16], provide local alignments rather than the input format that we assumed previously. In order to restrict the problem only to MSAs that conform to these local pairwise alignments, we must convert them to an SMG. In particular, we need to segment the sequences, and allow matches only between equal-length segments.

Starting with the set of sequences, we add breakpoints onto them based on the local alignments. This way, we progressively build the SMG, stopping when all local alignments are properly described. The ends of an alignment define breakpoints in the two aligned sequences. If the segments between those breakpoints have the same length, we simply add a connecting edge (or edges) to the SMG. However, the segments lengths may differ due to two reasons: First, gapped alignments match segments of unequal length; we solve this by adding breakpoints at the gap ends. Second, regions of the sequences corresponding to different alignments may overlap; we solve this by adding breakpoints at the ends of the overlapping region (or regions). Notice that if we add a breakpoint inside a segment that already has an edge associated with it, we must split the edge (and a corresponding breakpoint must be added to the connected segment).

Table 1. Number of table updates for three sets of human proteins. We compare full DP (Version 0), full DP on the Segment Matching Graph (Version 1), and the actual number of table updates when considering only relevant directions (Version 2); the SMG is generated using all significant gapped/un-gapped BLAST alignments. We see that in all cases, the actual work is several orders of magnitudes faster than the DP calculation.

		gapped BLAST		un-gapped BLAST	
Human proteins	full DP	Version 1	Version 2	Version 1	Version 2
MATK,SRC, ABL1,GRB2	6.65×10^{10}	$91 \cdot 98 \cdot 99 \cdot 89$ =78, 576, 498	7.20×10^{6}	$77 \cdot 84 \cdot 81 \cdot 74$ =38, 769, 192	1, 994, 813
PTK6,PTK7, RET, SRMS, DDR1	2.40×10^{14}	$92 \cdot 96 \cdot 106 \cdot 88 \cdot 125 = 1.03 \times 10^{10}$	281, 752	$60 \cdot 53 \cdot 66 \cdot 57 \cdot 58$=3, 736, 260	2, 980
GBAS, GBI1, GBT1, GB11, GB12	6.62×10^{12}	$148 \cdot 116 \cdot 113 \cdot 115 \cdot 120 = 2.68 \times 10^{10}$	270, 289	$61 \cdot 72 \cdot 68 \cdot 71 \cdot 70 = 4.3 \times 10^{8}$	145, 366

Example MSAs. We demonstrate the effectiveness of our algorithm by several examples of aligning human protein sequences. We align two sets of proteins from kinase cascades: (1) MATK, SRC, ABL1, and GRB2 of lengths 507, 535, 1130, and 217 respectively. (2) PTK6, PTK7, RET, SRMS, DDR1 of lengths 451, 1070, 1114, 488, and 913 respectively. We also align five heterotrimeric G-protein (subunits alpha) GBAS, GBI1, GBT1, GB11, GB12 of lengths 394, 353, 349, 359, and 380 respectively. We chose these (relatively long) proteins because their "mix-and-match" modular components characteristic highlights the strengths of our method. We use gapped and ungapped BLAST with E-value threshold of 10^{-2} to find local alignments. Namely, in the optimal MSA two letters can be matched only if they are in a local BLAST alignment with E-value at most 10^{-2}.

Table 1 lists the number of table updates needed to find the optimal MSA for these alignments. The first column has the size of the full DP matrix, or the product of the sequences lengths (same for gapped and un-gapped). The second column lists the number of segments in each sequence in the SMG, which was calculated from the BLAST gapped or ungapped alignments, and the size of their DP matrix. The last column has the actual number of vertices visited, or equivalently, the number of table updates. The number of updates drops dramatically, in the best case from 10^{14} to less than 3000. Other alignments that we studied had similar properties to the ones shown. Complete figures of the cases listed in Table 1 are available at [1] in a format that allows zooming for exploring the details.

5 The Transitive MSAS

In this section we further restrict the problem by making the following two assumptions, which allow for additional "shortcuts" in the DP algorithm.

ASSUMPTION 1: The matches are transitive, in the sense that if $\{i, j\}$ is an allowed direction at x, and $\{i, k\}$ is an allowed direction at x, then $\{j, k\}$ is also allowed at x (and hence, $\{i, j, k\}$ as well).
ASSUMPTION 2: The scoring function is such that we seek to find an alignment of minimal width, or equivalently, the shortest path from $(n_1, \ldots, n_k)$ to $(0, \ldots, 0)$ in $\mathbb{G}_0$.

The assumption of transitivity may be too restrictive in many biological relevant cases. We study it here for two main reasons: (1) The assumption holds in special cases of aligning nucleotide sequences, where a match indicates (near) identity; and (2) this analysis illuminates additional properties of the combinatorial structure of the problem, by further limiting the search space. The missing proofs appear in the full version.

Assumption 2 is achieved by setting the scoring function (over the edges of $\mathbb{G}_1$) as $s(x, x - e_I) = |I| - 1$: The longest possible path from $(n_1, \ldots, n_k)$ to $(0, \ldots, 0)$ is of length $\sum n_i$. Each edge $(x, x - e_I)$ "saves" $|I| - 1$ steps in the path, exactly its score. Hence, a shortest path, or the one that "saves" the most steps, is the highest scoring one. Since this scoring function is so simple over $\mathbb{G}_1$, it is convenient to return the discussion from $\mathbb{G}_2$ to $\mathbb{G}_1$. At the end of this section we prove that the techniques developed here apply to $\mathbb{G}_2$ as well.

We call the problem of finding the highest scoring path from $(n_1, \ldots, n_k)$ to $(0, \ldots, 0)$ in $\mathbb{G}_1(M)$, with s and M as above, the *Transitive MSAS Problem*.

Maximal Directions. The first observation is that an optimal solution to the Transitive MSAS proceeds in "maximal" steps.

Definition 5. *An edge $(x, x - e_I) \in E(\mathbb{G}_1(M))$ is called* maximal, *and the subset I a* maximal direction *(at x), if for all $J \supsetneq I$, the pair $(x, x - e_J)$ is not an edge. We denote by $D(x)$ the collection of maximal directions at vertex x (note that by transitivity, this is a partition of $[k]$). A path in $\mathbb{G}_1(M)$ is called a* maximal path *if it consists solely of maximal edges.*

Lemma 1. *There is an optimal path in $\mathbb{G}_1(M)$ that is also maximal.*

Henceforth, by "optimal path" we refer to a maximal optimal path. As a corollary of Lemma 1, the DP algorithm for the transitive MSA problem needs not check *all* directions (or all those that intersect a subset of independent relevance), only maximal ones.

This reduces the time complexity of the algorithm to $O(k \prod l_i)$, with a data structure that allows finding the maximal directions at a given vertex in $O(k)$. Details will be provided in the full version.

Obvious Directions. The notion of "relevant directions" discussed in Section 3 can be strengthened in the transitive setting. Indeed, there is a simple characterization of vertices in $\mathbb{G}_1$ for which the first step in an optimal path is obvious, and there is no need for recursion.

Definition 6. *Let x be a vertex in $\mathbb{G}_1(M)$ and I a maximal direction at x. The set I is called an* obvious direction *(at x) if for all $y \in \mathbb{G}_1(M)$, $y < x$, such that $x|_I = y|_I$, I is a maximal direction at y. If $y = x - c \cdot e_I$ is extremal with respect to x, and I is an obvious direction at x, we say that y is an* obvious vertex *with respect to x.*

Lemma 2. *Let p be an optimal path, x a vertex in p and I an obvious direction at x. Then there is an optimal path p' that is identical to p up to x, and that proceeds to $x - e_I$ from x.*

Corollary 1. *There is an optimal path p, such that if x is an extremal vertex in p, and y is obvious with respect to x, then p proceeds from x to y.*

Intuitively, obvious directions are cases where all benefits to the scoring function can be gained in the first step, or equivalently, there are no tradeoffs to consider. Hence, as for relevant directions, the DP algorithm can be revised to immediately move to an obvious vertex, avoiding the recursion over all extremal vertices.

Special Vertices. In this section we extend the "leaps" that the DP algorithm performs. Once more, we start with a few definitions.

Definition 7. *We say that a vertex y is* special *with respect to a vertex x if the following four conditions hold: (1) x dominates y; (2) $D(x) \neq D(y)$; (3) there is a path from x to y consisting solely of maximal edges; and (4) no vertex y' satisfies all the above, and dominates y. Denote by $S(x)$ the set of vertices that are special with respect to x.*

We define the set of *special vertices* $S \subseteq \mathbb{G}_1(M)$ as the smallest one such that $(n_1, \ldots, n_k) \in S$, and for every $x \in S$, $S(x) \subset S$. We first show that instead of "leaping" from one extremal vertex to another, we can "leap" from one special vertex to another.

Definition 8. *Let $p = (p_0, \ldots, p_r)$ and $p' = (p'_0, \ldots, p'_r)$ be two paths. Let $I_1, \ldots, I_r$ be the sequence of directions that p moves in, and $I'_1, \ldots, I'_r$ be the sequence of directions that p' does. We say that p and p' are equivalent if $p_0 = p'_0$, $p_r = p'_r$ and there is some permutation $\sigma \in S_r$ such that $I_i = I'_{\sigma(i)}$ for $i = 1, \ldots, r$.*

Note that equivalent paths have the same length, and hence the same score. We also observe:

Lemma 3. *Let p be an optimal path. Let x be a vertex in p, and let y be the first vertex in p that is also in $S(x)$. Then all maximal paths from x to y are equivalent.*

Let $p = (p_1, \ldots, p_t)$ be an optimal path. Define $x_1 = p_1$ and x_{i+1} to be the first vertex in p that is also in $S(x_i)$. Lemma 3 says that we only need to specify the vertices $\{x_i\}$ to describe an optimal path — all maximal paths connecting these vertices in order are equivalent.

As a corollary, we can further restrict the search space of the DP algorithm. When computing the shortest path from a vertex x, rather than considering the relevant extremal vertices, it is enough to consider the special ones. As we shall soon show, this is indeed a subset of the extremal vertices.

Before describing the modified algorithm in detail, let us observe that special points have a very specific structure.

Definition 9. *Let $x, y \in \mathbb{G}_1(M)$ be such that y is special with respect to x. Let I and I' be maximal directions at x. We say that y is a* breakpoint of direction I, *if $y = x - c \cdot e_I$ for some natural c, and I is not allowed at y.*
We say that y is a straight junction of direction I *if $y = x - c \cdot e_I$ for some natural c, and I is allowed, but not maximal, at y.*
We say that y is a corner junction of directions I and I' *if $y = x - c \cdot e_I - c' \cdot e_{I'}$ for some natural c and c', and I and I' are allowed, but not maximal, at y.*

Theorem 3. *Let y be a special vertex with respect to x. Then y is one of the types in definition 9. Furthermore, if x is an extremal vertex, then so is y.*

Corollary 2. *All special vertices are extremal vertices.*

This suggests a further improved DP algorithm that runs on $\mathbb{G}_2$. We defer the pseudo code listing and a detailed analysis of the running time of the algorithm to the full version. The analysis shows that the running time is linear in the number of segments and the number of special vertices, and at most cubic in the number of sequences.

Acknowledgements

P.K.A is supported by National Science Foundation (NSF) grants CCR-00-86013, EIA-98-70724, EIA-01-31905, and CCR-02-04118, and by a grant from the U.S.–Israel Bi-national Science Foundation. Y.B. was supported by the Israeli Ministry of Science. R.K. is supported by NSF grant CCR-00-86013. We are grateful to Chris Lee and Nati Linial for enlightening discussions on these topics.

References

1. http://trantor.bioc.columbia.edu/~kolodny/MSA/
2. Altschul, F. Stephen, L.M. Thomas, A. A. Schaffer, J. Zhang, Z. Zhang, W. Miller, and D.J. Lipman. Gapped BLAST and PSI-BLAST: A new generation of protein database search programs *Nucleic Acids Res.*, 25:3389–3402, 1997.
3. P. Bonizzoni and G. Della Vedova. The complexity of multiple sequence alignment with sp-score that is a metric. *Theoretical Computer Science*, 259(1-2):63–79, 2001.
4. H. Carrillo and D. Lipman. The multiple sequence alignment problem in biology. *SIAM J. Applied Math.*, 48(5):1073–1082, 1988.

5. F. Corpet. Multiple sequence alignment with hierarchical-clustering. *Nucleic Acids Research*, 16(22):10881–10890, 1988.
6. D. Eppstein, Z. Galil, R. Giancarlo, and G. F. Italiano. Sparse dynamic-programming: I. linear cost-functions. *JACM*, 39(3):519–545, 1992.
7. D. Eppstein, Z. Galil, R. Giancarlo, and G. F. Italiano. Sparse dynamic-programming: II. convex and concave cost-functions. *JACM*, 39(3):546–567, 1992.
8. M. R. Garey and D. S. Johnson. *Computers and Intractability–A Guide to the Theory of NP-completeness*. Freeman, San Francisco, 1979.
9. S. Henikoff and J. G. Henikoff. Amino acid substitution matrices from protein blocks. *Proc. Natl. Acad. Sci.*, 89: 10915–10919, 1992.
10. T. Jiang and L. Wang. On the complexity of multiple sequence alignment. *J. Comp. Biol.*, 1(4):337–48, 1994.
11. G. M. Landau M. Crochemore and M. Ziv-Ukelson. A sub-quadratic sequence alignment algorithm for unrestricted cost matrices. *Proc. 13th Annual ACM-SIAM Sympos. Discrete Algo.*, 679–688, 2002.
12. C. Lee, C. Grasso, and M. F. Sharlow. Multiple sequence alignment using partial order graphs. *Bioinformatics*, 18(3):452–464, 2002.
13. W. J. Masek and M. S. Paterson. A faster algorithm computing string edit distances. *J. Comput. Sys. Sci.*, 20(1):18–31, 1980.
14. B. Morgenstern. Dialign 2: improvement of the segment-to-segment approach to multiple sequence alignment. *Bioinformatics*, 15(3):211–218, 1999.
15. B. Morgenstern. A simple and space-efficient fragment-chaining algorithm for alignment of DNA and protein sequences *Applied Math. Lett.*, 15(1), 11–16, 2002.
16. B. Morgenstern, A. Dress. and T. Werner. Multiple DNA and protein sequence alignment based on segment-to-segment comparison. *Proc. Nat. Acad. Sci.*, 93(22):12098–12103, 1996.
17. M. Murata, J. S. Richardson, and J. L. Sussman. Simultaneous comparison of 3 protein sequences. *Proc. Nat. Acad. Sci.*, 82(10):3073–3077, 1985.
18. G. Myers and W. Miller. Chaining multiple-alignment fragments in sub-quadratic time. *Proc. 6th Annual ACM-SIAM Sympos. Discrete Algo.*, 38–47, 1995.
19. S. B. Needleman and C. D. Wunsch. A general method applicable to the search for similarities in the amino acid sequences of two proteins. *J. Mol. Biol.*, 48:443–453, 1970.
20. C. Notredame. Recent progress in multiple sequence alignment: a survey. *Pharmacogenomics*, 3(1):131–144, 2002.
21. C. Notredame, D. G. Higgins, and J. Heringa. T-coffee: A novel method for fast and accurate multiple sequence alignment. *J. Mol. Biol.*, 302(1):205–217, 2000.
22. G. D. Schuler, S. F. Altschul, and D. J. Lipman. A workbench for multiple alignment construction and analysis. *Proteins-Structure Function And Genetics*, 9(3):180–190, 1991.
23. R. M. Schwartz and M. O. Dayhoff. Matrices for Detecting Distant Relationships. *Atlas of Protein Sequences and Structure, (M.O. Dayhoff, ed.)*, 5, Suppl. 3 (pp; 353-358), National Biomedical Research Foundation, Washington, D.C., USA.
24. T. F. Smith and M. S. Waterman. Comparison of biosequences. *Adv. Applied Math.*, 2(4), 482–489, 1981.
25. J. D. Thompson, D. G. Higgins, and T. J. Gibson. Clustal-W - improving the sensitivity of progressive multiple sequence alignment through sequence weighting, position-specific gap penalties and weight matrix choice. *Nucleic Acids Research*, 22(22):4673–4680, 1994.
26. W. J. Wilbur and D. J. Lipman. Rapid similarity searches of nucleic-acid and protein data banks. *Proc. Nat. Acad. Sci.*, 80(3):726–730, 1983.
27. W. J. Wilbur and D. J. Lipman. The context dependent comparison of biological sequences. *SIAM J. Applied Math.*, 44(3):557–567, 1984.

Ortholog Clustering on a Multipartite Graph

Akshay Vashist[1,*], Casimir Kulikowski[1], and Ilya Muchnik[1,2]

[1] Department of Computer Science
[2] DIMACS Rutgers, The State University of New Jersey, Piscataway, NJ 08854, USA
vashisht@cs.rutgers.edu, kulikows@cs.rutgers.edu
muchnik@dimacs.rutgers.edu

Abstract. We present a method for automatically extracting groups of orthologous genes from a large set of genomes through the development of a new clustering method on a weighted multipartite graph. The method assigns a score to an arbitrary subset of genes from multiple genomes to assess the orthologous relationships between genes in the subset. This score is computed using sequence similarities between the member genes and the phylogenetic relationship between the corresponding genomes. An ortholog cluster is found as the subset with highest score, so ortholog clustering is formulated as a combinatorial optimization problem. The algorithm for finding an ortholog cluster runs in time $O(|E| + |V| \log |V|)$, where V and E are the sets of vertices and edges, respectively in the graph. However, if we discretize the similarity scores into a constant number of bins, the run time improves to $O(|E| + |V|)$. The proposed method was applied to seven complete eukaryote genomes on which manually curated ortholog clusters, KOG (eukaryotic ortholog clusters, http://www.ncbi.nlm.nih.gov/COG/new/) are constructed. A comparison of our results with the manually curated ortholog clusters shows that our clusters are well correlated with the existing clusters. Finally, we demonstrate how gene order information can be incorporated in the proposed method for improving ortholog detection.

1 Introduction

One of the fundamental problems in comparative genomics is the identification of genes from different organisms that are involved in similar biological functions. This requires identification of orthologs which are homologous genes that have evolved through vertical descent from a single ancestral gene in the last common ancestor of the considered species [1]. In recent years, many genome-wide ortholog detection procedures have been developed [2,3,4,5,6,7], however, they suffer from limitations that present real challenges for addressing the problem in a large set of genomes. Some of them are limited to identifying orthologs in a pair of genomes [2,3,4]; [7] requires phylogenetic information and are not computationally efficient, and others [5,6] require expert curation. Known complete methods for finding ortholog clusters have at least two stages - automatic and

* The author was, in part, supported by the DIMACS Graduate Students Awards.

R. Casadio and G. Myers (Eds.): WABI 2005, LNBI 3692, pp. 328–340, 2005.

manual. The role of the latter is to correct th results of the first stage which is usually a clustering procedure. Although specific implementations of clustering procedures in different methods vary, most successful methods include critical steps such as building clusters based on a set of "mutually most similar pairs" of genes from different genomes. These pairs are called BBH (bi-directional best hits [4,5,6]). This preprocessing is not robust as small changes in data or in the set of free parameters can alter the results substantially. So, currently there are three bottlenecks in ortholog extraction: (a) the manual curation, (b) time complexity, and (c) the hypersensitivity of the automatic stage to parameter changes. We propose a combinatorial optimization based approach for ortholog detection in a large set of genomes that addresses these bottlenecks.

The proposed method assigns a score to any arbitrary subset of genes from multiple genomes to assess orthologous relationships between genes in the subset, finding an ortholog cluster as the subset with the highest score. Thus, an ortholog cluster is found as a global solution to a combinatorial optimization problem on multipartite graphs. Assigning a score that best reflects the ortholog relationships among genes in an arbitrary subset of genes is critical to our approach. When considering orthologs from multiple genomes, observed sequence similarities between a pair of orthologous genes depends on the time since divergence of the corresponding genomes. So, in addition to sequence similarities between genes we consider the phylogenetic relationship between genomes. We also describe how the gene order information can be incorporated into our method to improve ortholog detection.

The method is efficient for finding candidate ortholog clusters in a large number of genomes and automatically determines the number of candidate ortholog clusters. We have applied this method to find ortholog clusters in seven genomes on which the KOG database [5] is constructed.

In the following we present our ortholog model in section 2 and describe the algorithm for extracting ortholog clusters in section 3. The implementation details and techniques to speed up the ortholog extraction are in section 4. The experimental results on the 7 genomes and the comparison of the orthologs are presented in sections 5. Section 6 contains the conclusion and future work.

2 Problem Formulation: Ortholog Model

A challenge in ortholog cluster extraction is to avoid detection of paralogs, which are genes that have evolved through duplication of an ancestral gene [1]. From a gene-function point of view, correct identification of orthologs is particularly important since they usually perform very similar function whereas paralogs, although highly similar at the primary sequence level, functionally diverge to adapt to new functions. Paralogs are closely related to orthologs, because if a duplication event follows a speciation event, orthology becomes a relationship between a set of paralogs [5]. Due to this complex relationship, paralogs related to ancient duplications are placed in different ortholog clusters whereas recently duplicated genes are placed in the same ortholog cluster, as has been done in

COG [6], KEGG [2] and Inparanoid [4]. Such definition of ortholog clusters is justified from a gene function perspective because anciently duplicated genes are most likely to have adapted to new functional niches.

We address the above challenge by modeling the ortholog cluster as clusters in a multipartite graph. Orthologs, by definition, are present in different genomes [1]. We represent this in the multipartite graph, by letting different genomes correspond to partite sets and the genes in a genome correspond to vertices in a partite set. The multipartite graph considers the similarity relationships between genes from different genomes and ignores similarities between genes within a genome. Furthermore, the multipartite graph representation is suitable for discriminating between recently and anciently duplicated paralogs. Recently duplicated paralogs are confined to a genome and are very similar in primary sequence, so these copies share similarities to the same set of genes in other genomes. On the other hand, anciently duplicated paralogs are more similar to orthologs in other genomes compared to paralogs within the genome. So, the multipartite graph clustering, described below, places recently duplicated genes in the same ortholog cluster and the ancient paralogs in different clusters.

A challenge specific to ortholog detection in a large set of genomes is the variation in observed sequence similarity between orthologs from different pairs of genomes. Within an ortholog family, orthologous sequences belonging to anciently diverged genomes are relatively less similar in comparison to those from the recently diverged genomes [8]. So, automatic methods based on numerical measures of sequence similarity must correct for these observed sequence similarities [1]. To correct the observed sequence similarity between a pair from two genomes, we use the distance between corresponding genomes.

Most ortholog detection methods consider genomes as a bag of genes and find ortholog clusters solely based on sequence similarity [6,5,4]. However, the leverage gained by using auxiliary information such as order of genes in a genome is widely recognized [8]. In fact, studies [9] show that the order of genes in the genome can reliably determine the phylogenetic relationship between closely related organisms. We describe later how the gene-order information can be used in conjunction with the sequence similarity to find ortholog clusters.

2.1 Ortholog Clusters on a Multipartite Graph

Consider the ortholog clustering problem with k genomes, where V_i, $i \in \{1, 2, \ldots, k\}$ represents the set of genes from the genome i. Then, the similarity relationships between genes from different genomes can be represented by an undirected weighted multipartite graph $G = (V, E, W)$, where $V = \cup_{i=1}^{k} V_i$ and V_i is the set of genes from the genome i, and $E \subseteq \cup_{i \neq j} V_i \times V_j$ is the set of weighted, undirected edges representing similarities between genes.

The problem of finding an ortholog cluster could be modeled as finding maximum weight multipartite clique, but no efficient procedure exists for solving this

[1] The issue related to the correction is more complicated due to diverse evolutionary rates across lineages and protein families [8]. Our intention is not to correct for the absolute rates of evolution.

problem [10]. Moreover, cliques are simple models for an ortholog cluster which requires robust models that allow some incompleteness in subgraph extracted as a cluster. Due to this, clusters are often modeled as quasi-cliques or dense graphs [2].

To find a weighted multipartite quasi-clique as an ortholog cluster, we assign a score $F(H)$ to any subset H of V. The score function denotes a measure of proximity among genes in H. Then, our multipartite quasi-clique, or cluster, H^* is defined as the subset with largest score value, i.e.,

$$H^* = \arg \max_{H \subseteq V} F(H) \tag{1}$$

The subset H contains genes from multiple genomes, so according to (1) our approach finds an ortholog cluster as a set of genes from multiple genomes by simultaneously considering all the similarity relationships in H. This is novel since to our knowledge all sequence similarity based methods, such as [6,4], find an initial set of orthologs from two genomes and possibly extend them at later stages. The function $F(H)$ is designed using a linkage function $\pi(i, H)$ which measures the degree of similarity of the gene $i \in H$ to other genes in H.

$$F(H) = \min_{i \in H} \pi(i, H), \quad \forall i \in H \ \forall H \subseteq V \tag{2}$$

In other words, $F(H)$ is the $\pi(i, H)$ values of the least similar (outlier) gene in H. Then, according to (1), the subset, H^* contains genes such that similarity of the least similar gene in H is maximum.

Our linkage function considers the sequence similarity between genes within the ortholog cluster, their relationship to genes outside the cluster, and the phylogenetic distance between the corresponding genomes. Consider a subset H of V that contains genes from at least two genomes, so that H be decomposed as $H = \cup_{i=1}^{k} H_i$ where H_i is the subset of genes from V_i present in H. If m_{ij} (≥ 0) is the similarity value between gene i from genome $g(i)$ and gene j from another genome $g(j)$ and $p(g(i), g(j))$ represents the distance between the two genomes, then the linkage function is defined as

$$\pi(i, H) = \sum_{\substack{\ell=1 \\ \ell \neq g(i)}}^{k} p(g(i), \ell) \left\{ \sum_{j \in H_\ell} m_{ij} - \sum_{j \in V_\ell \setminus H_\ell} m_{ij} \right\} \tag{3}$$

Given the phylogenetic tree for the genomes under study, the distance, $p(g(i), g(j))$ (≥ 0), between the genomes is defined as the height of the subtree rooted at the last common ancestor of the genomes $g(i)$ and $g(j)$. This term is used to correct the observed sequence similarities by magnifying the sequence similarities corresponding to genomes which diverged in ancient times. The term $\sum_{j \in H_\ell} m_{ij}$ aggregates the similarity values between the genes i from genome $g(i)$ and all other genes in the subset H that do not belong to genome $g(i)$, while the second term, $\sum_{j \in V_\ell \setminus H_\ell} m_{ij}$, estimates how this gene is related to genes from genome ℓ that are not included in H_ℓ. A large positive difference between these two terms ensures that the gene i is highly similar to genes in H_ℓ and at the same time very dissimilar from genes not included in H_ℓ. From a clustering point of view, this ensures large values of intra-cluster homogeneity and inter-cluster separability for extracted clusters. Translated to ortholog clustering, such a design enables a separation of ortholog clusters related to anciently duplicated paralogs.

3 Multipartite Graph Clustering

We now give an algorithm to find the solution for the combinatorial optimization problem defined in (1) and study properties of the functions $\pi(i, H)$ and $F(H)$ which guarantee an efficient algorithm to find the optimal solution. The linkage function in (3) and the score function in (2) were designed such that they satisfy these properties.

Definition 1. *A linkage function, $\pi : V \times 2^V \to \Re$, is monotone increasing if*

$$\pi(i, H) \geq \pi(i, H_1) \quad \forall i, \forall H_1, \forall H \,:\, i \in H_1 \subseteq H \subseteq V \tag{4}$$

Claim 1. The linkage function $\pi(i, H)$ defined in (3) is monotone increasing.

Proof. Observe that the distance $p(\imath, \jmath)$ is merely a scaling factor for the observed similarities and does not impact the monotonicity. Consider the case when H is extended to $H \cup \{k\}$, and assume $k \in V_s$. If $i \in V_s$ then $\pi(i, H \cup \{k\}) = \pi(i, H)$, otherwise $\pi(i, H \cup \{k\}) - \pi(i, H) = 2m_{ik} \geq 0$, which proves the claim. □

Definition 2. *A set function, $F : 2^V \setminus \emptyset \to \Re$, is quasi-concave if it satisfies*

$$F(H_1 \cup H_2) \geq \min(F(H_1), F(H_2)) \qquad \forall H_1, H_2 \subseteq V \tag{5}$$

Proposition 1. *The set function $F(H)$ as defined in (2) is quasi-concave if and only if the linkage function is monotone increasing.*

Proof. [⇒] Let $H_1, H_2 \subseteq V$, and $i^* \in H_1 \cup H_2$ be such that $F(H_1 \cup H_2) = \pi(i^*, H_1 \cup H_2)$. Suppose, $i^* \in H_1$, then using (4) we get, $F(H_1 \cup H_2) = \pi(i^*, H_1 \cup H_2) \geq \pi(i^*, H_1) \geq \min_{i \in H_1} \pi(i, H_1) = F(H_1) \geq \min(F(H_1), F(H_2))$.

[⇐] The proof is by contradiction. For $i \in H_1 \subseteq H \subseteq V$, assume that $\pi(i, H_1 \cup H) < \pi(i, H_1)$ and (5) hold. From the assumption we get $\min_{i \in H_1} \pi(i, H_1 \cup H) < \min_{i \in H_1} \pi(i, H_1) = F(H_1)$. Further, $F(H_1 \cup H) = \min_{i \in H_1 \cup H} \pi(i, H_1 \cup H) \leq \min_{i \in H_1} \pi(i, H_1 \cup H)$. Combining these two inequalities we get, $F(H_1 \cup H) < F(H_1)$ which contradicts the quasi-concavity property (5) in the assumption. □

Proposition 2. *For a quasi-concave set function $F(H)$ the set of all its maximizers, as defined by (2), is closed under the set union operation.*

Proof. Follows from the quasi-concavity of $F(H)$. □

A maximizer of $F(H)$ that contains all other maximizers is called the $\cup$-*maxi mizer*, $\hat{H}$. It is obvious from proposition (2) that $\hat{H}$ is the unique largest maximizer. The algorithm to find the optimal solution $\hat{H}$ is described in Table 1.

Table 1. Pseudocode for extracting $\hat{H}$

```
Step 0: Set t := 1; H_1 := V; Γ := V;
Step 1: Find M_t := {i : π(i, H_t) = min_{j∈H_t} π(j, H_t)};
Step 2: if ((H_t \ M_t = ∅) ∨ (π(i, H_t) = 0 ∀i ∈ H_t))   STOP.
        else { H_{t+1} := H_t \ M_t; t := t + 1; }
             if (F(H_t) > F(Γ)) {Γ = H_t;}
             go to Step 1.
```

This iterative algorithm begins by calculating $F(V)$ and the set M_1 containing the subset of vertices that satisfy $F(V) = \pi(i, V)$, i.e., $M_1 = \{i \in V : \pi(i, V) = F(V)\}$. The vertices in the set M_1 are removed from V to get $H_2 = V \setminus M_1$. At the iteration t, it considers the set H_{t-1} as input, calculates $F(H_{t-1})$, identifies the subset M_t such that $F(H_{t-1}) = \pi(i_t, H_{t-1}), \forall i_t \in M_t$, and removes this subset from H_{t-1} to produce $H_t = H_{t-1} \setminus M_t$. The algorithm terminates at the iteration T when $H_T = \emptyset$ or $F(H_T) = 0$. It outputs $\hat{H}$ as the subset, H_j with smallest j such that $F(H_j) \geq F(H_\ell) \forall l \in \{1, 2, \ldots, T\}$. Thus, the algorithm finds the largest optimal solution that includes all other optimal solution.

This algorithm resembles the one for finding the largest subgraph with maximum minimum degree [11], however, our formulation is very general and applies whenever the function $F(H)$ is quasi-concave. Furthermore, by designing an appropriate linkage function various structures in a graph can be obtained [12].

Theorem 1. *The subset Γ output by the above algorithm is the $\cup$-maximizer for F in the set V.*

Proof. According to the algorithm, $F(\Gamma) = \max_{H_i \in \mathcal{H}} F(H_i)$, where $\mathcal{H} = \{H_1, H_2, \ldots, H_T\}$ and $H_T \subset \ldots \subset H_2 \subset H_1 = V$. We divide the proof into two cases.
Case (i) $[H \setminus \Gamma \neq \emptyset]$: Let H_i be the smallest set in the sequence $\mathcal{H}$ containing H, so that $H \subseteq H_i$ but $H \not\subseteq H_{i+1}$. Since $M_i = H_i \setminus H_{i+1}$, there is at least one element, say i_H, common to both M_i and H. By definition of $F(H_i)$, we have

$$F(H_i) = \min_{i \in H_i} \pi(i, H_i) = \pi(i^*, H_i) \tag{6}$$

By construction of M_i, $i^* \in M_i$, so $\pi(i^*, H_i) = \pi(i_H, H_i)$. Using (4) we get

$$\pi(i^*, H_i) = \pi(i_H, H_i) \geq \pi(i_H, H) \geq \min_{i \in H} \pi(i, H) = F(H) \tag{7}$$

Using (6) and (7) we obtain $F(H_i) \geq F(H)$. According to the algorithm $F(\Gamma) > F(H_i)$, $\forall H_i \supset \Gamma$, so we prove

$$F(\Gamma) > F(H), \; \forall H \setminus \Gamma \neq \emptyset \tag{8}$$

Case (ii) $[H \subseteq \Gamma]$: Similar to the previous case, there exists a smallest subset H_i in the sequence $\mathcal{H}$ that includes H. So, the inequalities in (7) hold here too, and we could write $F(H_i) \geq F(H)$. On the other hand, $F(\Gamma) \geq F(H_i)$, which in conjunction with the previous inequality implies

$$F(\Gamma) \geq F(H), \; \forall H \subseteq \Gamma \tag{9}$$

Further, a maximizer satisfying inequalities (8) and (9) is the $\cup$-maximizer. □

Table 2. Pseudocode extract for finding a series of multipartite clusters

```
Initialization: V^0 := V; m := 0; C = ∅;
Step 1: Extract Ĥ^m from V^m using (1); Add Ĥ^m to C;
Step 2: V^{m+1} := V^m \ Ĥ^m; m := m + 1;
Step 3: if ( (V^m = ∅) ∧ (m_ij = 0 ∀i, j ∈ V^m) )
                Output C, V^m as R, and m; STOP;
        else go to step 1
```

3.1 Partitioning of Data into Multipartite Clusters (MPC)

The algorithm in Table 1 outputs one multipartite cluster. However, many such clusters are likely to be present in the set V. If we assume that these clusters are unrelated, we can use a simple heuristic of iteratively applying the above procedure to extract all these clusters. To do this we remove the elements belonging to the first cluster $\hat{H}$ from V and extract another multipartite cluster in the set $V \setminus \hat{H}$. This procedure is formalized in Table 2 and produces an ordered set, $C = \{\hat{H}^0, \hat{H}^1, \ldots, \hat{H}^m\}$, of m ortholog clusters, and a set of residual elements $R = \{i : i \in G \setminus C\}$. The number, m, of non-trivial clusters (ortholog clusters) is automatically determined by the method. It must be remarked that every cluster in C contains genes from at least two genomes.

4 Analysis and Implementation

The run-time of the algorithm depends on the efficiency of evaluation of the linkage function. A linkage function which can be updated efficiently, instead of having to be evaluated from scratch at each iteration, is preferable. The linkage function described in (3) is additive and can be updated efficiently when vertices are removed from the set.

Theorem 2. *The algorithm for finding an ortholog cluster runs in time $O(|E| + |V| \log |V|)$ and space $O(|E| + |V|)$.*

Proof. Clearly, the step 0 of the algorithm in Table 1 takes a constant time. The initialization includes computing $\pi(i, V)\ \forall i \in V$. To compute $\pi(i, V)$ we must look at all edges incident on i, thus computing $\pi(i, V)\ \forall i \in V$ takes $O(|E|)$ time. At subsequent iterations, due to the additive property of the linkage function (3), efficient updates are possible without recomputing from scratch. As each edge is deleted once, all linkage function updates together requires $O(|E|)$ time.

Step 1 involves determining the set M_t by finding the vertices with minimum value of the linkage function. Observe that in a sparse multipartite graph, only a few edges are deleted at each iteration implying that only a few linkage function values are updated. Consequently, the order of vertices determined by the linkage function values remains approximately fixed. We use Fibonacci heaps [13] (which work irrespective of the sparsity in the input graph) to store vertices according to their linkage function values. So, elements in the set M_t can be found in $O(1)$

time using the `find-min` operation. In Step 2, the set $H_{t+1} = H_t \setminus M_t$ can be found in $O(\log |H_t|)$ time using the `delete-min` operation and each update to linkage function value can be performed in $O(1)$ using the `decrease-key` operation. Thus, using Fibonacci heaps each iteration takes $O(\log |H_t|)$ time. The maximum number of iteration is $|V|$, and each iteration takes at most $O(\log |V|)$. Thus, the algorithm runs in $O(|E| + |V| \log |V|)$ time. Using the adjacency list representation for the graph, the algorithm requires $O(|E| + |V|)$ space. □

Theorem 3. *If the values in the similarity matrix are discretized into c different values, the algorithm runs in* $O(|E| + |V|)$ *time.*

Proof. The initialization step, as in Theorem 2, takes $O(|E|)$ time. To reduce the time complexity of subsequent iterations, we sort and store the vertices in the order of the initial linkage function values, $\pi(i, V)$. We assume that the graph is represented in the adjacency list format, and sort the vertices using the bucket sort algorithm [14]. Within each bucket, the vertices are stored using a linked list to accommodate multiple vertices with the same value. The edge weights can take c different values, so the initial linkage function values are bounded i.e., $0 \leq \pi(i, V) \leq |V|.c$. So, sorting the $|V|$ values of $\pi(i, V)$ takes $O(|V|)$ time.

At each iteration, finding the vertex with minimum value of the linkage function. This takes $O(1)$ time as vertices are sorted according to the linkage function values. Deletion of a vertex entails updates to the linkage function values for the neighboring vertices. The cost for updating linkage function values is already considered, but to preserve the sorted order of vertices, we must find the new place for each updated vertex. Since the edge weights are discretized, the new place must lie at most c bins away from the current position (towards the minimum). Thus, the new place is found in at most c, or $O(1)$ time. Every iteration requires $O(1)$ time, and since the number of iterations is at most $|V|$, the total time is bounded by $O(|V|)$. Combining this with the total cost for computing the linkage function values, the run time of the algorithm is $O(|E| + |V|)$. □

As a result of Theorem 2, the procedure for finding all the m ortholog clusters runs in time $O(m(|E| + |V|))$. We now give some implementation details which do not improve the complexity of the algorithm but enable a speedup in practice.

Vertices in different connected components of a graph cannot come together to form a cluster. So, different connected components can be processed in isolation. Furthermore, the input multipartite graph between genes is large but very sparse, so when a dense subgraph is extracted as the optimal solution, the remaining graph becomes disconnected. As a consequence, in practice, a significant speedup is achieved when the procedure in Table 2 is run on individual connected components after extraction of an ortholog cluster. Also, finding an optimal solution in different connected components is amenable to parallelism.

The algorithm in Table 1 removes the vertices corresponding to the minimum of the linkage function, if, however, we could remove a larger set of vertices without affecting the correctness of the procedure, the algorithm would be faster as more vertices would be removed at each iteration. The following theorem determines such elements.

Theorem 4. *Define $Q = \{i \in V : \pi(i,V) < \theta, \theta > 0\}$ and let $\hat{H}_{V \setminus Q}$ be the $\cup$-maximizer in the set $V \setminus Q$, and $\hat{H}$ be the $\cup$-maximizer in the set V. Then,*

$$F(\hat{H}_{V \setminus Q}) > \theta \Rightarrow \hat{H}_{V \setminus Q} = \hat{H}. \tag{10}$$

Proof. The score-value of $\hat{H}_{V \setminus Q}$, obtained from $V \setminus Q$ (a subset of V) can be at most the score-value of $\hat{H}$ obtained from V, i.e., $F(\hat{H}) \geq F(\hat{H}_{V \setminus Q}) > \theta$. Then using (2) it follows that $\forall i \in \hat{H}, \pi(i,\hat{H}) > \theta$. Further, by the monotonicity property we get $\pi(i,V) \geq \pi(i,\hat{H}) > \theta$. But, according to the definition of Q, $\pi(i,V) > \theta \Rightarrow i \in V \setminus Q$. This proves that $\forall i \in \hat{H}, i \in V \setminus Q$, in other words, $\hat{H} \subseteq V \setminus Q$. But, by the definition of $\cup$-maximizer, we have $F(\hat{H}_{V \setminus Q}) \geq F(H) \forall H \subseteq V \setminus Q$ i.e., $F(\hat{H}_{V \setminus Q}) \geq F(\hat{H})$. But we already had $F(\hat{H}) \geq F(\hat{H}_{V \setminus Q})$, so $F(\hat{H}_{V \setminus Q}) = F(\hat{H})$. Further, uniqueness of $\cup$-maximizer implies $\hat{H}_{V \setminus Q} = \hat{H}$. □

According to Theorem 4, we can remove all vertices whose linkage function value is less than the current estimate of the score value of $\cup$-maximizer. In the algorithm in Table 1, $F(\Gamma)$ is the current estimate of the $F(\hat{H})$, so we can remove all vertices, $M_t = \{i : \pi(i,H_t) < F(\Gamma)\} \cup \{i : \pi(i,H_t) = \min_{j \in H_t} \pi(j,H_t)\}$. This reduces the number of iterations as more vertices are likely to be removed at each iteration.

Theorem 4 can also improve the run-time of the procedure given in Table 2. This procedure, at the iteration t, finds a cluster as the optimal set $\hat{H}^t$, removes this optimal set from the current set to produce a resulting set, $V^{t+1} = V^t \setminus \hat{H}^t$, in which the optimal set, $\hat{H}^{t+1}$ is found at the next iteration $t+1$. For the ortholog clustering, we empirically found that $F(\hat{H}^{t+1})/F(\hat{H}^t) \geq 0.8$, so we apply Theorem 4 with $\theta = 0.8 * F(\hat{H}^t)$ for finding $\hat{H}^{t+1}$. Such preprocessing removes more than 95% of vertices that do not belong to the optimal set and thus leads to significant performance gains.

5 Experimental Results

We have applied the proposed method for constructing ortholog clusters to the complete genome data on which the manually curated eukaryotic orthologous groups (KOGs) [5] are constructed. The 4,852 KOGs contain 60,759 sequences from 112,920 sequences present in the seven eukaryotic genomes (*A. thaliana*, *C. elegans*, *D. melanogaster*, *E. cuniculi*, *H. sapiens*, *S. cerevisiae* and *S. pombe*).

For this study, we used the linkage function described in (3). The species tree in [5] was used to calculate the phylogenetic distance function. We used the pair-wise sequence similarity scores (bit-scores) computed using Blast [15] available at the KOG website. We considered only a subset of the top hits for any given sequence. To be precise, if the best hit for gene $i \in V_s$ in another genome V_t is $j \in V_t$ with bit-score m_{ij}, then we consider all genes from V_t which have bit-score value larger than $m_{ij}/2$. The idea behind such selection is to avoid low-scoring spurious hits for a given gene. The values of sequence similarity scores between genes within an ortholog family vary across ortholog families

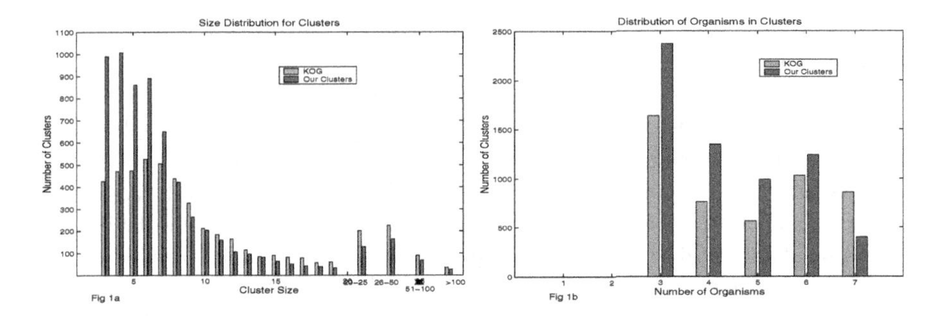

Fig. 1. A comparison of distribution of size and number of organisms in KOGs and our ortholog clusters

and one cannot use a constant threshold across all ortholog families. Such gene-specific and genome-specific cutoff [3] avoids spurious matches without filtering out potential orthologs.

Our method produced 36,034 clusters including 25,434 singletons, 2,870 clusters of size 2 and 7,830 clusters that contain at least 3 sequences. A comparative distribution of size and the number of organisms in our clusters and the KOGs is shown in Figure 1a and 1b, respectively. In comparison to KOGs, our ortholog clusters are relatively smaller in size and contain sequences from fewer genomes.

A KOG cluster, by construction, contains sequences from at least three genomes, so for the purpose of comparison, we divided the 7,830 clusters with at least three sequences into 1,488 clusters containing sequences from 2 genomes and 6,342 ortholog clusters that contain sequences from at least 3 genomes. The 6,342 clusters contain 61,272 sequences of which 47,458 are common with the 60,759 sequences in the KOG ortholog clusters. Of the 13,301 sequences from KOGs that are not covered by these clusters, 9,078 sequences are grouped into 1,566 clusters that contain sequences from at most two genomes while the remaining are classified as singletons. In comparison to KOGs our clusters contain fewer paralogs - a desirable feature obtained by ignoring similarities between genes within a genome. Although desirable from an ortholog clustering perspective, this also means that our ortholog clusters are smaller in size as the method avoids detection of paralogs in an ortholog cluster. This is consistent with the distribution of sizes (see Figure 1a) and the following statistical comparison.

To estimate the association between KOGs and our ortholog clusters, we used several statistical parameters. We used the rand index [16] to quantify the relatedness of the two clusterings, and obtained a value of 0.792 implying that approximately 80% of all pairs of sequences in input data are co-clustered in KOGs and in our results. The average number of KOGs that overlap with an ortholog cluster is 1.031, which indicates that most of our clusters contain sequences from a single KOG. Conversely, the average number of candidate clusters that overlap with a KOG is 3.012, however, in most cases a KOG completely contains the candidate clusters which suggests that our clusters are homogeneous with respect to the KOGs. These statistics along with the size distribution are a confirmation of our observation that our ortholog clusters are usually subsets of single KOGs.

A set-theoretic comparison of our clusters with the KOGs shows 844 KOGs exactly match our clusters. There are 611 KOGs that are divided into 2 clusters, each of which contains a sequence only from the corresponding KOG. Carrying out this analysis further, we found that 2,472 KOGs (51% of all KOGs) can be partitioned into our clusters whose members belong to a single KOG. The remaining 2,380 KOGs overlap with at least one mixed ortholog cluster, i.e., cluster that either contains sequences from multiple KOGs, or contains some sequences that do not belong to any KOG . Among 6,342 clusters produced by our method, there are 1,857 clusters that contain sequences from multiple KOGs. Using Pfam [17] annotations to assess the homogeneity of these 1,857 clusters, we found that all members in 952 of these clusters were annotated with the same Pfam families while no member in 436 clusters could be annotated with any Pfam family. In summary, the statistical coefficients and the set-theoretic comparison of our clusters with the manually curated ortholog clusters in KOG shows that the two clusterings to be very well correlated.

6 Conclusion

We have modeled the problem of finding orthologous clusters in a large number of genomes as clustering on a multipartite graph. The proposed method is efficient and finds an ortholog cluster in time $O(|E|+|V|\log|V|)$. To further speedup the method, we presented implementation choices that lead to significant speedups in practice. The proposed ortholog clustering method was applied to the seven eukaryote genomes on which KOG ortholog clusters are constructed. The analysis of the results shows that clusters obtained using the proposed method show a high degree of correlation with the manually curated ortholog clusters.

This method extracts an ortholog cluster by ignoring similarities between genes within a genome while emphasizing orthologous relationships between genes from different genomes. Since observed sequence similarity scores are higher for recently diverged orthologs compared to those for the anciently diverged orthologs, we used the species tree to correct for these differences (3).

Corrections to observed sequence similarity using phylogenetic trees assume the correctness of the given phylogenetic tree. However, there are instances when multiple hypothesis about the phylogenetic relationship between a group of organisms exist, in such cases the confidence in those hypotheses is low. We hope that the proposed method can be modified to resolve such conflicts by constructing the gene tree for each ortholog cluster and deriving support for the species tree(s) from these gene trees (personal communication with Roderic Guigo and Temple Smith [18]). When the phylogenetic information is controversial, an iterative process of finding ortholog clusters can resolve the ambiguities in the phylogenetic tree.

Recently the gene order has been used to find ortholog clusters in a pair of genomes [19]. The conserved gene order between a pair of genomes can be inferred using programs like DiagHunter [19]. This additional information about the orthologous relationships also can be incorporated in our method. Indeed, let us consider a new similarity coefficient between genes i and j which belong to different genomes, but are conserved in gene order as determined by methods such as DiagHunter. Then, we can design a new linkage function

$$\pi'(i,H) = \pi(i,H)\left(\sum_{\substack{\ell=1\\ \ell\neq g(i)}}^{k}\sum_{j\in H_\ell} e_{ij}\right) \tag{11}$$

where $\pi(i, H)$ is linkage function defined in (3). The first term inside the parentheses aggregates gene order similarity coefficients between gene i and genes in other genomes, while the second term aggregates gene order similarity coefficients between the gene i and genes in H_ℓ, the subset of genes from genome l present in H. Thus, $\pi'(i, H)$ incorporates three diverse components: sequence similarity, species tree, and the gene order information, critical for ortholog clustering.

References

1. Fitch, W.M.: Distinguishing homologous from analogous proteins. Syst Zool. **19** (1970) 99–113
2. Fujibuchi, W., Ogata, H., Matsuda, H., Kanehisa, M.: Automatic detection of conserved gene clusters in multiple genomes by graph comparison and P-quasi grouping. Nucleic Acids Res. **28** (2002) 4096–4036
3. Kamvysselis, M., Patterson, N., Birren, B., Berger, B., Lander, E.: Whole-genome comparative annotation and regulatory motif discovery in multiple yeast species. In: RECOMB. (2003) 157–166
4. Remm, M., Strom, C., Sonnhammer, E.: Automatics clustering of orthologs and in-paralogs from pairwise species comparisons. J Mol Biol. **314** (2001) 1041–1052
5. Koonin, E.V. *et al.*: A comprehensive evolutionary classification of proteins encoded in complete eukaryotic genomes. Genome Biol. **5** (2004)
6. Tatusov, R., Koonin, E., Lipmann, D.: A genomic perspective on protein families. Science **278** (1997) 631–637
7. Zmasek, C., Eddy, S.: RIO: Analyzing proteomes by automated phylogenomics using resampled inference of orthologs. BioMed Central Bioinformatics **3** (2002)
8. Huynen, M.A., Bork, P.: Measuring genome evolution. Proc. Natl. Acad. Sci. USA **95** (1998) 5849–5856
9. Tang, J., Moret, B.: Phylogenetic reconstruction from gene rearrangement data with unequal gene content. In: Proc. 8th Workshop on Algorithms and Data Structures (WADS'03). (2003) 37–46
10. Dawande, M., Keskinocak, P., Swaminathan, J.M., Tayur, S.: On bipartite and multipartite clique problems. J. Algorithms **41** (2001) 388–403
11. Matula, D.W., Beck, L.L.: Smallest-last ordering and clustering and graph coloring algorithms. J. ACM **30** (1983) 417–427
12. Mirkin, B., Muchnik, I.: Induced layered clusters, hereditary mappings, and convex geometries. Appl. Math. Lett. **15** (2002) 293–298
13. Fredman, M.L., Tarjan, R.E.: Fibonacci heaps and their uses in improved network optimization algorithms. J. ACM **34** (1987) 596–615

14. Cormen, T.H., Leiserson, C.E., Rivest, R.L., Stein, C.: Introduction to Algorithms, Second Edition. The MIT Press (2001)
15. Altschul, S. *et al.*: Gapped BLAST and PSI-BLAST: a new generation of protein database search programs. Nucleic Acids Res. **25** (1997) 3389–3402
16. Rand, W.M.: Objective criterion for the evaluation of clustering methods. J. Am. stat. Assoc. **66** (1971) 846–850
17. Bateman, A. *et al.*: The Pfam protein families database. Nucleic Acids Res. **32** (2004) 138–141
18. Guigo, R., Muchnik, I., Smith, T.: Reconstruction of ancient molecular phylogeny. Mol Phylogenet Evol. **6** (1996) 189–213
19. Cannon, S.B., Young, N.D.: OrthoParaMap: Distinguishing orthologs from paralogs by integrating comparative genome data and gene phylogenies. BMC Bioinformatics **4** (2003)

Linear Time Algorithm for Parsing RNA Secondary Structure

Extended Abstract

Baharak Rastegari and Anne Condon

Department of Computer Science, University of British Columbia

Abstract. Accurate prediction of pseudoknotted RNA secondary structure is an important computational challenge. Typical prediction algorithms aim to find a structure with minimum free energy according to some thermodynamic ("sum of loop energies") model that is implicit in the recurrences of the algorithm. However, a clear definition of what exactly are the loops and stems in pseudoknotted structures, and their associated energies, has been lacking.

We present a comprehensive classification of loops in pseudoknotted RNA secondary structures. Building on an algorithm of Bader et al. [2] we obtain a linear time algorithm for parsing a secondary structures into its component loops.

We also give a linear time algorithm to calculate the free energy of a pseudoknotted secondary structure. This is useful for heuristic prediction algorithms which are widely used since (pseudoknotted) RNA secondary structure prediction is NP-hard. Finally, we give a linear time algorithm to test whether a secondary structure is in the class handled by Akutsu's algorithm [1]. Using our tests, we analyze the generality of Akutsu's algorithm for real biological structures.

1 Introduction

RNA molecules play diverse roles in the cell: as carriers of information, catalysts in cellular processes, and mediators in determining the expression level of genes [8]. The structure of an RNA molecule is often the key to its function with other molecules. In particular, the *secondary structure*, which describes which bases of an RNA molecule bond with each other, can provide much useful insight as to the function of the molecule. If the RNA molecule is viewed as an ordered sequence of n bases (Adenine (A), Guanine (G), Cytosine (C), and Uracil (U)), indexed starting at 1 from the so-called 5' end of the molecule, then its secondary structure is a set of pairs $i \cdot j$, $1 \leq i < j \leq n$ with each index in at most one pair.

Most well known are *pseudoknot free* secondary structures in which no base pairs overlap - that is, there do not exist two base pairs $i \cdot j$ and $i' \cdot j'$ in the structure with $i < i' < j < j'$. Because of their biological importance, there has been a huge investment in understanding the thermodynamics of pseudoknot free secondary structure formation. For example, it is well understood that in a pseudoknot free secondary structure, the base pairs together with unpaired bases

R. Casadio and G. Myers (Eds.): WABI 2005, LNBI 3692, pp. 341–352, 2005.

form hairpin loops, internal loops (of which stacked pairs and bulge loops are special cases), external loops, or multiloops, with every unpaired base in exactly one loop and every base pair in exactly two loops. Parameters for estimating the free energies of such loops have been determined experimentally. The standard thermodynamic model posits that the free energy of a pseudoknot free secondary structure is the sum of the energies of its loops. A pseudoknot free secondary structure can be conveniently represented as a string in dot-parenthesis format, a generalization of a string of balanced parentheses in which matching parentheses denote base pairs and dots denote unpaired bases. It is straightforward to parse a pseudoknot free secondary structure represented in dot-parenthesis notation in linear time, in order to determine its loops and calculate its free energy. Finally, dynamic programming algorithms can find the minimum free energy (mfe) pseudoknot free secondary structure in $O(n^3)$ time; the mfe structure is the most stable of the possibly exponentially many structures that a molecule may form, according to current models.

In contrast, there has been no classification of loops in pseudoknotted secondary structures, though some examples of structural motifs, such as kissing hairpins, have been named. Since pseudoknotted secondary structure prediction is NP-hard, several polynomial time algorithms have been proposed for predicting the mfe secondary structure from restricted classes of structures that may contain pseudoknots. Of these, the $O(n^6)$ algorithm of Rivas and Eddy [12] handles (i.e. finds the mfe structure from) the most general class of structures. However, the loop types and thermodynamic model underlying the Rivas and Eddy and other algorithms are specified only implicitly in the recurrence equations of the algorithms. There is not a one-to-one correspondence between loops and terms in the recurrence equations, making it difficult to infer the loop types directly from the recurrences. The underlying energy models are unclear; there has been no algorithm to calculate the energy of a structure, and no way to compare the quality of thermodynamic models proposed by different authors.

In this work we present the first classification of loops that arise in pseudoknotted secondary structures. Our classification is derived from the algorithm of Rivas and Eddy, and allows us to formulate the thermodynamic models underlying the Rivas and Eddy and other dynamic programming algorithms as sum-of-loop-energies models. With this description, it becomes possible to evaluate the strengths and weaknesses of current thermodynamic models for pseudoknotted structures.

By extending an algorithm of Bader et al. [2], it is possible to parse a given secondary structure into its component loops in linear time. We present two applications of this parsing algorithm. First, we show how to calculate the free energy of a pseudoknotted secondary structure in linear time. This can be useful in heuristic algorithms, which hold promise since pseudoknotted secondary structure prediction is NP-hard [11].

The second application of our parsing algorithm is in assessing the trade-off between generality and running time of dynamic programming algorithms for RNA secondary structure prediction. Each dynamic programming algorithm in the literature only predicts structures from a restricted class. Usually, the more general the class, the higher the running time of the algorithm. An outstanding challenge is to design efficient dynamic programming algorithms that can predict biologically important structures. For example, Akutsu [1] proposed an algorithm that runs in $O(n^5)$ time, can in theory handle more secondary structures than the $O(n^5)$ algorithm of Dirks and Pierce [9], though less than the $O(n^6)$ algorithm of Rivas and Eddy. As another example, Uemura [14] proposed an algorithm that runs in $O(n^5)$ time, similar to Akutsu's algorithm in time complexity, but in theory handle more secondary structures than Akutsu's algorithm, though it is much more harder to understand and analyse. Let U, A, D&P, and R&E denote the classes of structures handled by the Uemura, Akutsu, Dirks and Pierce, and Rivas and Eddy algorithms, respectively. The question we address is: does A contain more biologically meaningful structures than does D&P and perhaps as many as U and/or R&E?

To help answer this question, we apply the parsing algorithm to give linear time test for membership in class A. In previous work [7], we obtained linear time tests for membership in the D&P and R&E classes. We provide a comparison of all four algorithms on a set of 1439 biological structures; the result shows that exactly 2 of the structures are in class A but not in class D&P.

The paper is organized as follows. In Sec. 2, we define what is a closed region in an RNA secondary structure. (The parsing algorithm, based on an algorithm of Bader et al. [2], is not shown due to the lack of space). In Sect. 3 we present our loops classification and our algorithm for enumerating the loops of a secondary structure. We briefly describe how to calculate the free energy of a secondary structure in Sect. 4. Our algorithm for testing membership in Akutsu's class is in Sect. 5, and conclusions are in Sect. 6.

We should note that some details of the algorithms and most of the details of the proofs are eliminated in this extended abstract.

2 Closed Regions

Here we first introduce *closed regions* of a secondary structure, which are important throughout the paper. Examples are shown in Fig. 1, where a secondary structure is represented as an arc diagram, in which base indices are shown as vertices on a straight line (backbone), ordered from the $5'$ end, and arcs (always above the straight line) indicate base pairs. Intuitively, a closed region is a "minimal" set of contiguous base indices - corresponding to a region of the line - with the property that no arcs leave the region and there is at least one arc in the region. The definitions in this and the following sections are with respect to a fixed non-empty secondary structure R for an RNA sequence of length n.

We denote the set of indices $i, i+1, ..., j$ by $[i;j]$ and call this set a region if $i \leq j$. We say that region $[i;j]$ is **weakly closed** if it contains at least one base pair and for all base pairs $i' \cdot j'$ of R, $i' \in [i;j]$ if and only if $j' \in [i;j]$. We say that $[i;j]$ is **closed**, and write $i;j$, if either (i) $i = 1$ and $j = n$ or (ii) $[i;j]$ is weakly closed and for all l with $i < l < j$, $[i;l]$ and $[l;j]$ are not weakly closed (Fig. 1(a)).

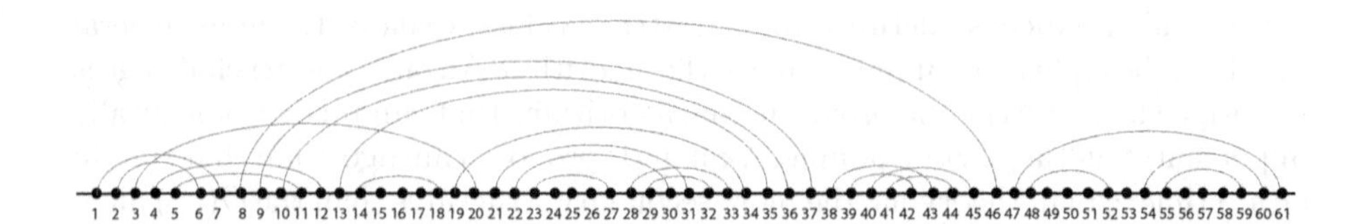

Fig. 1. Arc diagram representation of an RNA secondary structure R. **(a)** $[1;46]$, $[38;45]$, and $[47;61]$ are closed regions. $[48;60]$ is weakly closed but it is not closed as $[48;52]$ is weakly closed. $[38;45]$ is a pseudoknotted closed region. 38.43 and 40.45 are its external base pairs, and 38 and 45 are its left and right borders respectively. 38.43, 39.42, 40.45 and 41.44 are all pseudoknotted pairs and $38, 39, ..., 45$ are all pseudoknotted bases. **(b)** $[48;52]$ and $[54;60]$ are disjoint closed regions and both are nested in $[47;61]$. **(c)** $[8;10] \cup [36;46]$ is a band of pseudoknotted closed region $[1;46]$, and 8.46 and 10.36 are the band's outer and inner closing pairs. $[8;10]$ and $[36;46]$ are the band's regions, and 8 and 46 are the left and the right border of the band. 8.46, 9.37 and 10.36 span the band. **(d)** 47.61 is a multiloop external base pair with $(48, 52)$ and $(54, 60)$ as tuples. $(1, 46)$ and $(47, 61)$ are the tuples of an external loop. **(e)** $[1;46]$ is a pseudoknotted loop with bands $[1;2] \cup [6;7]$, $[3;3] \cup [20;20]$, $[4;5] \cup [11;12]$, $[8;10] \cup [36;46]$ and $[13;19] \cup [34;35]$. $[21;33]$ is the closed region nested in $[1;46]$. **(f)** 1.7 and 2.6 are the external and internal base pairs of an interior-pseudoknotted loop. **(g)** 8.46 is the external base pair of a multi-pseudoknotted loop with $(9, 37)$ and $(38, 45)$ as tuples. **(h)** $[38;45]$ is an in-Band loop, $[21;33]$ is an out-Band loop, and 8.46 is the external base pair of a span-Band loop (multi-pseudoknotted loop).

Let $i;j'$. If i' and j are such that $i.j$ and $i'.j'$ then we say that $i.j$ and $i'.j'$ are the *external* base pairs of $[i;j']$. If $i.j'$ then the region has just one external base pair; otherwise we call $[i;j]$ a **pseudoknotted closed region**. We also refer to i and j' as $[i;j']$'s left and right borders respectively.

Pair $i.j$ is **pseudoknotted** if there exists $i'.j'$ with $i < i' < j < j'$ or $i' < i < j' < j$. We also refer to i and j as **pseudoknotted** base indices.

2.1 Closed Regions Tree

Let $i;j$ and $i';j'$ with $i < i'$. If $j < i'$ we say that $[i;j]$ and $[i';j']$ are *disjoint*; otherwise we say that $[i';j']$ is *nested* in $[i;j]$ (Fig. 1(b)).

We say that closed region $[i';j']$ is a child of closed region $[i;j]$ if $[i';j']$ is nested in $[i;j]$ and is not nested in any closed region $[i'';j'']$ with $i < i''$. We say that $[i;j]$ and $[i';j']$ are siblings if they are children of the same closed region and $i \neq i'$. So the closed regions form a tree structure.

A tree $T(R)$ in which the children of a node are ordered is called the closed regions tree of R if: (i) there is a 1-1 correspondence between nodes of the tree and closed regions of R, and (ii) if node V corresponds to closed region C then V is the parent of all the nodes whose corresponding closed regions are nested in C. The children of each node are ordered by the left index of the closed region.

Building on an algorithm by Bader et al. [2], parsing algorithm (not shown) builds the closed region tree in linear time. (Details will be in full paper - omitted in this extended abstract)

3 Loops

In this section we describe the loops that comprise a pseudoknotted secondary structure, and how these can be enumerated in linear time. Models underlying the algorithm of Rivas and Eddy [12] and the algorithm of Dirks and Pierce [9] can be expressed by sum of the loops that we describe here. We need one important definition, that of a *band.*

3.1 Bands

Loosely speaking, a band is a pseudoknotted stem, which may contain internal loops or multi loops (Fig. 1(c)). We next define a band formally.

Let $i_2.j_2$ be a pseudoknotted base pair. We say that $i_2.j_2$ is *directly* banded in $i_1.j_1$ if (i) $i_1 \leq i_2 < j_2 \leq j_1$, and (ii) $[i_1+1, i_2-1]$ and $[j_2+1, j_1-1]$ are weakly closed. Note that the "is directly banded in" relation is reflexive. We let "are banded" be the symmetric and transitive closure of the "is directly banded in" relation. Let B be an equivalence class under the "are banded" relation. That is, B is a set of base pairs such that every two base pairs in B are banded and every base pair in B is pseudoknotted. B has outer and inner closing base pairs $i_1 \cdot j_1$ and $i'_1 \cdot j'_1$ respectively, such that for every base pair $i.j$ in B, $i_1 \leq i \leq i'_1$ and $j'_1 \leq j \leq j_1$. Note that $i_1 \cdot j_1$ may equal $i'_1 \cdot j'_1$.

We call the union of two non-overlapping regions a *gapped region.* A gapped region $[i_1; i'_1] \cup [j'_1; j_1]$ is a *band* if for some equivalence class B, $i_1 \cdot j_1$ and $i'_1 \cdot j'_1$ are the closing pairs of B. We refer to i_1 and j_1 as the left and the right border of the band respectively (Fig. 1(c)).

We refer to $[i_1; i'_1]$ and $[j'_1; j_1]$ as the band regions, which have borders i_1, i'_1 and j'_1, j_1 respectively. Closed region $i; j$ is *contained in* band $[i_1; i'_1] \cup [j'_1; j_1]$, if and only if $i; j$ is in a band region - that is, $i, j \in [i_1; i'_1]$ or $i, j \in [j'_1; j_1]$ - and there is no p, q with $p; q$, $p < i < j < q$, such that $p, q \in [i_1; i'_1]$ or $p, q \in [j'_1; j_1]$. Base pair $i.j$ *spans* band $[i_1; i'_1] \cup [j'_1; j_1]$ if $i_1 \leq i \leq i'_1$ and $j'_1 \leq j \leq j_1$.

We say that $[i_1; i'_1] \cup [j'_1; j_1]$ is a *band of closed region* $[i; j]$ if $i \leq i_1 \leq j_1 \leq j$ and there is no $p; q$ with $i < p \leq i_1 < j_1 \leq q < j$.

Lemma 1. *Let $i_1.j_1, i_2.j_2, ..., i_n.j_n$, $i_1 < i_2 < ... < i_n$, be the base pairs that span band $[i_1; i'_1] \cup [j'_1; j_1]$. Then $j_n < .... < j_2 < j_1$.*

3.2 Loop Types

Our definitions of hairpin and interior loops are standard for pseudoknot free structures so we do not include them here. The definitions of multiloop and external loop are generalized (Fig. 1(d)):

Multiloop: contains an external base pair $i.j$ and k tuples (i_1, j_1), (i_2, j_2), ..., (i_k, j_k), for some $k \geq 1$, along with the bases in $[i+1; j-1] - \cup[i_l; j_l], 1 \leq l \leq k$ all of which must be unpaired, where $i_l; j_l, 1 \leq l \leq k$, $i < i_1 < j_1 < i_2 < j_2 < ... < i_k < j_k < j$. Also, if $i_l.j_l$, $1 \leq l \leq k$, then k should be at least 2.

External loop: contains $k > 0$ tuples (i_1, j_1), (i_2, j_2), ..., (i_k, j_k) along with the bases in $[1; n] - \cup_{1 \leq l \leq k}[i_l; j_l]$, all of which must be unpaired, where $i_l; j_l$, $1 \leq l \leq k$, and $i_1 < j_1 < i_2 < j_2 < ... < i_k < j_k$.

We next introduce further types of elementary structures which are the consequence of having pseudoknotted base pairs and pseudoknotted regions.

Pseudoknotted loop: Let $[i; j']$ be a pseudoknotted closed region. Let the bands of $[i; j']$ be: $[i_1; i'_1] \cup [j'_1; j_1]$, $[i_2; i'_2] \cup [j'_2; j_2]$, ..., $[i_m; i'_m] \cup [j'_m; j_m]$. Let $[p_1; q_1]$, $[p_2; q_2]$, ..., $[p_k; q_k]$ be children of $[i; j']$ which are nested in $[i; j] - (\cup_{l=1}^{m}[i_l; i'_l] \cup_{l=1}^{m} [j'_l; j_l])$. The *pseudoknotted loop* corresponding to $[i; j']$ is the set: $\{(i_l, j_l), (i'_l, j'_l) | 1 \leq l \leq m\} \cup \{(p_l, q_l) | 1 \leq l \leq k\}$, along with the bases in: $[i; j'] - \cup_{l=1}^{k}[p_l; q_l] - \cup_{l=1}^{m}[i_l; i'_l] - \cup_{l=1}^{m}[j'_l; j_l]$ all of which must be unpaired (Fig. 1(e)).

Interior-pseudoknotted loop: contains two base pairs $i.j$ and $i'.j'$ where $i < i' < j' < j$, along with the bases in $[i+1, i'-1] \cup [j'+1, j-1]$ all of which must be unpaired. Moreover, there is a band $[bi; bi'] \cup [bj'; bj]$ such that $bi \leq i < bi'$ and $bj' < j \leq bj$. We refer to $i.j$ and $i'.j'$ as the interior-pseudoknotted loop external and internal base pairs respectively (Fig. 1(f)).

Multi-pseudoknotted loops: contains an external base pair $i.j$ and k tuples (i_1, j_1), (i_2, j_2), ..., (i_k, j_k), for some $k > 1$, along with the bases in $[i+1; j-1] - \cup_{1 \leq l \leq k}[i_l; j_l]$, all of which must be unpaired, where (i) there is a band $[bi; bi'] \cup [bj'; bj]$ such that $bi \leq i < bi'$ and $bj' < j \leq bj$, (ii) $i_l; j_l$, for all $1 \leq l \leq k$ except for exactly one tuple (i_{l_0}, j_{l_0}) for which $i_{l_0}; j_{l_0}$ is not true (i.e $[i_{l_0}; j_{l_0}]$ is not a closed region) and $i_{l_0}.j_{l_0}$ spans the band ($bi \leq i_{l_0} \leq bi'$ and $bj' \leq j_{l_0} \leq bj$), and (iii) $i < i_1 < j_1 < i_2 < j_2 < ... < i_k < j_k < j$ (Fig. 1(g)).

3.3 Enumerating Loops

We can enumerate the loops of a secondary structure in linear time. Each loop is fully specified by its list of tuples; thus an enumeration algorithm should list the tuples of each loop, with the external tuple first and the others in order.

Each node (closed region) of the tree corresponds to a hairpin loop, internal loop, multiloop, external loop, or pseudoknotted loop. A simple traversal of the tree suffices to enumerate such loops: when visiting a node, its closed region and the closed regions of its children (in order) are the needed tuples.

However, interior- and multi-pseudoknotted loops are not closed as their external base pair is pseudoknotted and spans a band. To enumerate these types of loops, two steps are needed:

Band finding: For each pseudoknotted closed region, construct the list of its bands regions, ordered by the left border index.

Loop finding: Identify all multi-pseudoknotted and interior-pseudoknotted loops, which are "nested" in the bands of the structure.

Algorithm 1 finds the bands of a pseudoknotted closed region $[i;j]$ of structure R. Loop finding is somewhat similar (details omitted).

Let L be a linked list representation of a secondary structure R for a strand of length n. In this representation, list elements are the base indices, with bidirectional links between adjacent elements and additionally bidirectional links between paired indices. In this algorithm, $bp(i)$ denotes j if $i.j$ or $j.i$, and 0 if i is unpaired. Algorithm 1 takes as input a sublist BL of L starting from index i to index j, in which unpaired base indices, and base indices corresponding to nested closed regions, are removed. Sublist BL can be generated using the closed region tree in time proportional to the number of closed regions that are nested in $[i;j]$. Thus, BL is a linked list representation of spanning band base pairs in $i;j$. Inspired by Lemma 1, Algorithm 1 scans list BL from left to right to identify bands and their region's borders.

algorithm Band-Finding
input: BL, a linked list representation of spanning band base pairs in $[i;j]$
output: ordered linked list of band regions in $[i;j]$
1 $b_i := i$;
2 **repeat**
3 $b_j := bp(b_i)$; // $b_i.b_j$ is the outer closing pair of a band, B
4 $b'_i := b_i$;
5 $b'_j := b_j$;
7 **while** $\text{Next}(b'_i, BL) = bp(\text{Prev}(b'_j, BL))$ **do**
8 $b'_i := \text{Next}(b'_i, BL)$;
9 $b'_j := \text{Prev}(b'_j, BL)$;
// $b'_i.b'_j$ is the inner closing pair of the band B So $B = [b_i; b'_i] \cup [b'_j; b_j]$ is a band of $i;j$
10 Add-Band-Region(BL, b_i, b'_i);
11 Add-Band-Region(BL, b'_j, b_j);
12 $b_i :=$Next-leftBase(b'_i, BL);
13 **until** $b_i = j + 1$;
14 **return** BL

Algorithm 1. Find bands of a pseudoknotted closed region

$\text{Next}(b'_i, BL)$ returns the index right after b'_i in BL and $\text{Prev}(b'_j, BL)$ returns the index right before b'_j in BL. Next-leftBase(b'_i, BL) returns l, the first index after b'_i in BL for which $bp(l) > l$. $l.bp(l)$ will be the outer closing pair of the next band.

Add-Band-Region(BL, b, b') (i) replaces index b in BL with a list element containing the band region borders (b and b') and (ii) removes from BL all other base indices that lie within the region $[b; b']$. At the end, BL is an ordered list of band regions.

By traversing the closed regions tree and applying the above algorithm to each pseudoknotted closed region, all lists of band regions can be constructed in time linear in the number of base pairs in R.

4 Energy Model

In the standard thermodynamic model for pseudoknot free secondary structures, the energy of a loop is a function of (i) loop type, (ii) an ordered list of its base pairs or tuples, (iii) the bases forming each base pair, and (iv) the bases in the loop (if any) that are adjacent to each base pair. The energy of a secondary structure is then calculated by summing the free energy of its component loops.

For pseudoknotted structures, the standard thermodynamic model is extended so that the energy of a loop depends additionally on (v) the *location status* of the loop, which shows its position relative to pseudoknotted loops in the structure. The location status can be one of the following (Fig. 1(h)).

span-Band: Interior-pseudoknotted and multi- pseudoknotted loops are called span-Band loops, since their external base pair spans a band.

Each of the remaining loop types corresponds to a closed region. Suppose that such a loop, $\mathcal{L}$, with corresponding closed region $[i_{\mathcal{L}}; j_{\mathcal{L}}]$, is a child of pseudoknotted closed region $[i; j]$. Then $\mathcal{L}$ can have one of the following two location statuses:

in-Band: If $[i_{\mathcal{L}}; j_{\mathcal{L}}]$ is contained in a band region of $[i; j]$, then $\mathcal{L}$ is an in-Band loop.

out-Band: Otherwise $\mathcal{L}$ is an out-Band loop.

standard: Loops that are not of the three types above are called standard loops. Such loops do not span bands and are not children of pseudoknotted loops.

4.1 Energy Calculation

It is straightforward to extend the loop enumeration algorithm so that the loop's type and location status is output in addition to its list of tuples. For example, the type of a loop corresponding to a closed region can be determined from the number and types of its children (e.g. if the closed region is not pseudoknotted and has no children, it must be a hairpin loop; if it has one child which is not a pseudoknotted closed region then it must be an internal loop). The location status of a loop can be determined using additionally the ordered list of band regions of its parent (if any). Then the free energy of the structure can be calculated by adding up the free energy of all loops.

4.2 Discussion

In the Rivas-Eddy model [12], the energy of a loop is exactly as in the standard model (for pseudoknot free structures) if the loop does not span a band. The standard model is generalized in the case of multiloops, which may now contain pseudoknotted regions, as follows: the energy is of the form $a + bu + ch + dm$, where a, b, c, and d are constants independent of the loop, u is the number of unpaired bases of the loop, h is the number of tuples (i, j) of the multiloop with $i \cdot j \in R$, and m is the number of tuples (i, j) of the multiloop with $i \cdot j \notin R$.

For multi-pseudoknotted loops, the constants a, b, c, d are replaced by distinct constants a', b', c', d'. In contrast, in the D&P model [9], the energy of a multiloop and multi-pseudoknotted loop are calculated using the same constants. In both models, the energy of a pseudoloop is the sum of terms, with one term depending on the total number of unpaired bases, one term per tuple of the pseudoloop, and one term that depends on the location status of the pseudoloop; however the dependence on the location status is different for both models. An interesting direction for future work would be to establish which method is most biologically plausible (neither paper provides justification for their choice of model).

The notion of what is a multiloop in the Rivas-Eddy is perhaps unnaturally restrictive. An (artificially small) example lies in the structure $\{1 \cdot 4, 2 \cdot 9, 3 \cdot 5, 6 \cdot 8, 7 \cdot 10\}$. Here, the base pairs $2 \cdot 9, 3 \cdot 5$, and $6 \cdot 8$ could be considered to form a "multiloop", but it is not recognized as such by the Rivas-Eddy algorithm, and thus also not by our classification. (We note that the Dirks-Pierce model, being less general, does not handle such loops.) We expect that the Rivas-Eddy algorithm could be reformulated to assign multiloop energies to such loops.

5 Akutsu's Structure Class

Akutsu's dynamic programming algorithms for RNA secondary structure prediction handles a restricted class of pseudoknotted RNA structures, called secondary structures with recursive pseudoknots [1]. We present a concise characterization of the class of structures Akutsu's algorithm can handle.

In this section, we will represent secondary structures as patterns, in which information about unpaired bases and base indices is lost but the pattern of nesting or overlaps among base pairs is preserved. To define patterns precisely, we use ϵ to denote the empty string and N_n to denote the natural numbers between 1 and n (inclusive).

Patterns: A string P (of even length) over some alphabet Σ is a pattern, if every symbol of Σ occurs either exactly twice, or not at all, in P. We say that secondary structure R for a strand of length n corresponds to pattern P if there exists a mapping $m : N_n \to \Sigma \cup \{\epsilon\}$ with the following properties: (i) if $i.j \in R$ then $m(i) \in \Sigma$ and $m(i) = m(j)$, (ii) if $i.j$ and $j.i \notin R$ for all $j \in N_n$, then $m(i) = \epsilon$, and (iii) $P = m(1)m(2)...m(n)$.

We refer to the index of the first and the second occurrence of any symbol σ in P by $L(P, \sigma)$ and $R(P, \sigma)$ respectively (L for Left and R for Right). When P is understood, we use $L(\sigma)$ and $R(\sigma)$. For example, pattern $P = abccdebaed$ corresponds to the closed region $[21; 33]$ in Fig. 1, and $L(a) = 1$ and $R(a) = 8$.

In what follows, let P be a pattern of size $2n$ over an alphabet Σ of size n.

5.1 Definitions

Definition 1. Our definition: *P is a* simplest pseudoknot *if and only if either:*

B1: $P = a_1 a_1$ *(for some a_1), or*

B2: *Either* $P = a_1 a_i P_1 a_i a_1 P_2$ *or* $P = a_1 P_1 a_i a_1 a_i P_2$*, where* $a_1 P_1 a_1 P_2$ *is a* simplest pseudoknot.

P is a B&C simple pseudoknot *if and only if either it is a* simplest pseudoknot *or for some* $a_1, a_i, \ldots a_r \in \Sigma$ *it is equal to* $a_1 P_1 a_1 a_i\ a_{i+1} \ldots a_r a_r \ldots a_{i+1} a_i P_2$*, where* $a_1 P_1 a_1 P_2$ *is a* simplest pseudoknot.

Theorem 1. *B&C simple pseudoknot is equivalent to Akutsu's definition of simple pseudoknot.*

Therefore, in what follows, we will simply refer to simple pseudoknots. The following definition is derived from Akutsu [1].

Definition 2. *Pattern P is a* recursive pseudoknot *if and only if P is a* simple pseudoknot *or* $P = P_1 P_2 P_1'$ *where* P_2 *is a nonempty* simple pseudoknot *and* $P_1 P_1'$ *is a* recursive pseudoknot.

We say that an RNA secondary structure R is a secondary structure with recursive pseudoknots *or conveniently* recursive pseudoknot structure *if its corresponding pattern P is a* recursive pseudoknot.

Assume that C is the closed region corresponding to node V and $C_1, \ldots, C_m$ are the closed regions correspond to the children of V. Then we say that the pattern corresponding to C also corresponds to node V. Also, $C' = C - \cup_{i=1}^{m} C_i$ is called the *private region* corresponding to V and we refer to the pattern corresponding to C' as the *private pattern* of V.

Theorem 2. *R is an Akutsu (i.e. recursive pseudoknot) structure if and only if all of the private patterns corresponding to the nodes in $T(R)$ are simple pseudoknots.*

5.2 Akutsu Tests

Our algorithm for testing whether a pattern P is a simple pseudoknot has two steps. In the first step it deals with the $a_i a_{i+1} \ldots a_r a_r \ldots a_{i+1} a_i$ subpattern and removes it from P, making the pattern a *simplest pseudoknot.* This can be done in linear time by scanning the symbols of P, starting from the symbol after the second occurrence of a_1, and removing the subpattern $a_i a_{i+1} \ldots a_r a_r \ldots a_{i+1} a_i$ if any.

Next the algorithm determines if P is a simplest pseudoknot, building on both cases in the definition of simplest pseudoknot. We define two *simplify* operations according to **B2**: **(i)** $a_1 a_i S_1 a_i a_1 S_2$ is converted to $a_1 S_1 a_1 S_2$, and **(ii)** $a_1 S_1 a_i a_1 a_i S_2$ is converted to $a_1 S_1 a_1 S_2$. We define one more operation, *final* operation, according to **B1**: **(iii)** $a_1 a_1$ is converted to ϵ. In these cases we say that a simple/final operation is applicable to a_1.

The linear time algorithm for testing whether the pattern P is a simplest pseudoknot (1) applies one of the simplify operations, **i** or **ii**, on the first symbol, a_1, if applicable, repeatedly (2) does the *final* operation, **iii**, on a_1 if it is applicable. (3) return true if the pattern is empty and false otherwise.

Thus, using Theorem 2, to test whether a secondary structure R is an Akutsu (i.e. recursive pseudoknot) structure, it is sufficient to check whether the private pattern corresponding to each node of $T(R)$ is a simple pseudoknot. It is straightforward to generate the private pattern for all nodes in linear time; thus the overall algorithm is a linear time algorithm.

5.3 Classification of Biological Structures

Condon et al. [7] provide linear time algorithms to test if an input structure is in the R&E and D&P classes. To compare the generality of Akutus's algorithm with those of R&E and D&P, we applied our algorithms for membership in Akutsu's recursive class along with those of Condon et al.[7] to classify biological structures from several sources [3,10,6,4,5,13,15]. As results show (Table 1), exactly 2 of the structures are in class A but not in class D&P.

Table 1. Structure classification. Columns 2-8 present data for each RNA data set. For each data set (column), the entry in the first row lists the number of structures in the data set. The second row lists the average number of base pairs in the structures. The remaining rows list the number of structures of the data set that are in D&P, A, and R&E classes.

	PBase	Pseudo Viewer	Gutell	RCSB	RNase	SR PDB	tm RNA
# Strs	240	15	426	279	468	4	7
Avg. #Bps	14.2	144	970.6	35.5	198.4	92	85.71
D&P	232	11	354	244	95	3	5
A	232	11	354	246	95	3	5
R&E	240	15	369	274	468	4	7

6 Conclusions

In this work we present a precise definition of the structural elements in a secondary structure, and a comprehensive way to classify the type of loops that arise in pseudoknotted structure. Based on an algorithm of Bader et al. [2], we also introduced a linear time algorithm to parse a pseudoknotted secondary structure to its component loops, and to calculate its the free energy. Finally, we applied our algorithm to compare the generality of Akutsu's algorithm with those of Dirks and Pierce and Rivas and Eddy on a large test set of biological structures.

Our work can be continued in future in several directions. First, heuristic algorithms commonly use a procedure to calculate the free energy for a given sequence and structure. Incorporating our linear time free energy calculation algorithm into heuristic algorithms may cause improvements in their efficiency. Second, it would be interesting to investigate the structures which are in Akutsu's class but not in D&P class. Third, there is no linear time characterization of Uemura's [14] algorithm and having one makes it possible to figure out about the differences between Uemura's class of structures and other classes of structures (A, D&P, and R&E). Fourth, the parsing algorithm can be used to analyse known biological RNA structures, in order to find out what structures occur more frequently in biology. Finally, it would be useful to refine the thermodynamic model presented in this paper, to obtain mfe predictions of better quality.

Acknowledgement. We would like to thank Satoshi Kobayashi for his useful comments and pointing out an error in an earlier version of the paper.

References

1. Akutsu, T.: Dynamic programming algorithms for RNA secondary structure prediction with pseudoknots. *Discrete Applied Mathematics* **104** (2000) 45–62.
2. Bader, D. A., Moret, B. M.E., Yan, M.: A linear-time algorithm for computing inversion distance between signed permutations with an experimental study. *Journal of Computational Biology* **8** (2001) 483–491.
3. Batenburg, F. H. D. van et al.: Pseudobase: a database with RNA pseudoknots. *Nucl. Acids Res.* **28** (2000) 201–204.
4. Berman, H.M. et al.: The Nucleic Acid Database; A Comprehensive Relational Database of Three-Dimensional Structures of Nucleic Acids. *Biophys. J.* **63** (1992) 751–759.
5. Brown, J.W.: The Ribonuclease P Database. *Nucl. Acids Res.* **27** (1999) 314.
6. Cannone, J.J. et al.: The Comparative RNA Web (CRW) Site; an online database of comparative sequence and structure information for ribosomal, intron, and other RNAs. *BMC Bioinformatics* **3** (2002)
7. Condon, A., Davy, B., Rastegari, B., Zhao, S. and Tarrant, T.: Classifying RNA pseudoknotted structures. *Theor. Comput. Sci.* **320** (2004) 35–50.
8. Dennis, C.: The brave new world of RNA. *Nature* **418** (2002) 122–124.
9. Dirks, R. M., Pierce, N. A.: A partition function algorithm for nucleic acid secondary structure including pseudoknots. *J. Comput. Chem.* **24** (2003) 1664–1677.
10. Han, K., Byun, Y.: PseudoViewer2: visualization of RNA pseudoknots of any type, *Nucl. Acids Res.* **31** (2003) 3432–3440.
11. Lyngsø R. B., Pedersen, C. N.: RNA pseudoknot prediction in energy-based models. *J. Computational Biology* **7** (2000) 409–427.
12. Rivas, E., Eddy, S. R.: A dynamic programming algorithm for RNA structure prediction including pseudoknots. *J. Molecular Biology* **285** (1999) 2053–2068.
13. Rosenblad, M.A., Gorodkin, J., Knudsen, B., Zwieb, C., Samuelsson, T.: SRPDB: Signal Recognition Particle Database. *Nucl. Acids Res.* **31** (2003) 363–364.
14. Uemura, Y., Hasegawa, A., Kobayashi, S., Yokomori, T.: Tree adjoining grammars for RNA structure prediction. *Theor. Comput. Sci.* **210** (1999) 277–303.
15. Zwieb, C., Gorodkin, J., Knudsen, B., Burks, J., Wower, J.: tmRDB (tmRNA database). *Nucl. Acids Res.* **31** (2003) 446-447.

A Compressed Format for Collections of Phylogenetic Trees and Improved Consensus Performance

Robert S. Boyer, Warren A. Hunt Jr, and Serita M. Nelesen

Department of Computer Sciences,
The University of Texas, Austin, TX 78712, USA
{boyer, hunt, serita}@cs.utexas.edu

Abstract. Phylogenetic tree searching algorithms often produce thousands of trees which biologists save in Newick format in order to perform further analysis. Unfortunately, Newick is neither space efficient, nor conducive to post-tree analysis such as consensus. We propose a new format for storing phylogenetic trees that significantly reduces storage requirements while continuing to allow the trees to be used as input to post-tree analysis. We implemented mechanisms to read and write such data from and to files, and also implemented a consensus algorithm that is faster by an order of magnitude than standard phylogenetic analysis tools. We demonstrate our results on a collection of data files produced from both maximum parsimony tree searches and Bayesian methods.

1 Introduction

Producing a phylogeny for a set of taxa involves four major steps. First, comparative data for the taxa must be collected. This data often takes the form of DNA sequences, other biomolecular information or matrices of morphological data. Second, this data is aligned to ensure that comparable information is considered as input to the tree producing step. The third step is to produce candidate trees. There are many techniques for doing this including optimizing maximum parsimony or maximum likelihood criteria, or, more recently, by using Bayesian methods [7] [10]. These techniques rarely result in a single optimal tree. Instead, there are often many trees that a phylogeneticist would like to save for further processing such as consensus analysis, which is used to summarize the collection of trees. These post-tree analyses are the final step.

We have developed methods for storing and retrieving phylogenetic tree data, and using these methods we have implemented a consensus algorithm. Our approach permits very large data sets to be compactly stored and retrieved without any loss of precision. Also, our implementation of our consensus algorithm provides greatly increased performance when performing strict and majority consensus computations as compared to PAUP [18] and TNT [8].

Our system is called the Texas Analysis of Symbolic Phylogenetic Information (TASPI), and it is an experimental system, written from scratch. It is a stand alone tool for a few kinds of phylogenetic data manipulation. TASPI is written in the ACL2 [12] formal logic, where all operations are represented as pure functions. Using ACL2's associated mechanical theorem prover, it is possible to prove assertions about the TASPI system.

R. Casadio and G. Myers (Eds.): WABI 2005, LNBI 3692, pp. 353–364, 2005.

In this paper, we explain our representation of phylogenetic trees, and how this format reduces the storage requirement for a collection of trees. We also give an algorithm for computing strict and majority consensus trees that exhibits improved performance as compared to currently available software. Finally, we include an empirical study confirming our results.

2 Representation

Newick format [6] is the standard way of storing a collection of phylogenetic trees. Adopted in 1986, Newick is a parenthetical notation that uses commas to separate sibling subtrees, parentheses to indicate children, and a semicolon to conclude a tree. Newick outlines each tree in its entirety whether storing one tree, or a collection of trees.

On the other hand, TASPI capitalizes on common structure within a collection of trees. TASPI stores a common subtree once, and then each further time the common subtree is mentioned, TASPI references the first occurrence. This saves considerable space since potentially large common subtrees are only stored once, and the references are much smaller (for empirical results see Section 5).

There are two layers to the TASPI representation of trees. At a high-level, trees are represented as Lisp lists, similar in appearance to Newick, but without commas and semicolons. This is the format presented to the user of TASPI and on which user functions operate. At a low-level, the data are instead represented in a form that uses hash-consing [9] to achieve decreased storage requirements and improved accessing speeds. For ease of reference in Section 5, we call this the Boyer-Hunt compression.

Consider the following set of rooted trees in Newick format:

```
(a,((b,(c,d)),e));
(a,((e,(c,d)),b));
(a,(b,(e,(c,d))));
((a,b),(e,(c,d)));
```

The format of these trees presented to the user of TASPI is straightforward:

```
(a ((b (c d)) e))
(a ((e (c d)) b))
(a (b (e (c d))))
((a b) (e (c d)))
```

Notice that storing this set of trees involves restoring the subtree containing taxa c and d once for every tree. The Boyer-Hunt compression instead stores the c-d clade once, the first time it is encountered. If, subsequently, the c-d clade is encountered again, the first time is marked with "#n=" for the current value of a counter n that is incremented each time it is used. Then, instead of re-storing the c-d clade, a reference in the form "#n#" is stored in its place. This compression has parallels to the Lempel-Ziv data compression which is based only on characters seen so far [20]. The compressed version of the trees above is given below:

```
((A ((B #1=(C D )) E ))
(A (#2=(E #1#) B))
(A (B #2#))
((A B)#2#))
```

We use a technique sometimes called hash-consing, which ensures that no object is ever stored twice. In the context of phylogenetic trees, an object is a subtree, and consing is a tree constructor that joins subtrees. Hashing, put simply, is a technique that creates a table that allows for fast searches. In this case, hashing is used to quickly determine if a subtree was previously encountered. The format, using "#n=" and "#n#", is a standard read dispatch macro from Lisp programming [17].

Two subtleties remain to be addressed. First, though we will be presenting rooted trees in this paper, trees are not all rooted. In fact, most tree searching algorithms return unrooted trees since determining the root of a tree may itself be a computationally intensive problem [7]. Newick format does not distinguish between rooted and unrooted trees except through the use of auxiliary flags. By placing [&R] and [&U] just before the beginning of a tree, rooted and unrooted trees, respectively, are indicated. Without these flags, the onus is on the user to interpret the trees appropriately.

Second, Newick does not give a unique representation for a tree. Consider the tree on the right. There are many representations for this tree in both Newick and TASPI. Possible TASPI representations include:

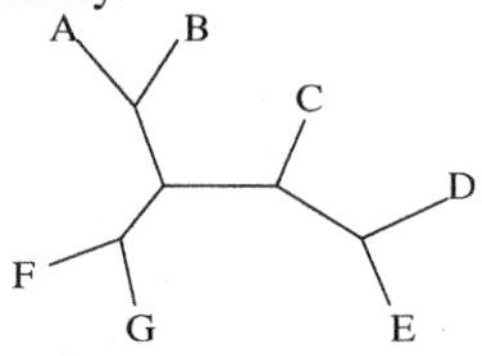

`((F G) ((A B) (C (D E))))` and
`((C (E D)) ((B A) (G F)))`.

To ensure a unique answer in our computations, we order the output with respect to an ordering on the taxa. As far as we can tell, PAUP also does this. Thus, given an alphabetical ordering, we would order the tree above as `(A B ((C (D E)) (F G)))`.

3 Consensus Analysis

3.1 Background

Consensus trees are defined by Felsenstein as "trees that summarize, as nearly as possible, the information contained in a set of trees whose tips are all the same species" [7]. The idea of a consensus tree was first proposed by Day in 1972 [1], and quickly followed by other criteria for agreement between trees. In 1981, Margush and McMorris defined the majority rule trees as we know them today. They proposed this form of consensus as following best the "dictionary definition of consensus as 'general agreement' or 'majority of opinion'" [13]. It was also around this time that Sokal and Rohlf coined the term "strict consensus" [16].

Consensus methods return a single tree, or an indication that no tree meeting that method's requirements exists. The types of consensus include Adams, maximum agreement subtree, semi-strict, also called loose, combinable component or Bremer [7], greedy, local, and Nelson-Page. See Bryant [4] for an overview of various consensus methods and their interrelationships.

Two of the most common types of consensus trees are strict and majority. Both of these decide which branches in the input trees to keep, and then build a tree from the resulting branches. Strict consensus requires that any branch in the consensus tree be a branch in every input tree, while a majority tree only requires that any branch in the consensus tree be a branch in at least some majority of the input trees. A threshold is a parameter to majority consensus that determines what percentage is to be used as a cutoff. Strict consensus is a special case of majority consensus; that is, it is a majority consensus with a threshold of 100%. Strict and majority consensus algorithms always return a tree, and have optimal O(kn) algorithms as described by Day [5] and Amenta et al. [2] (where k is the number of trees and n is the number of taxa).

3.2 Our Algorithm

We compute a consensus through a sequence of steps. We first read the source file containing the trees for which a consensus is to be computed. During the read process, we identify every subtree for which we have already read an identical subtree; thus, instead of creating a new data structure for the subtree just read, we reference the previously created subtree. We next create a mapping from all subtrees to every parent in which a subtree is referenced. Using this information, we compute the occurrence frequency of every subtree. Finally, after we have selected the subtrees that match our selection criteria, we construct the consensus answer. We give an example computation in Subsection 3.3.

In the following explanation, we use the notion of a "multiset", which is, intuitively speaking, a kind of set in which the number of occurrences count. More formally, one may regard a multiset as a function to the set of positive integers. If **A** and **B** are multisets, then **A** is a multisubset of **B** if and only if for each **x** in the domain of **A**, **x** is in the domain of **B** and **A(x)** <= **B(x)**.

For example, suppose **u**, **v**, and **w** are all distinct objects. Let **A** = <**u, 1**>, <**v, 2**> and let **B** = <**u, 2**>, <**v, 4**>, <**w, 5**>, then **A** is a multiset with one occurrence of **u** and two of **v**. Thus, **A(v) = 2**. **A** is a multisubset of **B** because **A(u)** <= **B(u)** and **A(v)** <= **B(v)**.

One way to represent multisets is with lists in which the number of occurrences of an element in a list represents the number of times that the element is in the corresponding multiset. So for example, we may represent the example **A** above with the Lisp list **(u v v)**.

Several definitions will be useful.

- **tip**: a symbol or integer.
- **tree**: a tip or, recursively, a list of one or more trees.
- **fringe**: a list of all tips in a tree.
- **subtree**: If a and b are trees, then **a** is a *subtree* of **b** if and only if either (1) **a** is **b** or (2) **b** is a list and **a** is a subtree of a member of **b**.
- **proper subtree** If **a** and **b** are trees, **a** is a *proper subtree* of **b** iff **a** is a subtree of **b** and **a** is not **b**.
- **domain**: The *domain* of an association list (a list of key-value pairs) is the set of the keys of the members of the association list.
- **replete** An association list **db** is *replete* if and only if for all **t1** in the domain of **db**, (1) **t1** is a nontip tree and (2) if **t2** is a nontip proper subtree of **t1**, then **db(t2)**

is a list representing the multiset of all trees in the domain of **db** that have **t2** as a member, including **t1**. Note that the multiset ((**a**) (**b**) (**a**)) has the tree (**a**) as a member twice.
- **top level** A tree in the domain of a replete **db** is said to be *top level* if and only if it is a proper subtree of no member of the domain of **db**.

To compute the consensus, our algorithm proceeds by:

1. Producing a replete association list of all of the subtrees in the original input,
2. Counting the frequencies of the non-tip subtrees,
3. Collecting the subtrees that appear as often as the designated majority threshold, and finally,
4. Constructing the consensus tree.

Step one is accomplished by our function **replete-trees-list-top** which converts the original input list of trees into a replete association list (database). This replete database is a mapping from subtrees to every parent tree containing the subtree in question. Step two is performed by the function **fringe-frequencies** which counts the frequencies of every subtree fringe in the replete database by iterating through the replete database. Step three is collecting the subtrees that occur as often as the threshold. Finally, using this collection of subtrees, function **build-term-top** constructs the consensus answer.

Our function **replete-trees-list-top** takes a list **l** of non-tip trees no member of which is a proper subtree of another, such as a list of trees all with the same set of taxa. **replete-trees-list-top** returns a replete association list **db** such that (1) **x** is a member of the domain of **db** if and only if **x** is a member of **l** or is a non-tip proper subtree of a member of **l** and (2) if **x** is in the domain of **db**, then **db(x)** is an integer if and only if **x** is a member of **l** and **db(x)** is the number of times **x** occurs in **l**. For an example execution of **replete-trees-list-top**, see Subsection 3.3.

Function **fringe-frequencies** takes a list **l** of nontip trees such that no member of **l** is a proper subtree of any other member of **l** (such as that produced by **replete-trees-list-top**). **fringe-frequencies** returns a minimal length association list that pairs the fringe **fr** of each nontip subtree of each member of **l** with the number of occurrences in **l** of non-tip subtrees of members of **l** that have fringe **fr**.

By scanning through the resulting association list, we just pick out the subtrees that appear as often as the desired threshold. We have no need to store the actual number of times any specific subtree appears, we simply collect the desired subtrees (fringes) into a list.

The function **build-term-top** takes two arguments. The first argument is a sorted list **l** of the subtrees' fringes; **l** is sorted using a lexicographic (normalization) order that is based both on the internal tips and the size of the elements in each subtree. All the subtrees in **l** must appear in the consensus answer. The second argument is a normalization taxa list **tx**, that is used by our lexicographic ordering function so we can produce a unique representation of any subtree that itself includes more than one subtree. Remember, we represent each subtree as a list of subtrees, so to make the representation unique we sort members of each subtree. **build-term-top** constructs a consensus answer tree recursively by first building an answer of the first subtree of **l**. Once the first answer subtree is computed for the first element in **l**, any (sub-)subtrees required to build the first subtree are "crossed out" from **l** that remain to be processed, and we continue with the next remaining element of **l** until no entries remain.

3.3 Example

Consider the five trees in Figure 1. The TASPI representation of these trees is the input to the function **replete-trees-list-top**. This function returns the following association list, where keys are in boldface:

```
((A B)((A B) C))
((((A B) C) ((D E) (F G))) . 1)
((D E)((D E) F G)
      ((D E) (F G)))
(((D E) F G) ((A B) C) ((D E) F G)))
((((A B) C) ((D E) F G)) . 1)
(((A B) C)(((A B) C) (D (E (F G))))
           (((A B) C) ((D E) F G))
           (((A B) C) ((D E) (F G))))
((F G) (E (F G))
       ((D E) (F G)))
((E (F G)) (D (E (F G))))
((D (E (F G))) (((A B) C) (D (E (F G)))))
((((A B) C) (D (E (F G)))) . 1)
((B C) (A (B C)))
((D E F) ((D E F) G))
(((D E F) G) ((A (B C)) ((D E F) G)))
(((A (B C)) ((D E F) G)) . 1)
((A (B C))((A (B C)) ((D E) (F G)))
           ((A (B C)) ((D E F) G)))
(((D E) (F G))((A (B C)) ((D E) (F G)))
               (((A B) C) ((D E) (F G))))
(((A (B C)) ((D E) (F G))) . 1)
```

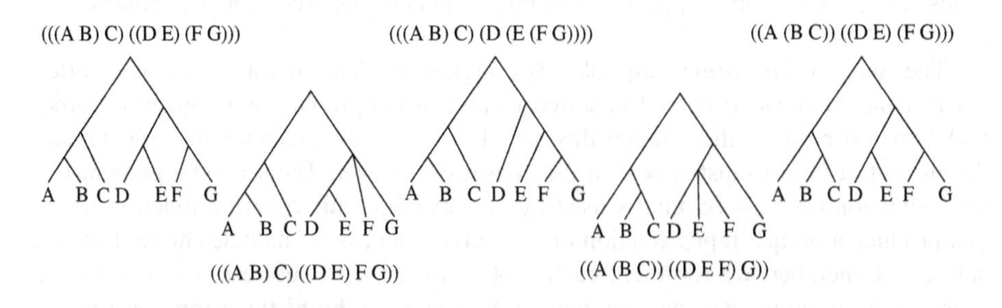

Fig. 1. A collection of trees together with their TASPI representations

A subtree is the key for each element of the list, and the remainder of each entry (the values) is either (1) trees or subtrees in which the key appears, or (2) an integer representing the number of times this top level tree occurs in the input collection. Thus, this is a replete association list. This association list is now the input to the function **fringe-frequencies**, which produces this list:

```
((A B) . 3)        ((D E F). 1)
((D E) . 3)        ((A B C) . 5)
((F G) . 3)        ((D E F G) . 5)
((E F G) . 1)      ((A B C D E F G) . 5)
((B C) . 2)
```

This frequency list has each fringe from our replete association list, together with an integer. Remember, a fringe is simply a list of the tips in a tree, so we do not distinguish between the fringe from `(A (B C))` and the fringe from `((A B) C)`. The integer gives the number of trees that have a subtree with this fringe.

We are now prepared to sweep through this list and record the fringes that occur at least as often as the threshold for both a strict and majority consensus. In this example, for the strict majority we collect those fringes that occur 5 times, and for the majority, we collect those that occur at least 3 times. This gives us:

```
((A B C D E F G) . 5)
((D E F G) . 5)
((A B C) . 5)
```

and

```
((A B C D E F G) . 5)
((D E F G) . 5)
((A B C) . 5)
((F G) . 3)
((D E) . 3)
((A B) . 3)
```

Finally, the function **build-term-top** uses either the strict or majority fringes together with a normalization list such as `(A B C D E F G)` to create the strict and majority consensus trees. In this case we create `((A B C) (D E F G))` and `(((A B) C) ((D E) (F G)))`.

4 Experiments

4.1 Data Sets

We first obtained collections of phylogenetic trees from Dr. Usman Roshan and Dr. Tiffani Williams. These trees were created by PAUP and TNT performing maximum parsimony searches using biomolecular data sets. We have analyzed hundreds of these collections though we only present the results from ten collections. The results presented are representative of the full set. We also generated sets of trees using MrBayes [11] that had more taxa than either PAUP or TNT can even read; these data sets were created by the third author.

Table 1 gives characteristic information for each collection we present, namely, the numbers of taxa per tree, the number of trees in the collection, and the source of the collection.

Table 1. Data set statistics

Data Set Number	Data Set Name	Number of Taxa	Number of Trees	Source
1	Dom_2org	8506	47	Roshan
2	sRNA_mito	2587	369	Roshan
3	Will_Euk	2000	537	Roshan
4	Three567	567	2505	Williams
5	Actino	4583	301	Roshan
6	Ocho854	854	2505	Williams
7	John921	921	2505	Williams
8	t10000	500	10000	Roshan
9	Will2000	2000	2505	Williams
10	Mari2594	2594	2505	Williams
11	20000seqs	20000	1001	Nelesen
12	50000seqs	50000	1001	Nelesen

4.2 Methods

The files we obtained often contained comments about how the trees were generated, parsimony scores, or other output from their production. TASPI does not store this information, so we began by creating files that contained only the topological tree information so that we could accurately assess our compression.

Next, we created a suite of Perl scripts that take these original files and generate appropriate input files for PAUP and TNT. In each case, the taxa list is created from the first tree in the file, and the trees themselves are collected. Then, for PAUP, a Nexus file is produced with the taxa list, the trees, and a PAUP block containing the commands to compute consensus. Similarly for TNT, an appropriate input file is created with the taxa list, trees, and commands to compute consensus.

TASPI reads the source files directly. As with PAUP and TNT, TASPI can be run both interactively, where we submit one command at a time, or using an input file containing all commands needed for the desired computation.

Using PAUP, TNT and TASPI, we measured the time it took the software to read each collection, and the time needed to compute both a strict and majority consensus tree. For PAUP, we produced a strict consensus tree using its majority consensus command with percent set to 100 since the strict consensus command took considerably longer to do the same calculation. Also, by default, TNT does not include branches that are not well supported by the data used to create trees. However, we were not including any initial data other than the trees themselves, so we turned this feature off using the command *collapse notemp*.

Our experiments, where we were able to compare PAUP, TNT and TASPI, were all performed on an Intel Pentium 4 CPU 3.4 Ghz computer. However, for the two largest data sets, we used an AMD Opteron CPU 2.4 Ghz computer, which has similar computational performance, but more physical memory. Either computer produces the same compressed files. The largest files are too large to be read by either PAUP or TNT due to internal limitations on the number of taxa allowed in a tree.

5 Results

The first major contribution of TASPI is the condensed format in which trees can be stored, while maintaining structural information. Figure 2 shows four sets of sizes for each of our benchmark data sets. The Newick data represents the size of the trees as they were given to us, after removing information that TASPI does not currently store (e.g. comments and branch lengths) and Newick.bz2 illustrates the size of the file after compression using the algorithm implemented in bzip2 [15]. TASPI.bhz displays the size of the file after compression using the Boyer-Hunt method. Notice that this file is still in ASCII, but with redundancies removed. Unlike most compression methods, all the information present in the original files is still immediately accessible, without a decompression step. Finally, TASPI.bhz.bz2 shows the size of the file if it is compressed using the Boyer-Hunt method and then bzip2 is applied.

Using the compressed TASPI format saves considerable memory space. For the data sets we present, the storage requirement for the TASPI format ranges from 2% of the storage requirement of Newick for the t10000 (data set 8) collection, up to 26% for the Dom_2org (data set 1) collection. Over all data sets, the compressed TASPI format uses just 5% of the storage requirement of the Newick format.

The amount of storage space saved is dependent on the amount of similarity between input trees. The more similarity between input trees (i.e. the greater the number of common subtrees) the more effective the compression. It is known in the phylogenetic community that trees derived from independent data sets are unlikely to have common structure [4]. However, it appears that collections of trees such as those we are presenting do have common structure since our compression was able to reduce the storage requirement for these collections of trees. Further, the greater the number of trees in the collection, the more likely there will be common structure.

It is readily apparent that bzip2 produces smaller files than the Boyer-Hunt compression on the smaller collections of trees, but for the very large data sets, the Boyer-Hunt compression produces smaller files than bzip2. Further, the Boyer-Hunt files are ASCII,

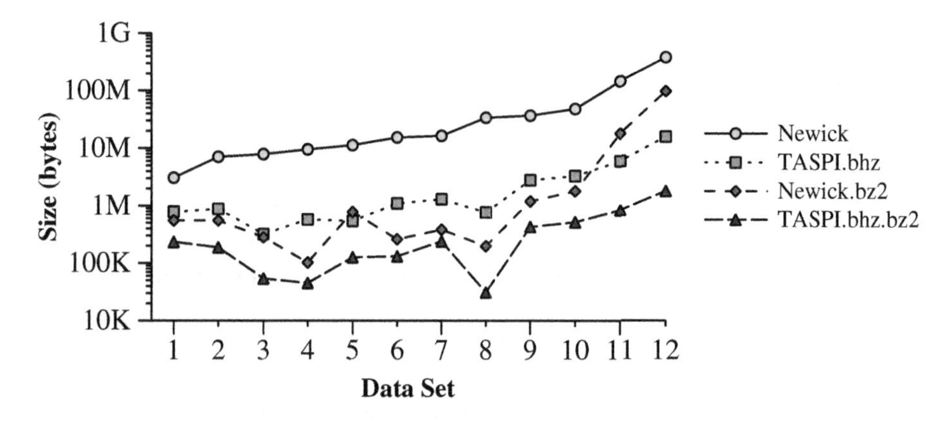

Fig. 2. Storage requirements

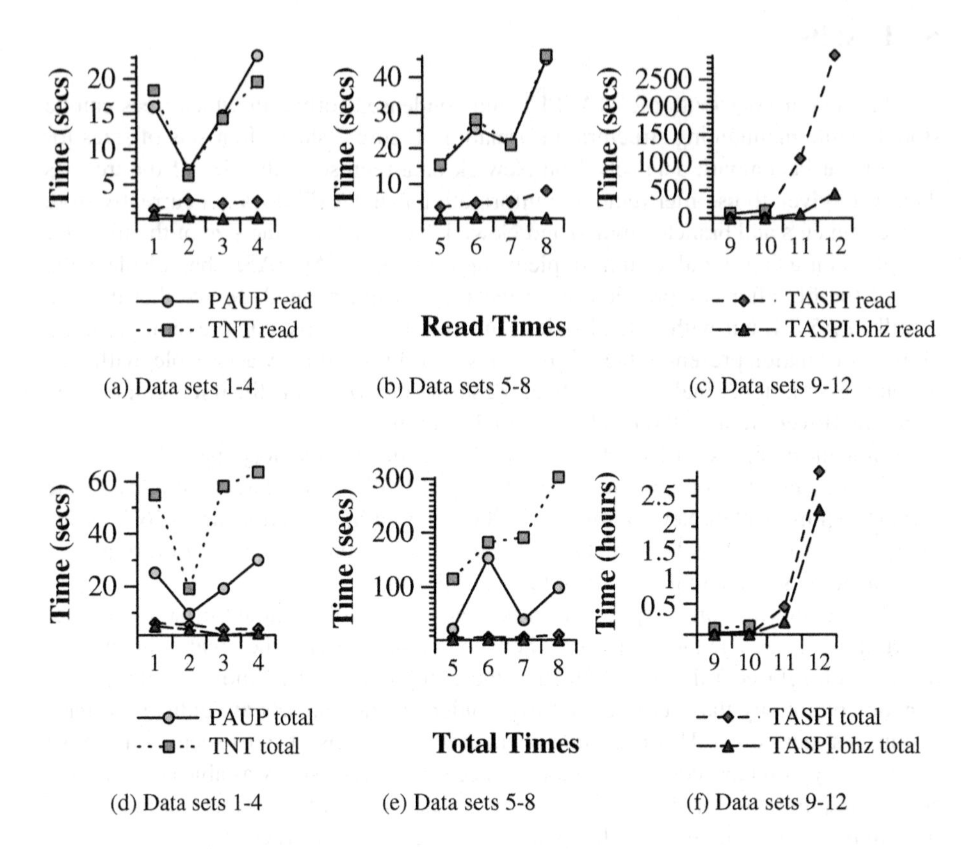

Fig. 3. Time to read a collection of trees (a-c) and compute strict and majority consensus trees with PAUP, TNT and TASPI (d-f)

and thus are ready to be used as input to analysis, such as consensus. If the data are not currently required as input to a post-tree analysis, compressed TASPI is even more useful. Boyer-Hunt files can be further compressed using bzip2 to produce even smaller files than those produced by using bzip2 on the original Newick files for sharing and transmission purposes. For our data sets, using the Boyer-Hunt compression together with bzip2 produces files that require 1% of the storage space of Newick.

The second major contribution of TASPI is its ability to read collections of trees quickly. Figure 3(a-c) shows average read times in seconds for each of our benchmark collections of trees. Notice that while reading trees with TNT or PAUP requires comparable times, reading the Boyer-Hunt compressed trees with TASPI is by far the fastest time for any collection. In fact, neither PAUP nor TNT is able to read the last two data sets. For the data sets which PAUP and TNT can read, reading the compressed TASPI format takes just 2% of the time to read the Newick files with PAUP. This means that loading these files takes more than 48 times longer when read with PAUP or TNT rather than using TASPI to read their compressed counterpart. Even reading the source files is faster in TASPI than it is in either PAUP or TNT – using TASPI to read the Newick files takes just 16% of the time needed to read the same files with PAUP or TNT.

The third major contribution of TASPI is a consensus implementation with improved performance. Figure 3(d-f) shows the time to compute consensus with each of TASPI, TNT and PAUP. In each case, both a strict consensus tree and a majority consensus tree are computed. Notice that the time to compute consensus includes the time to read the collection of trees since the trees are the input to a consensus calculation. Thus, we show both the time to compute consensus when reading compressed trees and also the time when reading Newick trees.

In all cases, the result TASPI produces is identical to that produced by PAUP (when PAUP is able to read the input), but TASPI is faster. For the data sets PAUP and TNT can process that we present, using TASPI to compute consensus with input trees in compressed TASPI format requires 5% of the time it takes PAUP to compute consensus with input trees in Newick format. If we factor out the improved reading time, TASPI computes these consensus trees in about 10% of the time it takes PAUP to do the same computation.

6 Conclusion

In phylogenetics, the ability to store large numbers of trees is increasingly important. Bayesian methods, which use Monte Carlo Markov Chains, are visiting more trees than previous methods, and are growing in popularity. Biologists are also choosing to retain additional trees visited during a search. We have shown that our format provides decreased storage requirements, while maintaining data accessibility for further processing. Further, our format together with techniques like memoization allows for improved performance in post-tree analysis. We showed this using strict and majority consensus.

The use of post-tree analyses are also becoming more prevalent. Williams et al. propose using the rate of change of a consensus tree as a stopping criterion for heuristic maximum parsimony searches, which requires the computation of a consensus tree multiple times over the course of an analysis [19]. We have given a new format for collections of phylogenetic trees that would make this feasible. In addition, our replete database, the output of the first step in our consensus algorithm, provides a possible starting point for phylogenetic databases such as those proposed in [14].

In the future we hope to investigate the changes necessary to make our consensus algorithm incremental. This would allow online consensus analysis as proposed in [3]. We would also like to look at even larger collections of trees (larger both in number of trees and number of taxa) and consider application of our techniques to supertree methods.

Acknowledgment

This work was funded in part by an ITR from the National Science Foundation (EF-0331453).

References

1. E. N. Adams. Consensus techniques and the comparison of taxonomic trees. *Systematic Zoology*, 21:390–397, 1972.
2. Nina Amenta, Katherine St. John, and Frederick Clarke. A linear-time majority tree algorithm. In Gary Benson and Roderic D. M. Page, editors, *Proc. of the 3rd International Workshop on Algorithms in Bioinformatics (WABI 2003)*, volume 2812 of *Lecture Notes in Computer Science*, pages 216–227. Springer-Verlag, 2003.
3. Tanya Y. Berger-Wolf. Online consensus and agreement of phylogenetic trees. In I. Jonassen and J. Kim, editors, *Proc. of the 4th International Workshop on Algorithms in Bioinformatics (WABI 2004)*, volume 3240 of *Lecture Notes in Computer Science*, pages 216–227. Springer-Verlag, 2004.
4. David Bryant. A classification of consensus methods for phylogenetics. In M. Janowitz, F.J. Lapointe, F. McMorris, B. Mirkin, and F. Roberts, editors, *Bioconsensus*, DIMACS Series in Discrete Mathematics and Theoretical Computer Science. DIMACS-AMS, 2001.
5. William H. E. Day. Optimal algorithms for comparing trees with labeled leaves. *Journal of Classification*, 2(1):7–28, 1985.
6. J. Felsenstein. The newick tree format. `http://evolution.genetics.washington.edu/phylip/newicktree.html`, 1986.
7. Joseph Felsenstein. *Inferring Phylogenies*. Sinauer Associates, Inc., 2004.
8. P.A. Goloboff, J.S. Farris, and K.C. Nixon. TNT (Tree analysis using new technology) (BETA) ver. 1.0. Published by the authors, Tucumán, Argentina, 2000.
9. E. Goto, T. Soma, N. Inade, T. Ida, M. Idesawa, K. Hiraki, M. Suzuki, K. Shimizu, and B. Philpov. Design of a lisp machine - flats. In *LFP '82: Proceedings of the 1982 ACM Symposium on LISP and functional programming*, pages 208–215, 1982.
10. David M. Hillis, Craig Moritz, and Barbara K. Mable, editors. *Molecular Sytematics*. Sinauer Associates, Inc., Sunderland, Massachusetts, 2nd edition, 1996.
11. J. P Huelsenbeck and F. Ronquist. MRBAYES: Bayesian inference of phylogeny. *Bioinformatics*, 17:754–755, 2001.
12. Matt Kaufmann, Pete Manolios, and J. S. Moore. *Computer-Aided Reasoning: An Approach.* Kluwer Academic Publishers, 2000.
13. T. Margush and F.R. McMorris. Consensus n-trees. *Bulletin of Mathematical Biology*, 43(2):239–244, 1981.
14. L. Nakhleh, D. Miranker, F. Barbancon, W.H. Piel, and M.J. Donoghue. Requirements of phylogenetic databases. In *Proceedings of the Third IEEE Symposium on Bioinformatics and Bioengineering (BIBE 2003)*, pages 141–148. IEEE Press, 2003.
15. Julien Seward. bzip2. `http://sources.redhat.com/bzip2/`, 2002.
16. Robert R. Sokal and F. James Rohlf. Taxonomic Congruence in the Leptopodomorpha Re-Examined. *Systematic Zoology*, 30(3):309–325, 1981.
17. Guy L. Steele. *Common Lisp the Language*, chapter 22.1.4. Digital Press, 2nd edition, 1990.
18. D. L. Swofford. *PAUP*: Phylogenetic Analysis Using Parsimony (and Other Methods) 4.0 Beta*. Sinauer Associates, Sunderland, Massachusetts, 2002.
19. Tiffani Williams, Tanya Berger-Wolf, Bernard Moret, Usman Roshan, and Tandy Warnow. The relationship between maximum parsimony score and phylogenetic tree topologies. Personal Communication.
20. J. Ziv and A. Lempel. A universal algorithm for sequential data compression. *IEEE Transactions on Information Theory*, 23:337–342, 1977.

Optimal Protein Threading by Cost-Splitting

P. Veber[1], N. Yanev[1,*], R. Andonov[1], and V. Poirriez[2]

[1] IRISA, Campus de Beaulieu, 35042 Rennes Cedex, France
{pveber, nyanev, randonov}@irisa.fr
[2] University of Valenciennes, 59313 Valenciennes, France
vpoirriez@univ-valenciennes.fr

Abstract. In this paper, we use integer programming approach for solving a hard combinatorial optimization problem, namely protein threading. For this sequence-to-structure alignment problem we apply cost-splitting technique to derive a new Lagrangian dual formulation. The optimal solution of the dual is sought by an algorithm of polynomial complexity. For most of the instances the dual solution provides an optimal or near-optimal (with negligible duality gap) alignment. The speed-up with respect to the widely promoted approach for solving the same problem in [17] is from 100 to 250 on computationally interesting instances. Such a performance turns computing score distributions, the heaviest task when solving PTP, into a routine operation.

1 Introduction

Protein folding is one of the most extensively studied problems in computational biology. The problem can be simply stated as follows: given a protein sequence, which is a string over the 20-letter amino acid alphabet, determine the positions of each amino acid atom when the protein assumes its 3D folded shape. Although simply stated, this problem is extremely difficult to solve and is widely recognized as one of the most important challenges in computational biology today [10,16,6].

In case of remote homologs, one of the most promising approaches to the above problem is protein threading, i.e., one tries to align a query protein sequence with a set of 3D structures to check whether the sequence might be compatible with one of the structures. Fold recognition methods based on threading are complex and time consuming computational techniques consisting of the following components:

1. a database of known 3D structural templates;
2. an objective function which evaluates any alignment of a sequence to a template structure;
3. a method for finding the best (with respect to the score function) possible sequence-3D structure alignment;
4. a statistical analysis of the raw scores allowing the detection of the significant sequence-structure alignments.

* on leave from University of Sofia, 5, J. Bouchier str., 1126 Sofia, Bulgaria.

R. Casadio and G. Myers (Eds.): WABI 2005, LNBI 3692, pp. 365–375, 2005.

The third point above is related to the problem of finding the optimal sequence-to-structure alignment and is referred to as protein threading problem (PTP). From a computer scientist's viewpoint this is the most challenging part of the threading methods. Until recently, it was the main obstacle to the development of efficient and reliable fold recognition methods. In the general case, when variable-length alignment gaps are allowed and pairwise amino acid interactions are considered in the score function, PTP is NP-hard [8]. Moreover, it is MAX-SNP-hard [1], which means that there is no arbitrary close polynomial approximation algorithm, unless P = NP. In this context the progress done by the computational biology community in solving PTP during the last few years is really remarkable [12,20,2,17,18,3]. The empirical results clearly illustrate that PTP is easier in practice than in theory and that it is possible to solve real-life (biological) instances in a reasonable amount of time. These results also show that one of the most promising approaches in solving this problem is using advanced mathematical programming (Mixed Integer Programming, MIP) models for PTP [19,20,2,17]. The most amazing observation is that for almost all (more than 95%) of the instances, the LP relaxation of the MIP models is integer-valued, thus providing optimal threading. This is true even for polytopes with more than 10^{46} vertices. Moreover, when the LP relaxation is not integer, its value is a relatively good approximation of the integer solution. However, to the best of our knowledge, this observation has not been practically used before the current paper. Other successful Integer Programming approaches for solving combinatorial optimization problems originated in molecular biology are discussed in the recent survey [11].

The main drawback of mathematical programming approaches is that the corresponding models are often very large (over 10^6 variables). Even the most advanced MIP solvers need prohibitively large running time for solving such instances. For example, the authors in [17] find out 30 templates for which it takes about 15 hours to thread one target onto them on a Silicon Graphics Origin 3800 system, which has 40400 MHz MIPS R12000 CPUs and 20 GB of RAM. Different divide-and-conquer methods and parallel algorithms can be used to overcome this drawback [18,19,20].

A further step in solving the huge MIP models is the development of special-purpose algorithms based on advanced combinatorial optimization techniques like Lagrangian relaxation. Such an algorithm has been recently designed by S. Balev in [3] and computationally compared with the B&B algorithm from [8] and a heuristic used in [9]. The computational results are very impressive and clearly show that the Lagrangian relaxation (LR) significantly outperforms both other algorithms. However, comparisons with MIP solver are not provided in [3].

In this paper we continue the same direction of research and propose a new dedicated algorithm for solving protein threading MIP models. It is as well based on Lagrangian relaxation. But, both our Lagrangian dual formulation and the optimization technique that we use for solving it (the so-called cost-splitting [13]), differentiate from those described in [3]. Extensive computational results prove that: (i) our algorithm is in most cases faster than the one in [3]; (ii) both

Lagrangian relaxation algorithms significantly outperform solving MIP models by LP relaxation. To the best of our knowledge the only other impressive application of LR to an alignment problem is discussed in [5].

Another contribution of the current paper concerns the 4th point above. When aligning a given query sequence to a set of 3D structures it is not possible to directly use the raw scores to rank the 3D structures. The reason is that these scores strongly depend on the query and template lengths and also, in a complicated way, on the particular features of the 3D structures. In addition, the query sequence may correspond to none of the existing folds. Therefore one must have means to evaluate the significance of an alignment score. This can be done as a preprocessing stage, by empirically calculating a distribution of scores for each template, using a set of sequences not related to it[1]. The underlying score normalization procedure involves threading a large set of queries against each template and requires solving millions of PTP. For example the package FROST (Fold Recognition-Oriented Search Tool) [9], uses a database of about 1,200 known 3D structures, each one associated with empirically determined score distributions. Computing these distributions is extremely time consuming: it requires solving about 1,200,000 sequence-to-structure alignments and takes about 40 days on a 2.4 GHz computer and about 3 days on a cluster of 12 PCs [14]. Accelerating computations involved in this component is crucial for the development of efficient fold recognition methods.

Based on extensive comparisons we observe that the approximated solutions obtained by any one of the three algorithms considered in this paper can be successfully used when computing scores distributions. Since these approximated solutions are obtained by polynomial algorithms, we experimentally prove that this heavy stage can be polynomially computed.

The organization of the paper is as follows. In section 2 we introduce a formal presentation of PTP, and then study some special cases in the section 3. Section 4 presents the cost-splitting technique. Last section is dedicated to experimental results.

2 Protein Threading Problem

For the sake of brevity, in this paper we stick to the network optimization problem formulation proposed in [19,2]. Formally, for a given sequence (query) of R characters and an ordered sequence (template) $\{s_1, \ldots, s_m\}$ of m blocks (structural elements), the ith one of length $|s_i|$, an alignment could be defined by insertion of gaps g_i between the blocks, s.t. $|g_1 s_1 g_2 s2 \ldots s_m g_{m+1}| = R$. Obviously, $\sum |g_i| = n' = R - \sum |s_i| \geq 0$ and $0 \leq |g_i|, i = 1, \ldots, m+1$. Now, we can substitute the gaps by s.c. relative positions defined by $r_k = \sum_{i=1}^{k} |g_i| + 1, k = 1, \ldots, m$ and to determine an alignment by choosing a point $(r_1, \ldots, r_m) \in Z_+^m$ with $1 \leq r_1 \cdots \leq r_m \leq n = n' + 1$.

[1] More justifications for this phase the interested reader can find in [9].

Let us introduce binary variables y_{ij} for modeling the event $r_i = j$, and denote by Y the set of feasible threading, defined by the following:

$$\sum_{k=1}^{n} y_{ik} = 1 \qquad i = 1, \ldots, m \tag{1}$$

$$\sum_{l=1}^{k} y_{il} - \sum_{l=1}^{k} y_{i+1,l} \geq 0 \qquad i = 1, \ldots, m-1,\ k = 1, \ldots, n-1 \tag{2}$$

$$y_{ik} \in \{0, 1\} \qquad i = 1, \ldots, m,\ k = 1, \ldots, n \tag{3}$$

To facilitate the interpretation of this set, we will many times refer to a m-partite graph G with a vertex set V given by the grid of points, whose coordinates $(i, j), i = 1, \ldots, m; j = 1, \ldots, n$ correspond to the indices of y variables. Let G be the digraph obtained by adding arcs from each grid point (i, j) in the ith column (layer), $i = 1, \ldots, m-1$, of the grid to all points $(i+1, k), k = j, \ldots, n$ in the $i+1$th column.One could easily check that the set Y is equivalent to the set of all paths in G of length $m - 1$. Thus the feasible threading could be regarded either as a point in Y either as a path in G. In the sequence-to-structure alignment context each layer corresponds to a block, and each vertex in a layer corresponds to a positioning of this block on a query protein. Let $L \subseteq \{(i, k) \mid 1 \leq i < k \leq m\}$ be a given set of inter-layers links. This is the so-called *contact graph*: a link between layers i and k means that the corresponding structural elements are in contact (close) in the 3D structure. As for the graph G, each such link adds the arcs (called z-arcs) $((i, j), (k, l))l = j, \ldots, n$ for each vertex $(i, j), j = 1, \ldots, n$. In the optimization problem, given below, the feasible set consist of previously defined paths together with all z-arcs having their both ends on such a path.

Let A_{ik} be the $2n \times \frac{n(n+1)}{2}$ node-arc incidence matrix for the subgraph spanned by the layers i and k, $(i, k) \in L$. The submatrix A_i, the first n rows of A_{ik}, (resp. A_k, the last n rows) corresponds to the layer i (resp. k). To avoid added notation we will use vector notation for the variables $y_i = (y_{i1}, \ldots y_{in}) \in B^n$ where B^n is the set of n-dimensional binary vectors, with assigned costs $c_i = (c_{i1}, \ldots c_{in}) \in R^n$ and $z_{ik} = (z_{i1k1}, \ldots, z_{i1kn}, z_{i2k1}, \ldots, z_{inkn}) \in B^{\frac{n(n+1)}{2}}$ for $(i, k) \in L$ with assigned costs $d_{ik} = (d_{i1k1}, \ldots, d_{i1kn}, d_{i2k1}, \ldots, d_{inkn}) \in R^{\frac{n(n+1)}{2}}$. In the sections below the vector d_{ik} will be considered as a $n \times n$ upper triangular matrix, having arbitrarily large coefficient below the diagonal. This slight deviation from the standard definition of an upper triangular matrix is used only for formal definition of some matrix operations.

Now the protein threading problem $PTP(L)$ is defined as:

$$v_{ip}^L = v(PTP(L)) = \min\{\sum_{i=1}^{m} c_i y_i + \sum_{(i,k) \in L} d_{ik} z_{ik}\} \tag{4}$$

$$\text{subject to: } y = (y_1, \ldots, y_m) \in Y, \tag{5}$$

$$y_i = A_i z_{ik},\ y_k = A_k z_{ik} \qquad (i, k) \in L \tag{6}$$

$$z_{ik} \in B^{\frac{n(n+1)}{2}} \qquad (i, k) \in L \tag{7}$$

The shortcut notation $v(.)$ will be used for the optimal objective function value of a subproblem obtained from $PTP(L)$ with some z variables fixed. Throughout the next section, vertex costs c_i are assumed to be zero. We study three sorts of contact graph that make PTP polynomially solvable.

3 Special Cases

3.1 Contact Graph Contains No Crossing Edges

Two links (i_1, k_1) and (i_2, k_2) such that $i_1 < i_2$ are said to be crossing when k_1 is in the open interval (i_2, k_2). The case when the contact graph L contains no crossing edges has been mentioned to be polynomially solvable for the first time in [1]. Here we present a different sketch for $O(n^3)$ complexity of PTP in this case.

If L contains no crossing edges, then $PTP(L)$ can be recursively divided into independent subproblems. Each of them consists in computing all shortest paths between the vertices of two layers i and k, discarding links that are not included in (i, k). Thus the result of this computation is a distance matrix D_{ik} such that $D_{ik}(j, l)$ is the optimal length between vertices (i, j) and (k, l). Note that for $j > l$ as there is no path in the graph, $D_{ik}(j, l)$ is an arbitrarily large coefficient. Finally, the solution of $PTP(L)$ is the smallest entry of D_{1m}.

We say that an edge $(i, k), i < k$ is included in the interval $[a, b]$ when $[i, k] \subseteq [a, b]$. Let us denote by $L_{(ik)}$ the set of edges of L included in $[i, k]$. Then, an algorithm to compute D_{ik} can be sketched as follows:

1. if $L_{(ik)} = \{(i, k)\}$ then the distance matrix is given by

$$D_{ik} = \begin{cases} d_{ik} \text{ if } (i, k) \in L \\ \tilde{0} \quad \text{otherwise} \end{cases} \tag{8}$$

   where $\tilde{0}$ is an upper triangular matrix in the previously defined sense (arbitrary large coefficients below the main diagonal) and having only zeros in its upper part.
2. otherwise as $L_{(ik)}$ has no crossing edges, there exists some $s \in [i, k]$ such that any edge of $L_{(ik)}$ but (i, k) is included in $[i, s]$ or in $[s, k]$. Then

$$D_{ik} = \begin{cases} D_{is}.D_{sk} + d_{ik} \text{ if } (i, k) \in L \\ D_{is}.D_{sk} \qquad\quad \text{otherwise} \end{cases} \tag{9}$$

   where the matrix multiplication is computed by replacing $(+, \times)$ operations on reals by $(\min, +)$.

Remark 1. If the contact graph has m vertices, and contains no crossing edges, then the problem is decomposed into $O(m)$ subproblems. For each of them, the computation of the corresponding distance matrix is a $O(n^3)$ procedure (matrix multiplication with $(\min, +)$ operations). Overall complexity is thus $O(mn^3)$. Typically, n is one or two orders of magnitude greater than m, and in practice, this special case is already expensive to solve.

3.2 All Edges Have Their Left End Tied to a Common Vertex

A set of edges $L = \{(i_1, k_1), \dots, (i_r, k_r)\}, k_1 < k_2 < \dots k_r$ is called a *star* if it has at least two elements and $i_t = i_1, t \leq r$. The arc costs corresponding to the link (i, k_s) are given by the upper triangular matrix d_{ik_s}.

The following algebra is used to prove the $O(n^2)$ complexity of the corresponding PTP.

Definition 1. *Let A, B be two matrices of size $n \times n$. $M = A \otimes B$ is defined by* $M(i,j) = \min_{i \leq r \leq j} A(i,r) + B(i,j)$

In order to compute $A \otimes B$, we use the following recursion: let M' be the matrix defined by $M'(i,j) = \min_{i \leq r \leq j} A(i,r)$, then

$$M'(i,j) = \min\{M'(i,j-1), A(i,j)\}, \text{ for all } j \geq i$$

Finally $A \otimes B = M' + B$. From this it is clear that $\otimes$ multiplication for $n \times n$ matrices is of complexity $O(n^2)$.

Theorem 1. *Let $L = \{(i, k_1), \dots, (i, k_r)\}$ be a star.*
Then $D_{ik_r} = (\dots (d_{ik_1} \otimes d_{ik_2}) \otimes \dots) \otimes d_{ik_r}$

Proof. The proof follows the basic dynamic programming recursion for this particular case: for the star $L = \{(i, k_1), \dots, (i, k_r)\} = L' \bigcup \{(i, k_r)\}$, we have $v(L : z_{ijk_rl} = 1) = d_{ijk_rl} + \min_{j \leq s \leq l} v(L' : z_{ijk_{r-1}s} = 1)$

3.3 Sequence of Independent Subproblems

Given a contact graph $L = \{(i_1, k_1), \dots, (i_r, k_r)\}$, $PTP(L)$ can be decomposed into two independent subproblems when there exists an integer $e \in (1, m)$ such that any edge of L is included either in $[1, e]$, either in $[e, m]$. Let $I = \{i_1, \dots, i_s\}$ be an ordered set of indices, such that any element of I allows for a decomposition of $PTP(L)$ into two independent subproblems. Suppose additionally that for all $t \leq s - 1$, one is able to compute $D_{i_t i_{t+1}}$. Then we have the following theorem:

Theorem 2. *Let $p = (p_1, p_2, \dots, p_n)$ be obtained by the following matrix-vector multiplication $p = D_{i_1 i_2} D_{i_2 i_3} \dots D_{i_{s-1} i_s} \overline{p}$, where $\overline{p} = (0, 0, \dots, 0)$ and the scalar product in the matrix-vector multiplication is defined by changing "+" with "min" and "." with "+". Then for all i, $p_i = v(PTP(L : y_{1i} = 1)$, and $v(PTP(L)) = \min\{p_i\}$.*

Proof. Each multiplication by $D_{i_k i_{k+1}}$ in the definition of p is an algebraic restatement of the main step of the algorithm for solving the shortest path problem in a graph without circuits.

Remark 2. With the notations introduced above, the complexity of $PTP(L)$ for a sequence of such subproblems is $O(sn^2)$ plus the cost of computing matrices $D_{i_t i_{t+1}}$.

From the last two special cases, it can be seen that if the contact graph can be decomposed into independent subsets, and if these subsets are single edges or stars, then there is a $O(srn^2)$ algorithm, where s is the cardinality of the decomposition, and r the maximal cardinality of each subset, that solves the corresponding PTP.

4 Cost Splitting

In order to apply the results from the previous section, we need to find a suitable partition of L into $L^1 \bigcup L^2 ... \bigcup L^t$ where each L^s induces an easy solvable $PTP(L^s)$, and to use the s.c. cost-splitting variant of the Lagrangian duality. Now we can restate (4)- (7) equivalently as:

$$v_{ip}^L = \min \left\{ \sum_{s=1}^{t} \left(\sum_{i=1}^{m} c_i^s y_i^s + \sum_{(i,k)\in L^s} d_{ik} z_{ik} \right) \right\} \quad (10)$$

$$\begin{aligned} \text{subject to: } & y_i^1 = y_i^s, & & s = 2, t & & (11) \\ & y^s = (y_1^s, .. y_m^s) \in Y, & & s = 1, \dots, t & & (12) \\ & y_i^s = A_i z_{ik}, \ y_k^s = A_k z_{ik} & & s = 1, \dots, t \quad (i,k) \in L^s & & (13) \\ & z_{ik} \in B^{\frac{n(n+1)}{2}} & & s = 1, \dots, t \quad (i,k) \in L^s & & (14) \end{aligned}$$

Taking (11) as the complicating constraints, we obtain the Lagrangian dual of $PTP(L)$:

$$v_{csd} = \max_{\lambda} \min_{y} \sum_{s=1}^{t} \left(\sum_{i=1}^{m} c_i^s(\lambda) y_i^s + \sum_{(i,k)\in L^s} d_{ik} z_{ik} \right) = \max_{\lambda} \sum_{s=1}^{t} v_{ip}^{L^s}(\lambda) \quad (15)$$

subject to (12), (13) and (14).

The Lagrangian multipliers λ^s are associated with the equations (11) and $c_i^1(\lambda) = c_i^1 + \sum_{s=2}^{t} \lambda^s$, $c_i^s(\lambda) = c_i^s - \lambda^s, s = 2, \dots, t$. The coefficients c_i^s are arbitrary (but fixed) decomposition (cost-split) of the coefficients c_i, i.e. given by $c_i^s = p_s c_i$ with $\sum p_s = 1$. From the Lagrangian duality theory follows $v_{lp} \leq v_{csd} \leq v_{ip}$. This means that for each PTP instance s.t. $v_{lp} = v_{ip}$ holds $v_{csd} = v_{ip}$. By applying the subgradient optimization technique ([13]) in order to obtain v_{csd}, one need to solve t problems $v_{ip}^{L^s}(\lambda)$ (see the definition of $v_{ip}^{L^s}$) for each λ generated during the subgradient iterations. As usual, the most time consuming step is $PTP(L^s)$ solving, but we have demonstrated its $O(n^2)$ complexity in the case when L^s is a union of independent stars and single links.

5 Experimental Results

The numerical results presented in this section were obtained on an Intel(R) Xeon(TM) CPU 2.4 GHz, 2 GB RAM, RedHat 9 Linux. The behavior of the algorithm was tested by computing the distributions used in FROST (Fold Recognition Oriented Search Tool) software [9]. The MIP models were solved using CPLEX 8.1.1 solver [7].

We pursued three main objectives. First: to compute the distributions with approximated values (found in a polynomial time) instead of the exact values, and to study the quality of these "approximated" distributions. Second: to compare the running times obtained by the cost-split algorithm with the ones of the CPLEX LP solver. Third: to compare the running times obtained by both Lagrangian algorithms.

We first focused on the phase of computing score distributions. Five distributions are associated to any 3D template in the FROST database, and any distribution requires solving about 200 sequence-to-template alignments. Only the 1st and at the 3rd quartile values from any of these distribution (q_{25} and q_{75}) are needed to compute the normalized score which is used for final evaluation [9,14].

We conducted the following experiment. We chose a set of 12 non-trivial templates, which are the same as given in [3], plus few extra-large instances based on real-life data generated by FROST. 60 distributions are associated to them. We first computed these distributions using an exact algorithm for solving the underlying alignment problem. The same distributions have been afterwords computed using approximated values obtained by any of the three PTP algorithms here considered. By approximated value we respectively mean: i) for a MIP model, this is the solution given by the LP relaxation; ii) for Stefan Balev's Lagrangian Relaxation algorithm (SB-LR) [3], this is the solution obtained for 500 iterations; iii) for the Cost-Splitting Lagrangian Relaxation algorithm (CS-LR) this is the solution obtained either for 300 iterations, or when the relative duality gap becomes less than 0.001.

For a MIP model we used the s.c. MYZ model introduced in [2] which has been proved faster than the model used in the package RAPTOR [17]. Because MIP models are relatively slow, we present here results from only 10 distributions which required solving 2000 alignments. We observed that in the 1st quartile the relative error between the exact and approximated (LP) distribution was 3×10^{-3} in only two cases and less than 10^{-6} for all other cases. In the 3rd quartile, the relative error was 10^{-3} in two cases and less than 10^{-6} for all other cases.

With LR algorithms we were able to compute all 12125 alignments for the selected set of 60 templates. We observed that in the 1st quartile, the exact and approximated values were equal for all cases for both (SB-LR and CS-LR) algorithms. In the 3rd quartile the exact solution of SB-LR algorithm was equal to the approximated one in all, but two cases, in which the relative error was respectively 10^{-3} and 10^{-5}. As for the CS-LR algorithm the exact value was equal to the approximated one in 12119 instances and the relative error was 7×10^{-4} in only 6 cases.

Obviously, such loss of precision is negligible and does not degrade the quality of the prediction. For this reasons we didn't explore a few nodes in a Branch and Bound tree just to improve insignificantly the heuristic solution in one out of 2000 cases.

We therefore conclude that the approximated values given by any of above mentioned algorithm can be successfully used in order to compute distributions.

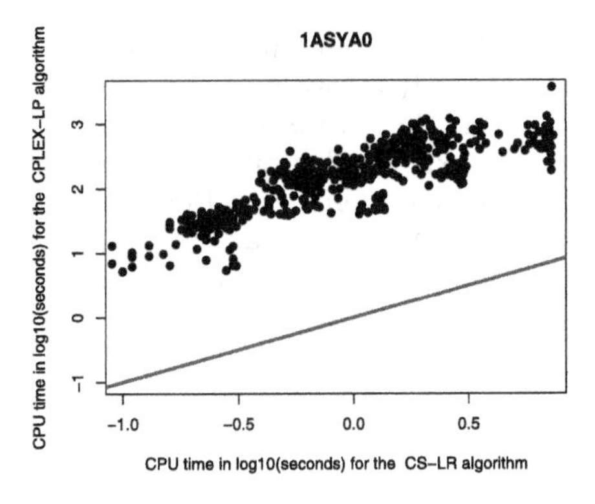

Plot of time in seconds with CS-LR algorithm on the x-axis and the LP algorithm from [2] on the y-axis. Both algorithms compute approximated solutions for 962 threading instances associated to the template 1ASYA0 from the FROST database. The linear curve in the plot is the line $y = x$. What is observed is a significant performance gap between the algorithms. For example in a point $(x, y) = (0.5, 3)$ CS-LR is $10^{2.5}$ times faster than LP relaxation.

Fig. 1. Cost-Splitting Lagrangian Relaxation versus LP Relaxation

Table 1. CS-LR versus SB-LR running time comparison. For cores given respectively in the first and the seventh column, and from left to right, the meaning of the columns is as follows: PDB code, number of blocks, number of query sequences that were aligned to the core, the minimum and maximum size of the solution space. The columns g_mean give the geometric mean of the ratio between SB-LR and CS-LR execution time.

Core	m	ali.	s_{min}	s_{max}	g_mean	Core	m	#	s_{min}	s_{max}	g_mean
1KEVA	27	28	1.7e+33	1.7e+33	0.928	1DIK	57	15	7.85e+48	7.85e+48	0.957
1MUCA	25	25	3.96e+31	3.96e+31	0.925	1ECEA	20	73	6.21e+29	6.21e+29	1.006
1AJSA	25	600	1.27e+25	5.06e+33	0.920	1FCDA	27	37	1.8e+35	1.8e+35	1.044
1AK5	23	979	9.1e+23	3.82e+37	1.084	1HRDA	26	17	1.74e+33	1.74e+33	1.194
1ASYA	26	752	3.48e+20	2.07e+37	1.242	1JDC	29	69	5.53e+39	5.53e+39	0.858
1LYLA	26	591	5.29e+28	1.89e+37	1.202	2TYSB	25	600	7.52e+22	3.49e+32	1.066
1SESA	26	593	5.18e+23	8.95e+33	1.300	2NACA	28	600	2.81e+28	8.4e+36	1.285
1PHP	29	600	4.19e+26	5.54e+36	1.033	2PHLA	23	600	3.22e+22	9.85e+30	1.386
1AFWB	24	600	3.89e+23	1.14e+32	1.309	1CG2A	26	600	5.19e+20	2e+32	0.998
1BGLA	55	480	5.39e+27	6.65e+77	0.936	1PBE	25	400	8.31e+25	3.64e+30	1.035
1CXSA	56	77	1.75e+68	1.75e+68	0.682						

Our second numerical experiment compared running times for computing "approximated" distribution by LP and CS-LR algorithm. The obtained results, given on Fig. 1 clearly show that CS-LR algorithm significantly outperforms the LP relaxation.

Our third test concerned CS-LR versus SB-LR running time comparison. For this purpose we selected a set of 8837 hard instances based on computations from [14]. These instances are related to 21 cores listed in table 1. In order to evaluate the performance of our algorithm, we computed the ratio of execution time between SB-LR and CS-LR algorithms for each one of these instances. We then computed the geometric mean of this ratio, for each core as shown in table 1. The overall mean is 1.12 showing that both methods are comparable, with a slight advantage for CS-LR algorithm. Fig. 2 visualizes the obtained results.

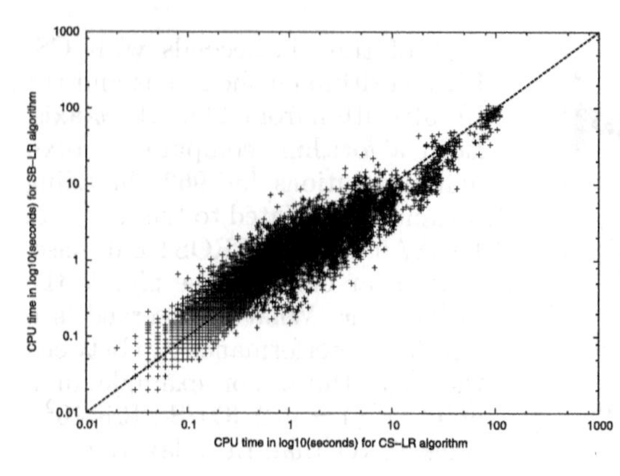

Plot of time in seconds with CS-LR algorithm on the x-axis and the SB-LR algorithm on the y-axis. Each point corresponds to one alignment. Both the x-axis and y-axis are in logarithmic scales. The linear curve in the plot is the line $y = x$.

Fig. 2. CS-LR versus SB-LR : recapitulation plot concerning 8337 alignments

6 Conclusion

The results in this paper confirm once more, that integer programming approach is well suited to solve protein threading problem. Here, we proposed a cost-splitting approach, and derived a new Lagrangian dual formulation for this problem. This approach compares favorably with the Lagrangian relaxation proposed in [3]. It allows to solve huge PTP instances[2] (of size above to 10^{77}) within a few minutes.

References

1. T. Akutsu and S. Miyano. On the approximation of protein threading. *Theoretical Computer Science*, 210:261–275, 1999.
2. R. Andonov, S. Balev and N. Yanev, Protein Threading Problem: From Mathematical Models to Parallel Implementations, *INFORMS Journal on Computing*, 2004, 16(4): 393-405
3. Stefan Balev, Solving the Protein Threading Problem by Lagrangian Relaxation, 4th Workshop on Algorithms in Bioinformatics, WABI, 2004
4. D. Fischer, http://www.cs.bgu.ac.il/ dfishcer/CAFASP3/, Dec. 2002
5. A. Caprara, R. Carr, S. Israil, G. Lancia and B. Walenz, 1001 Optimal PDB Structure Alignments: Integer Programming Methods for Finding the Maximum Contact Map Overlap *Journal of Computational Biology*, 11(1), 2004, pp. 27-52
6. H. Greenberg, W. Hart, and G. Lancia, Opportunities for combinatorial optimization in computational biology, *INFORMS Journal on Computing*, 16(3), 2004.
7. Ilog cplex. http://www.ilog.com/products/cplex
8. R. Lathrop, The protein threading problem with sequence amino acid interaction preferences is NP-complete, Protein Eng., 1994; 7: 1059-1068
9. A. Marin, J.Pothier, K. Zimmermann, J-F. Gibrat, FROST: A Filter Based Recognition Method, Proteins, 2002 Dec 1; 49(4): 493-509

[2] Solution space size of 10^{40} corresponds to a MIP model with 4×10^4 constraints and 2×10^6 variables [20].

10. T. Lengauer. Computational biology at the beginning of the post-genomic era. In R. Wilhelm, editor, *Informatics: 10 Years Back - 10 Years Ahead*, volume 2000 of *Lecture Notes in Computer Science*, pp. 341–355. Springer-Verlag, 2001.
11. G. Lancia. Integer Programming Models for Computational Biology Problems. *J. Comput. Sci. & Technol.*, Jan. 2004, Vol. 19, No.1, pp.60-77
12. R.H. Lathrop and T.F. Smith. Global optimum protein threading with gapped alignment and empirical pair potentials. *J. Mol. Biol.*, 255:641–665, 1996.
13. G. L. Nemhauser and L. A. Wolsey. *Integer and Combinatorial Optimization.* Wiley, 1988.
14. V. Poirriez, A. Marin, R. Andonov, J-F. Gibrat. FROST: Revisited and Distributed, HiCOMB 2005, Fourth IEEE International Workshop on High Performance Computational Biology, April 4, 2005, Denver, CO
15. R: A language and environment for statistical computing, R Foundation for Statistical Computing, Vienna, Austria, 2004, http://www.R-project.org
16. J.C. Setubal, J. Meidanis, Introduction to computational molecular biology, 1997, Chapter 8: 252-259, Brooks/Cole Publishing Company, 511 Forest Lodge Road, Pacific Grove, CA 93950
17. J. Xu, M. Li, G. Lin, D. Kim, and Y. Xu. RAPTOR: optimal protein threading by linear programming. *Journal of Bioinformatics and Computational Biology*, 1(1):95–118, 2003.
18. Y. Xu and D. Xu. Protein threading using PROSPECT: design and evaluation. *Proteins: Structure, Function, and Genetics*, 40:343–354, 2000.
19. N. Yanev and R. Andonov. Solving the protein threading problem in parallel. In *HiCOMB 2003 – Second IEEE International Workshop on High Performance Computational Biology*, 2003, Avril, Nice, France
20. N. Yanev and R. Andonov, Parallel Divide and Conquer Approach for the Protein Threading Problem, Concurrency and Computation: Practice and Experience, 2004; 16: 961-974

Efficient Parameterized Algorithm for Biopolymer Structure-Sequence Alignment

Yinglei Song[1], Chunmei Liu[1], Xiuzhen Huang[2], Russell L. Malmberg[3], Ying Xu[4], and Liming Cai[1,⋆]

[1] Dept. of Computer Science, Univ. of Georgia, Athens GA 30602, USA
cai@cs.uga.edu
[2] Dept. of Computer Science, Arkansas State Univ., State University, AR 72467, USA
[3] Dept. of Plant Biology, Univ. of Georgia, Athens GA 30602, USA
[4] Dept. of Biochemistry and Molecular Biology, Univ. of Georgia, Athens, GA 30602, USA

Abstract. Computational alignment of a biopolymer sequence (e.g., an RNA or a protein) to a structure is an effective approach to predict and search for the structure of new sequences. To identify the structure of remote homologs, the structure-sequence alignment has to consider not only sequence similarity but also spatially conserved conformations caused by residue interactions, and consequently is computationally intractable. It is difficult to cope with the inefficiency without compromising alignment accuracy, especially for structure search in genomes or large databases.

This paper introduces a novel method and a parameterized algorithm for structure-sequence alignment. Both the structure and the sequence are represented as graphs, where in general the graph for a biopolymer structure has a naturally small tree width. The algorithm constructs an optimal alignment by finding in the sequence graph the maximum valued subgraph isomorphic to the structure graph. It has the computational time complexity $O(k^t N^2)$ for the structure of N residues and its tree decomposition of width t. The parameter k, small in nature, is determined by a statistical cutoff for the correspondence between the structure and the sequence. The paper demonstrates a successful application of the algorithm to developing a fast program for RNA structural homology search.

1 Introduction

Structure-sequence alignment plays the central role in a number of important computational biology methods. For instance, protein threading, an effective method to predict protein tertiary structure, is based on the alignment between the target sequence and structure templates in a template database [3,5,37,19,36]. Structure-sequence alignment is also essential to RNA structural homology search, a viable approach to annotating (and identifying new) non-coding RNAs [10,12,29,22]. Structure-sequence alignment also finds applications in other bioinformatics tasks where structure plays an instrumental role, such as in the identification of the structure of intermolecular interactions [25,27], and in the discovery of the structure of biological pathways through comparative genomics [8].

⋆ Corresponding author.

R. Casadio and G. Myers (Eds.): WABI 2005, LNBI 3692, pp. 376–388, 2005.

The structure-sequence alignment is to find an optimal way to "fit" the residues of a target sequence in the spatial positions of a structure template. To be able to identify the structure of remote homologs, the alignment has to consider not only sequence similarity but also spatially conserved conformations caused by sophisticated interactions between residues, and consequently is computationally intractable. For example, it is both NP-hard for protein threading with amino acid interactions [18] and for thermodynamic determination of RNA secondary structure including pseudoknots [23].

The alignment problem has often been formulated as integer programming that characterizes residue spatial interactions with (a large number of) linear inequality constraints [36,20]. Commercial software packages for linear programming are usually used to approximate the integer programming and to reduce the computation time. More sophisticated techniques, such as branch-and-cut, can be used to dynamically include only needed linear constraints [20,28]. Moreover, a divide-and-conquer method based on the notion of "open-links" has also been devised to address the residue-residue interaction issue [37]. For RNA structure-sequence alignment, dynamic programming has been extended to include crossing patterns of RNA nucleotide interactions [32,7]. The above algorithmic techniques cope with the alignment intractability, however, most of them still require computation time polynomial of a high-degree.

In this paper, we introduce an efficient structure-sequence alignment algorithm. Both structure and sequence are represented as mixed graphs (with directed and undirected edges); the optimal alignment corresponds to finding in the sequence graph the maximum valued subgraph isomorphic to the structure graph. In addition, we introduce an integer parameter k to constrain the correspondence between the graphs. A dynamic programming algorithm is developed over a tree decomposition of the structure graph. For each value of k, the optimal alignment can be found in time $O(k^t N^2)$ for each structure template containing N residues given a tree decomposition of tree width t.

Our algorithm is a parameterized algorithm [11], in which the naturally small parameter k determined by a statistical cutoff reflects the accuracy of the alignment. The new algorithm with the time complexity $O(k^t N^2)$ is more efficient than previous algorithms, for example, of the time complexity $O(N^k)$ [37]. This is also because the tree width t of the graph for a biopolymer structure is small in nature. For example, the tree width is 2 for the graph of any pseudoknot-free RNA and the width can only increase slightly for all known pseudoknot structures (see Figure 5). Our experiments also show that among 3890 protein tertiary structure templates compiled using PISCES [33], only 0.8% of them have tree width $t > 10$ and 92% have $t < 6$, when using a 7.5 $\mathring{A}$ C_β-C_β distance cutoff for defining pair-wise interactions (Figure 2(a)).

The alignment algorithm has been applied to the development of a fast RNA structure homology search program [31]. With a significantly reduced amount of computation time, the new search method achieves the same accuracy as searches based on the widely used Covariance model (CM) [13]. The new algorithm yields about 24 to 50 times of speed up for the search of pseudoknot-free RNAs with 90 to 150 nucleotides; it gains even more significant advantage for larger RNAs or structures including pseudoknots. In addition, for all the conducted tests, including the searches of medium to large RNAs in bacteria and yeast genomes, parameter $k \leq 7$ has been sufficient for the accurate identification.

2 Problem Formulation

We formulate structure-sequence alignment as a *generalized* subgraph isomorphism problem. Graphs used here are *mixed* graphs contaning both undirected and directed edges. Let $V(G)$, $E(G)$, and $A(G)$ denote the vertex set, the undirected edge set, and the directed edge (arc) set of graph G, respectively.

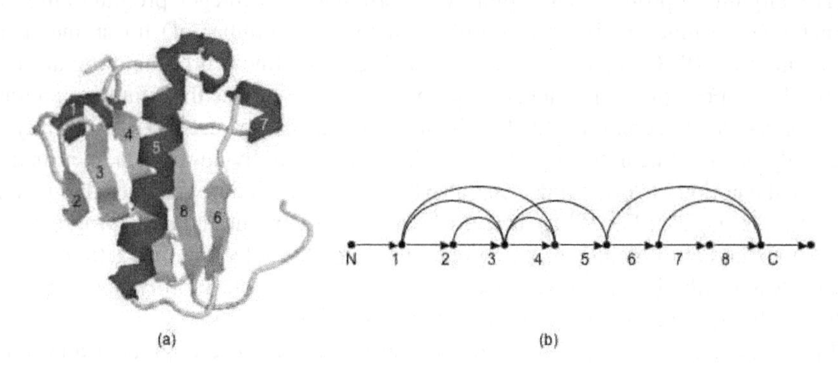

Fig. 1. (a) Folded ChainB of Protein Kinase C interacting protein with 8 cores (the PDB-file corresponding to PDB-ID 1AV5); (b) its corresponding structure graph

Definition 1. A *structural unit* in a biopolymer sequence is a stretch of contiguous residues (nucleotides or amino acids). A non-structural stretch, between two consecutive structural units, is called a *loop*.

A structure of the sequence is characterized by interactions among structural units. For example, structural units in a tertiary protein are α helices and β strends, called *cores*. Figure 1(a) shows a protein structure with 8 structural units. In the RNA secondary structure, a structural unit is a stretch of nucleotides, one half of a stem formed by a stack of base pairings.

Given a biopolymer sequence, a *structure graph* H can be defined such that each vertex in $V(H)$ represents a structural unit, each edge in $E(H)$ represents the interaction between two structural units, and each arc in $A(H)$ represents the loop ended by two structural units. Figure 1(b) shows the structure graph for the folded protein in 1(a). Figure 5 shows the graph for bacterial tmRNAs.

The alignment between a structure template and a target sequence is to place residues of the sequence in the spatial positions of the template. Instead of placing individual residues to the spatial positions, the method we introduce in this paper allows us to put a stretch of residues as a whole in the position of some structural unit of the template. The sequence to be aligned to the structure is preprocessed so that all *candidates* in the sequence are identified for every structural unit in the template.

By representing each candidate as a vertex, the target sequence can also be represented as a mixed graph G, called a *sequence graph*. Each edge in $E(G)$ connects a pair of candidates that may possibly interact but do not overlap in sequence positions, and an arc in $A(G)$ connects two candidates that do not overlap.

Based on the graph representations, the structure-sequence alignment problem can be formulated as the problem of finding in the sequence graph G a subgraph isomorphic to the structure graph H such that the objective function based on the alignment score achieves the optimum. For this, we first introduce a mechanism to parameterize (and to scrutinize) the mapping between H and G.

Definition 2. A *map scheme* M between graphs H and G is a function: $V(H) \rightarrow 2^{V(G)}$ that maps every vertex in H to a subset of vertices in G. The maximum size of such a subset, $k = \max_{v \in V(H)}\{|M(v)|\}$, is called the *map width* of the map scheme.

A map scheme can be obtained in the preprocessing step that finds all candidates of every structural unit. The qualification of these candidates can usually be quantified by a statistical cutoff of the degree to which a candidate is aligned to a structural unit. One may simply choose the top k candidates for each structural unit. More sophisticated map schemes are possible (see section 4), in which ideally, the parameter k reflects the accuracy of alignment results. We define the following parameterized problem:

GENERALIZED SUBGRAPH ISOMORPHISM:
INPUT: mixed graphs H and G, and map scheme M of width k;
OUTPUT: a subgraph G' of G and an isomorphic mapping $f : V(H) \rightarrow V(G')$, constrained by $f(x) \in M(x)$ for any x, such that the objective function

$$\sum_{u \in V(H)} S_1(u, f(u)) + \sum_{(u,v) \in E(H)} S_2((u,v),(f(u),f(v))) + \sum_{\langle u,v \rangle \in A(H)} S_3(\langle u,v \rangle, \langle f(u), f(v) \rangle) \qquad (1)$$

achieves the optimum (i.e., maximum or minimum).

Functions S_1, S_2, and S_3 are application dependent, scoring respectively three different alignments between the structure template and the target sequence: the alignment between a structural unit u and its candidate $f(u)$, the alignment between the interaction of two structural units (u, v) and the interaction of the corresponding candidates $(f(u), f(v))$, and the alignment between a loop (connecting two neighboring structural units u and v) and its correspondence loop in the sequence.

This problem generalizes the well-known NP-hard subgraph isomorphism decision problem. Efficient algorithms for subgraph isomorphism may be obtained on constrained instances. However, algorithms of this kind only exist for the cases where H is small, fixed, and G is planar or of a small tree width [1,14,24]. None of these conditions can be satisfied by the application in structure-sequence alignment, where the structure can be large and the sequence graph is often arbitrary.

We conclude this section by noting that the parameterization introduced on the map width does not trivialize the problem under investigation. In fact, one can transform NP-hard problem 3-SAT to (a decision version of) this problem when k is fixed to be 3, leading to the following theorem (the proof details are omitted).

Theorem 1. The problem GENERALIZED SUBGRAPH ISOMORPHISM remains NP-hard on map schemes of map width $k = 3$.

3 Parameterized Alignment Algorithm

Definition 3. [30] Pair (T, X) is a *tree decomposition* of a mixed graph H if

1. T is a tree,
2. $X = \{X_i | i \in V(T), X_i \subseteq V(H)\}$, and $\bigcup_{X_i \in X} X_i = V(H)$,
3. $\forall u, v, (u, v) \in E(H)$ or $\langle u, v\rangle \in A(H)$, $\exists i \in V(T)$ such that $u, v \in X_i$, and
4. $\forall i, j, k \in V(T)$, if k is on the path from i to j in tree T, then $X_i \cap X_j \subseteq X_k$.

The *tree width* of (T, X) is defined as $\max_{i \in V(T)}\{|X_i|\} - 1$. The *tree width of the graph* is the minimum tree width over all possible tree decompositions of the graph.

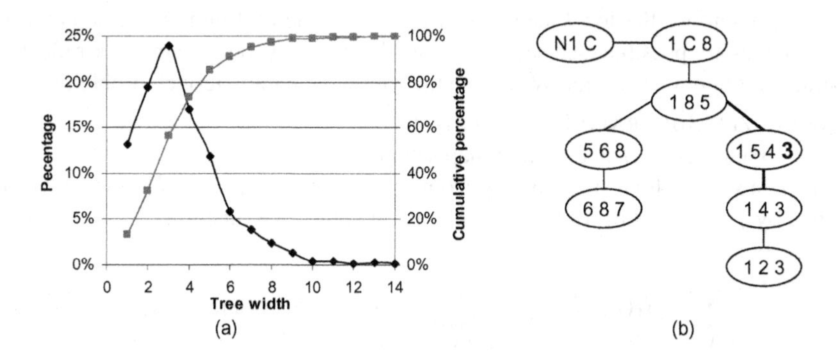

Fig. 2. (a) Tree width distribution of the graphs for 3,890 protein structure templates compiled using PISCES [33,34]. (b) A tree decomposition for the structure graph in Figure 1(b).

Biopolymer structure graphs in general have small tree width. For instance, the tree width of the structure graphs for pseudoknot-free RNAs is 2, and it can only increase slightly for all known pseudoknots. Figure 2(a) gives a statistics on the tree width of about 3,890 protein structure templates compiled using PISCES [33,34]. Figure 2(b) shows a tree decomposition for the protein structure graph in Figure 1(b).

3.1 Parameterized Algorithm for Subgraph Isomorphism

We now describe a tree decomposition based parameterized algorithm for the problem GENERALIZED SUBGRAPH ISOMORPHISM formulated in section 2. Our algorithm assumes a given tree decomposition (T, X) of width t for structure graph H. Our algorithm follows the basic idea of the tree decomposition based techniques in [1,2].

To simplify our discussion, we assume that T for the tree decomposition is a binary tree. The following notations will also be useful. Let $U \subseteq V(H)$ and $Y \subseteq V(G)$ such that $|U| = |Y|$. Then a mapping $f : U \to Y$ is a *valid mapping for* U if f is a subgraph isomorphism between the graph induced by U and the graph induced by Y. If $W \subseteq U$, then $f|_W$ is f *projected onto* W, therefore a valid mapping for W. A *partial isomorphism for* H *with respect to* X_i is a valid mapping f for $U = X_i \cup \bigcup_{k \in D(i)} X_k$, where $D(i)$ is the set of i's descendent nodes in the tree.

In a bottom up fashion, the algorithm establishes one table for each tree node. Let $X_i = \{u_0, u_1, \ldots, u_t\}$. Table m_i for tree node i consists of $|X_i| + 3$ columns, one for every vertex in X_i. Rows are all possible mappings for X_i restricted by the map scheme M; each row is of the form $\langle x_0, x_1, \ldots, x_t \rangle$ representing the mapping f, $f(u_l) = x_l$, $l = 0, 1, \ldots, t$. There are three additional columns in the table: V, S, Opt (see Figure 3). $V(f) =$ '$\surd$' if and only if mapping f is valid for X_i. $S(f)$ is the optimal score over all the partial isomorphism e for H with respect to X_i such that $f = e|_{X_i}$. $Opt(f)$ indicates whether $S(f)$ is the optimal over all valid mapping f' for X_i, where $f'|_{X_i \cap X_p} = f|_{X_i \cap X_p}$ for p, the parent node of i.

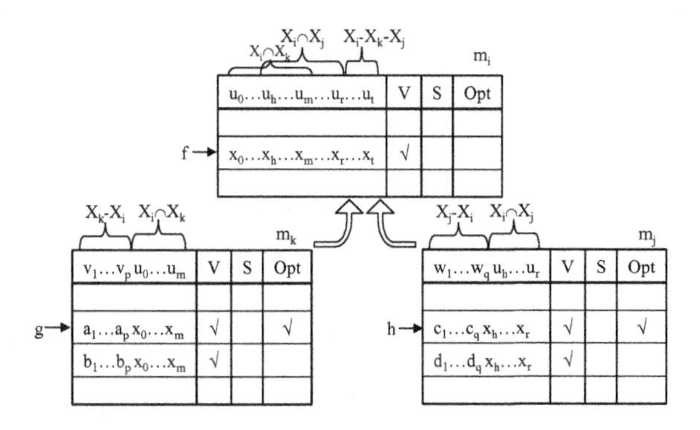

Fig. 3. Computing dynamic programming tables over a tree decomposition in which tree node i has two children k and j

If i is a leaf node, the score $S(f)$ is simply the value computed based on formula (1) (given in section 2) for vertices in X_i only. If i is an internal node with children nodes k and j, $S(f)$ is the sum of the following three value :

1. The value computed for f with formula (1) for vertices in X_i only,
2. The maximum S value over all valid mappings g in table m_k such that $g|_{X_i \cap X_k} = f|_{X_i \cap X_k}$, and
3. The maximum S value over all valid mappings h in table m_j such that $h|_{X_i \cap X_j} = f|_{X_i \cap X_j}$.

Figure 3 illustrates the computation for row f in table m_i of the internal node i that has two children nodes k and j. The formal algorithm, GENSUBGISOMO, is outlined as a recursive process in Figure 4. The optimal score computed in the table for the root of the tree T is the best isomorphism score. A recursive routine can be used to trace back the corresponding optimal isomorphism. Details are omitted here.

We need to prove that the (bottom up) dynamic programming always produces correct partial isomorphisms. Since the algorithm automatically validates the isomorphism for locally involved vertices, it suffices to ensure that for every $u \in X_i$, the mapping from u to x for some $x \in M(u)$ does not conflict with an earlier mapping from v to x, for some vertex $v \in X_k$, where k is a descendent of i. Interestingly enough, for

ALGORITHMGENSUBGISOMO (T, X_i, M, i, m_i)
If i has left child k, GENSUBGISOMO(T, X_k, M, k, m_k);
If i has right child j, GENSUBGISOMO(T, X_j, M, j, m_j);
For every every mapping f for X_i, constrained by M
    If i has left child k in T
        Find in m_k a valid mapping g, such that $g|_{X_i \cap X_k} = f|_{X_i \cap X_k}$ of $Opt(g)$ being '$\surd$';
    If i has right child j in T
        Find in m_j a valid mapping h, such that $h|_{X_i \cap X_j} = f|_{X_i \cap X_j}$ of $Opt(h)$ being '$\surd$';
    Compute score $score(f)$ with formula (1) for X_i only;
    Let $S(f) = score(f) + S(g) + S(h)$;
    If i has parent p in T, and $S(f)$ maximizes over all f' with $f'|_{X_i \cap X_p} = f|_{X_i \cap X_p}$
        Let $Opt(f) =$ '$\surd$';
Return (m_i);

Fig. 4. An outline for the tree decomposition based recursive algorithm GENSUBGISOMO that solves the problem GENERALIZED SUBGRAPH ISOMORPHISM. The algorithm assumes the input of a tree decomposition (T, X) and a map scheme M; it returns table m_i for every node i in T.

mixed graphs H constructed from biopolymer structures, the non-conflict property is also automatically guaranteed. The following is a brief justification for this claim.

Note that the directed edges in graph H form the total order relation $(V(H), \preceq)$ defined as follows: $v \preceq u$ if (i) either $\langle u, v \rangle \in A(H)$, or (ii) $\exists w, v \preceq w$ and $\langle u, w \rangle \in A(H)$. This relation needs to be satisfied by any (partial) isomorphism. Assume vertices $u \in X_i$, $v \in X_k$, and k is one of i's descendants in the tree. Assume $v \preceq u$ (the case of $u \preceq v$ is similar). Then in general there exists j on the path from i to k, such that $\exists w \in X_j, v \preceq w$ and $w \preceq u$. An induction on the distance of the chain from u to v can assert that the mapping conflict cannot occur between u and v so long as $v \preceq u$.

Theorem 2. GENSUBGISOMO correctly solves the GENERALIZED SUBGRAPH ISOMORPHISM problem for every given tree decomposition and every given map scheme.

Corollary 3. Parameterized algorithm GENSUBGISOMO computes the optimal structure-sequence alignment for every given map scheme of width k.

3.2 Tree Decomposition and Total Alignment Time

For graphs with tree width t, theoretical algorithms [4] can find an optimal tree decomposition in time $O(c^t n)$ for some (possibly large) constant c. We introduce a simple greedy algorithm for tree decomposition that practically runs fast on structure graphs.

Given a structure graph H, undirected edges are selected such that removals of these edges from the graph result in an outerplanar graph. The removals of these edges are done by first removing an edge (but not the endpoints) that *crosses* with the maximum number of other edges, and then repeating the same process until the resulting graph contains no crossing edges. Note that two edges (u, v) and (u', v') in H *cross* each other if either $v' \preceq v \preceq u' \preceq u$ or $v \preceq v' \preceq u \preceq u'$ (see section 3.1 for the definition of the partial order $(V(H), \preceq)$).

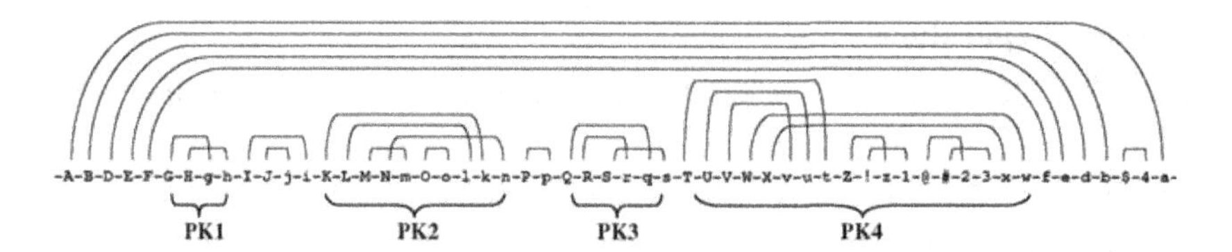

Fig. 5. Diagram of the pairing regions on the tmRNA gene. Upper case letters indicate base sequences that pair with the corresponding lower case letters. The four pseudoknots constitute the central part of the tmRNA gene and are called Pk1, Pk2, Pk3, Pk4 respectively.

A simple recursive algorithm can find a tree decomposition of tree width 2 for the remaining outerplanar graph. Then for each removed edge (u, v), in the tree we place v in every node on the (shortest) path from a node containing v to a node containing u. The tree decompositon shown in Figure 2(b) is obtained by first removing crossing edge $(3, 5)$. Then a tree decomposition for the remaining outerplanar graph is built, which is extended to the tree decomposition for the original graph by placing vertex 3 (in the bold font) in node $\{1, 5, 4\}$ on the path from node $\{1, 4, 3\}$ to node $\{1, 8, 5\}$. This strategy produces a tree decomposition of size at most $2 + c$ if there are c crossing edges removed. In reality, the obtained tree decomposition has much smaller tree width. For example, for the structure graph constructed from the bacterial $tmRNA$ structure (Figure 5), our strategy shall yield a tree decomposition of tree width 4 instead of 9. This algorithm is of linear time $O(|E(H) + |A(H)| + |V(H)|)$.

The running time for algorithm GENSUBGISOMO is $O(k^t t^2 n)$, for map width k, tree width t, number of vertices n in H. For each row in the table, the compliance with subgraph isomorphism needs to be validated and a score computed according to formula (1) (by looking up pre-computed values of functions S_1, S_2, S_3). The former step needs $O(t^2)$ and the latter $O(t^2 + 2t \log_2 k)$ (note that the rows of a table can be ordered to facilitate binary search by the computation for its parent node).

It takes $O(knN)$ time to preprocess the target sequence of length N to construct the sequence graph. Simultaneously, this step pre-computes the values of functions S_1, S_2. The values of function S_3 can then be pre-computed, using time $O(k \sum_{i=1}^{l} l_i^2) = O(knN)$, where l_i is the length of ith loop and l is the number of loops in the structure. Summing up the times needed by the preprocessing, tree decomposition, and ALGORITHM GENSUBGISOMO gives us a loose upper bound $O(k^t nN)$, or $O(k^t N^2)$, for the total time for the structure-sequence alignment.

4 Applications in Fast RNA Structural Homology Search

To evaluate the performance of our method and algorithm for structure-sequence alignment, we have applied them to the development of a fast program that can search for RNA structural homologs. We have also conducted extensive tests on finding medium to large RNA secondary structures (including pseudoknots) in both random sequences and biological genomes (bacteria and yeasts) [31]. We summarize our test results in the following.

4.1 Data Preparations

The tests on RNA structure searches that we conducted can be grouped into three categories:

1. On 8 RNA pseudoknot-free structures, of medium size (61 - 112 nucleotides), inserted in random sequences of length 10^5,
2. On 6 RNA pseudoknot structures, of medium size (55 - 170 nucleotides), inserted in random sequences of length 10^5, and
3. On 3 RNA pseudoknot structures, of medium to large size (61 - 755), in a variety of genomes of lengths range from 2.7×10^4 to 1.1×10^7.

Each homologous RNA family is modelled with a structure graph. Each undirected edge in the graph represents a stem that is profiled with a simplified Covariance Model (CM) [13]. Each arc in the graph represents a loop (5' to 3') that is profiled with a profile Hidden Markov Model (HMM). In the first two categories of searches, for each family we downloaded from the Rfam database [16] 30 RNA sequences with their mutual identities below 80%. We used them to train the CMs and profile HMMs in the model.

For each family we downloaded from Rfam another 30 sequences with their mutual identities below 80% and use them for search. They were inserted in a random background of 10^5 nucleotides generated with the same base compositions. Using a method similar to the one used in RSEARCH [17], we computed the statistical distribution for the alignment scores with a random sequence of 3,000 nucleotides generated with the same base composition as the sequences to be searched. An alignment score with a Z-score exceeding 5.0 was reported as a hit. Both random sequences and genomes were scanned through with a window of a size correlated with the structure model size. The segment of the sequence falling within the window was aligned to the model with the structure-sequence alignment algorithm presented in the earlier sections.

For the tests of the third category, we searched for three RNA pseudoknot structures: the pseudoknot structure in the 3' UTR in the corona virus family [15], the bacterial tmRNA structure (see Figure 5) that contains 4 pseudoknots [26], and yeast telomerase RNA consisting of up to 755 nucleotides [9]. The structures for these RNAs were trained with 14, 85, and 5 available sequences respectively. The searched genomes for the 3' UTR pseudoknot were Bovine corona virus, Murine hepatitus virus, Porcine diarrhea virus, and Human corona virus, with the average length 3×10^4. The two searched bacteria genomes for the tmRNA were *Haemophilus influenzae* and *Neisseria meningitidis*, with the average length 2×10^6. Yeast genomes,*Saccharomyces cerevisiae* and *Saccharomyces bayanus* of the average length 11×10^6, were used to search for the telomerase RNA.

To obtain a reasonably small value for the parameter k, the map scheme between the structure and the sequence was designed with the constraint that candidates of a given stem were restricted in certain region in the target sequence. For this, we assumed that for homologous sequences, the distances from each pairing region of the given stem to the 3' end follow a Gaussian distribution, whose mean and standard deviation were computed based on the training sequences. For training sequences representing distant homologs of an RNA family, we could effectively divide data into groups so that a different but related structure model was built for each group and used for searches. This method ensures a small value for the parameter k in search models.

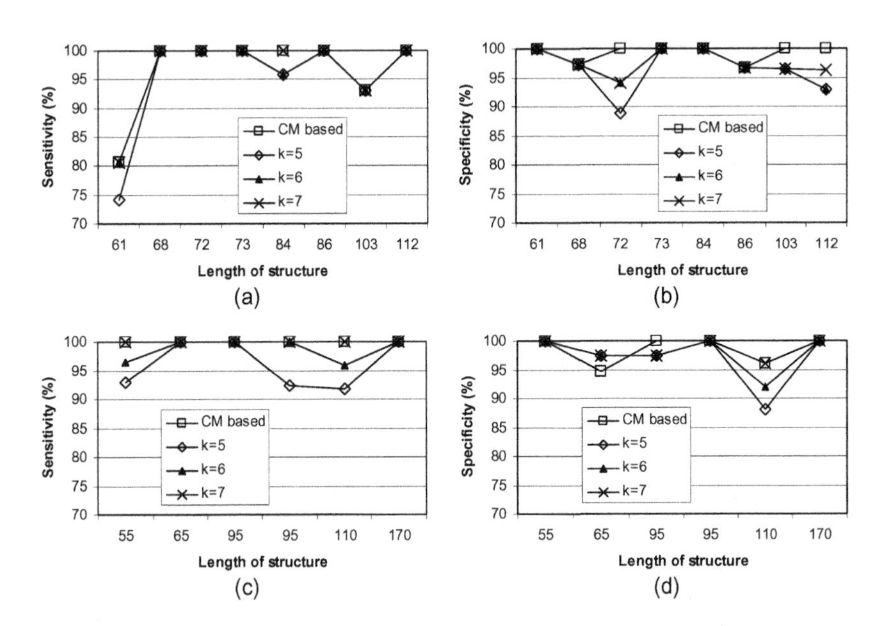

Fig. 6. Performance comparison between the tree decomposition based method and the CM based method on search for RNA structures, (a) and (b) for pseudoknot-free structures, (c) and (d) for pseudoknots

4.2 Performance Evaluations

We conducted the tests on the tree decomposition based search program and on a Covariance Model (CM) based search system[1] and compared the performances of the two. The tests results showed that, on all three categories, parameter $k = 7$ was sufficient for our new search program to achieve the same accuracy as the CM based search system does. But the computation time used by the new method was significantly reduced.

Figure 6(a) and (b) respectively show the sensitivity comparison and specificity comparison between the two search methods on pseudoknot-free RNA structures. These structures were from eight RNA families: Entero_CRE, SECIS, Lin_4, Entero_OriR, Let_7, Tymo_tRNA-like, Purine, and S_box, in the increasing order of their length. The tree decomposition based algorithm performed quite well for $k = 6$ and larger values.

Figure 6(c) and (d) respectively show the sensitivity comparison and specificity comparison between the two search methods on RNA pseudoknot structures. These were from six RNA families: Antizyme_FSE, corona_pk3, HDV_ribozyme, Tombus_3_IV, Alpha_RBS, and IFN_gamma, in the increasing order of their lengths. As for pseudoknot-free structures, the tree decomposition based searches for pseudo-

[1] We developed this CM based system [21] in the same spirit of Brown and Wilson's work [6] that profiles pseudoknots with intersection of CMs. CM was first introduced by Eddy and Durbin [13] and has proved very accurate in profiling for search of pseudoknot-free RNA structures.

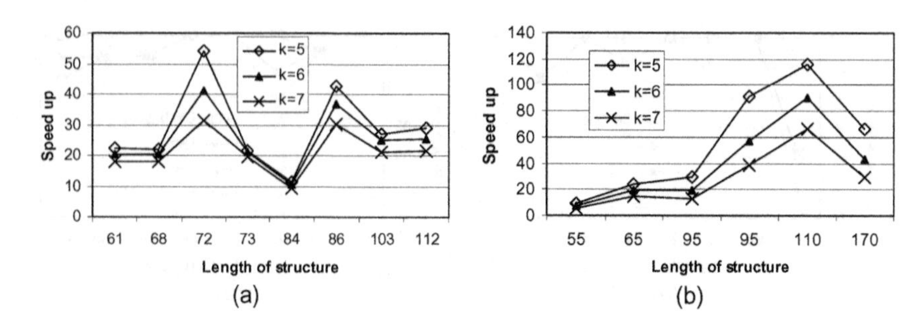

Fig. 7. The speed up of the tree decomposition based method over the CM based method: (a) on pseudoknot-free structures, and (b) on pseudoknot structures

	ncRNA	Real location		Tree decomposition based			CM based			Genome length
		Left	Right	Left offset	Right offset	Time	Left off	Right off	Time	
3'PK	BCV	30798	30859	0	0	0.053	0	0	1.24	3.1×10^4
	MHV	31792	31153	0	0	0.053	0	0	1.27	3.1×10^4
	PDV	27802	27882	0	0	0.048	0	0	1.17	2.8×10^4
	HCV	27063	27125	0	0	0.047	0	0	1.12	2.7×10^4
tmRNA	HI	472209	472574	-1	-1	44.0	0	0	1700	1.83×10^5
	NM	1241197	1241559	0	0	52.9	0	0	2044	2.2×10^5
TLRNA	SC	307688	308429	-3	-1	492.3	-	-	-	1.03×10^7
	SB	7121529	7122284	-3	2	550.2	-	-	-	1.15×10^7

Fig. 8. Performance comparison between the tree decomposition based method and the CM based method on RNA structure searches on genomes. Offset is between the annotated and the real positions. Time unit is hour.

knots achieved the same performance as the CM based method for parameter values $k \leq 7$.

Figure 7 shows the speed up by the new method over the CM based method, for (a) pseudoknot-free and (b) pseudoknot structures. It is evident that for $k = 7$ the new method was about 20 to 30 times faster than the other method on pseudoknot-free structures. On the pseudoknot structures, typically on Alpha_RBS and Tombus_3_IV containing more than 100 nucleotides, the new method was 66 and 38 times faster, suggesting its advantage in the search of larger and more complex structures.

Figure 8 compares the search results obtained by the two methods on three types of RNA pseudoknots in virus, bacteria, and yeast genomes. Parameter $k = 7$ is used for the parameterized algorithm. Both methods achieve 100% sensitivity and specificity. It clearly shows that the new method had a speed-up of about 30 to 40 times over the other method for searches in virus and bacteria genomes. With the new method, searching genomes of a moderate size for structures as complex as tmRNA gene (see Figure 5) only took days, instead of months. Searching a larger genome such as yeast for larger structure like telomerase RNAs was also successful, a task not accomplishable by the CM based system within a reasonable amount of time.

5 Conclusions

We introduced a novel method and an efficient parameterized algorithm for the structure-sequence alignment problem by exploiting the small tree width of biopolymer structure graphs. The algorithm was applied to the development of a fast search program that is capable of accurately identifying complex RNA secondary structure including pseudoknots in genomes [31]. Our method provides a new perspective on structure-sequence alignment that is important in a number of bioinformatics research areas where structure plays an instrumental role. In particular, we expect the tree decomposition based method, together with one for protein side-chain packing [35], to yield efficient and accurate protein threading algorithms.

References

1. S. Arnborg and A. Proskurowski, "Linear time algorithms for NP-hard problems restricted to partial k-trees", *Discrete Applied Mathematics*, 23: 11-24, 1989.
2. S. Arnborg, J. Lagergren, and D. Seese, "Easy problems for tree-decomposable graphs", *Journal of Algorithms* 12: 308-340, 1991.
3. J. Bowie, R. Luthy, and D. Eisenberg, "A method to identify protein sequences that fold into a known three-dimensional structure", *Science* 253: 164-170, 1991.
4. H. L. Bodlaender, "A linear time algorithm for finding tree-decompositions of small treewidth", *SIAM Journal on Computing* 25: 1305-1317, 1996.
5. S.H. Bryant and S.F. Altschul, "Statistics of sequence-structure threading", *Curr. Opinion Struct. Biol.* 5: 236-244, 1995.
6. M. Brown and C. Wilson, "RNA pseudoknot modeling using intersections of stochastic context free grammars with applications to database search", *Pacific Symposium on Biocomputing*, 109-125, 1995.
7. L. Cai, R. Malmberg, and Y. Wu, "Stochastic Modeling of Pseudoknot Structures: A Grammatical Approach", *Bioinformatics*, 19, $i66 - i73$, 2003.
8. T. Dandekar, S. Schuster S, B. Snel, M. Huynen, and P. Bork, "Pathway alignment: application to the comparative analysis of glycolytic enzymes", *Biochemical Journal.* 1: 115-24, 1999.
9. A. T. Dandjinou, N. Lévesque, S. Larose, J. Lucier, S. A. Elela, and R. J. Wellinger, "A phylogenetically based secondary structure for the yeast telomerase RNA.", *Current Biology*, 14: 1148-1158, 2004.
10. J.A. Doudna, "Structural genomics of RNA", *Nature Structural Biology* 7(11) supp. 954-956, 2000.
11. R. Downey and M. Fellows, *Parameterized Complexity*, Springer, 1999.
12. S.R. Eddy, "Computational genomics of non-coding RNA genes", *Cell* 109:137-140, 2002.
13. S. Eddy and R. Durbin, "RNA sequence analysis using covariance models", *Nucleic Acids Research*, 22: 2079-2088, 1994.
14. D. Eppstein, "Subgraph isomorphism in planar graphs and related problems", *Journal of Graph Algorithms and Applications*, 3.3: 1-27, 1999.
15. S. J. Geobel, B. Hsue, T. F. Dombrowski, and P. S. Masters, "Characterization of the RNA components of a Putative Molecular Switch in the 3' Untranslated Region of the Murine Coronavirus Genome.", *Journal of Virology*, 78: 669-682, 2004.
16. S. Griffiths-Jones, A. Bateman, M. Marshall, A. Khanna, and S. R. Eddy, "Rfam: an RNA family database", *Nucleic Acids Research*, 31: 439-441, 2003.

17. R. J. Klein and S. R. Eddy, "RSEARCH: Finding Homologs of Single Structured RNA Sequences.", *BMC Bioinformatics*, 4:44, 2003.
18. R.H. Lathrop, "The protein threading problem with sequence amino acid interaction preferences is NP-complete", *Protein Engineering* 7: 1069-1068, 1994.
19. R.H. Lathrop, R.G. Rogers Jr, J. Bienkowska, B.K.M. Bryant, L. J. Buturovic, C. Gaitatzes, R.Nambudripad, J.V. White, and T.F. Smith, "Analysis and algorithms for protein sequence-structure alignment", in *Computational Methods in Molecular Biology*, Salzberg, Searls, and Kasif ed., Elsevier, 1998.
20. H-P. Lenhof, K. Reinert, and M. Vingron. "A polyhedral approach to RNA sequence structure alignment", *Journal of Computational Biology* 5(3): 517-530, 1998.
21. C. Liu, Y. Song, R. Malmberg, and L. Cai, "Profiling and searching for RNA pseudoknot structures in genomes", *Lecture Notes in Computer Science* 3515, 968-975, 2005.
22. T. M. Lowe and S. R. Eddy, "tRNAscan-SE: A Program for improved detection of transfer RNA genes in genomic sequence", *Nucleic Acids Research*, 25: 955-964, 1997.
23. S.B. Lyngso and C.N. Pedersen, "RNA pseudoknot prediction in energy-based models", *Journal of Computational Biololgy* 7(3):409-427, 2000.
24. J. Matousek and R. Thomas, "On the complexity of finding iso- and other morphisms for partial k-trees", *Discrete Mathematics*, 108: 343-364, 1992.
25. E.M. Marcotte, P. Matteo, HL. Ng, D.W. Rice, T.O. Yeates, and D. Eisenberg, "Detecting protein function and protein-protein interactions from genome sequences", *Science* 285: 751-753, 1999.
26. N. Nameki, B. Felden, J. F. Atkins, R. F. Gesteland, H. Himeno, and A. Muto, "Functional and structural analysis of a pseudoknot upstream of the tag-encoded sequence in E. coli tmRNA", *Journal of Molecular Biology*, 286(3): 733-744, 1999.
27. D.D. Pervouchine, "IRIS: Intermolecular RNA Interaction Search", *Genome Informatics* 15(2): 92-101, 2004.
28. K. Reinert, H-P. Lenhof, P. Mutzel , K. Mehlhorn , and J.D. Kececioglu, "A branch-and-cut algorithm for multiple sequence alignment", *Proceedings of the first annual international conference on Computational molecular biology*, 241-250, 1997.
29. E. Rivas and S. R. Eddy, "Noncoding RNA gene detection using comparative sequence analysis", *BMC Bioinformatics*, 2:8, 2001.
30. N. Robertson and P. D. Seymour, "Graph Minors II. Algorithmic aspects of tree-width", *Journal of Algorithms* 7: 309-322, 1986.
31. Y. Song. C. Liu, R. Malmberg, F. Pan, and L. Cai, "Tree decomposition based fast search of RNA secondary structures in genomes", *Proceedings of 2005 IEEE Computational Systems Biology Conference*, in press.
32. Y. Uemura, A. Hasegawa, Y. Kobayashi, and T. Yokomori, "Tree adjoining grammars for RNA structure prediction", *Theoretical Computer Science*, 210: 277-303, 1999.
33. G. Wang and R.L. Dunbrack, Jr. "PISCES: a protein sequence culling server", *Bioinformatics* 19: 1589-1591, 2003.
34. D. Xu, M.A. Unseren, Y. Xu, and E.C. Uberbacher, "Sequence-structure specificity of a knowledge based energy function at the secondary structure level", *Bioinformatics* 16:257-268, 2000.
35. J. Xu, "Rapid side-chain packing via tree decomposition", In *Proceedings of 2005 International Conference on Research in Computational Biology*, to appear.
36. J. Xu, M. Li, D. Kim, and Y. Xu. "RAPTOR: optimal protein threading by linear programming", *Journal of Bioinformatics and Computational Biology*, 1(1):95-113, 2003.
37. Y. Xu, D. Xu, and E.C. Uberbacher, "An efficient computational method for globally optimal threading", *Journal of Computational Biology* 5(3):597-614.

Rotamer-Pair Energy Calculations Using a Trie Data Structure

Andrew Leaver-Fay[1], Brian Kuhlman[2], and Jack Snoeyink[1]

[1] Department of Computer Science, University of North Carolina at Chapel Hill
[2] Department of Biochemistry, University of North Carolina at Chapel Hill

Abstract. Protein design software places amino acid side chains by precomputing rotamer-pair energies and optimizing rotamer placement. If the software optimizes by rapid stochastic techniques, then the precomputation phase dominates run time. We present a new algorithm for rapid rotamer-pair energy computation that uses a trie data structure. The trie structure avoids redundant energy computations, and lends itself to time-saving pruning techniques based on a simple geometric criteria. With our new algorithm, we compute rotamer-pair energies nearly 4 times faster than the previous approach.

1 Introduction

Researchers have recently achieved notable success in computational protein design. Homme Hellinga's lab redesigned the active site of Ribose Binding Protein to bind TNT [1,2]. David Baker's lab designed a more stable protein-L and created a novel protein fold [3,4]. The two labs solve a common subproblem: with a fixed protein backbone as scaffold, they search for a side chain placement that packs them snuggly without collisions. The snugness-of-fit is captured by an energy function. The problem of minimizing the energy function over all side chain conformations is known as the side chain placement problem.

Side chain conformational flexibility is typically modeled by creating many possible atom placements. Each side chain may be modeled with bond lengths and bond angles fixed to standard or experimentally determined values; selected torsional (or dihedral) angles that are variable give the flexibility. These angles have preferred values that have been observed within the Protein Databank (PDB) [5] and confirmed through quantum mechanical calculations [6]. Designers sample the continuous torsional space near these torsional angles' preferred values to generate "rotamers:" conformational iso*mers* that differ by torsional *rota*tions. Scientists have collected preferred side chain conformations into rotamer libraries [7,8,9,10].

Designers divide the side chain placement problem into two phases. In phase 1, they precompute all possible rotamer-pair interaction energies for their rotamer library, and in phase 2, they search for the (globally) optimal side chain placement. Significant work has gone into exact algorithms for the side chain placement problem [11,12,13,14,15]. Still, the problem is NP-Complete [16], and

R. Casadio and G. Myers (Eds.): WABI 2005, LNBI 3692, pp. 389–400, 2005.

many researchers choose fast stochastic optimization techniques [17,18,19,20,21]. The rotamer-pair energy computation of phase 1 can be a significant fraction of the running time for both techniques, and usually dominates the running time of stochastic techniques. Thus, in this paper, we address rotamer-pair energy computation.

The interaction energy between two rotamers, A and B, is the sum of the atom/atom interaction energies over all atoms of A with all atoms of B. When a pair of rotamers on the same residue share torsional angles, they share atoms. Repeated atoms imply repeated atom/atom energy evaluations when computing all rotamer-pair energies

The obvious way to avoid repeating atom/atom energy computations is to store in a table the result of atom/atom energy computations for unique atom pairs. When a unique atom pair is encountered for the first time, calculate the pair's interaction energy and store it. When a unique atom pair is encountered any subsequent time, simply look up the old result. However, with a moderately large rotamer library of 2K rotamers, which we use as our running example, a single residue can generate ~10K unique atoms. A unique-atom by unique-atom table with 10K x 10K entries would occupy 400 MB. This table does not fit in a processor's cache (~512 KB). Although storing energies avoids repeated computation, retrieving the table entries incurs cache misses, eroding any savings in running time.

We use a trie to represent all the rotamers on a single residue. With a pair of these "rotamer tries" we can rapidly compute the rotamer-pair energies, while reducing our memory usage. We have implemented our algorithm within the Rosetta molecular modeling software [17]. Because our memory use is minimal, and because we reuse atom/atom energy computations, our algorithm runs nearly 4 times faster than Rosetta's existing method.

2 Methods

2.1 Rotamers and Tries

As mentioned in the introduction, rotamer libraries are usually built by sampling certain torsional (or dihedral) angles; these are denoted by $\chi_1, \chi_2, \ldots$ in order from the protein backbone. The most flexible amino acids, lysine and arginine, have four χ dihedrals. Rotamers of the same amino acid on the same residue that share a prefix of χ dihedrals place many of their atoms in the same position. For instance, two leucine rotamers that share a χ_1 dihedral place their C_β, $1H_\beta$, $2H_\beta$ and C_γ atoms identically. If we order atoms by distance from the backbone as well, then the shared atoms are also a prefix. Trie data structures are perfect for capturing shared prefixes.

A "trie" is a rooted tree. Each node in the trie contains an object. Each root-to-leaf path in the trie represents a string of objects. Tries are often used to represent dictionaries. In a dictionary trie, each node represents a letter. Each root-to-leaf path represents a word. For example, consider a dictionary containing just two words: 'apple' and 'apply.' The root would be the letter 'a'.

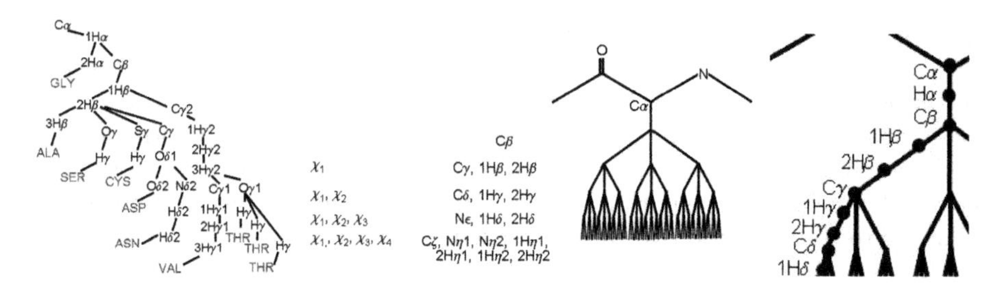

Fig. 1. Two example tries. a) One rotamer from each of the small amino acids and three for threonine. b) A set of arginine rotamers showing the branching pattern for arginine's four χ dihedrals. c) Angle χ_1 determines the coordinates of $1H_\beta$, $2H_\beta$ and C_γ, so they lie together along a path.

The shared prefix 'appl' would lie along an unbranched path of the trie. The 'l' node would have two children, 'e' and 'y'. The path to the leaf node 'y' from the root spells out the word 'apply'.

In a rotamer trie, each node contains an atom. Each root-to-leaf path represents a rotamer. We depict a few rotamer tries in Figure 1. Note that the trie connectivity does not reflect the amino acids' chemical structure. Note also that the three threonine rotamers depicted differ only in their hydrogen position: their substantial shared prefix lets us save space. The trie for arginine in Figure 1(b&c) shows the trie branching produced by its four χ dihedrals. A trie for a complete rotamer set would be too large to display.

Tries have proven useful in a number of other string problems in computational biology [22,23,24]. Homme Hellinga previously introduced representing rotamer sets in a trie-like structure to weed out rotamers colliding with the background [25] but does not use tries to compute energies.

It is with a pair of rotamer tries that we compute rotamer-pair energies. We have implemented our algorithm to be compatible with the energy function from Rosetta. We describe the details of Rosetta's energy function, and the existing rotamer-pair energy subroutine in the next section.

2.2 Rosetta's Energy Function

Rosetta has four terms that apply on an atom-by-atom basis. Between all heavy atom pairs, Rosetta includes three terms: a van der Waal's attractive term, a van der Waal's repulsive term, and a Lazaridis-Karplus implicit solvation [26] term. Each of these terms depend on the atom types of the two heavy atoms, and their distance. For speed, Rosetta uses a maximum distance threshold of 5.5 Å: if two heavy atoms are further than 5.5 Å apart, then their interaction energy is zero.

Between hydrogen/other atom pairs two terms apply: a van der Waal's repulsive term, and a statistically derived hydrogen bonding term [27]. The hydrogen bonding term is usually described by four atoms acting simultaneously: the donor hydrogen, the donor heavy atom, the acceptor and the acceptor-base. The term depends on one distance and the cosine of two angles: the hydrogen-acceptor distance, the cosine of the donor heavy atom—hydrogen—acceptor angle and the cosine of the hydrogen—acceptor—acceptor-base angle. We reformulate the hydrogen bond function to depend on only two atoms, the hydrogen and the acceptor, by including orientation vectors with each atom. The orientation vectors allow us to compute the two cosines needed.

Because Rosetta's terms involving hydrogens are so short-ranged, Rosetta's developers use a distance threshold between two heavy atoms to determine if their attached hydrogen atoms could be close enough to interact. If two heavy atoms are further than 4.6 Å apart, then all hydrogen/other atom pairs for the attached hydrogens have zero interaction energy.

Rosetta's existing rotamer-pair energy function, get_energies(), takes two rotamers sets, R and S, and outputs their rotamer-pair energies into a rotamer/rotamer energy table (*rot_rot_E*). We describe get_energies() with the following pseudocode:

```
get_energies(R, S)
  for i = 1 : R.num_rotamers
    for j = 1 : S.num_rotamers
      if cbeta_dis( i, j) > threshold(amino_acid(i), amino_acid(j) )
        continue;
      energy_sum = 0
      for k = 1 : num_heavy_atoms(i)
        for l = 1 : num_heavy_atoms(j)
          energy_sum += atom_atom_energy( R.atom(i,k), S.atom(j,l) );
          if dis( R.atom(i,k), S.atom(j,l) ) < 4.6
            energy_sum += calc_attached_h_energies( R.atom(i,k), S.atom(j,l) );
      rot_rot_E[ i, j ] = energy_sum;
  return;
```

where calc_attached_h_energies(k, l) iterates over the hydrogen/heavy atom pairs and hydrogen/hydrogen atom pairs calling atom_atom_energy() for the heavy atoms k and l and their attached hydrogen atoms. Because hydrogens make up roughly half of the atoms in a rotamer, it would be roughly 4 times more expensive to evaluate all atom/atom energies as it would be to evaluate all heavy atom/heavy atom energies, descending into the hydrogens only as needed.

Rosetta does not compute rotamer-pair energies between rotamers if their C_β atoms are so distant that it is impossible for any pair of rotamers of those two amino acid types to interact. Rosetta uses its 5.5 Å heavy atom distance cutoff to calculate these thresholds.

2.3 Trie Node

The trie data structure stores everything we need for evaluating Rosetta's energy function. It also stores a number of variables needed to prune energy computations, which we describe after the algorithm.

We represent our trie as an array. We store the nodes in their preorder traversal order. In the recursive description of the algorithm that we give below (Sec. 3.1), we refer to child pointers as if they were explicit. However, we store the depth of each node in the tree instead of explicit child pointers. The preorder/depth representation is sufficient to completely describe the trie structure.

We store the atom type for each atom, its xyz coordinate, and its orientation vector. In a redundant, but time-saving extension of the atom type, we keep several boolean flags: is_backbone, is_heavy_atom, is_acceptor, is_donor_h, etc. Each of these flags is stored as a single bit. Each bit value is determined by the atom type and is therefore redundant. However, the logic is somewhat complex to convert between atom type and these boolean values. Instead of evaluating the conversion functions during the trie traversal $O(n^2)$ times, we evaluate the value of these flags outside of the main loop and store them compactly in the trie node.

When we prune, we use 40 bytes per node. The last three variables in the trie_node are needed only for pruning. The "no pruning" implementation does not allocate space for these three variables, and so the cost per trie_node drops to 32 bytes. We have found it especially important to make sure our trie nodes align with the 32-bit memory boundaries.

```
struct trie_node
   float[3]       xyz;                  //12 bytes
   float[3]       o_vector;             //12 bytes
   unsigned char  atom_type;            // 4 bytes
   unsigned char  depth;
   unsigned char  hv_depth;
   unsigned char  flags;
   unsigned short flags2;               // 4 bytes
   unsigned short hybridization;
   unsigned short rotamers_in_subtree;  // 4 bytes
   unsigned short sibling;
   float          subtree_radius;       // 4 bytes
```

3 Interaction Energy Between Two Rotamer Tries

We now give the algorithm to calculate the rotamer-pair energies between two tries, R and S. The idea is simple, we perform a preorder traversal of R, and for each atom $r \in R$, we perform a preorder traversal of S. We evaluate atom_atom_energy(r, s) for each pair of atoms we encounter ($s \in S$). (We refer to the preorder traversal order when we use the words 'before', 'preceed' and 'after' below.)

To calculate the rotamer-pair energies, we use two recursive functions: atom_vs_trie() and trie_vs_trie(). For clarity we describe these functions recursively; for speed we implement them iteratively.

- atom_vs_trie$(r, s, ancestral_E)$ recursively computes the interaction energy between atom r and all the rotamers in the subtree of S rooted at node s. It stores these energies in a global variable, AREnergies, a stack of arrays. atom_vs_trie() calls atom_atom_energy() and is called by trie_vs_trie(r, S).
- trie_vs_trie(r, S) recursively computes the interaction energy between the rotamers in the subtree of R rooted at node r and the rotamers in the trie S. It stores these energies in a global variable, RREnergies, the table of rotamer/rotamer energies. An invocation of trie_vs_trie$(R.root, S)$ calculates all rotamer-pair energies. trie_vs_trie() invokes atom_vs_trie().

3.1 Functions in Detail

Global Variables. We use three global variables in these recursive functions:

- RREnergies. Rotamer/Rotamer Energies. This table has (R.num_rotamers $\times$ S.num_rotamers) entries, one for each rotamer pair.
- AREnergies. Atom/Rotamer Energies. This is a stack of arrays. Each array contains S.num_rotamers entries and holds r's ancestors' interaction energies with rotamers of S. The stack height is limited to the maximum number of ancestors with siblings of any leaf in a rotamer tree.
- ARStackTop. Top of stack pointer for AREnergies.

atom_vs_trie(r, s, ancestral_E) in Detail. Precondition: r is an atom of R, s is an atom of S. ancestral_E holds the sum of the interaction energies r has with all ancestors of s. AREnergies[ARStackTop] contains the sum of the interaction energies of all of r's ancestors with the rotamers of S that terminate at or after s, and contains the sum of r's ancestors' and r's interaction energies for all rotamers of S that terminate before s.

Postcondition: AREnergies[ARStackTop] contains the sum of interaction energies of r and its ancestors with the rotamers of S that terminate before s or terminate in s's subtree. AREnergies[ARStackTop] contains the sum of the interaction energies of all other rotamers in S with r's ancestors only.

Pseudocode:

```
atom_vs_trie(r, s, ancestral_E)
   ancestral_E += atom_atom_energy(r, s);
   if (s.terminal_rotamer_id != -1)
      AREnergies[ARStackTop][s.terminal_rotamer_id] += ancestral_E;
   for (int i = 0; i < s.num_children; i++)
      atom_vs_trie(r, s.child[i], ancestral_E);
   return;
```

trie_vs_trie(r, S) in Detail. Precondition: r is an atom of R. AREnergies[ARStackTop] contains the sum of the interaction energies of r's ancestors with the rotamers of S. If r is the root, then ARStackTop must be zero and each entry in AREnergies[0] is zero.

Postcondition: RREnergies contains the interaction energies for all rotamers of S and the rotamers of R that terminate in the subtree of R rooted at r.

Pseudocode:

```
trie_vs_trie(r, S)
  atom_vs_trie(r, S.root, 0);
  if (r.terminal_rotamer_id != -1)
    //copy entire AREnergies row
    RREnergies[r.terminal_rotamer_id] = AREnergies[ARStackTop];
  if (r.num_children > 0)
    ARStackTop++;
    for (int i = 0; i < r.num_children-1, i++)
      //copy stack top for children with siblings
      AREnergies[ARStackTop] = AREnergies[ARStackTop - 1];
      trie_vs_trie(r.child[i], S);
    ARStackTop--;
    //last child doesn't need its own stack copy
    trie_vs_trie(r.child[r.num_children - 1], S);
  return;
```

Because we traverse S repeatedly, it is critical that S fit inside the processor's cache. The size of each node in the trie is 40 bytes. In our example rotamer set with 10K unique atoms, S would occupy 400KB. Since most cache sizes are 512KB, S fits comfortably. AREnergies's size would be 4 rows $\times$ 2K rotamers/row $\times$ 4 bytes/float = 32KB. There are only 4 rows in AREnergies since the most flexible amino acids have only 4 χ dihedrals.

3.2 Pruning Computations

We can use the tree structure of the two tries R and S to avoid performing many of the atom/atom energy computations. Suppose we are somewhere in the middle of the trie traversals, examining atoms $r \in R$ and $s \in S$. Beneath r is a subtree containing some number of atoms, beneath s is another subtree containing some number of atoms. If r is a heavy atom, then there are some number of hydrogen atoms bound to r in the subtree beneath r. We have three conceptual entities: atoms, heavy atoms (including their associated hydrogen atoms), and subtrees. We may prune calculations for any combination of entities.

We decide how to prune based on r and s's distance. In the following sections we describe the additional data structures we maintain. Briefly, here is a sketch of our pruning options.

1. atom/atom: If the distance between r and s exceeds a threshold, we can assign their interaction energy to zero without doing a more detailed calculation. After this prune, we still must continue calculating interactions between the atoms in r's and s's subtrees. Rosetta already includes this prune within its atom_atom_energy() function. Because we use this function as well, we get this prune for free.
2. atom/subtree: If the distance between r and s is so great that r must be too far to interact with any atom in s's subtree, then we can make an atom/subtree prune.
3. subtree/subtree: If the distance between r and s is so great that all atoms in r's subtree must be too far to interact with any atom in s's subtree, then we can make a subtree/subtree prune. Rosetta makes a similar prune using C_β atoms (see Section 2.2 above).
4. heavy atom/heavy atom: If the distance between heavy atoms r and s exceeds 4.6 Å, we can skip calculating interactions among their bound hydrogens. Rosetta already employs this prune, so we must too, if we hope to improve upon the running time. After this prune, we still must continue calculating interactions between the atoms in r's and s's subtrees.
5. heavy atom/subtree: Much like the atom/subtree pruning, we may see that a heavy atom r and all of its attached hydrogens are too far to interact with all the atoms in s's subtree and then perform this skip.

Heavy Atom/Heavy Atom Pruning. As we described above, if a pair of heavy atoms are further apart than $\lambda = 4.6$ Å, then all the interactions between the hydrogen/heavy atom pairs and the hydrogen/hydrogen pairs is zero. We prune computations based on this cutoff by 1) restricting the atom ordering within the trie and 2) including another global variable in our algorithm, skipH.

We order the atoms in our trie so that the closest ancestral heavy atom for a hydrogen is the heavy atom to which it is chemically bound. For instance, the ordering of atoms for alanine would be: C O N H CA HA CB 1HB 2HB 3HB. (We include backbone atoms as part of the trie. This leaves room for later incorporating backbone flexibility as part of a protein redesign task.) We store each atom's "heavy atom depth" (hv_depth). The heavy atom depth for a heavy atom is its position in a list of an amino acid's heavy atoms. Alanine's CA heavy atom depth is 4. The heavy atom depth for a hydrogen atom is the depth of its parent heavy atom. Alanine's HA heavy atom depth is also 4.

Our new global variable, skipH, is a stack of booleans, represented as a table. It has MAX_HEAVY rows (the largest number of heavy atoms for a single amino acid, which in tryptophan is 14). Each row has S.num_heavyatoms entries. This table is a stack in that its contents describes properties for ancestor atoms of our currently focused r atom. In essence, the top-of-stack pointer is stored within each atom of R by its heavy atom depth.

Now we'll describe how we use skipH. If a heavy atom r at heavy atom depth d is at least λ away from heavy atom s of S, then we set skipH$[d][s]$ to 'true.' Later, if we want to know if the parent heavy atom for a hydrogen atom of R at heavy atom depth, d, and the heavy atom s of S are greater than λ apart, then skipH$[d][s]$ tells us. To capture this formally, we revise the atom_vs_trie($r, s, ancestral\_E$) pre- and postconditions.

Additional Precondition: For all heavy atom ancestors r' of r, skipH$[r'.\text{hv_depth}][s']$ holds 'true' iff s' is further than λ from r' for all heavy atoms $s' \in S$. Additionally if r is a heavy atom, then for all s'' that precede s, skipH$[r.\text{hv_depth}][s'']$ holds 'true' iff s'' is further than λ from r.

Additional Postcondition: For all heavy atom ancestors r' of r, (including r if r is a heavy atom), skipH$[r'.\text{hv_depth}][s']$ holds 'true' iff s' is further than λ from r' for all heavy atoms $s' \in S$.

skipH scales in size with the number of heavy atoms in S, unlike some of our other global variables that scale with the number of rotamers in S. It is a rather large data structure. In our example rotamer trie with 10K atoms, roughly 5K would be heavy atoms. In this case, skipH would occupy 70KB.

Subtree/Subtree Pruning. In a subtree/subtree prune, we avoid calculating atom/atom energies for all pairs of atoms in the subtrees of r and s. We prune based on a sphere overlap test. The "interaction sphere" of heavy atom r is the sphere centered at r that has a radius of one half of the threshold distance for heavy atom/heavy atom interaction; in our case, one half of 5.5 Å. For two heavy atoms to interact, their interaction spheres must overlap. The "subtree-interaction sphere" of heavy atom r is centered at r, and is large enough that, for any atom s to interact with an atom in r's subtree, s's interaction sphere must overlap with r's subtree-interaction sphere. Equivalently, the radius of the subtree-interaction sphere is the greatest distance between r and all heavy atoms in r's subtree + (5.5 Å/2).

When two subtree-interaction spheres do not overlap, we may make a subtree/subtree prune. The non-overlapping condition is met when the squared distance between r and s exceeds the square of the sum of r and s's subtree-interaction-sphere radii. This comparison is very fast and we can afford to make it at each heavy atom pair we encounter.

When we decide to skip computations involving the subtrees rooted at r and s we immediately add ancestral_E to AREnergies for all rotamers of S that terminate in the subtree rooted at s. We then skip to the first node of S, that is not in s's subtree. To make this jump, we must know the sibling of s's closest ancestor (which may be s itself if s has a sibling).

We also skip over s's subtree for all of r's descendants in later calls to atom_vs_trie(). We maintain another global array, trim_depth, to record which subtrees should be skipped. This array has one entry for each heavy atom of S. The values stored in each entry are small (< 14) so we can get away with using a single byte per entry. In our example rotramer trie with 5K heavy atoms, trim_depth occupies only 5KB.

We describe the functionality of this variable with another pre- and postcondition pair for atom_vs_trie($r, s, ancestral\_E$).

Additional Precondition: If r is a heavy atom (hydrogen), then trim_depth[s] is less than (less than or equal to) r.hv_depth if 1) the subtree-interaction sphere of the heavy atom ancestor of r at depth trim_depth[s] (call this ancestor, r') does not overlap with s's subtree-interaction sphere, and 2) s is the only atom amongst it and its ancestors whose subtree-interaction sphere does not overlap with r''s subtree-interaction sphere. The value of trim_depth[s] is undefined for those atoms of S for which condition 1, but not condition 2, holds. If s's interaction sphere overlaps with the subtree-interaction spheres for all ancestors of r, then trim_depth[s] is greater than or equal to r.hv_depth when r is a heavy atom, or strictly greater than r.hv_depth when r is a hydrogen atom.

Additional Postcondition: If r is a heavy atom (hydrogen) and trim_depth[s] was less than (less than or equal to) r.hv_depth, then trim_depth[s] remains the same, and the values in trim_depth for atoms in the subtree rooted at s are undefined. If r and s are heavy atoms, and r and s's subtree-interaction spheres do not overlap, then trim_depth[s] is r.hv_depth. Otherwise, trim_depth[s] is MAX_HEAVY + 1.

We also make subtree/subtree prunes when we encounter colliding atoms. Collisions reflect physically impossible situations, and an exact representation of a collision's energy is unnecessary. We prune when we find an atom/atom energy that exceeds 20 kcal/mol.

Heavy Atom/Subtree Pruning. If r's interaction sphere and s's subtree-interaction sphere do not overlap we may skip past s's subtree. In order to repeat this subtree-skip for the hydrogen atoms attached to r, we maintain another global variable, skipSubtree. skipSubtree, like skipH, is a stack of boolean arrays represented as a table. Each array has S.num_heavyatoms entries. There are

MAX_HEAVY rows. We do not provide additional pre- and postconditions as they are so similar to those describing skipH. skipSubtree would occupy 70KB in our example 10K atom rotamer trie.

4 Results

We wrote our algorithm in C++ and verified that it generates the same energies as Rosetta's existing rotamer-pair energy function, get_energies(). We compared the running time of our algorithm using six pruning options against get_energies() in 57 complete protein redesign tasks (Fig. 2). All six variants included heavy atom/heavy atom pruning. We measured running times on Intel Xeon 2.8 GHz processors each with 2.5 GB RAM. Our algorithm runs 3.87 times faster than get_energies().

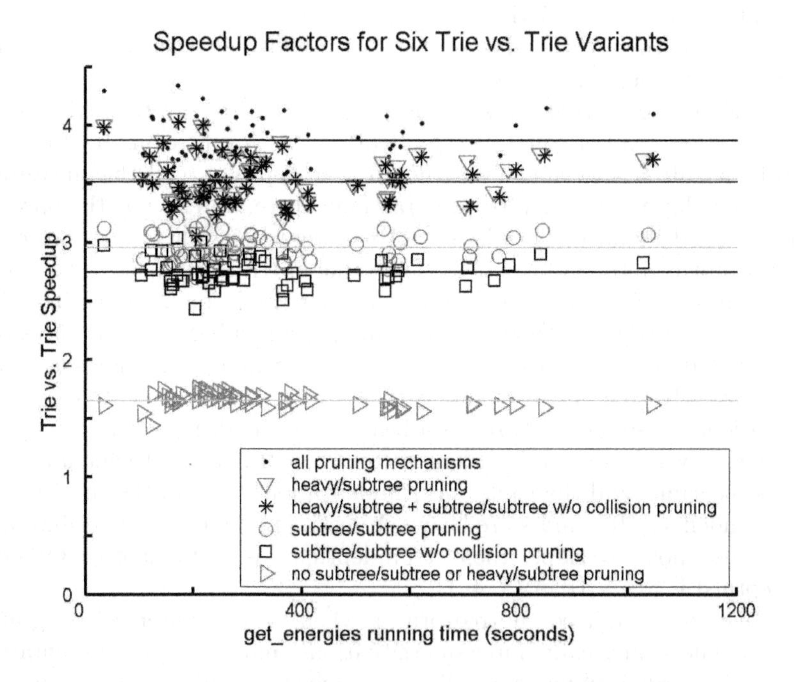

Fig. 2. Comparing trie_vs_trie() and get_energies() for 57 entire-protein-redesign energy computations. Mean speedup factors for the six pruning combinations were 1.65, 2,75, 2.96, 3.52, 3.54, and 3.87.

5 Discussion

We have sped up the bottleneck stage of Rosetta's protein design module. There are a few direct consequences of our trie structure we would like to highlight.

Hydroxyl Hydrogens. Fine grained sampling of dihedral space for terminal hydroxyl groups now comes at a reduced cost. The shared atomic prefix for two rotamers that differ in their hydroxyl hydrogen placement spans all atoms but the last two: the hydroxyl oxygen and hydrogen. Because the orientation vectors on hydroxyl oxygens point at the hydrogen, each oxygen is distinct. For tyrosine the shared atomic prefix includes 13 side chain atoms.

Uniform Rotamer Libraries. Common dihedral angles can improve trie performance by 23%. Currently Rosetta selects its rotamers using Dunbrack's backbone dependent rotamer library. In this library, very few χ_1 dihedrals agree. The overlap that buys us our performance boost comes from the additional rotamer samples Rosetta takes at χ_2 that surround ($\pm\sigma$) Dunbrack's rotamers. We measured a 10% decrease in the number of unique atoms in the rotamer trie when we construct our rotamers using a new rotamer library built from rounding Dunbrack's rotamers to the nearest 10°.

Flexible Backbone Design. Imagine sampling a few backbone conformations for a pair of residues [28] and attaching hundreds of rotamers to each sample. This setup for flexible backbone design promotes the backbone from the role of static background into the role of structural variable. With flexible backbone design, side chain/backbone energies must be included in the rotamer-pair energy calculations. We incorporated backbone atoms into our tries so that when we begin using flexible backbone design, we can make effective reuse of side chain/backbone computations.

References

1. Looger, L.L., Dwyer, M.A., Smith, J.J., Hellinga, H.W.: Computational design of receptor and sensor proteins with novel functions. Nature **423** (2003) 185–190
2. Dwyer, M., Looger, L., Hellinga, H.: Computational design of a biologically active enzyme. Science **304** (2004) 1967–1971
3. Kuhlman, B., O'Niell, J.W., Kim, D.E., Zhang, K.Y., Baker, D.: Accurate computer-based design of a new backbone conformation in the second turn of protein L. J. Mol. Bio. **315** (2002) 471–477
4. Kuhlman, B., Dantas, G., Ireton, G., Varani, G., Stoddard, B., Baker, D.: Design of a novel globular protein fold with atomic-level accuracy. Science **302** (2003) 1364–1368
5. Berman, H.M., Westbrook, J., Feng, Z., Gilliland, G., Bhat, T., Weissig, H.N.S.I., Bourne, P.: The protein data bank. Nucleic Acids Research **28** (2000) 235–242
6. Butterfoss, G., Hermans, J.: Boltzmann-type distribution of side-chain conformation in proteins. Protein Science **12** (2003) 2719–2731
7. Ponder, J.W., Richards, F.: Tertiary templates for proteins. Use of packing criteria in the enumeration of allowed sequences for different structural classes. J. Mol. Bio. **193** (1987) 775–791
8. R. L. Dunbrack, J., Karplus, M.: Backbone dependant rotamer library for proteins: application to side chain prediction. J. Mol. Bio. **230** (1993) 543–574
9. R. L. Dunbrack, J.: Rotamer libraries in the 21st century. Curr. Opin. Struct. Biol. **12** (2002) 431–440

10. Lovell, S.C., Word, J.M., Richardson, J.S., Richardson, D.C.: The penultimate rotamer library. Proteins: Structure Function and Genetics **40** (2000) 389–408
11. Desmet, J., Maeyer, M.D., Hazes, B., Lasters, I.: The dead-end elimination theorem and its use in protein side-chain positioning. Nature **356** (1992) 539–541
12. Goldstein, R.F.: Efficient rotamer elimination applied to protein side-chains and related spin glasses. Biophysical Journal **66** (1994) 1335–1340
13. Looger, L.L., Hellinga, H.W.: Generalized dead-end elimination algorithms make large-scale protein side-chain structure prediction tractable: implications for protein design and structural genomics. J Mol Biol **307** (2001) 429–45
14. Gordon, D., Mayo, S.: Branch-and-terminate: a combinatorial optimization algorithm for protein design. Structure Fold Des **7** (1999) 1089–98
15. Leaver-Fay, A., Kuhlman, B., Snoeyink, J.: An adaptive dynamic programming algorithm for the side chain placement problem. In: Pacific Symposium on Biocomputing, 2005, The Big Island, HI (2005) 17–28
16. Pierce, N., Winfree, E.: Protein design is NP-hard. Protein Engineering **15** (2002) 779–82
17. Bradley, P., Chivian, D., Meiler, J., Misura, K., Rohl, C., Schief, W., Wedemeyer, W., Schueler-Furman, O., Murphy, P., anc C. Strauss, J.S., Baker, D.: Rosetta predictions in CASP5: Successes, failures, and prospects for complete automation. Proteins: Structure Function and Genetics **53** (2003) 457–68
18. Dahiyat, B.I., Mayo, S.L.: De novo protein design: fully automated sequence selection. Science **278** (1997) 82–87
19. Holm, L., Sander, C.: Fast and simple monte carlo algorithm for side chain optimization in proteins: application to model building by homology. Proteins **14** (1992) 213–23
20. Saven, J.G., Wolynes, P.G.: Statistical mechanics of the combinatorial synthesis and analysis of folding macromolecules. J. Phys. Chem. B **101** (1997) 8375–8389
21. Desjarlais, J.R., Handel, T.M.: De novo design of the hydrophobic cores of proteins. Protein Science **4** (1995) 2006–2018
22. Weiner, P.: Linear pattern matching algorithms. In: Proc. 14th IEEE Annual Symp. on Switching and Automata Theory. (1973) 1–11
23. McCreight, E.M.: A space-economical suffix tree construction algorithm. Jrnl. of Algorithms **23** (1976) 262–272
24. Ukkonen, E.: On-line construction of suffix trees. Algorithmica **14** (1995) 249–260
25. Hellinga, H., Richards, F.: Construction of new ligand binding sites in proteins of known structure. I: Computer-aided modeling of sites with pre-defined geometry. J. Mol. Bio. **222** (1991) 763–85
26. Lazaridis, T., Karplus, M.: Effective energy function for proteins in solution. Proteins: Structure Function and Genetics **35** (1999) 133–152
27. Kortemme, T., Morozov, A.V., Baker, D.: An orientation-dependent hydrogen bonding potential improves prediction of specificity and structure for proteins and protein-protein complexes. J. Mol. Bio. **326** (2003) 1239–1259
28. Noonan, K., O'Brien, D., Snoeyink, J.: Probik: Protein backbone motion by inverse kinematics. In: WAFR'04, Utrecht/Zeist, The Netherlands (2004)

Improved Maintenance of Molecular Surfaces Using Dynamic Graph Connectivity *

Eran Eyal and Dan Halperin

School of Computer Science,
Tel Aviv University, Israel
{eyaleran, danha}@tau.ac.il

Abstract. We present recent developments in efficiently maintaining the boundary and surface area of protein molecules as they undergo conformational changes. As the method that we devised keeps a highly accurate representation of the outer boundary surface and of the voids in the molecule, it can be useful in various applications, in particular in Monte Carlo Simulation. The current work continues and extends our previous work [10] and implements an efficient method for recalculating the surface area under conformational (and hence topological) changes based on techniques for efficient dynamic maintenance of graph connectivity. This method greatly improves the running time of our algorithm on most inputs, as we demonstrate in the experiments reported here.

1 Introduction

We study efficient techniques for dynamic maintenance of protein molecular surfaces as the molecules undergo conformational changes. Our techniques include: efficient detection of intersections of atoms, local update of the molecular surface, perturbation that allows for robust computation of the surface using floating-point arithmetic, and maintenance of the connectivity of the surface. Previously, our solution to this last step of connectivity maintenance, was rather straightforward. The major contribution of the current work is the application of an improved method for maintaining the connected components of the surface which is an adaptation to our setting of a fully dynamic algorithm for graph connectivity.

A common approach to modeling the three-dimensional geometric structure of molecules is to represent each atom as a sphere of fixed radius in a fixed placement relative to the other atoms. The radius assigned to each atom depends on the type of the atom. The spheres are allowed to penetrate one another.

* Work reported in this paper has been supported in part by the IST Programme of the EU as Shared-cost RTD (FET Open) Project under Contract No IST-2001-39250 (MOVIE - Motion Planning in Virtual Environments), by The Israel Science Foundation founded by the Israel Academy of Sciences and Humanities (Center for Geometric Computing and its Applications), and by the Hermann Minkowski – Minerva Center for Geometry at Tel Aviv University.

R. Casadio and G. Myers (Eds.): WABI 2005, LNBI 3692, pp. 401–413, 2005.

This model, called the *hard sphere model*, has proven useful in many practical applications, in spite of its approximate nature.

Molecular surfaces have many uses, such as drug design, studies of solvation and hydrophobicity, the protein folding problem, and more. One type of molecular surfaces is simply the outer boundary of the union of the spheres in the hard sphere model. This type uses the van der Waals radii, and is often referred to as the van der Waals surface. There are two closely related types of surfaces: The *solvent accessible surface* introduced by Lee and Richards [19] and the *smooth molecular (solvent excluded) surface* introduced by Richards [24]. See also [6,7,21] and the survey by Mezey [23] for an extensive discussion on molecular surfaces.

The study of the conformations adopted by proteins is an important topic in structural molecular biology. Some of the methodologies used for this study are Monte Carlo Simulation (MCS) [3,14] and Molecular Dynamics Simulation (MDS) [1,18]. In the context of molecular simulations, the surface area of a molecule is required when calculating the energy of the molecule (see [4] for a discussion and more references). Therefore fast methods to maintain the surface area of a molecule dynamically during conformation changes are desired.

Several algorithms and their software implementation for calculation of the various surfaces mentioned above have been designed in the last two decades [6,20,21,27]. Halperin and Shelton [13] used *controlled perturbation* to calculate the van der Waals and the solvent accessible surfaces robustly.

Bajaj *et al* [2] maintain molecular surfaces dynamically as the radius of the solvent probe-atom changes continuously. Edelsbrunner *et al* [5] developed an algorithm for maintaining an approximating triangulation of a deforming surface in $\mathbb{R}^3$. Bryant *et al* [4] calculate the area derivatives of molecular surfaces in motion, for a molecular dynamics simulation. Sanner and Olson [25] presented surface reconstruction for moving molecular fragments when only a small number of atoms move in each step. Lotan *et al* [22] introduced a fast implementation of MCS of proteins where a large number of atoms move in each step. They exploit the fact that proteins are long kinematic chains.

Several algorithms for dynamic graph connectivity have been designed in the last two decades. The first non-trivial fully-dynamic connectivity algorithm was presented by Frederickson [11] and supported $O(\sqrt{m})$ time updates (where m is the number of edges) and constant time queries. Eppstein *et al* [9] improved the update time to $O(\sqrt{n})$ (where n is the number of vertices). Henzinger and King [15] presented a randomized algorithm supporting updates in $O(\log^3 n)$ expected amortized time and $O(\log n/\log\log n)$ time queries. Holm *et al* [16] presented a deterministic fully dynamic algorithm with $O(\log^2 n)$ amortized time updates and $O(\log n/\log\log n)$ time queries. Both poly-logarithmic algorithms use $O(m+n\log n)$ space. Thorup [26] further improved these bounds to $O(\log n(\log\log n)^3)$ expected amortized time updates, $O(\log n/\log\log\log n)$ time queries and $O(m)$ space.

In our previous work [10] we maintain the boundary and surface area of proteins as they undergo conformational changes. We exploit the fact that proteins are long kinematic chains (and not an arbitrary collection of spheres). As the conformations change, we update the torsion angles of the protein backbone, instead of updating the Cartesian coordinates of the atoms. This allows us to modify the boundary of the molecule quickly even when a large number of atoms move, as is usually the case in conformation changes of proteins. The update time of the boundary depends on the number of intersecting pairs of atom spheres whose intersection pattern changed, which is relatively small when just a few torsion angles are changed in each step of the simulation. Maintaining a highly accurate[1] representation of the outer boundary surface and of the voids of the molecule allows us to keep track of the surface area of the molecule and the contribution of each atom to the outer boundary and to the voids, which can be useful in various applications such as MCS. Our use of controlled perturbation ensures the robustness of our implementation even while using floating-point arithmetic.

In [10] we also suggested an alternative method based on efficient maintenance of graph connectivity. Here we present its implementation, heuristic improvement and experimental results. The new method yields an amortized update time of $O(p \log^2 n)$ for each accepted conformational change — here and throughout the paper n is the total number of atoms in the molecule and p is the number of atom spheres whose intersection pattern with the other atom spheres was affected by a conformational change. For Monte Carlo simulation, the number p is typically much smaller than the number of moving atoms.

The implementation of this method improved our running time by up to 55% compared to the original method, which itself was improved to run up to 30% faster than the original implementation in [10]. The graph connectivity implementation can yield even better results when using general heuristics to improve the basic graph connectivity algorithm (up to 57% faster than the original method). In our best experimental results (for a molecule with 5614 atoms) we managed to update the molecular surface under conformational changes in 1% of the total time it would take to construct that surface from scratch. Our results indicate that our algorithm gives better gains for larger molecules. The algorithm is useful in particular for MCS, where in each step of the simulation few degrees of freedom are modified, and therefore p is small.

2 An Overview of the Algorithm

Before we describe the novel contribution of the current work (in Section 3), we review the overall algorithm for dynamic maintenance of the boundary surface of molecules.

We compute a highly accurate representation of the boundary of a molecule (both the outer boundary and the voids), and the surface area of each connected component of the boundary. The contributions in terms of surface area of each atom to the outer boundary of the molecule and to the voids are also calculated.

[1] We use the description *highly accurate* rather than *exact* to avoid confusion with exact geometric computing, since we are using floating point arithmetic.

Initially this information in computed when the molecule is first loaded. For that purpose we construct the *spherical arrangement* for each atom sphere (a subdivision of the atom sphere induced by the collection of intersection circles of that atom sphere with the other atom spheres — see Figure 1) and connect these spherical arrangements of intersecting atoms to form a subset of the 3D *arrangement of the spheres* of the atoms (the subdivision of $\mathbb{R}^3$ induced by the atom spheres), which is traversed in order to find the two-dimensional faces (*regions*) of the arrangement that form the boundary of the molecule. This construction is based on work by Halperin and Shelton [13], which uses a perturbation scheme, controlled perturbation, that overcomes degeneracies and precision problems in computing spherical arrangements while using floating-point arithmetic. The initial construction of the boundary takes $O(n)$ time (recall that n is the number of atoms in the molecule) due to favorable properties of molecules [12] and a careful calculation of the perturbation parameters [13].

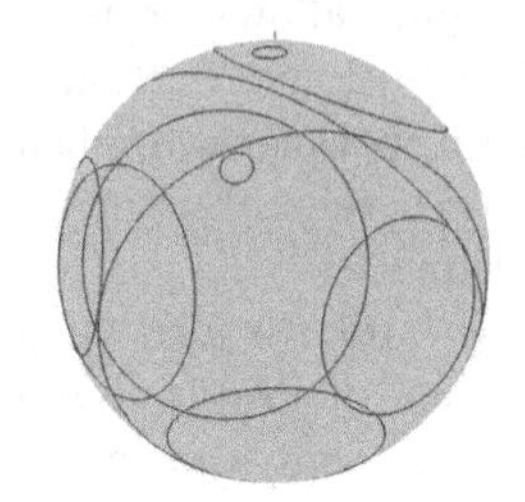

Fig. 1. A spherical arrangement

Next we describe the dynamic maintenance of the molecular surface under conformation changes, which includes updating the spherical arrangements and updating the connectivity of the surface. A more detailed description of this algorithm can be found in [10].

2.1 Updating the Spherical Arrangements

When we allow the atoms of the molecule to move, it is practically expensive to reconstruct the data structure used for detecting intersections in [13] as well as the arrangements and the boundary surface from scratch, and may be prohibitively slow for large molecules.

In [22] Lotan *et al* introduced a novel data structure called the *ChainTree* (CT), which takes advantage of the fact that proteins are long kinematic chains (and not an arbitrary collection of spheres) and that few degrees of freedom (DOFs) are changed at each step of the simulation. They represent the protein as a kinematic chain [8] of rigid links (each link consists of a group of atoms with no DOFs between them) separated by torsion angle DOFs. See [22] for more details. In [10] we use the CT to detect self-collisions and to find the modified pairs of intersecting atoms after performing DOF changes. When a DOF change is accepted (when it incurs no self-collisions), we have to modify some of the spherical arrangements and portions of the arrangement of spheres in order to compute the new boundary surface of the molecule and its area.[2] To find these pairs we introduced a data structure called the *IntersectionsTree* (IT) [10]. We summarize the worst-case performance of the CT and IT in the following theorem, which is proven in [22].

[2] In order to use the surface area in energy calculations for the acceptance criterion, these calculations will have to be done in each step of the simulation, and in rejected steps will be reversed.

Theorem 1. *The overall cost of finding all the modified intersections (including the maintenance of the CT and IT) is $O(n^{\frac{4}{3}})$.*[3]

The CT and IT tell us which old circles need to be removed from the spherical arrangements and which new circles should be added. Additional pairs of intersecting atoms for which the intersection circle has not changed may have to be removed and re-added to the arrangements due to our extensions of the controlled perturbation scheme for the dynamic case [10].

Lemma 1. *The overall cost of updating the spherical arrangements is $O(p)$, where p is the number of atoms whose spherical arrangement is involved in a change.*
The proof is given in [10].

2.2 Updating the Connectivity of the Surface

After the modification of the spherical arrangements, we have to reconstruct the outer boundary and void boundaries of the molecule and to calculate their areas, as well as the contribution of each atom to the outer boundary and to the voids.

The outer boundary of the molecule is constructed by starting from the bottommost region (of the bottommost atom), and traversing the arrangement of spheres, adding regions to the surface as we move along. Each time we reach an arc that connects two intersecting atoms, we move from the spherical arrangement of the current atom to that of the other. For each visited region of the outer boundary we calculate its area and sum the areas to get the total surface area. Later we calculate the void boundaries. This is done by finding the set of exposed regions, and excluding from this set all the regions on the already computed outer boundary. Then we traverse the remaining regions and construct the void boundaries, in the same way that we construct the outer boundary.

The computation of the exposed regions of a given atom was recently improved. Instead of subtracting from the set of all regions the regions buried within each of the atoms that intersect it, we do a single traversal of the regions of the atom a, and find for each region how many atoms cover it. We start with an arbitrary region which we assume to be exposed (covered by 0 atoms). Whenever we cross an arc to a new region, we determine if we are entering or leaving an intersection circle, and update the number of atoms covering the newly visited region accordingly. During the traversal we maintain a list of the regions covered by the minimum number of atoms (this number is 0 if the initial region is really exposed, and negative if not). After we finish the traversal, this list holds all the exposed regions of a, unless the entire atom is buried within other atoms (which can be determined by checking a single region from this list against the atoms that intersect a to see if any of them cover it).

[3] The $O(n^{\frac{4}{3}})$ The bound in Theorem 1 is a worst-case bound, but the typical practical performance is much better and constitutes a negligible portion of the overall time of an update step (see Figure 4).

This construction takes $\Theta(n)$ time, since we traverse the entire boundary, which has an overall $\Theta(n)$ complexity in the worst case. However, a great deal of the required calculations depend on the number p of modified atoms in the current step.

3 Dynamic Connectivity

Avoiding the traversal of the spherical arrangements that have not changed requires some more care in terms of identifying connected components of the boundary. The main difficulty is that in general there can be topological changes to the boundary and connected components of the boundary may merge, split, newly appear or disappear. We now present an efficient approach that despite the topological changes can accurately recompute the surface area of every boundary component in total time $O(p \log^2 n)$. For that purpose we adapt tools from dynamic maintenance of graph connectivity.

3.1 The Algorithm

We define the following graph: Each exposed region of the spherical arrangements becomes a vertex of the graph; two vertices of the graph are connected by an edge if their respective regions are adjacent on the boundary of the union of all spheres. See Figure 2 for an illustration. As the molecule undergoes DOF changes, some regions are modified, some regions are deleted and new regions are created. These changes are reflected in the graph by deleting the vertices of deleted and modified regions and adding the vertices of new and modified regions. For each deleted region, all the edges incident to its vertex in the graph are deleted.

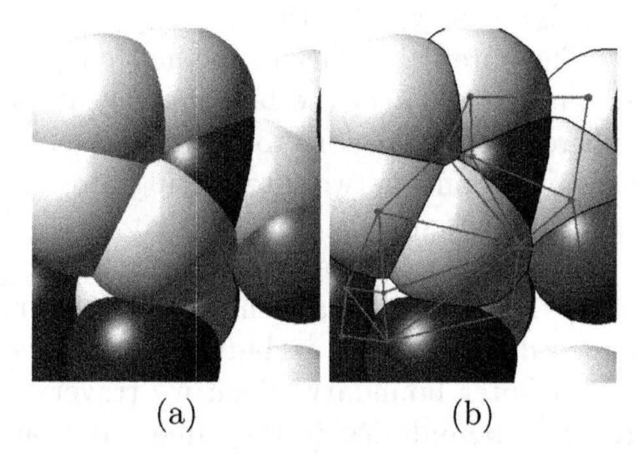

Fig. 2. A portion of the union of all spheres (a) and the subgraph induced by it (b)

In order to maintain the connected components of the boundary of the molecule, we simply need to maintain the connected components of this graph as the molecule undergoes DOF changes. One connected component of the graph represents the outer boundary of the molecule and the rest of the components represent the voids.

In [16] Holm *et al* present a poly-logarithmic deterministic fully-dynamic algorithm for graph connectivity. Their algorithm maintains a spanning forest of a graph, answers connectivity queries in $O(\log n)$ time in the worst case and uses $O(\log^2 n)$ amortized time per insertion or deletion of an edge. Here n, the number of vertices of the graph, is assumed to be fixed as edges are added and removed. In our case the vertices are not fixed, since we create and delete regions during the DOF changes. However, the number of vertices throughout the simulation remains $O(n)$ [12,22], and therefore the algorithm still works with the same amortized time bound. We next describe the original algorithm and our extension of it that efficiently maintains the *surface area* of the boundary of the molecule without traversal of the entire boundary.

The connectivity algorithm in [16] maintains a spanning forest F of the input graph G, and uses for this purpose a data structure called ET-tree. An *ET-tree* is a dynamic balanced binary tree over some *Euler tour* around a tree T. An Euler tour around a tree is a maximal closed walk over the graph obtained from the tree by replacing each edge by a directed edge in each direction. If we merge two trees or split a tree, the new Euler tours can be constructed by at most two splits and two concatenations of the original Euler tours, which take $O(\log n)$ time while maintaining the balance of the ET-tree(s). Each vertex of the tree may occur several times in the Euler tour, and one of these occurrences is chosen arbitrarily as a representative. Each ET-node represents the set of representative leaves below it, and may hold data that represent these leaves. See Figure 3 for an illustration. For more details cf. [15,16].

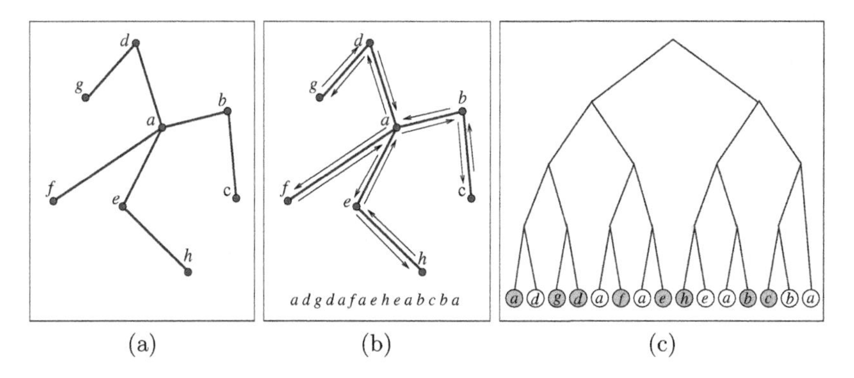

Fig. 3. A tree (a), an Euler Tour of that tree (b), and an ET-tree of that Euler Tour with its representative occurrences marked (c)

The edges of the graph are split into $\ell_{\max} = \lfloor \log_2 n \rfloor$ levels, and a hierarchy $F = F_0 \supseteq F_1 \supseteq \ldots \supseteq F_{\ell_{\max}}$ of spanning forests is maintained, where F_i is the sub-forest of F induced by the edges of level $\geq i$.

Inserting an edge to the graph as well as removing a non-tree edge are simple. Removing a tree edge $e = (v, w)$ requires finding a replacement edge, reconnecting the two trees T_v and T_w created by the removal of e. Such an edge can only be found in levels $\leq l(e)$ [16]. The replacement edge is searched recursively in the levels $\leq l(e)$ starting with level $l(e)$. The amortization argument of the algorithm is based on increasing the levels of the edges (since the level of each edge can be increased at most $\ell_{\max}$ times).

We add to each representative node of each ET-tree the area of its respective region. Each internal node of the ET-tree will hold the sum of the areas of the representative leaves in its sub-tree. The root of each tree of F will hold the total surface area of that connected component. Maintaining the area information in the ET-trees takes $O(\log n)$ time per each split or merge of the ET-trees, the same time required by the original data structure [16]. Maintaining this information in the spanning forest F takes $O(\log^2 n)$ amortized time when an edge is inserted or deleted.

To summarize:

Theorem 2. *(i) The amortized cost of recalculating the surface area of the outer boundary and voids of the molecule is $O(p \log^2 n)$, where p is the number of atoms whose spherical arrangement is involved in a change. (ii) The cost of computing the contribution of an atom to the boundary and all the voids is $O(\log n)$.*

Proof. (i) The number of inserted and deleted regions involved in a change is $O(p)$, as the complexity of each spherical arrangement is bounded by a constant. Since each insertion or deletion of an edge of G takes $O(\log^2 n)$ amortized time, the overall amortized cost is $O(p \log^2 n)$. (ii) The number of regions in an atom is bounded by a constant. Given any region of the atom, we can find the connected component it belongs to in $O(\log n)$ time by finding the root of its tree in the spanning forest F. Therefore we can compute the contribution of the atom to the surface area of all the components in $O(\log n)$ time. □

3.2 Implementation Details

Our implementation of the dynamic graph connectivity algorithm is based on the implementation by Iyer, Karger, Rahul and Thorup [17] of the algorithm by Holm *et al* [16].

Creating the Boundary Graph. After the initial construction of the spherical arrangements, we find for each atom its exposed regions. Each such region will be represented by a vertex of our graph. Then for each such region we create an edge from its vertex to the vertices of its adjacent exposed regions. The vertices and the edges are then passed on to the dynamic graph connectivity structure, and the initial spanning forest of the graph is constructed. We maintain a list of the connected components of the graph, for easier access to the outer boundary and voids. Each component is represented by the root of its tree, which holds its surface area.

Updating the Boundary Graph. During each simulation step, we mark all the regions that were modified (regions which are split into smaller regions or merged into larger regions, due to updates of the spherical arrangements). The vertices of these regions will be removed from the graph. After the spherical arrangements are updated we find the exposed regions of each modified atom, and collect the newly created exposed regions. Those regions will be added as vertices to the graph and their areas will be calculated. For each of these vertices we find their adjacent exposed regions and create edges corresponding to the adjacencies.

Next we remove all the vertices of the modified regions and their adjacent edges from the graph. Note that whenever we remove an edge that belongs to a spanning tree of some connected component (a *tree edge*), we search for a replacement edge, and this search is the most costly part of the algorithm. Since each deleted edge is adjacent to some vertex all whose edges are deleted, if we remove those edges in an arbitrary order, the algorithm is likely to replace deleted tree edges with edges about to be deleted, and thus work harder than is necessary. The solution to this problem is simply to first remove all the *non-tree edges* and then remove the tree edges.

The original implementation [17] does not handle deletion of graph vertices. Therefore, whenever we want to delete a vertex, we simply store that vertex in a list of vertices to be recycled. When new vertices will be added to the graph, the recycled vertices will be reused.

After the modified vertices and their adjacent edges are removed from the graph, the new vertices and edges are added. At the end of this addition process, we have a spanning forest of the new graph, and each connected component of this graph holds the area of a boundary component of the molecule.

Whenever we require to find the contribution of an atom to the outer boundary of the molecule and to the voids, we simply go over the exposed regions of the atom, and for each such region find the component it belongs to in $O(\log n)$ time, by finding the root of its tree in the spanning forest.

Heuristics. The implementation by Iyer *et al* has some heuristics that may run faster than the original algorithm of Holm *et al* on certain inputs. These heuristics are aimed to reduce the cost of searching for a replacement edge for a deleted tree-edge: (i) *Sampling*: Searches for a replacement edge within the first s (the *sampling threshold*) non-tree edges of the smaller tree created by the removal of the tree-edge, without promoting any edges. To keep the $O(\log^2 n)$ time of the operation, s can be at most $O(\log n)$. (ii) *Truncating Levels*: At a high level of the hierarchy, where the trees are guaranteed to be small, it is no longer worth doing anything sophisticated. Therefore it may be more efficient to simply check *all* the non-tree edges. For that purpose we choose a *base size* b, and for trees with less than b nodes we perform this simple search. We experimented with various values of s and b and briefly report the results in Section 4.

4 Experimental Results

The experiments described in this section were all executed on a 1 GHz Pentium III machine with 2 GB of RAM.

Table 1 describes the proteins used in our experiments reported here. In PDB files that contain more than one backbone chain, we handle only the first chain. We also show in this table the initial size of the boundary graph. We can see

Table 1. Proteins used in the experiments. The numbers of vertices and edges are of the initial boundary graph (induced by the boundary of the molecule at the original conformation).

Input File	# of Atoms	# of Amino Acids	# of Links	# of Vertices	# of Edges
4PTI.pdb	454	58	117	3405	10553
1BZM.pdb	2034	260	521	15254	47266
2GLS.pdb	3636	468	937	29385	90820
1JKY.pdb	5614	748	1497	45558	138818
1KEE.pdb	8181	1058	2117	62308	191317
1EAO.pdb	11180	1452	2905	84536	260096

that both the number of vertices and the number of edges are proportional to n (the number of atoms) which verifies the proof [12] that the complexity of the boundary of a molecule is linear. The ratio between the number of edges and the number of vertices is similar for all the tested inputs — about 3 — which means that the average degree of each vertex is about 6. Due to the linear bound on the complexity of a molecule boundary, the overall size of the boundary graph remains bounded by $O(n)$ during conformation changes.

Each simulation consisted of 1,000 steps. At each step the changed DOFs were picked uniformly at random and the magnitude of the change was chosen uniformly at random between $-1°$ and $1°$ (we chose small angle changes in order to increase the number of accepted steps). The results, reported in the following figures and tables, refer only to accepted simulation steps whose number was usually several hundreds (the time taken by rejected simulation steps is negligible compared to accepted steps).

We improved the original implementation (as described in [10]). Some of the improvements are algorithmic (as described in Section 2.2) while others are the result of technical code tuning. These improvements reduced the average running time of the original algorithm by up to 29%.

In Tables 2 and 3 we compare the time it takes to update the surface after a k-DOF change to the time it takes to reconstruct the surface from scratch. The reconstruction time is the time it takes to construct the static surface (not including the time spent on the construction of the CT and IT). The update time is the average time (for accepted steps) it takes to update the CT, the IT, the spherical arrangements and the surface. We made this comparison for several values of simultaneous DOF changes (k). For each update time, we give the percentage of that time from the reconstruction time. Table 2 gives the results of the naïve connectivity algorithm while Table 3 gives the results of the dynamic connectivity algorithm. Note that the static construction times are different for the two implementations, since the initial construction of the surface is different (the initial construction of the connectivity graph is slightly slower than the naïve construction of the surface). However, for the dynamic updates of the surface, the dynamic algorithm is faster in most cases.[4] The dynamic

[4] For 50-DOF simulations the dynamic connectivity algorithm runs a little slower (up to 11%) than the naïve algorithm.

Table 2. Time (in seconds) of static reconstruction vs. dynamic modification of the surface, using the naïve connectivity algorithm

Input File	# of Atoms	static	1-DOF	5-DOFs	20-DOFs
4PTI.pdb	454	1.95	0.11 (5.5%)	0.48 (24.4%)	0.83 (42.6%)
1BZM.pdb	2034	8.79	0.61 (7%)	1.49 (16.9%)	2.24 (25.5%)
2GLS.pdb	3636	18.25	0.57 (3.1%)	1.45 (7.9%)	2.65 (14.5%)
1JKY.pdb	5614	27.31	0.61 (2.3%)	1.43 (5.2%)	2.81 (10.3%)
1KEE.pdb	8181	36.48	1.10 (3%)	2.29 (6.3%)	3.51 (9.6%)
1EAO.pdb	11180	53.53	1.29 (2.4%)	2.91 (5.4%)	4.79 (8.9%)

Table 3. Time (in seconds) of static reconstruction vs. dynamic modification of the surface, using the dynamic connectivity algorithm

Input File	# of Atoms	static	1-DOF	5-DOFs	20-DOFs
4PTI.pdb	454	2.05	0.09 (4.7%)	0.51 (24.9%)	0.92 (44.7%)
1BZM.pdb	2034	9.27	0.56 (6%)	1.57 (17%)	2.46 (26.6%)
2GLS.pdb	3636	19.27	0.37 (1.9%)	1.39 (7.2%)	2.82 (14.6%)
1JKY.pdb	5614	28.91	0.27 (1%)	1.18 (4.1%)	2.81 (9.7%)
1KEE.pdb	8181	38.62	0.65 (1.7%)	2.03 (5.3%)	3.55 (9.2%)
1EAO.pdb	11180	56.49	0.64 (1.1%)	2.54 (4.5%)	4.95 (8.8%)

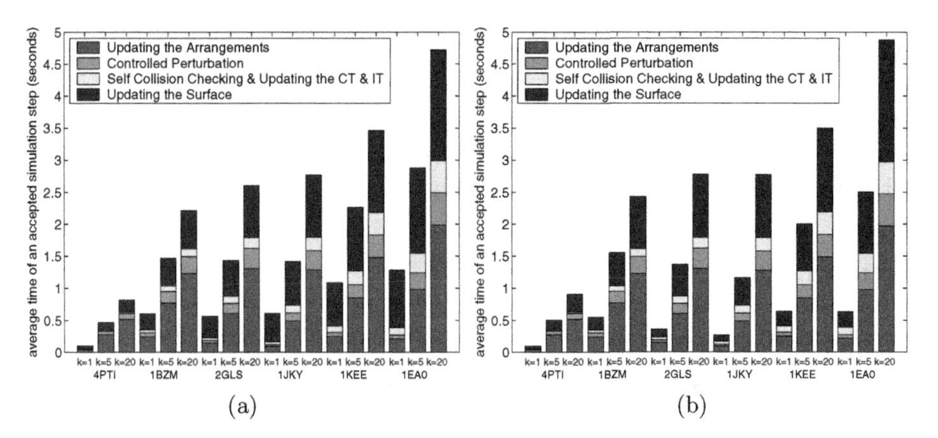

Fig. 4. Average breakdown of the running time of the main components of our application in a single accepted simulation step for different k values using the naïve algorithm (a) and the dynamic connectivity algorithm (b)

connectivity algorithm works better for small numbers of simultaneous DOF changes, but as the size of the molecules grows, it becomes faster than the naïve algorithm even for larger numbers of simultaneous DOF changes. The dynamic connectivity algorithm runs up to 55% faster compared to the naïve algorithm.

Figure 4 shows the fractions of the average running time taken by the main components of our application for both the naïve algorithm (a) and the dynamic connectivity algorithm (b). It can be seen that for most inputs updating the surface is faster with the dynamic connectivity algorithm, and takes a smaller fraction of the total time.

We also experimented with the heuristic parameters added by Iyer *et al* to the dynamic connectivity algorithm (See Section 3.2). We tested different values of s and b (the sampling threshold and the base size). Most values of s and b tested gave better running times than the original algorithm, by speeding up the deletion of tree edges in up to 30% for the larger molecules. The best heuristic we found was for $s = 10000$ and $b = 5000$.

5 Future Work

The graph connectivity algorithm used in our work was designed for general graphs. It may be possible to develop a more efficient algorithm that better suits the graph used in our application, in which all vertices have a low degree bounded by a constant. Our implementation may also be improved by detecting small changes in the molecular surface that do not affect the topology of the graph. Finally, we observe that in a typical scenario of protein motion simulation there is one very big component of the molecular surface (the outer boundary) and several much smaller components (the voids). It would be interesting to use this imbalance of component sizes to improve their maintenance.

Acknowledgments

Our implementation is based on code by Itay Lotan (the CT) [22], by Christian Shelton (the static construction of a molecular surface) [13] and by Iyer *et al* (the dynamic graph connectivity algorithm) [17].

References

1. B. J. Alder and T. E. Wainwright. Phase transition for a hard sphere system. *J. Chem. Phys.*, 27:1208–1209, 1957.
2. C. L. Bajaj, V. Pascucci, A. Shamir, R. J. Holt, and A. N. Netravali. Dynamic maintenance and visualization of molecular surfaces. *Discrete Applied Mathematics*, 127(1):23–51, 2003.
3. K. Binder and D. Heerman. *MCS in Statistical Physics*. Springer Verlag, Berlin, 2nd edition, 1992.
4. R. Bryant, H. Edelsbrunner, P. Koehl, and M. Levitt. The area derivative of a space-filling diagram. *Discrete & Computational Geometry*, 32:293–308, 2004.
5. H. L. Cheng, T. K. Dey, H. Edelsbrunner, and J. Sullivan. Dynamic skin triangulation. *Discrete & Computational Geometry*, 25:525–568, 2001.
6. M. L. Connolly. Analytical molecular surface calculation. *J. of Applied Crystallography*, 16:548–558, 1983.
7. M. L. Connolly. Solvent-accessible surfaces of proteins and nucleic acids. *Science*, 221:709–713, 1983.

8. J. J. Craig. *Introduction to Robotics: Mechanics and Control.* Prentice Hall, 3rd edition, 2005.
9. D. Eppstein, Z. Galil, G. F. Italiano, and A. Nissenzweig. Sparsification — a technique for speeding up dynamic graph algorithms. *Journal of the ACM*, 44(5):669–696, 1997.
10. E. Eyal and D. Halperin. Dynamic maintenance of molecular surfaces under conformational changes. In *Proceedings of the 21st ACM Symposium on Computational Geometry (SoCG)*, pages 45–54, 2005.
11. G. N. Frederickson. Data structures for on-line updating of minimum spanning trees, with applications. *SIAM J. Computing*, 14(4):781–798, 1985.
12. D. Halperin and M. H. Overmars. Spheres, molecules, and hidden surface removal. *Computational Geometry: Theory and Applications*, 11(2):83–102, 1998.
13. D. Halperin and C. R. Shelton. A perturbation scheme for spherical arrangements with application to molecular modeling. *Comput. Geom. Theory Appl.*, 10:273–287, 1998.
14. H. Hansmann and Y. Okamoto. New Monte Carlo algorithms for protein folding. *Current Opinion in Structural Biology*, 9(2):177–183, 1999.
15. M. R. Henzinger and V. King. Randomized fully dynamic graph algorithms with polylogarithmic time per operation. *Journal of the ACM*, 46(4):502–516, 1999.
16. J. Holm, K. De Lichtenberg, and M. Thorup. Poly-logarithmic deterministic fully-dynamic algorithms for connectivity, minimum spanning tree, 2-edge, and biconnectivity. *Journal of the ACM*, 48(4):723–760, 2001.
17. R. Iyer, D. Karger, H. Rahul, and M. Thorup. An experimental study of polylogarithmic, fully dynamic, connectivity algorithms. *J. Exp. Algorithmics*, 6:4, 2001.
18. A. R. Leach. *Molecular modeling: Principles and applications.* Addison Wesley Longman Limited, 1996.
19. B. Lee and F. M. Richards. The interpretation of protein structure: Estimation of static accessibility. *J. of Molecular Biology*, 55:379–400, 1971.
20. S. M. LeGrand and K. M. Merz. Rapid approximation to molecular surface area via the use of boolean logic and lookup tables. *Comput. Chem.*, 14:349–352, 1993.
21. J. Liang, H. Edelsbrunner, P. Fu, P. V. Sudhakar, and S. Subramaniam. Analytical shape computation of macromolecules: I. molecular area and volume through alpha shape. *Proteins: Structure, Function, and Genetics*, 33:1–17, 1998.
22. I. Lotan, F. Schwarzer, D. Halperin, and J.-C. Latombe. Algorithm and data structures for efficient energy maintenance during Monte Carlo simulation of proteins. *Journal of Computational Biology*, 11(5):902–932, 2004.
23. P. Mezey. Molecular surfaces. In K. B. Lipkowitz and D. B. Boyd, editors, *Reviews in Computational Chemistry*, volume I, pages 265–294. VCH Publishers, 1990.
24. F. M. Richards. Areas, volumes, packing, and protein structure. *Annual Reviews of Biophysics and Bioengineering*, 6:151–176, 1977.
25. M. F. Sanner and A. J. Olson. Real time surface reconstruction for moving molecular fragments. In *Pacific Symposium on Biocomputing '97*, Maui, Hawaii, 1997.
26. M. Thorup. Near-optimal fully-dynamic graph connectivity. In *STOC '00: Proceedings of the thirty-second annual ACM symposium on Theory of computing*, pages 343–350, New York, NY, USA, 2000. ACM Press.
27. A. Varshney, F. P. Brooks Jr., and W. V. Wright. Computing smooth molecular surfaces. *IEEE Computer Graphics and Applications*, 14:19–25, 1994.

The Main Structural Regularities of the Sandwich Proteins

Alexander Kister

Department of Health Informatics, SHRP, University of Medicine and Dentistry of New Jersey, Newark, NJ, 07107, USA

Abstract. The examination of the arrangement of the strands in beta-sandwich proteins reveals strict rules, which constrain the folding of a polypeptide chain. These structural rules allowed us to investigate the main principles of the packing of strands in the sandwich-like proteins and place severe restrictions on the number of allowed ways these proteins can fold. It was found that dissimilar sequences from different protein families and superfamilies, which share the same sandwich-like architecture, have 8 common key positions in sequences, whose residues govern the similar protein folding. These structural determinants can serve for protein classification.

1 Introduction

As with any text, the genetic code requires both reading and understanding. Much progress has been made with regard to the first task, while the thrust of work in molecular biology today is directed toward understanding the relationships between sequence and structure, between sequence and function, between sequence and the location of binding site, and so on.

The seminal insight into sequence-structure relationship of protein chains is due to Anfinsen [1]. He has shown that all information about the native structure of a protein is coded in the amino acid sequence. It follows that similar sequences fold into similar structure in the same solution environment. This conclusion has been confirmed for many proteins.

What about dissimilar sequences? Would they necessarily fold into dissimilar 3D structures? Although, proteins from different superfamilies do tend to fold differently, it is also true that non-homologous proteins often have very similar folds [2-5]. Good examples of dissimilar sequences forming similar structures are the proteins from different superfamilies with the same architecture, which is defined in CATH database by relative orientations of secondary structures (barrel, 2-layer sandwich, or alpha four helix bundle [6].

To explain how non-homologous sequences may share similar architecture we put forth a two-part hypothesis that can be formulated as follows:

1. dissimilar sequences, which form similar structures possess common essential elements at the sequence level – to be referred to as *structure determinants*;
2. *structure determinants* govern protein folding.

R. Casadio and G. Myers (Eds.): WABI 2005, LNBI 3692, pp. 414–422, 2005.

Each structure determinant is a key position within sequence occupied by chemically similar amino acids with similar structural roles. Structure determinants are interspersed within a sequence. Note, that the same structural determinants, which are responsible for similar chain folding may be located at widely different sites within the sequences of different protein families / superfamilies. By this reason the popular algorithms such as PSI-BLAST or Hidden Markov models [7-8], for uncovering common sequence features on the basis of sequence similarity cannot be used when dealing with a group of such diverse proteins. A principally different approach is called for.

The starting point of our approach is the statement that proteins from different families/superfamilies with similar architecture share supersecondary structure features - common regularities in the arrangement of strands and helices. Once these common supersecondary structural elements have been identified, their constituent secondary structure elements with analogous structural roles can be compared. Analogous secondary structure elements can be aligned among themselves to yield invariant sequence features for a group of proteins. The proposed approach involves projecting, as it were, common 3D supersecondary structural features onto the 1D sequence to reveal sequence similarity.

The subject of our investigation is a large group of beta proteins - 'sandwich-like proteins' (SPs) - from different folds and superfamilies. The underlying architectural motif of these proteins consists of two beta sheets packed against each other. And yet sandwich-like proteins from different superfamilies have no detectable sequence similarity; the number of strands and arrangement of the strands in the beta sheets may vary considerably; they have different biological functions. The goal of this research is to find the constraint structural rules and use these rules for uncovering the structural determinants in the sequences of sandwich proteins.

Our analysis has shown that despite a seemingly unlimited number of arrangements of strands resulting in sandwich-like structure, there exist a rigorously defined constraints on supersecondary structure that apply to almost all SPs. Knowledge of these constraints makes it possible to carry out multiple sequence alignment of SPs and to find positions in sequences that are occupied by residues with similar chemical and structural natures. Our analysis revealed eight hydrophobic positions conserved across all SPs that fulfill the criteria for structure determinants.

In perspective we suppose that the knowledge of the structural and sequence features of seemingly completely dissimilar sequences arising from different ancestors will shed light on the fundamental question of how sequence determines structure.

2 Results and Discussion

2.1 Construction and Analysis of Supersecondary Structures

The concept of *strandon* is essential for our analysis. We define strandon as *a set of all consecutive strands connected by hydrogen bonds among main chain atoms*. If a strand is not hydrogen-bonded to a consecutive strand it by itself constitutes a strandon.

Consider, for example, the NGF binding domain of trkA receptor (the PDB code: 1he7, chain A). According to the PDBSum database the domain has 7 strands. Our calculations of the inter-strand hydrogen bonds, represented by a dash, '-', reveal the following arrangement of the 7 strands in the two main beta sheets, termed A and B:

A: 1-5-4

B: 7-6-2-3

According to our definition, **strandon I** includes only strand 1. **Strandon II** consists of two strands: 2 and 3: **strandon III** is made of strands 4 and 5, and **strandon IV** from strands 6 and 7. Thus, if we denote the strandons by Roman Numerals, their arrangement in the two beta sheets A and B that form the main motif of the trkA receptor can be expressed as:

A: I III V

B: IV II

An arrangement of strandons in a structure will be referred to as supermotif.

Following the SCOP hierarchical structural classification, we determined the supermotifs for 303 protein structures from 21 folds, 46 superfamilies, 76 families [9]. The examination of the supersecondary structures of these proteins revealed that six most popular supermotifs, which describe about 90% of all beta-sandwich domains

2.2 The Constraints on the Arrangement of Strandons Within Supermotifs

Analysis of strandon arrangements within the supermotifs showed that strandons are combined in such a way so as to satisfy three following constraints:

Rule 1. Two consecutive strandons are always located in different sheets. It follows that the odd-numbered strandons are to be found in one sheet, and the even-numbered strandons in the other.

Rule 2. Two strandons located on the same edge of two sheets are always consecutive. For example, in supermotif #1 (see fig. 1) two pairs of consecutive strandons; pairs I/II and IV/V are located at the left and right edge of

1) **A: I III V**
B: II VI IV

2) **A: I III**
B: IV II

3) **A: I III**
B: IV II

4) **A: I V III**
B: VI II IV

5) **A: III I V**
B: II IV VI

6) **A: I III VII V**
B: II VIII IV VI

Fig. 1. The six supermotifs that describe 90% of sandwich proteins

the beta sheets, respectively. In this research we assume cyclic ordering of the strandons, i.e. the ?rst strandon of the domain follows the last strand. For example, the strandons VI and I are considered as consecutive neighbors in the supermotif # 4.

Rule 3. For any pair of consecutive strandons, $i$and$i$+1, where at least one strandon is not at the edge of a sheet, there is always another pair of consecutive strandons, k and k+1, such that the arrangement of these two pairs have the following characteristics (Fig. 2):

a) The strandons $i$and $k$are the neighbors in one sheet and the strandons i+1 and k+1 are neighbors in the other sheet;

b) If strandon i is the right (left) of$k$, then i+1 is the left (right) of k+1.

In effect, this rule describes an invariant substructure within all SPs, shown on Fig.2, which we call the '*strandon interlock*'.

Fig. 2. Schematic representation of a *strandon interlock* in the supermotif. '•'- denotes a strandon.

2.3 Collaries of the Constraint Rules

Supersecondary structures of sandwich proteins are governed by few specific rules, which place severe restrictions on the number of allowed ways these proteins can fold. Nearly all sandwich structures can be described by just a handful of supermotifs.

Let us consider in more detail how the constraint rules delimit the number of allowed supermotifs for SPs. Clearly, supermotifs made of 3 strandons could not exist in the sandwich proteins as existence of the interlock requires at least 4 strandons within SP structure. It can be easily shown that from the rules follows only two permissible four-strandon supermotifs ## 2 and 3 in Fig. 1. Both of them are found to represent protein structures

Analysis of the arrangements of five strandons in the structures shows that the three supermotifs shown below are not valid, because they either do not satisfy the rule stipulating that odd and even strandons be in opposite sheets, or the rule requiring consecutive strandons at the edge of the sheets, or do not conform with strandon interlock. These supermotifs are "illegal":

'

A: II V III	**A: I III V**	**A: I III V**
B: I IV	**B: IV II**	**B: II IV**

Consideration of all possible arrangements of five strandons that would satisfy the constraint rules leads one to the conclusion that only the following two supermotifs are possible:

A: III V I **A: V I III**
B: IV II **B: IV II**

These two supermotifs do in fact represent all currently known five-strandon SP structures.

Similar analysis can be carried out to discover all the permissible supermotifs with six and more strandons.

2.4 Two Main Constraints on Strand Arrangement Within Strandons

Having described the rules of arrangement of strandons (supersecondary elements) in the structures, we proceed to investigate the regularities of the arrangement of strands (secondary elements) within the strandons.

Here the analysis yields two rules that will be illustrated using an example in Fig. 3.

a)

	→	←
A:	**(1–2) –**	**(6–5)**
B:	**(8–7) –**	**(3–4)**
	←	→

b)

i	k
A: I –	**III**
B: IV –	**II**
k+1	i+1

Fig. 3. Schematic representation of the arrangement of strands and strandons in domain 4kbp, chain A: 9-120; a) Arrangement of the strands is the motif of the domain. The arrows point in the direction of increasing sequential strand number within the strandons. b) Arrangement of the strandons is the supermotif of the domain.

1. The arrangement of the strands in the strandons i and k, which take part in an interlock, can be represented using "→" and "←" arrows. The arrows point toward each other to represent the fact that strands in the two hydrogen-bonded strandons are lined up in sequentially increasing and decreasing order, respectively. The arrangement of strands in the two other strandons of the interlock, $k+1$ and $i+1$, can be represented by arrows pointing in opposite direction: "←" and "→". In the example shown in fig. 3, strands 1 and 2 within strandon i = I, are bound to strands 6 and 5 within strandon k =III in such a way that the 'later' strands of each strandon form hydrogen bonds. By contrast, the arrangement of strands in strandons II ($i+1$) and IV ($k+1$) in the structure 4kbp is such that the numerically 'earlier' strands are bonded to each other.
2. The sequential numbering of strands in the 'edge strandons' runs in antiparallel directions and may be represented by arrows pointing in opposite directions: "←" and "→". Thus, in Fig. 3, each edge strandon pairs: i/ $k+1$ and $i+1$ / k, is marked by two arrows pointing in opposite directions.

In summary: analysis at the level of secondary structure revealed two rigorous constraints on the arrangement of the strands within all strandons. In fact, from Rule 3, which describes strandon interlock follows that there are only two possibilities of the strandons in the structure to be at the edge of the sheet or participate in the interlock.

2.5 Structural Determinants of Sandwich Proteins

The main idea of structure-based sequence alignment is to find and align amino acids with similar structural properties in different proteins. Selection of residues in corresponding strands in a group of homologous proteins (protein family) with the same number and similar arrangement of strands in space (same motif) is a straightforward problem. Structurally similar strands can be readily selected in all proteins of the group and aligned with each other.

In contrast to the analysis of homologous proteins, the determination of the corresponding strands in a group of non-homologous proteins can be a rather complicated problem. Fortunately, in case of sandwich proteins, it was possible to delineate an invariant supersecondary substructure, the interlock, common for all proteins. Various strands that play analogous role in the formation of interlock can then be identified and aligned with each other.

Let us consider the arrangement of the strands, presented in Fig. 3. In the strandons i and k, the last strands are 2 and 6, respectively, whereas in the strandons i+1 and k+1, the first strands are 3 and 7, respectively. Let us call these four strands 2, 3, 6 and 7 as J, J+1, M and M+1 respectively. These strands form *strand interlock*, which can be defined by analogy to the *strandon interlock* in the following way: If two pairs of the strandons ***i***, ***i***+1 and ***k***, ***k***+1 form strandon interlock, then there are always two pairs of consecutive strands J, J+1 and M, M+1 in these strandons such that strand J in strandon ***i*** and strand M in strandon ***k*** are hydrogen bonded in one sheets, while strand J+1 in strandon ***i***+1 and strand M+1 in strandon ***k***+1 are hydrogen-bonded in the other sheet. Strand J can be either to the left of M or to the right of M. If J is to the left of M, then J+1 is to the right of M+1, and vice versa.

A sandwich-like protein contains four strands - J, J+1, M and M+1, each of which has similar structural properties across all SP. Thus, we were able to identify and collect all J strands from SP structures and aligned with each other, then all J+1 strands, etc. The essential feature of our method of structure-based sequence alignment is that it involves an alignment not of whole sequences, but of structurally analogous strands. The multiple alignment is carried out separately for each set of the corresponding strands. It is important to note that for purposes of alignment no gaps within strands are allowed, because strands are viewed as indivisible structural units. Adjacent residues within a strand are always assigned sequential position numbers.

In order to find conserved positions in strands J, J+1, M and M+1, we characterized each residue with respect to its (i) residue-residue contacts, (ii) hydrogen bonds, (iii) residue surface exposure, and (iv) structural superposition of strands. Since strand alignment is based on structural properties of residues, the first residue in J^{th} strand of one sequence can possess similar structural properties, and be aligned with, for example, 3rd residue of Jth strand of another sequence.

Structure-based alignment of SP sequences revealed that in each of the strands that make up the strand interlock there are only three positions that have the similar structural properties and are occupied by a residue in all known SPs. The remainder of the positions in the four strands can only be assigned a residue in a subset of SP sequences. Thus, there is a total of 12 positions, which are occupied by residues with structurally-similar properties in their respective SP structures.

Inspection of amino acid frequencies in these 12 positions showed that two of three positions in each strand are *conserved hydrophobic positions* of SPs: they are occupied by either aliphatic (A, V, L and I), aromatic (W, Y and F) or non-polar residues (M and C). Residues at these 8 conserved positions were termed the *SP structure determinants*: they are the structurally and chemically conserved positions of the sandwich proteins. Eighty percent of all *structure determinants* were occupied by residues V, L, I and F. [10]. Thus the first hypothesis about the common sequence features in dissimilar sequences is proved.

2.6 Role of the Structural Determinants in Protein Folding

Identification of a distinct set of structural determinants in a group of proteins as diverse as sandwich-like proteins may shed light on how the folding of this type of structures may be determined by the primary structure. The question is: what are the structural roles of these residues in the folding process. Recently, the analysis of folding kinetics by using fluorescence and far-UV CD detection showed that the structural determinants discovered by us play very important role for structure formation [11]. Half of the structure determinants participate in the folding nucleus with little affect on native-state stability, whereas the other half governs high native state stability without participating in the folding transition state. It follows from this observation that similar folding behavior of all SPs is largely due to structure determinants. Thus we proved the second hypothesis that non-similar sequences, which form similar structures have common sequence features – *structural determinants,* which govern the protein folding.

2.7 Role of the Structural Determinants for Protein Classification

For the structural classification we explored spatial distribution of eight C_α atoms of the structure determinants. Calculations of distances between C_α atoms revealed similar substructure in approximately one half of protein domains designated as 'sandwiches' in SCOP database [12]. Thus, this substructure can be used for an automatic or semiautomatic protein classification procedures. We suppose that in the other sandwich-like domains there is another type of interlock or very few geometrically different types of interlock. This analysis is the goal of our further research.

Discovery of a small set of the structure determinants furnishes us with characteristic amino acid patterns for proteinfamilies, superfamilies or groups of superfamily and allows us to develop a computer algorithm for classificationof proteins. To assign a query sequence to its proper group of proteins, weneed to find a match between residues at positions in the querysequences and the residues of the structural determinants of the given group of proteins. In fact, we need not know residues at all positionsin the query sequence. The advantage of our approach is thatit is not necessary to compare all residues in a query sequence: findings characteristic set of residues at the defining positions is sufficient for proper assignment of a sequence.

The algorithm for assignment (classification) of a query sequence to the group of proteins can be presented as anorderly search for structural determinants within the sequence from N to C end. Each residue of the sequence is analyzed in terms of its chemical properties and interval distances (a possible rangeof residues) to neighboring sequence determinants to determine whether it fits the profile of any structure determinant. Results of the analysis are formulated simply as the number of structural determinants found within a query sequence. If the sequence inquestion contains all or almost all structure determinants of a particular protein group, then the sequence is consideredto belong to that protein group. This approach has been successfully tested for the cadherin family [13].

In future we plan to extend our approach to classification of genomic sequences with no known homologues. They, too, can be checked out against any known set of structure determinants. Once a genomic sequence is found possess particular set of determinants, it can be assigned to its proper protein group and a number of testable predictions about its structure and function can be made.

References

1. Anfinsen, C.B.: Principles that govern the folding of protein chains. Science. (1973) **181**, 223-230
2. Chothia, C. and Lesk, A. M.: The relation between the divergence of sequence and structure in proteins. *Embo* J., (1986) **5**, 823-826
3. Chothia, C.: Proteins. One thousand families for the molecular biologist. Nature (1993).**357**, 543-544
4. Russell R.B.: Classification of Protein Folds. Molecular Biotechnology (2002) 20,17-28 **5.** Thornton, J. M., Orengo, C. A., Todd, A. E. and Frances, M. G. and Pearl, F. M. G.: Protein folds, functions and evolution. J. Mol. Biol. (1999) **293**: 333-34
5. Orengo, C.A., Michie, A.D., Jones, S., Jones, D.T., Swindells, M.B., and Thornton, J.M. : A hierarchic classification of protein domain structures. Structure. (1997) **5**. 1093-1108.
6. Altschul SF, Madden TL, Schaffer AA, Zhang J, Zhang Z, Miller W, Lipman DJ. : Gapped BLAST and PSI-BLAST: a new generation of protein database search programs. Nucleic Acids Res. (1997) **25(17),** 3389-402.
7. Eddy, SR, Mitchison G, Durbin R : Maximum Discrimination Hidden Markov Models of Sequence Consensus. *J Comput Biol.* (1995) **2(1)**:9-23.
8. Murzin A. G., Brenner S. E., Hubbard T., Chothia C.: SCOP: a structural classification of proteins database for the investigation of sequences and structures. J. Mol. Biol. (1995), **247**, 536-540

9. Kister, A. Finkelstein, A and Gelfand, I Common features in structures and sequences of sandwich-like proteins. Proc. Natl. Acad. Sci. USA (2002) **99**, 14137-14141
10. Corey J. Wilson, C.J and Wittung-Stafshede, P.: Role of structural determinants in folding of the sandwich-like protein *Pseudomonas aeruginosa* azurin. Proc. Natl. Acad. Sci. USA (2002) **102,** 3984-3987
11. Aksianov E., Alexeevsky A., Kister, A., Gelfand I : Interlock structural motif is widely spread in sandwich-like protein domains (in preparation)
12. Kister, A.E, Roytberg, M.A., Chothia, C., Vasiliev, Y.M. & Gelfand, I.M. The sequence determinants of cadherin molecules. Prot. Sci. (2001) 10, 1801-1810.

Discovery of Protein Substructures in EM Maps

Keren Lasker[1], Oranit Dror[1], Ruth Nussinov[2,3], and Haim Wolfson[1]

[1] School of Computer Science, Raymond and Beverly Sackler Faculty of Exact Sciences, Tel Aviv University, Tel Aviv 69978, Israel
[2] Sackler Inst. of Molecular Medicine, Sackler Faculty of Medicine, Tel Aviv University, Tel Aviv 69978, Israel
[3] Basic Research Program, SAIC-Frederick, Inc, Lab. of Experimental and Computational Biology, Bldg. 469, Rm. 151, Frederick, MD 21702, USA

Abstract. Cryo-EM has become an increasingly powerful technique for elucidating the structure, dynamics and function of large flexible macromolecule assemblies that cannot be determined at atomic-resolution. A major challenge in analyzing EM maps of complexes is the identification of their subunits. We propose a fully automated highly efficient method for discovering high-resolution subunits of a complex, given as an intermediate resolution map, without prior knowledge of their boundaries and content. The method extracts helices from an EM map and uses their spatial arrangement to detect candidate subunits. The method was tested successfully on several simulated 8.0Å resolution maps. The obtained spatial helix arrangement was sufficient for the discovery of the correct subunits from a dataset of 887 SCOP representatives.

Keywords: Structural bioinformatics, intermediate resolution cryo EM maps, 3D alignment of secondary structures, macromolecular assemblies.

1 Introduction

Structure determination of large macromolecular assemblies is one of the main challenges in structural genomics. To date, only 1.5% of the structures in the Protein Data Bank (PDB) are of large macromolecular complexes [1]. The reason is that X-ray crystallography, the most prolific and accurate technique for structure determination, has difficulties in the crystallization process of large and unstable assemblies such as membrane proteins and viruses.

In the absence of crystals, cryo-electron microscopy (Cryo-EM) is a valuable source of structural information. It is well suited for studying both the structure and dynamics of large macromolecule assemblies [2,3]. Its main limitation is the relatively low resolution of the data, ranging between 6 to 30Å. The predominant resolution criterion is the Fourier shell correlation, which is calculated between the 3D Fourier transforms of two independent 3D reconstructions [4]. At low resolution (15-30Å) only the overall shape and, possibly, subunit boundaries can be revealed. At 7-9Å , often referred to as *intermediate resolution* [5], secondary structure elements (SSEs) become apparent. Helices appear as cylinders and β-sheets appear as planar regions. Due to major improvements in the cryo-EM

R. Casadio and G. Myers (Eds.): WABI 2005, LNBI 3692, pp. 423–434, 2005.

technique, 10Å resolution maps become available [6]. Thus, developing methods for analyzing intermediate resolution EM data is of utmost importance.

Cryo-EM methods can be synergistically combined with atomic resolution methods for structure determination to overcome the limitations of either method alone. For large flexible complexes that cannot be crystallized, it is possible to fit the atomic structures of the individual subunits into a low or intermediate resolution EM map of the entire complex. The resulting quasi atomic model of the complex may provide crucial information about the interactions of its subunits.

Indeed, several hybrid approaches have been developed for fitting atomic resolution domains into cryo-EM data. Some of them are based on manual placement with visualization tools [7,8], while others are fully automated. The majority of the automated methods apply a 6D search of a predefined domain in the EM map using variants of cross correlation as a measure of fitness [9,10,11,12].

One major drawback of the domain fitting methods is the assumption that the searched domain is present in the map and its conformation is *a-priori* known. For intermediate resolution maps, a different methodology has been suggested [11]. The methodology exploits the fact that the scaffold of a domain is defined by the spatial arrangement of its secondary structure elements (SSEs). First, helices in the given EM map are identified using Helixhunter and then known homologous folds are revealed using DEJAVU [13] or COSEC [14].

In our preliminary study presented here, we adopt the methodology introduced in [11] for intermediate resolution maps. We suggest a combined new approach both for helix extraction and fold alignment (Figure 1). The hybrid method detects helices in an EM map and uses the 3D arrangement of the identified helices to query a dataset of high-resolution folds to find potential structural homologues. The method is highly efficient and suitable for exploring macromolecular assemblies. Another important feature of the method is its ability to detect 'partial alignments' between the extracted set of helices and the database folds. Thus, the method is tolerant to errors in the helix extraction stage and capable of detecting non-predefined motifs.

2 Method

2.1 Helix Extraction

Problem Definition. The *input* is an EM map at intermediate resolution given as a three-dimensional (3D) grid, in which each voxel is associated with a density value. The *output* is a set of 3D undirected segments $\{s_i = (p_i, q_i) \mid p_i, q_i \in 3D\}$, where segment s_i represents the central axis of the ith predicted helix.

Outline. In intermediate resolution EM maps helices are usually characterized as highly dense long regions [2,11]. We exploit this observation and define a *helix-like region* as a region in the EM grid with the following properties: (i) the region is highly dense (compared to the average grid density); (ii) the region is homogeneous (its density standard deviation is below a predefined threshold);

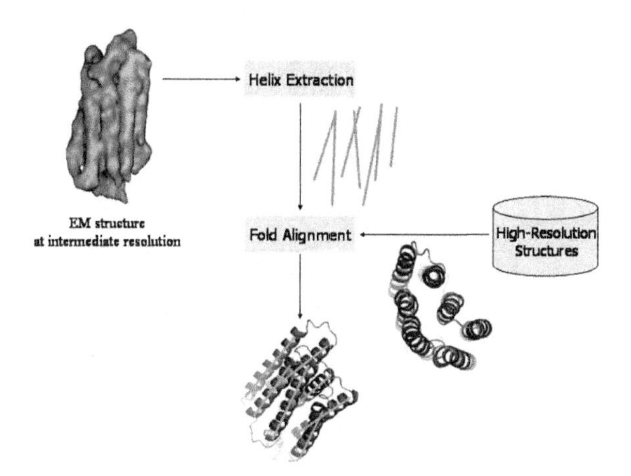

Fig. 1. Substructure Discovery Flow. The hybrid method detects helices in an EM map and uses the 3D arrangement of the identified helices to query a dataset of high-resolution structures to find potential homologous folds.

(iii) the 2D slices of the region that are perpendicular to its central axis behave roughly as a 2D Gaussian [15]; and (iv) the region's shape is a thin cylinder (formally defined below as a *helix predicate*).

The Helix extraction method consists of four main stages (Figure 2). The objective of the first two stages is to enhance helical regions and suppress non-helical ones by thresholding and fitting techniques. In the next two stages, the goal is to identify *helix-like* regions and represent them as 3D undirected segments using a segmentation procedure followed by a linkage procedure for linking small fractions of the same helix. Below is a detailed description of each stage.

Threshold Filtering. We apply an image processing technique for filtering noise (non-helical regions in our case) [15].

Helix Fitting. We use cross correlation for matching a helix template in the EM grid. The template is a 3D electron density pattern of a blurred ideal two-turn α-helix. We construct this template by first interpolating its atoms onto a grid with the same sampling as the searched EM grid. Then, for each atom we convolute the helix's grid with a Gaussian mask defined by $\sqrt{2\pi}\sigma A e^{-2\pi^2\sigma^2 r^2}$, where A is the number of electrons in the atom and σ is its influence radius, which depends on the map's resolution. Similar blurring methods are also used in [12,16,17].

The helix template is exhaustively correlated with the EM grid. All possible orientations and positions (up to six degrees of freedom) of the template are searched to find the optimal match for each voxel. The optimal match for a voxel is defined as the orientation of the helix template with maximum cross correlation coefficient (CCC). The result is a new grid in which each voxel stores the optimal orientation and its normalized CCC. Note that the highest CCCs are found along the helices' central axes and decrease as we reach their boundaries.

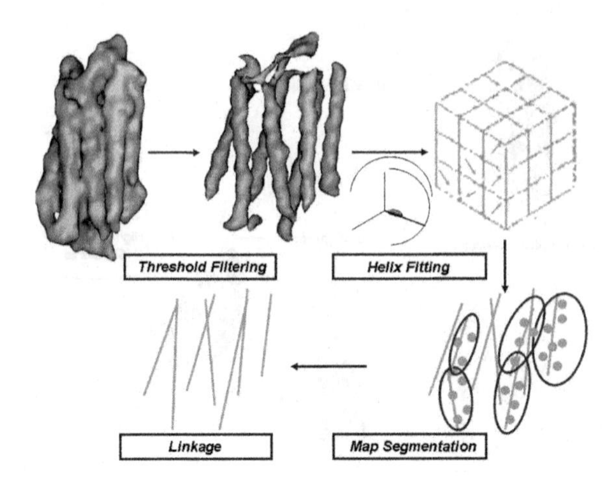

Fig. 2. Helix Extraction Flow. In the filtering and fitting stages helical regions are enhanced and non-helical ones are suppressed. In the next two stages, helix-like regions are identified and represented as 3D undirected segments.

The orientation of the helix template is determined by 3 rotation angles (azimuth, elevation and tilt). Utilizing cylinder shape symmetry, all tilt rotations and half of the azimuth angles can be disregarded. This leaves us with $(\pi/\rho)^2$ orientations of the search object, where ρ is the rotational sampling interval ($\pi/12$ by default). Once an orientation has been determined, we rapidly scan the translation space by utilizing the advantages of the fast Fourier transform (FFT) [18]. For a grid of n voxels each translation scan takes $O(n \log n)$ time and, thus, the total time complexity of the exhaustive search is $O((\pi/\rho)^2 \cdot n \log n)$.

Segmentation. A helix is defined by its orientation and length. Even though the fitting procedure reveals strongly correlated orientations and potential helices' center points, the lengths of the helices remain unresolved. Furthermore, inherent difficulties due to resolution problems may lead to false identification (strands as helices or two parallel close helices as one helix) and to failures in distinguishing between short helices and noise. To solve these problems we have developed the following *helix-like region segmentation* method.

We use a graph-theoretic approach. Let $G = (V, E)$ be an undirected graph. Each vertex $v_i \in V$ represents a voxel in the new EM grid and stores the orientation and the CCC of the best match with the helix template, denoted as $dir(v_i)$ and $score(v_i)$ respectively. The scores of the vertices in G satisfy $score(v_i) \geq threshold$. A pair of vertices v_i and v_j are connected by an edge if the following conditions are satisfied: (i) v_i and v_j represent neighboring voxels in the EM grid; and (ii) $dir(v_i)$ and $dir(v_j)$ are ϵ-parallel, where two vectors are considered as *ϵ-parallel* if the angle between them is not higher than ϵ. Note that the first condition bounds the number of edges in the graph to $O(|V|)$.

A *helix-like region segmentation* S is a partition of V into regions $R_1, ..., R_k$ so that the following conditions are satisfied: (i) $\cup R_i = V$; (ii) $\forall i \neq j$ $R_i \cap R_j = \emptyset$;

(iii) R_i is a connected region, meaning that each pair of vertices in R_i is connected by a path in R_i; and (iv) each R_i satisfies a *helix predicate* (D), which prefers thin cylinder-like regions. Formally, a region R is said to satisfy the predicate $D(R)$ if $|pc_1(R)|$,$|pc_2(R)| \leq 3.5\mathring{A}$ (roughly a helix radius) and for each $v \in R$, $dir(v)$ is ϵ-parallel to $pc_1(R)$, where $pc_i(R)$ is the ith principal components of the region. Note that the predicate is valid only if the region contains a sufficient number of voxels. One can show that given a region R with pre-calculated principal components, we can answer whether $R \cup \{u \in V\}$ satisfies the predicate in $O(1)$ time. The principal components of a set S of points in R^3 are the eigen vectors of its 3×3 covariance matrix. Assuming we have already calculated the average point and covariance matrix of S, we can calculate the average point and the covariance matrix of $S \cup p$ in $O(1)$, for any $p \in R^3$. Hence, calculation of the principal components of $S \cup p$ is done in constant time.

For a graph G a helix-like region segmentation is not uniquely defined. Following the approach introduced in [19], a helix-like region segmentation is *satisfactory* if (i) it is *not too fine* - $\forall i \neq j, R_i \cup R_j$ does not satisfy D; and (ii) *not too coarse* - any *refinement* S' of S, where $S \neq S'$ is too fine. A *refinement* S' of S is such that any region in S' is contained in (or equal to) a region in S. Similarly to the proof given in [19] it can be shown that there is at least one helix-like region segmentation that is not too fine and not too coarse.

To find a satisfactory helix-like region segmentation we apply a greedy approach. We construct a set of helix-like regions from a number of seed vertices using a variant of the BFS algorithm [20]. First, we sort the vertices by their scores and add them to a *seed queue* in descending order. Then, the vertex at the top of the seed queue is given as a seed vertex for BFS. The initial helix-like region R consists of the seed vertex. During the search we iteratively add to R newly discovered vertices v that satisfy $D(R \cup \{v\})$. If the newly discovered vertex cannot be added to R, then its neighbors are not further explored. When no vertex can be added to R we define R as a new helix-like region. The vertices of R are marked as visited and removed from the seed queue. We repeat the BFS procedure until the seed queue is empty. The time required for building the graph is linear in the number of the EM grid voxels, that is $O(n)$. Sorting the vertices costs $O(|V| \log |V|)$. The BFS costs $O(|V| + |E|)$ since the validation of the helix predicate for each vertex costs only $O(1)$. Since E is linear in V, the overall running time is $O(n + |V| \log |V|)$. Although theoretically $|V|$ equals n, in practice the thresholding and fitting procedures dramatically decrease $|V|$.

Linkage. The input is a set of helix-like regions $S = (R_1, R_2, ...R_k)$ in the EM grid such that each satisfies the helix predicate. We apply a liberal set of conditions to allow fragments of the same helix to be connected. The output is a coarse set $S' = (R'_1, R'_2, ...R'_l)$ $l \leq k$. Each region is represented as a segment. The segment's direction is the first principal component (pc_1) of the region (for helices the direction is along its central axis). The segment's endpoints are determined by projecting each of the region's voxels onto its direction and choosing the extreme projected points. Two regions can be linked if the angle between them is below a threshold ($\pi/9$ by default), and the minimal distance between the two segments' endpoints is below another threshold (6Å by default). Finding the refined set is equivalent to finding connected components [20] in an undirected graph $G' = (V', E')$, where each $v'_i \in V'$ represents a region's segment and two vertices are connected if their regions can be linked. The output helices are S' regions' segments with pc_1 higher than the length of a two-turn helix.

For each vertex we can find all the neighboring vertices in $O(1)$ by querying the EM grid. Thus, E' is linear in V' and the time required to construct the graph is $O(|V'|)$. Finding connected components costs $O(|V'|)$ and the overall time complexity of the stage is $O(|V'|)$, where $|V'|$ is bounded by $|S|$.

Complexity. For an EM grid of size n, the overall complexity is $O((\pi/\rho)^2 \cdot n \log n + n + |V| \log |V| + |V'|)$, where both $|V|$ and $|V'|$ are at most n. Thus, the total time complexity is $O(n \log n)$.

2.2 Fold Alignment

The fold alignment algorithm is partially based on the MASS method for multiple 3D alignment of proteins by their secondary structures [21,22]. While MASS compares high resolution structures, the fold alignment method was designed to compare proteins for which there are no atomic structures and the 3D arrangement of their SSEs is the only available information. The input for the fold alignment method is a set of SSEs extracted from cryo-EM data and the output is a set of folds that share the largest common configuration of SSEs with the input. This is carried out by comparing the spatial arrangement of the input SSEs with a dataset of known folds. Below is a description of the method for a single comparison with one dataset structure.

Problem Definition. We are given two protein structures, A and B. Each structure is characterized by the set of its SSEs, where each SSE is represented as an undirected line segment in 3D space. A segment is represented by the two endpoints of its central axis, that is $A = \{a_i = (a_{i1}, a_{i2})\}$ and $B = \{b_i = (b_{i1}, b_{i2})\}$. The task is to find a rigid transformation (rotation and translation) T and two corresponding subsets $A' = \{a'_i\}_{i=1}^k \subseteq A$ and $B' = \{b'_i\}_{i=1}^k \subseteq B$ of maximal cardinality (k) so that the corresponding segments of $T(A')$ and B' are coincident up to a predefined error threshold ϵ. The *coincidence error* between two segments is defined as a pair (e_1, e_2), where e_1 is the Euclidean distance between their midpoints and e_2 is the angle between them. Given an error threshold $\epsilon = (\epsilon_1, \epsilon_2)$ two segments with coincidence error (e_1, e_2) are said to be *ϵ-coincident* if $e_i < \epsilon_i$ for $i = 1, 2$. By default, $\epsilon = (6.0Å, 0.5\ radians)$. The transformation T defines the *alignment* between the two structures and the corresponding subsets A' and B' are its *core*.

Outline. The algorithm exploits the observation that a biologically interesting alignment consists of at least two common SSEs. We define a *basis* as an ordered pair of nonlinear SSEs and represent it by a 5D vector *fingerprint*. The fingerprint is invariant to a 3D rigid transformation and composed of: (i-ii) the types of the SSEs (helix or strand) (iii) the angle between them; (iv) their midpoint distance;

and (v) their line distance, which is the closest distance in space between the lines on which the segments are located. Two bases are said to have a *similar configuration* if the type of the corresponding SSEs is the same and the differences between the other attributes of their fingerprints are below predefined thresholds.

Construction of Basis Alignments. We use the hashing approach described in [21] to efficiently detect all pairs of bases with similar configuration between the two structures. Then, for each pair, $(a_i, a_j) \in A$ and $(b_k, b_l) \in B$, we compute the two possible transformations for superimposing the basis of A onto the basis of B in $O(1)$ time. Specifically, for each basis we uniquely define a Cartesian reference frame as follows: (i) the first SSE segment of the basis is defined as the X-axis and its direction is arbitrary defined; (ii) the line distance segment between the two SSE segments is considered as the Y-axis and its direction is from the first to the second SSE; and (iii) the Z-axis is the cross product of the X-axis with the Y-axis. There are two possible transformations for superimposing the reference frames of the bases, one for each possible direction for the X-axis of the first basis. For each transformation, we compute the coincidence error for the two pairs of corresponding segments, (a_i, b_k) and (a_j, b_l), and if the error is below ϵ the transformation is considered as a potential alignment between the two structures with a core of at least two SSEs.

Global Extension. In this stage we extend the basis alignments. Specifically, for a given basis alignment with a transformation T, we are interested to find two corresponding subsets $A' = \{a'_i\}_{i=1}^k \subseteq A$ and $B' = \{b'_i\}_{i=1}^k \subseteq B$ of maximal cardinality (k) so that the corresponding segments of $T(A')$ and B' are ϵ-coincident. This optimization problem can be solved by an exact algorithm for finding maximal matching in a bipartite graph [23], where the graph is $G = (T(A) \cup B, E)$ and an $e = ((T(a_i), b_k) \in E$ if and only if the corresponding segments $T(a_i)$ and b_k are ϵ-coincident. The time required for constructing the graph and finding maximal bipartite matching is $O(|G| + \sqrt{|A| + |B|} \cdot |E|)$. For $n = max(|A|, |B|)$ there are $O(n)$ edges in the graph, since a segment can ϵ-coincide with a bounded number of segments. Thus, the total time complexity is $O(n^{1.5})$. Due to efficiency considerations, we have solved the problem by a greedy approach. We store the segments' midpoints of $T(A)$ in a 3D grid. Then, for each segment $b_k \in B$, we use its midpoint to access the grid and examine a ball of radius ϵ_1. The segment whose for which the midpoint is in this ball and ϵ-coincides with b_k with the smallest error is matched to b_k. Since different helices cannot be too close in space, the number of segment midpoints in an ϵ_1-radius ball is bounded for a small ϵ_1. Thus, for each segment of B we examine a constant number of possible alignments and the total time runtime is $O(n)$. Finally, after extending a basis alignment, we refine its transformation by applying the Least-Squares Fitting technique [24] on the corresponding segment midpoints.

Scoring, Clustering and Ranking. The extended alignments are sorted by their core size and the RMSD between the midpoints of the corresponding segments. Since two structures may share more than one common substructure, we report the t top-ranking non-redundant alignments. For this purpose, we apply an iterative RMSD clustering procedure. In each iteration, we pop the top-ranking alignment from the sorted list and add it to a non-redundant list of top-ranking alignments if its transformation differs from the transformations of all the current alignments in this list. Two transformations are considered different if the RMSD distance between their images on the same reference set of points is above a threshold. Specifically, given a reference set $\{p_1, ..., p_r\}$ and two transformations T_1 and T_2, their RMSD distance is defined as $((\sum_{i=1}^{r} ||T_1(p_i) - T_2(p_i)||^2)/r)^{0.5}$. For a constant-size reference set it takes $O(1)$ time to check if two transformations are different. Thus, the total time is $O(m \log m + m \cdot t)$, where m is the number of initial alignments. This is equal to $O(m \log m)$ for a small t.

Complexity. For $n = max(|A|, |B|)$ the number of possible bases for each structure is $O(n^2)$. In the theoretical worst case, we will construct, extend and score $O(n^4)$ alignments. This takes an overall $O(n^4) \cdot O(1) + O(n^4) \cdot O(n) + O(n^4 \log n) = O(n^5)$ time. In practice, the runtime is significantly lower due to the usage of a hash table [21].

3 Results

We tested our application on simulated 8Å resolution cryo-EM maps (with 1.5Å voxel spacing) of several proteins. Among them are the four proteins used to validate the results of Helixhunter (PDB codes: 1c3w, 1irk, 1tim:A, 1bvp:1) [11]. The simulated maps were constructed using the pdb2mrc utility [16]. For each simulated map we applied our two-tier approach (Figure 1). Specifically, a set of helices was extracted from the map and used to query a SCOP representative dataset consisting of 887 domains, where each domain is a fold representative [25]. The alignments between the set of predicted helices and each domain in the dataset were obtained and sorted by the core size and the RMSD between the midpoints of their central axes. All experiments were performed on a standard PC (Pentium© 4, 2.60 GHz with 2GB RAM). Below is a detailed description of some results. A summary is given in Table 1.

Photosynthetic Reaction Center Complex. The photosynthetic reaction center is a transmembrane protein complex that converts light into chemical energy. A simulated EM map of the whole complex from *Rhodopseudomonas viridis* (PDB:1r2c) was given as input to the method. The complex consists of five subunits, where the mostly helical subunits, L (light) and M (medium), provide the scaffold. A set of 26 helices was detected in six minutes. Most of these helices are located in the L and M subunits (Figure 3a). 37 out of the 68 helices of the high resolution structure were not detected due to their short size (less than 2.5 turns). Also notice that although the complex contains an all-β domain, there were no false positives, meaning that none of the strands was falsely identified as a helix. Despite the large size of the complex, comparing it with all the 887 SCOP domains of the dataset took less than one hour and the top-ranking solution was the fold representative of the L subunit domain

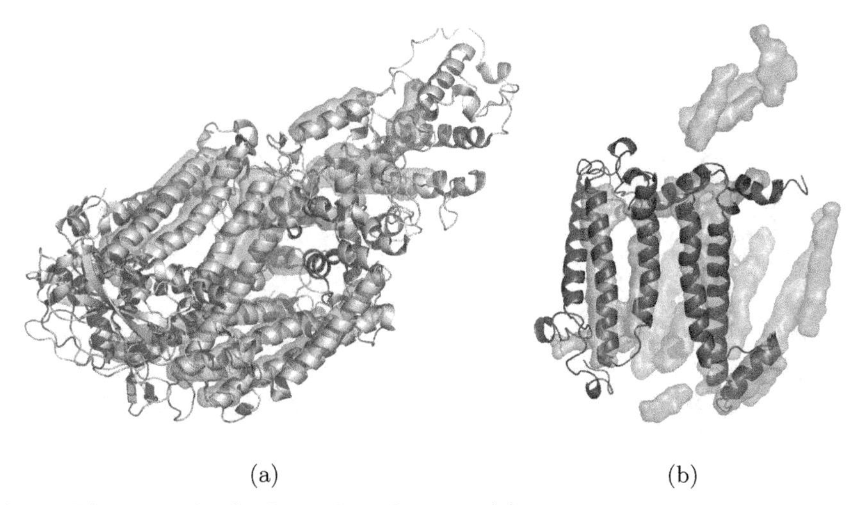

(a) (b)

Fig. 3. Photosynthetic Reaction Center. **(a)** The alignment between the high resolution complex (PDB:1r2c, gray) and the predicted set of helices from the EM-map (orange); **(b)** The alignment between the detected set of helices (orange) and an L subunit domain (PDB:1dxr, chain L, green). This figure and the subsequent one were prepared using PyMOL (`http://www.pymol.org`).

(PDB:1dxr, chain L). Figure 3b presents the alignment with a core of six helices and an RMSD of 2.9Å between their midpoints. This example demonstrates the ability of our method to identify partial alignments without prior knowledge of subunit boundaries and content.

TIM-barrel Fold. The TIM-barrel fold is a common fold observed in many different superfamilies. A simulated EM map of the fold from the *Triosephosphate isomerase* superfamily (PDB:1tim, chain A) was given as input to our method. Nine helices were identified in less than seven minutes. Also, despite the fact that the structure belongs to the α/β class and contains a β-sheet of eight strands, there were no false-positives. Comparing the detected set of helices with the 887 SCOP domains took less than 13 minutes. The TIM-barrel fold representative (PDB:1thf) was the top-ranking domain despite the fact that it belongs to a different superfamily (the *Ribulose-phoshate binding barrel* superfamily). Figure 4a presents the alignment of the two structures. The core consists of seven helices with an RMSD of 3.7Å between the midpoints of their central axes.

Endocytic AP2 Complex. AP2 is a heterotetrameric clathrin adaptor complex that plays a key role in many vesicle trafficking pathways in the cell. A simulated EM map of an endocytic AP2 core (PDB:1gw5) was given as input to our method. The AP2 core is a compact assembly of five subunits, where two of them have an α-α superhelix fold. Despite the size of the assembly, a set of 80 helices was detected in less than ten minutes. Also, although the assembly contains a large all-β subunit, there were no false positives. When we compared the identified set of helices with the structures of the SCOP dataset, the top-ranking

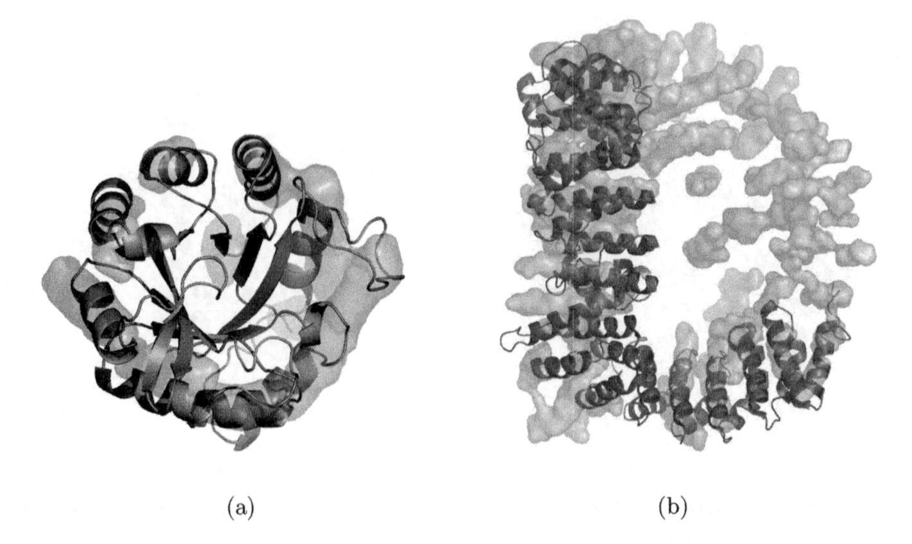

(a) (b)

Fig. 4. (a) TIM-barrel. The alignment between the detected set of medium-resolution helices of a TIM-barrel domain from the *Triosephosphate isomerase* superfamily (PDB:1tim, orange) and a high resolution TIM-barrel domain from the *Ribulose-phoshate binding barrel* superfamily (PDB:1thf, green). **(b) AP2 Core.** The alignment between the detected set of medium-resolution helices for an endocytic AP2 core (PDB:1gw5, orange) and a high resolution structure of one of its α-α superhelix domain (PDB:1gw5, chain B, green).

was one of the α-α superhelix domains of the assembly (PDB:1gw5, chain B). Figure 4b presents the alignment for which the core consists of nine helices with an RMSD of 3.8Å between the midpoints of their central axes.

4 Conclusions

We have described a new fully automated method for discovering subunits in intermediate resolution maps of macromolecular assemblies without *a-priori* knowledge of the subunits boundaries and content. The method reveals the spatial arrangement of helices in a given EM map of a complex and exploits this information for identifying all subunits appearing in a fold database. The method is highly efficient and the preliminary experimental results are encouraging. The results demonstrate the capability of the method to extract almost all longer than two-turn helices without false positives as well as its capability to discover the correct folds even when the helical information is partial. We are currently testing the application on real data taken from the EBI database [26]. We consider this method as the first of a larger set of tools for structure interpretation of molecular assemblies at intermediate resolution. Future challenges include β-sheet extraction from the EM data, RNA structure analysis, handling conformational flexibility of the various subunits, and the integration of the presented method with advanced multi-molecular docking methodologies [27].

Table 1. Results: For each query the data appearing in the columns are: (1) the PDB code followed by the chain identifier of the modeled 8Å resolution structure. In brackets are the number of identified helices and the number of helices in the original structure; For some structures the number of predicted helices is significantly lower than the number of helices in the original structures. The main reason is that the structures contain many short helices between loops (*e.g.* 37 in PDB:1r2c and 9 in PDB:1hno). The other reason is that parallel helices that are a turn apart are merged in the linkage step; (2) The PDB code and the chain identifier of the top-ranking SCOP domain and the number of the helices in the structure; (3) the core size of the alignment; (4) the RMSD between the axial midpoints of the corresponding SSEs; (5) the running times for the helix extraction and fold alignment vs. the full database respectively.

Query	Fold Homologue	Matched Helices	RMSD	Runtime (hh:mm:ss)
1c3w (7/8)	1bm1 (7)	7	1.2	00:06:23 00:13:11
1tim:A (9/12)	1thf (11)	7	3.7	00:06:54 00:12:26
1irk (7/9)	1gz8:A (13)	6	2.9	00:04:44 00:11:13
1bvp:1 (6/10)	2btv:P (9)	6	1.9	00:04:48 00:13:37
1s0p:A (7/8)	1s0p (8)	5	1.9	00:04:38 00:12:31
1hn0 (16/27)	1gai (18)	6	3.5	00:04:13 00:11:27
1r2c (26/68)	1dxr:L (17)	6	2.9	00:06:06 00:54:15
1gw5 (80/84)	1gw5:B (36)	9	3.7	00:09:53 07:43:13

Acknowledgments

We thank Maxim Shatsky for stimulating discussions. The research of H.J.W. is partially supported by the Hermann Minkowski-Minerva Center for Geometry at TAU. The research of R.N. has been funded in whole or in part with Federal funds from the NCI, NIH, under contract number NO1-CO-12400. The content of this publication does not necessarily reflect the view or policies of the Dep. of Health and Human Services, nor does mention of trade names, commercial products, or organization imply endorsement by the U.S. Government.

References

1. Dutta, S., Berman, H.M.: Large macromolecular complexes in the protein data bank: A status report. Structure **13** (2005) 381–388
2. Chiu, W., Baker, M.L., Jiang, W., Dougherty, M., Schmid, M.F.: Electron cryomicroscopy of biological machines at subnanometer resolution. Structure **13** (2005) 363–372
3. Baumeister, W., Steven, A.: Macromolecular electron microscopy in the era of structural genomics. Trends. Biochem. Sci. **25** (2000) 624–631
4. Harauz, G., Van Heel, M.: Exact filters for general geometry three dimensional recontruction. Proceedings of the IEEE Computer Vision and Pattern Recognition Conf. **73** (1986) 146–156
5. Auer, M.: Three-dimentional electron cryo-microscopy as a powerful structural tool in molecular medicine. J. Mol. Med. **78** (2000) 191–202

6. Rossmann, M.G., Morais, M.C., Leiman, P.G., Zhang, W.: Combining x-ray crystallography and electron microscopy. Structure **13** (2005) 355–362
7. Jones, T., Zou, J., Cowan, S., Kjeldgaard, M.: Improved methods for building protein models in electron density maps and the location of errors in these models. Acta Crystallogr. **A47** (1991) 110–119
8. Humphrey, W., Dalke, A., Schulten, K.: VMD: visual molecular dynamics. J.Mol. Graph. **14** (1996) 33–38
9. Volkmann, N., Hanein, D.: Docking of atomic models into reconstruction from electron microscopy. Methods Enzymol **374** (2003) 204–225
10. Chacon, P., Wriggers, W.: Multiresolution contour-based fitting of macromolecular structure. J. Mol. Biol. **317** (2002) 375–384
11. Jiang, W., Baker, M.L., Ludtke, S.J., Chiu, W.: Bridging the information gap: Computational tools for intermediate resolution structure interpretation. J. Mol. Biol. **208** (2001) 1033–1044
12. Rossmann, M.G.: Fitting atomic models into electron-microscopy maps. Acta Crystallogr. **D56** (2000) 1341–1349
13. Kleywegt, G.J., Jones, T.: Detecting folding motifs and similarities in protein structures. Methods Enzymol **277** (1997) 525–545
14. Mizuguchi, K., Go, N.: Comparison of spatial arrangements of secondary structural elements in proteins. Protein Eng. **8** (1995) 353–362
15. Gonzalez, R.C., Woods, R.E.: Digital Image Processing, second edition. Prentice Hall, Upper Saddle River N.J (2002)
16. Ludtke, S.J., Baldwin, P.R., Chiu, W.: EMAN: Semiautomated software for high-resolution single-particle reconstructions. J. Struct. Biol. **128** (1999) 82–97
17. Wriggers, W., Birmanns, S.: Using situs for flexible and rigid-body fitting of multiresolution single-molecule data. J. Struct. Biol. **133** (2001) 193–202
18. Brigham, E.: The Fast Fourier Transform and its Applications. Prentice Hall, Upper Saddle River N.J (1988)
19. Felzenszwalb, P.F., Huttenlocher, D.P.: Efficient graph-based image segmentation. Int. J.Comput. Vision **59** (2004) 167–181
20. Cormen, T.H., Leiserson, C.E., Rivest, R.L.: Interoduction to Algorithms. The MIT Press, Cambridge ,Massachusetts London England (1990)
21. Dror, O., Benyamini, H., Nussinov, R., Wolfson, H.: MASS: multiple structural alignment by secondary structures. Bioinformatics **19 Suppl. 1** (2003) i95–i104
22. Dror, O., Benyamini, H., Nussinov, R., Wolfson, H.: Multiple structural alignment by secondary structures:Algorithm and applications. Protein Sci. **12** (2003) 2492–2507
23. Mehlhorn, K.: The LEDA Platform of Combinatorial and Geometric Computing. Cambridge University Press, United Kingdom (1999)
24. Kabsch, W.: A discussion of the solution for the best rotation to relate two sets of vectors. Acta Crystallogr. **A 34** (1978) 827–828
25. Murzin, A., Brenner, S., Hubbard, T., Chothia, C.: SCOP: a structural classification of proteins database for the investigation of sequences and structures. J. Mol. Biol. **247** (1995) 536–540
26. Tagari, M., Newman, R., Chagoyen, M., Carazo, J., Henrick, K.: New electron microscopy database and deposition system. Trends. Biochem. Sci. **27** (2002) 589
27. Inbar, Y., Benyamini, H., Nussinov, R., Wolfson, H.: Prediction of multimolecular assemblies by multiple docking. J. Mol. Biol. **349** (2005) 435–447

Author Index

Lecture Notes in Bioinformatics

Vol. 3695: M.R. Berthold, R. Glen, K. Diederichs, O. Kohlbacher, I. Fischer (Eds.), Computational Life Sciences. XI, 277 pages. 2005.

Vol. 3692: R. Casadio, G. Myers (Eds.), Algorithms in Bioinformatics. X, 436 pages. 2005.

Vol. 3678: A. McLysaght, D.H. Huson (Eds.), Comparative Genomics. VIII, 167 pages. 2005.

Vol. 3615: B. Ludäscher, L. Raschid (Eds.), Data Integration in the Life Sciences. XII, 344 pages. 2005.

Vol. 3594: J.C. Setubal, S. Verjovski-Almeida (Eds.), Advances in Bioinformatics and Computational Biology. XIV, 258 pages. 2005.

Vol. 3500: S. Miyano, J. Mesirov, S. Kasif, S. Istrail, P. Pevzner, M. Waterman (Eds.), Research in Computational Molecular Biology. XVII, 632 pages. 2005.

Vol. 3388: J. Lagergren (Ed.), Comparative Genomics. VII, 133 pages. 2005.

Vol. 3380: C. Priami (Ed.), Transactions on Computational Systems Biology I. IX, 111 pages. 2005.

Vol. 3370: A. Konagaya, K. Satou (Eds.), Grid Computing in Life Science. X, 188 pages. 2005.

Vol. 3318: E. Eskin, C. Workman (Eds.), Regulatory Genomics. VII, 115 pages. 2005.

Vol. 3240: I. Jonassen, J. Kim (Eds.), Algorithms in Bioinformatics. IX, 476 pages. 2004.

Vol. 3082: V. Danos, V. Schachter (Eds.), Computational Methods in Systems Biology. IX, 280 pages. 2005.

Vol. 2994: E. Rahm (Ed.), Data Integration in the Life Sciences. X, 221 pages. 2004.

Vol. 2983: S. Istrail, M.S. Waterman, A. Clark (Eds.), Computational Methods for SNPs and Haplotype Inference. IX, 153 pages. 2004.

Vol. 2812: G. Benson, R.D. M. Page (Eds.), Algorithms in Bioinformatics. X, 528 pages. 2003.

Vol. 2666: C. Guerra, S. Istrail (Eds.), Mathematical Methods for Protein Structure Analysis and Design. XI, 157 pages. 2003.